Methods in Cell Biology

VOLUME 97

Microtubules: In Vivo

CONTRIBUTORS

Numbers in parentheses indicate the pages on which the authors' contributions begin.

Anna Akhmanova, (91) Department of Cell Biology, Erasmus Medical Center, 3000 CA Rotterdam, The Netherlands

Andreas Anders, (147) Wellcome Trust Centre for Cell Biology, School of Biological Sciences, University of Edinburgh, Edinburgh EH9 3JR, United Kingdom

Ammar Azioune, (133) Systems Cell Biology of Cell Division and Cell Polarity, UMR144, Institut Curie, CNRS, Paris 75248, France

Monica Bettencourt-Dias, (223) Instituto Gulbenkian de Ciência, Rua da Quinta Grande, P-2780-156 Oeiras, Portugal

Margaret E. Bisher, (415) Department of Molecular Biology, Princeton University, Princeton, New Jersey 08544

Gary J. Brouhard, (497) Department of Biology, McGill University, Montréal, Québec, Canada H3A 1B1

Rebecca D. Burdine, (415) Department of Molecular Biology, Princeton University, Princeton, New Jersey 08544

Henrik Buschmann, (373) Department of Cell and Developmental Biology, John Innes Centre, Norwich NR4 7UH, United Kingdom

Grant Calder, (373) Department of Cell and Developmental Biology, John Innes Centre, Norwich NR4 7UH, United Kingdom

Giuliano Callaini, (223) Department of Evolutionary Biology, University of Siena, I-53100 Siena, Italy

Frédérique Carlier-Grynkorn, (185) Institut Curie, UMR 144 CNRS, Paris 75005, France

Nicolas Carpi, (133) Systems Cell Biology of Cell Division and Cell Polarity, UMR144, Institut Curie, CNRS, Paris 75248, France

Stéphane Chiron, (203) INSERM, U974, Université Pierre et Marie Curie-Paris, UMR-S974, CNRS, UMR-7215, Institut de Myologie, IFR14, Paris, F-75013, France

Judite Costa, (185) Cell & Developmental Biology, University of Pennsylvania, Philadelphia, Pennsylvania 19104

Maitreyi Das, (203) Department of Molecular and Cellular Pharmacology (R-189), University of Miami Miller School of Medicine, Miami, Florida 33101

Scott C. Dawson, (307) Department of Microbiology, One Shields Avenue, University of California – Davis, Davis, CA 95616

Jennifer G. DeLuca, (53) Department of Biochemistry and Molecular Biology, Colorado State University, Fort Collins, Colorado 80523

Nick P. Ferenz, (81) Precept Medical Communications, Berkeley Heights, New Jersey 07922

Timur Gatanov, (475) Department of Physics, Harvard University, Cambridge, Massachusetts 01238

Sarah Gierke, (15) Department of Cell & Tissue Biology, University of California, San Francisco, California 94143-0512

Gohta Goshima, (259) Division of Biological Science, Graduate School of Science, Nagoya University, Nagoya 464-8602, Japan

Ralph Gräf, (341) Department of Cell Biology, Institut for Biochemistry and Biology, University of Potsdam, Potsdam-Golm 27708, Germany D-14476

Ilya Grigoriev, (91) Department of Cell Biology, Erasmus Medical Center, 3000 CA Rotterdam, The Netherlands

Casper C. Hoogenraad, (111) Department of Neuroscience, Erasmus Medical Center, 3015 GE, Rotterdam, The Netherlands

Susan A. House, (307) Department of Microbiology, One Shields Avenue, University of California – Davis, Davis, CA 95616

Yumi Iida, (359) Laboratory for Developmental Genomics, RIKEN Center for Developmental Biology, Kobe 650-0047, Japan

Kazuho Ikeda, (401) Department of Cell Biology, R. D. Berlin Center for Cell Analysis and Modeling, University of Connecticut Health Center, Farmington, Connecticut 06032-1507

Kimberly M. Jaffe, (415) Department of Molecular Biology, Princeton University, Princeton, New Jersey 08544

Mary Ann Jordan, (1) Department of Molecular, Cellular, and Developmental Biology and the Neuroscience Research Institute, University of California Santa Barbara, Santa Barbara, California 93106

Kathy Kamath, (1) Department of Molecular, Cellular, and Developmental Biology and the Neuroscience Research Institute, University of California Santa Barbara, Santa Barbara, California 93106

Lukas C. Kapitein, (111) Department of Neuroscience, Erasmus Medical Center, 3015 GE, Rotterdam, The Netherlands

Akatsuki Kimura, (437) Cell Architecture Laboratory, Center for Frontier Research, National Institute of Genetics, Mishima 411-8540, Japan

Roman I. Koning, (455) Department of Molecular Cell Biology, Section Electron Microscopy, Leiden University Medical Center, 2300 RC, Leiden, The Netherlands

Michael P. Koonce, (341) Division of Translational Medicine, Wadsworth Center, Albany, New York 12201-0509

Praveen Kumar, (15) Department of Cell & Tissue Biology, University of California, San Francisco, California 94143-0512

Clive W. Lloyd, (373) Department of Cell and Developmental Biology, John Innes Centre, Norwich NR4 7UH, United Kingdom

Pedro Machado, (223) Instituto Gulbenkian de Ciência, Rua da Quinta Grande, P-2780-156 Oeiras, Portugal

Helder Maiato, (243) IBMC—Instituto de Biologia Molecular e Celular, Universidade do Porto, 4150-180 Porto, Portugal

Nicola Maghelli, (173) Max Planck Institute of Molecular Cell Biology and Genetics (MPI-CBG), 01307 Dresden, Germany

Ana Rodrigues Martins, (223) Instituto Gulbenkian de Ciência, Rua da Quinta Grande, P-2780-156 Oeiras, Portugal

Irina Matos, (243) Laboratory of Cell and Molecular Biology, Faculdade de Medicina, Universidade do Porto, 4200-319 Porto, Portugal

Adeline Mayeux, (185) Institut Curie, UMR 144 CNRS, Paris 75005, France

Irene Meyer, (341) Department of Cell Biology, Institut for Biochemistry and Biology, University of Potsdam, Potsdam-Golm 27708, Germany D-14476

Sara Moutinho-Pereira, (243) IBMC—Instituto de Biologia Molecular e Celular, Universidade do Porto, 4150-180 Porto, Portugal

Elena Nazarova, (277) Department of Biology, McGill University, Montreal, Quebec, Canada H3G 0B1

Daniel J. Needleman, (475) Department of Molecular and Cellular Biology, School of Engineering and Applied Sciences, Center for Systems Biology, Harvard University, Cambridge, Massachusetts 01238

Daniyar Nurgaliev, (475) Department of Physics, Harvard University, Cambridge, Massachusetts 01238

Shuichi Onami, (437) Advanced Computational Sciences Department, RIKEN Advanced Science Institute, Yokohama 230-0045, Japan

Emin Oroudjev, (1) Department of Molecular, Cellular, and Developmental Biology and the Neuroscience Research Institute, University of California Santa Barbara, Santa Barbara, California 93106

Edouard Pesquet, (373) Department of Cell and Developmental Biology, John Innes Centre, Norwich NR4 7UH, United Kingdom

Matthieu Piel, (133) Systems Cell Biology of Cell Division and Cell Polarity, UMR144, Institut Curie, CNRS, Paris 75248, France

Alexander Rauch, (277) Institute of Biochemistry, ETH-Zurich, 8093 Zurich, Switzerland

Rania S. Rizk, (35) Department of Molecular Genetics and Cell Biology, University of Chicago, Chicago, Illinois 60637

Vladimir Rodionov, (401) Department of Cell Biology, R. D. Berlin Center for Cell Analysis and Modeling, University of Connecticut Health Center, Farmington, Connecticut 06032-1507

Adrian Sambade, (373) Department of Cell and Developmental Biology, John Innes Centre, Norwich NR4 7UH, United Kingdom

Itaru Samejima, (147) Wellcome Trust Centre for Cell Biology, School of Biological Sciences, University of Edinburgh, Edinburgh EH9 3JR, United Kingdom

Matthias Samereier, (341) Department of Cell Biology, Institut for Biochemistry and Biology, University of Potsdam, Potsdam-Golm 27708, Germany D-14476

Kenneth E. Sawin, (147) Wellcome Trust Centre for Cell Biology, School of Biological Sciences, University of Edinburgh, Edinburgh EH9 3JR, United Kingdom

Irina Semenova, (401) Department of Cell Biology, R. D. Berlin Center for Cell Analysis and Modeling, University of Connecticut Health Center, Farmington, Connecticut 06032-1507

Sidney L. Shaw, (35) Department of Biology, Indiana University, Bloomington, Indiana 47405

Hilary A. Snaith, (147) Wellcome Trust Centre for Cell Biology, School of Biological Sciences, University of Edinburgh, Edinburgh EH9 3JR, United Kingdom

Asako Sugimoto, (359) Laboratory of Developmental Dynamics, Graduate School of Life Sciences, Tohoku University, Sendai 980 8577, Japan

Manuel Théry, (133) Laboratoire de Physiologie Cellulaire et Végétale, iRTSV, CEA/CNRS/UJF/INRA, 38054 Grenoble, France

Stephan Y. Thiberge, (415) Lewis-Sigler Institute for Integrative Genomics, Princeton University, Princeton, New Jersey 08544

Iva M. Tolić-Nørrelykke, (173) Max Planck Institute of Molecular Cell Biology and Genetics (MPI-CBG), 01307 Dresden, Germany

Mika Toya, (359) Laboratory for Developmental Genomics, RIKEN Center for Developmental Biology, Kobe 650-0047, Japan

Phong T. Tran, (185) Institut Curie, UMR 144 CNRS, Paris 75005, France and Cell & Developmental Biology, University of Pennsylvania, Philadelphia, Pennsylvania 19104

Qingzong Tseng, (133) Laboratoire de Physiologie Cellulaire et Végétale, iRTSV, CEA/CNRS/UJF/INRA, 38054 Grenoble, France

U. Serdar Tulu, (81) Department of Biology, Duke University, Durham, North Carolina, 27708

Guilhem Velve-Casquillas, (185) Institut Curie, UMR 144 CNRS, Paris 75005, France

Fulvia Verde, (203) Department of Molecular and Cellular Pharmacology (R-189), University of Miami Miller School of Medicine, Miami, Florida 33101

Jackie Vogel, (277) School of Computer Science, McGill University, Montreal, Quebec, Canada H3A 2A7

Patricia Wadsworth, (81) Department of Biology, University of Massachusetts, Amherst, Massachusetts 01003

Patricia Wadsworth, (81) Department of Biology, University of Massachusetts, Amherst, Massachusetts 01003

Claire E. Walczak, (35) Medical Sciences, Indiana University, Bloomington, Indiana 47405

Torsten Wittmann, (15) Department of Cell & Tissue Biology, University of California, San Francisco, California 94143-0512

Kah Wai Yau, (111) Department of Neuroscience, Erasmus Medical Center, 3015 GE, Rotterdam, The Netherlands

Olga Zhapparova, (401) Department of Cell Biology, R. D. Berlin Center for Cell Analysis and Modeling, University of Connecticut Health Center, Farmington, Connecticut 06032-1507

PREFACE

Microtubules provide critical functions in cells, forming the mitotic spindle of dividing cells, the tracks for polarized vesicle and organelle movements, and the core structure of cilia and flagella. Microtubule assembly and turnover are essential for maintenance of the genome over generations (error-free mitosis) and for the cell shape changes and movements necessary for development. This volume of Methods in Cell Biology describes proper use of tools to measure microtubule assembly and turnover in living cells grown in culture or within living organisms. Twenty years ago were limited to injection into living cells of dye-labeled, purified mammalian brain tubulin as a means to follow microtubule assembly dynamics. These methods were amenable to a limited number of cell types, including mammalian cells in culture and some marine eggs/oocytes. The expanding use of green fluorescent protein and other fluorescent variants, expressed in model organisms and combined with numerous genetic backgrounds, has greatly expanded our understanding of the microtubule cytoskeleton structure and functions. Knowledge of how and when associated proteins bind to the tips of microtubules has also expanded the tools available for study of microtubules, providing a high-resolution view of microtubules in the polymerization stage. Photobleaching and photoactivation methods continue to provide measures of microtubule turnover and movement within the dense arrays of the mitotic spindle. High-resolution approaches, including electron microscopy tomography and single molecule analyses, continue to expand the tools available for accurate mechanistic insight.

This volume includes chapters from 26 research groups whose laboratories span the globe. Eight chapters describe methods for measuring microtubule dynamics and turnover in mammalian cells in culture. Fifteen chapters are devoted to model and nonmodel organisms, including budding yeast, fission yeast, Drosophila embryos and S2 cells, plant cells, Zebrafish, the intestinal parasite Giardia, fish melanophores, *Caenorhabditis elegans* embryos, and Dictyostelium. An additional three chapters describe methods for high-resolution imaging and analysis.

Volume 95, Microtubules, *in vitro*, edited by Leslie Wilson and John J. Correia was published earlier in 2010. Microtubules *in vitro* contain a number of methods for tubulin purification, determination of microtubule structure and dynamics, drug disruption of microtubule functions, microtubule interactions with motors and microtubule associated proteins (MAPs), and functional cell extracts for force measurements. Together the two volumes provide a wealth of techniques for study of the microtubule cytoskeleton, with advice from experienced researchers on how to avoid common errors and pitfalls. We expect these volumes to provide a valuable resource for those already working in this area and for others who are now beginning microtubule studies.

Thanks are due to all the authors who contributed to this volume of Methods in Cell Biology. Their efforts made the volume easy to assemble and a sure thing for success. Thanks also to Tara Hoey, Zoe Kruze, and Narmada Thangavelu at Elsevier for keeping us organized and close to meeting deadlines throughout the process.

Lynne Cassimeris
Department of Biological Sciences, Lehigh University,
Bethlehem, PA, USA

and

Phong Tran
Department of Cell & Developmental Biology, University of Pennsylvania School of Medicine, Philadelphia, PA, USA
and
Institut Curie
UMR 144 CNRS, Paris, France

CHAPTER 1

Determination of Microtubule Dynamic Instability in Living Cells

Kathy Kamath, Emin Oroudjev, *and* Mary Ann Jordan

Department of Molecular, Cellular, and Developmental Biology and the Neuroscience Research Institute, University of California Santa Barbara, Santa Barbara, California 93106

Abstract

The precise regulation of microtubules and their dynamics is critical for cell cycle progression, cell signaling, intracellular transport, cell polarization, and organismal development. For example, mitosis, cell migration, and axonal outgrowth all involve rapid and dramatic changes in microtubule organization and dynamics. Microtubule-associated proteins (MAPs) such as MAP2 and tau (Bunker *et al.*, 2004; Dhamodharan and Wadsworth, 1995) and microtubule-interacting proteins such as stathmin, the kinesin MCAK, and EB1 (Cassimeris, 1999; Moore and Wordeman, 2004; Ringhoff and Cassimeris, 2009; Rusan *et al.*, 2001) as well as numerous clinically approved or experimental anti-mitotic drugs including the taxanes, vinca alkaloids, and colchicine-like compounds modulate microtubule dynamic in cells (Jordan, 2002; Jordan and Kamath, 2007). In this chapter, we describe methods to

978-0-12-381349-7
DOI: 10.1016/S0091-679X(10)97001-5

analyze the dynamic instability of microtubules in living cells by microscopy of microinjected or expressed fluorescent tubulin, time-lapse microscopy, and analysis of time-dependent microtubule length changes.

I. Introduction

Microtubules are dynamic filamentous polymers assembled from the protein tubulin, a heterodimer composed of an α and a β subunit. Both *in vitro* and in cells, microtubules exhibit a nonequilibrium behavior called dynamic instability, in which microtubule ends alternate between periods of slow growth and rapid shortening (Mitchison and Kirschner, 1984). Plus ends of microtubules (the ends with β-tubulin exposed) are more dynamic, undergoing greater changes in length than minus ends (the ends with α-tubulin exposed). Although dynamic microtubule ends likely undergo continuous exchange with free tubulin dimers and continuously grow and shorten, when the changes in microtubule length are below the resolution of the light microscope, the microtubule is said to be in a state of attenuated or paused dynamic instability. Microtubule dynamic instability can be described by four main parameters: the rates of microtubule growth and shortening and the frequencies of "catastrophes" and "rescues." A "catastrophe" is a transition from a period of growth or pause to shortening, and a "rescue" is a transition from shortening to growth or pause (Walker *et al.*, 1988). The overall rate of exchange of tubulin with the microtubule end is called "dynamicity" (Toso *et al.*, 1993).

While much of our understanding of microtubule behavior comes from studies on microtubules assembled from purified tubulin *in vitro*, many of the basic principles of microtubule behavior gained from these studies apply also to microtubules in cells. However, in cells, microtubules often grow at a five- to ten-fold faster rate and transitions between growth and shortening occur ~10 times more frequently than with microtubules assembled from purified tubulin *in vitro*. The dramatic changes in microtubule organization and dynamics throughout the cell cycle reflect a high degree of spatial and temporal regulation of cellular microtubule behavior. Regulation is achieved by posttranslational modification of tubulin, by tubulin isotype expression, and by a large group of MAPs and other microtubule-interacting proteins that either stabilize or destabilize microtubules (Cassimeris, 1999; Desai and Mitchison, 1997; Walczak, 2000).

Both microinjection and expression of fluorescent tubulin in living cells have proven to be extremely powerful techniques for studying microtubule-based processes in real time in a cellular environment (Dhamodharan and Wadsworth, 1995b; Dhamodharan *et al.*, 1995; Faire *et al.*, 1999; Goncalves *et al.*, 2001; Landen *et al.*, 2002; Mikhailov and Gundersen, 1998; Saxton *et al.*, 1984; Shelden and Wadsworth, 1993, 1996; Wadsworth and Bottaro, 1996; Waterman-Storer and Salmon, 1997; Waterman-Storer *et al.*, 2000; Yvon *et al.*, 1999; Zhang *et al.*, 1990). Pioneering work in the 1980s demonstrated that when tubulin covalently modified with a fluorescent tag is introduced into cells it behaves like endogenous tubulin, is incorporated into all microtubule-related structures, and does not affect cell viability (Saxton *et al.*, 1984). Studies using microinjection of fluorescent tubulin into living cells and, more recently, using transfection of green fluorescent protein (GFP)-, EGFP (enhanced GFP)- or mCherry-tubulin have confirmed and refined the significance of *in vitro* work on the modulation of microtubule dynamics by proteins and biochemical agents such as drugs (Dhamodharan and Wadsworth, 1995; Dhamodharan *et al.*, 1995; Faire *et al.*, 1999; Landen *et al.*, 2002; Shelden and Wadsworth, 1996; Yvon *et al.*, 1999). In addition, these studies have provided novel critical insights into microtubule behavior in mitosis and other cellular processes such as migration that cannot be recapitulated *in vitro* (Mikhailov and Gundersen, 1998; Rusan *et al.*, 2001; Wadsworth and Bottaro, 1996; Waterman-Storer and Salmon, 1997; Yvon and Wadsworth, 1997).

Expression of fluorescent tubulin in cloned cells is often the method of choice for forming fluorescent microtubules and analyzing their behavior. Plasmids are available from CLONTECH (Palo Alto, CA) and can be transformed following the Superfect (Qiagen, Valencia, CA) protocol or any standard transfection protocol. This technique has the advantage of avoiding purification and fluorescent labeling of tubulin and of microinjecting cells. Microtubules formed of expressed GFP-tubulin in cells have very similar dynamics to those in cells microinjected with rhodamine-tubulin. However, one advantage of microinjection is that rhodamine-tubulin is more photostable than GFP-tubulin thus allowing approximately two to three times longer duration of time-lapse recording prior to photobleaching and a better signal-to-noise ratio. On the other hand, with fluorescent tubulin transfection one has more freedom to experimentally alter other cellular characteristics by microinjection of regulatory proteins or antibodies or by siRNA treatments. In this case, an additional fluorescent marker must be used with the microinjected protein to locate the microinjected cells. Two types of fluorescent markers are useful: lysine fixable markers and small molecular weight fluorescent dextrans. For example, Bunker *et al.* (2004) microinjected tau proteins into EGFP-tubulin-expressing cells along with 1.4 mM β-mercaptoethanol and 0.45 mg/ml RITC-dextran as a marker for injected cells (all centrifuged immediately before filling the needles, see detailed instructions below). Fluorescent dextran is washed out of the cells during the fixation process and thus lysine-fixable markers are necessary if the microinjected cells are to be fixed. This chapter focuses on optimization of cell choice and cell preparation, the microinjection of fluorescent tubulin, and visualization and analysis of microtubule dynamics in living cells.

II. Methods and Materials

A. Choice of Cell Line

The choice of cell line is perhaps the most important criterion for successful microinjection as well as for imaging microtubule dynamics, even after expression of fluorescent tubulin. The cell type must (1) adhere tightly to the substrate for successful microinjection and (2) have a flat, well-spread morphology for successful imaging of microtubules. Some cells adhere to coverslips under incubation conditions (37°C and 5% CO_2), but when they are brought to room temperature and air for microinjection, they begin to round up and detach. Cells that are not well attached tend to stick to the microneedle after injection and become detached from the coverslip when the needle is retracted. For example, NIH3T3 cells exhibit a neuronal-like morphology and their processes tend to be anchored to the substrate but the cell body does not adhere well, making them difficult to inject. BT549 breast carcinoma cells are also difficult to inject because, while they appear very flat in culture, they round up after being out of the incubator for a short period of time. DU145 prostate cancer cells attach well but exhibit a spindle-shaped morphology; they are easy to inject but the lamellar edge is not spread in such a way that allows for individual microtubules to be easily visualized. Both MCF7 breast carcinoma and A549 lung carcinoma can be microinjected and analyzed relatively easily. A list of cell lines that have been used for imaging microtubule dynamics is shown in Table I.

B. Optimization of Cell Growth Conditions

The type of coverslip and substrate upon which cells are grown, as well as the growth medium and conditions, all influence cell morphology and spreading which are important for microinjection and analysis.

1. Coverslips

While cells often appear flatter when grown on plastic, cells should be grown on glass coverslips for imaging, since many types of plastic coverslips generate background autofluorescence and interfere with the fluorescent tubulin signal. No. 1 or 1.5 thickness may be used. Glass coverslips etched with grid squares that are marked by an identifying

Table I
Cell Types Used for Analysis of Microtubule Dynamics

Non-mammalian cells	Species and cell type	Reference
Yeast	*Saccharomyces cerevisiae*	Tischfield *et al.* (2010)
Drosophila embryonic cells	*Drosophila*	Brust-Mascher and Scholey (2002)
Sea hare neurons	*Aplysia* bag cell neurons	Lee and Suter (2008) and Schaefer *et al.* (2002)
Mammalian non-tumor cell lines		
MDCK	Canine kidney	Wadsworth and Bottaro (1996)
CHO	Hamster ovarian	Shelden and Wadsworth (1993)
TC-7	Monkey kidney	Faire *et al.* (1999)
PTK1	Rat kangaroo kidney epithelial	Saxton *et al.* (1984)
PTK2	Rat kangaroo kidney epithelial	Landen *et al.* (2002)
NRK	Rat kidney	Mikhailov and Gundersen (1998)
LLCPK 1	Pig kidney	Faller and Brown (2009)
HUVEC	Human umbilical vein endothelial cells	Pasquier *et al.* (2005)
HMEC-1	Human dermal microvascular endothelial cells	Pasquier *et al.* (2005)
Mammalian tumor cell lines		
A498	Human kidney	Yvon *et al.* (1999)
A549	Human lung	Goncalves *et al.* (2001)
CaOV3	Human ovary	Yvon *et al.* (1999)
MCF7	Human breast	Galmarini *et al.* (2003)
BT549	Human breast	Balasubramani *et al.* (2010)
B16	Mouse melanoma	Ballestrem *et al.* (2000)
BSC-1	SV40-transformed monkey kidney	Dhamodharan and Wadsworth (1995b) and Saxton *et al.* (1984)

alphanumeric code (Electron Microscopy Sciences, Hatfield, PA, emsdiasum.com; MatTek Corp, Ashland, MA; and Millenium, Thermo Fisher Scientific) are useful for identifying individual cells that have been injected. Although this is generally not necessary if the microinjected protein is fluorescent, it can be of utmost importance if a nonfluorescent microtubule regulatory protein or antibody is microinjected.

2. Substrate

Cells grown on poly-L-lysine-treated coverslips may adhere and spread adequately. If not, a substrate such as laminin, collagen, or fibronectin may promote cell adherence and spreading. The optimal substrate and substrate concentration may vary with the cell type. We plate A549 and MCF7 cells on laminin, fibronectin, or both (described further below). In our experience, A549 and MCF7 cells spread optimally on a combination of laminin and fibronectin and do not spread well on collagen. It is worth keeping in mind that the particular substrate used may affect cellular microtubule dynamics.

3. Culture Media and Growth Conditions

Cells spread best when allowed to attach for at least 48 h prior to microinjection and analysis. Serum starvation also may promote cell spreading (Jordan-Sciutto *et al.*, 1997) and can be used alternatively or in addition to plating on specific substrates.

4. Cell Density

Dynamics of individual microtubules and cells are inherently variable, and care should be taken to eliminate variability due to conditions that are not relevant to the objective of the study. For example, cell–cell contact suppresses microtubule dynamic instability (Waterman-Storer *et al.*, 2000). In addition, the cell membranes of contiguous cells are less flat and microtubules are often more difficult to image in cells that are in contact with other cells on all sides. Cell colonies should be broken apart by pipetting and plated at a low density that yields single cells or cells with a large portion of their plasma membrane free of neighboring cells after 48 h of incubation. If cells are migrating, it is important to note that dynamic instability at the trailing edge is faster than at the leading edge (Wadsworth, 1999).

C. Detailed Protocol for Preparation of MCF7 Cells for Microinjection and Analysis of Microtubule Dynamic Instability

1. Solutions

a. Make a stock solution of poly-L-lysine (2.5 mg/ml) (Sigma) in sterile water and store at 4°C.
b. If needed, dissolve laminin and fibronectin (Gibco, Carlsbad, CA) in sterile water to concentrations of 0.33 mg/ml and 1.0 mg/ml, respectively, and store at –80°C.
c. Injection buffer: 50 mM potassium glutamate, 0.5 mM $MgSO_4$. Divide it into 6 μl aliquots and freeze.
d. Versene (137 mM NaCl, 2.7 mM KCl, 1.5 mM KH_2PO_4, 8.1 mM Na_2HPO_4, 0.5 mM EDTA, pH 7.2).

e. Cell culture media supplemented with 10% fetal bovine serum (FBS).
f. Media supplemented with 2% FBS.

2. Materials

Coverslips etched with identified grid squares, if necessary for locating experimentally treated cells (sources listed above).
4- or 6-well plates (Nunc, Rochester, NY).
Sterile water.
Silver foil Femtotips (Eppendorf).
Silane (hexamethyldisilazane, Pierce, Rockford, IL).

3. Preparation of Rhodamine-Labeled Tubulin for Microinjections

Rhodamine-tubulin is available from Cytoskeleton, Inc. (Denver, CO) and might conceivably be used successfully for labeling microtubules in living cells. In practice, we have found it necessary to label our own tubulin for successful microinjection and labeling of cellular microtubules, and we now routinely make a large batch as follows: Purify or otherwise obtain bovine brain tubulin (Miller and Wilson, 2010). Label with carboxyrhodamine succinimidyl ester (Molecular Probes, Eugene, OR) according to the method of Hyman and Mitchison (Hyman *et al.*, 1991). A detailed protocol can be found at http://mitchison.med.harvard.edu/protocols/label.html. Determine the concentrations of tubulin and carboxyrhodamine in the labeled tubulin by spectrophotometry at A280 and A525 (or the emission maximum for the specific rhodamine used), respectively. This will yield information on the stoichiometry of rhodamine:tubulin. Adjust the tubulin concentration to 10 mg/ml with injection buffer. Rhodamine tubulin concentrations <10 mg/ml appear to be less stable and denature when stored for long periods (>2 months). Make single-use aliquots (5–6 μl), flash freeze, and store in liquid nitrogen.

The concentration of labeled tubulin to use for microinjection depends on the stoichiometry of labeling. The optimal concentration can be determined by injecting cells with a range of concentrations between 2 and 10 mg/ml, allowing the cells to recover for 4 h and evaluating the microtubules by fluorescence microscopy. Use the lowest concentration that yields sufficiently bright microtubules. With carboxyrhodamine-tubulin, an injection concentration of 2–3 mg/ml is adequate to visualize microtubules when the stoichiometry is 0.5–0.8 rhodamine/tubulin dimer (final cellular concentration of ~2 μM). It is advisable to use the same batch of rhodamine-labeled tubulin for all experiments in a study to eliminate any variability due to differences in the final cellular tubulin concentration after addition of rhodamine-tubulin.

4. Detailed Protocol

There are at least three methods for cleaning and sterilizing coverslips. (1) Clean with ethanol, dry overnight in a vacuum oven, and then autoclave. (2) Clean as above, then flame them, holding with a sterile forceps, and rinse with sterile water. (3) Wash with Alconox, rinse extensively with running hot water followed by sterile water, store in 95% ethanol, autoclave, and store or flame immediately before use. Gridded coverslips are presterilized and do not need washing.

Incubate coverslips with poly-L-lysine at a working concentration of 0.05 mg/ml for 2 h at 37°C, followed by incubation with laminin (10 μg/ml) and fibronectin (20 μg/ml) for an additional 2 h at 37°C in 4- or 6- well plates. Rinse coverslips with sterile water and transfer to a new 4- or 6-well plate.

Detach MCF7 cells from flasks by trypsinization (0.5 mg/ml trypsin in versene) and resuspend them in growth media. Break up cell clumps by vigorous pipetting. Determine cell concentration by counting and adjust to a final concentration of 3×10^4 cells/ml in growth media supplemented with 10% FBS.

Plate cells (6×10^4/well) into each well of a 4- or 6-well plate containing pretreated coverslips and allow to adhere for 24 h.

Serum-starve the cells for 24 h prior to microinjection by replacing normal growth media with media supplemented with 2% serum.

Choose cells for microinjection and imaging that are either isolated or at the edge of a colony of cells.

D. Optimization of Microinjection

1. Preventing Tubulin Polymerization/Clogging of the Microneedle

One of the main obstacles to successful microinjection of tubulin is clogging of the microneedle. Clogging generally occurs from either tubulin polymerization/aggregation or debris. The two types of clogs can be distinguished by visual inspection of the needle with a 40× objective. Tubulin aggregation can be seen as a minute white speck at the very tip of the needle, which sometimes protrudes from the tip. Debris is usually located in the wider portion of the needle and appears black.

The following precautions will minimize clogging:

On the day before microinjection, pretreat the injection needles (silver foil Femtotips) with silane (hexamethyldisilazane). Silanize several needles together in a 100 mm glass Petri dish. To do this first press the large part of each needle distal to its tip into a thin linear rolled piece of modeling clay that has been pressed into the bottom of the dish. This will keep the tip well away from the bottom of the dish and prevent breakage. Using a small syringe, place approximately five drops of hexamethyldisilazane (Pierce, Rockford, IL) into the bottom of the Petri dish. The hexamethyldisilazane is toxic; cover the dish and leave it in a fume hood overnight.

Prior to microinjection, clarify the tubulin to remove any aggregates or denatured tubulin as follows: Adjust the solution to the appropriate concentration (2–3 mg/ml) with freshly filtered Injection Buffer and centrifuge at 4°C for 20 min at 65,000×*g* (35K RPM in a Beckman Ultra tabletop TLA 100 centrifuge using the TLA 100.3 rotor). Transfer the supernatant to a clean tube and store on ice.

Place the Femtotips and microloaders (Eppendorf) in the freezer and chill a small amount of medium on ice prior to use. Place the cold medium on the cells just before injection to prevent the tubulin from polymerizing in the needle during injection.

Cover the Femtotips, microloaders, pipette tips, Eppendorf tubes, and anything that comes in contact with the tubulin solution when not in use to prevent dust from getting into the tubulin and clogging the microneedle.

Use the "clean" function on the microinjection apparatus (such as Eppendorf 5246 Transjector and Injectman) frequently during injection to flush out the tip of the microneedle.

2. Microinjection Pressure

The optimal pressure settings vary with cell size. Set the pressure to deliver ~10% of the total cell volume. Injecting this volume creates a visible wave across the surface of the cell, but does not cause the cell to bleb, become misshapen or blow up. Cells should return to their normal size after injection. Set the "compensation pressure" high enough that the solution constantly flows out of the needle. For example, we routinely use 2.1 psi injection pressure and 1.5 compensation pressure for MCF7 cells with the Eppendorf 5246 Transjector and Injectman. We maintain the injection time at 0.3 s. Several methods have been developed to quantify the injection volume if this information is critical (Lee, 1989; Minaschek *et al.*, 1989).

3. Microinjection Protocol

Make a circular well about 3 mm in height on a large glass slide using VALAP (1/3 Vaseline, 1/3 lanolin, 1/3 paraffin). Place the coverslip with the cells to be injected on the slide in the circle and cover it with cold medium. Load 2 μl of clarified rhodamine-labeled tubulin using a chilled microloader into a chilled silanized Femtotip. Inject the cells under the 40× objective lens of an inverted phase-contrast microscope mounted with the microinjector. After microinjection, return the cells to the incubator for at least 3 h to allow for recovery and incorporation of labeled tubulin into microtubules before imaging. Microtubules can be expected to remain strongly labeled for 16–24 h after injection.

4. Addition of Drug

If the effects of a drug are being studied, it is important to allow the drug to be taken up in the cells to equilibrium. In the case of drugs such as taxanes and vinca alkaloids, we have found that uptake time is 4–6 h, depending on the concentrations used (Jordan and Wilson, 1999).

E. Preparation of Cells for Time-Lapse Fluorescence Microscopy

1. Recording Medium

Image MCF7 cells in Dulbecco's modified Eagles Medium lacking phenol red and supplemented with 25 mM Hepes, 4.5 g/l glucose, and 10–20 μl/ml OxyFluor (Oxyrase Inc., Mansfield, OH). The phenol red is excluded because it exhibits some autofluorescence. Hepes buffer is used rather than sodium bicarbonate to compensate for the lack of a CO_2 buffering system. The oxygen scavenger OxyFluor is included to minimize photobleaching and photodamage generated by fluorescence excitation. The extra glucose is included to counterbalance the effects of the oxygen scavenging system.

If a drug with a rapid efflux time, e.g., 2-methoxyestradiol, is being studied, it is important to include the drug in the recording medium (Kamath *et al.*, 2006). In order to more precisely maintain the equilibrium between the drug content of the medium and of the cells, recording medium from incubation of drug with a second identical coverslip of cells can be removed from that second coverslip and added to the cells that are to be imaged (and have already been incubated with drug for the requisite time).

2. Preparation of Cell-Containing Coverslips

Rinse a precoated (see above) coverslip containing attached cells in warm recording media and seal it in a Rose chamber (Rose *et al.*, 1958; Wadsworth, 2007). An updated version of the chamber with dimensions given is shown in Fig. 1. The chamber is essentially a sandwich of cells between two 25 mm diameter coverslips in recording media. Three parafilm or silicone rings are used; one ring is placed between the two coverslips and the other two are placed outside the coverslip sandwich, between the sandwich surfaces and the metal surface of the chamber. Make sure that the cells face the interior of the sandwich. Be careful to avoid including air bubbles.

F. Acquisition of Time-Lapse Images and Analysis of Microtubule Dynamic Instability

We image cells with a Nikon E800 upright microscope using a Nikon 100× 1.4 N.A. oil immersion lens. Others have used a Nikon Eclipse TE300 inverted microscope equipped

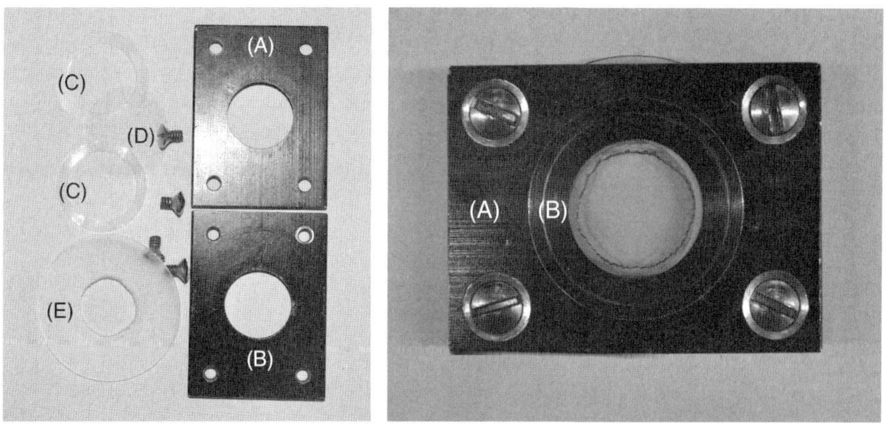

Fig. 1 A recent version of the Rose chamber. Left panel: the separate parts including the following: (A and B) the inner flat surfaces of the top and bottom metal plates, respectively; diameter of hole in each is 19 mm; (C) two Parafilm or silicone rings, inner diameter 19 mm and outer diameter 27 mm, that are placed between the 25 mm diameter coverslips and the metal plates; (D) one of the two 25 mm coverslips; (E) the single internal silicone or Parafilm spacer, inner diameter 16–19 mm, outer diameter > 25 mm, < 40 mm. The rings can be cut from sheets using a cork borer and used repeatedly. Right panel: the assembled chamber. The order of parts from top to bottom is metal plate, ring, coverslip, spacer, coverslip with cells attached, ring, and metal plate. (A) shows the flat outer surface and (B) shows the depressions in the outer surfaces that are machined to allow objective and condenser to touch the coverslip surfaces.

with a 100× 1.3 N.A. objective lens and coupled to a PerkinElmer spinning disc confocal scan head (PerkinElmer) (Salaycik *et al.*, 2005), an Axiovert 200 M (Carl Zeiss, Jena, Germany) equipped with an LSM510META scan head (Carl Zeiss) (Ringhoff and Cassimeris, 2009), a Leica DM-IRBE (Pasquier *et al.*, 2005), or an Axio Image MI (Zeiss) scope with a 63× Plan Fluor 1.4 N.A. objective (Photometrics) (Tischfield *et al.*, 2010).

Cell processes and microtubule dynamics are very sensitive to temperature. To eliminate temperature-induced variability in microtubule dynamics the stage should be enclosed in a Plexiglass or wooden box with an air- or stage-heating system or by some other means kept at a constant temperature of 37 °C. We use a forced air heating system that maintains the sample temperature within 36 ± 1 °C (the lower temperature being a precaution against any unexpected upward temperature fluctuation). Place a temperature probe on the stage next to the sample chamber.

We acquire time-lapse images with a CoolSNAP HQ2 camera (Roper Scientific GmbH, Ottobrunn, Germany) and a Uniblitz shutter system driven by Metamorph software (Universal Imaging, Media, PA). Others have used Princeton Instruments Micromax interline transfer cooled CCD camera (Roper Instruments, NJ, USA) (Salaycik *et al.*, 2005) or a CoolsnapFX CCD camera (Princeton Instruments, Trenton, NJ) (Pasquier *et al.*, 2005). We typically acquired 41 sequential images at 3-s intervals for each cell using approximately 300–1500 ms exposure time and a gain of 2–3. Time-lapse settings will vary depending on the sensitivity of the camera, the brightness of the specimen, the fluorophore used for labeling, and the amount of photobleaching. We sometimes open the aperture only slightly at the beginning of the time-lapse recording, and as the fluorescence photobleaches the aperture is slowly opened to generate a series of images that are more evenly exposed and thus are easier to analyze. Alternatively, the contrast and brightness of each frame can be normalized in Image J or other image-processing software.

Several frames from a time-lapse series of images of an MCF7 human breast cancer cell that has been transfected with GFP-tubulin are shown in Fig. 2. Arrows mark the sequential positions of the plus ends of three microtubules. The plus ends of microtubules can be tracked over time using the "Track Points" function of Metamorph Software (Universal Imaging, Brandywine, PA) or by MetaView imaging software (Molecular

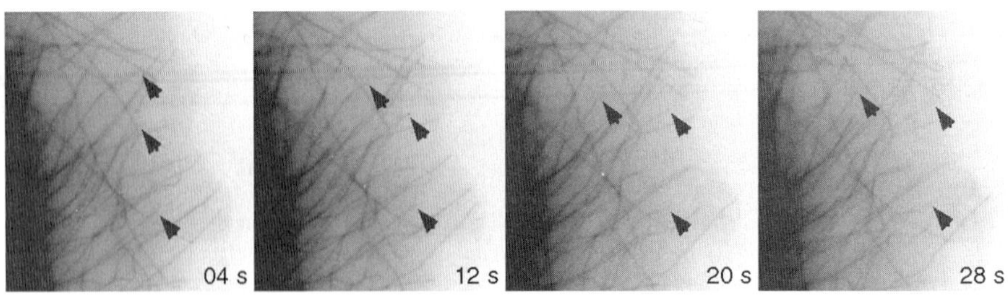

Fig. 2 Time lapse images of microtubules in an MCF7 breast cancer cell that expresses rhodamine-labeled tubulin. Sequential positions of the ends of three microtubules are shown by arrows. The upper microtubule shortens for three frames and has grown slightly in the fourth frame. The other two microtubules undergo growth throughout the sequence.

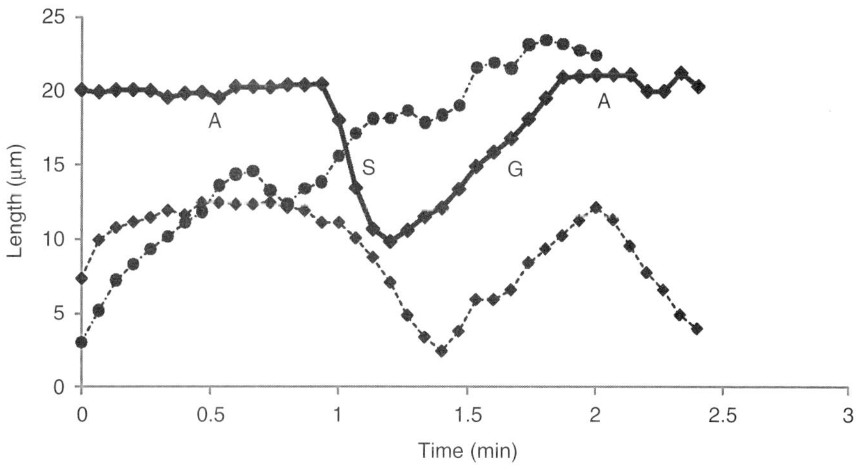

Fig. 3 A life history plot of length changes in three microtubules. One of them (diamonds and solid line) undergoes a phase of attenuation (A) for almost a minute, then undergoes a catastrophe and shortens (S), followed by a rescue and growth (G) and then further attenuation (A).

Devices, Sunnyvale, CA) or MT-LHAP software (see below) which has built-in tracking functions. Image interphase microtubules that have ends clearly discernible at the cell periphery and that persist for 16–40 frames. All clearly defined and trackable microtubules within an image series should be tracked. To keep a record of the microtubules that have been analyzed, print an image of each cell and mark the microtubules on the print with an identifier as they are tracked. The data from Track Points are sent to a Microsoft Excel spreadsheet and converted to Real-Time Measurement (RTM) using Track-to-RTM software (Walker *et al.*, 1988) or to a recently developed software that we now use to automate the tracking and analysis process [Microtubule Life History Analysis Package (MT-LHAP); Oroudjev (2010)]. A typical graph of the changes in microtubule length over time (microtubule life history traces) is shown in Fig. 3. The microtubule dynamics parameters are determined by linear regression using RTM software or by MT-LHAP.

1. Criteria for Analysis of Life History Traces

Analysis of life history traces can be approached in several ways. After careful comparison of several methods, we now use the following procedure: Determine the beginning and end of each event by the overall shape of the microtubule trace. If the beginning or end of an event cannot be determined clearly from the life history plot, retrace the microtubule and compare the two traces. If the two traces are significantly different, retrace the microtubule again. If the ambiguity is not resolved, then compare the traces while viewing the time-lapse sequence to determine which is most accurate.

Changes in microtubule length of less than 0.5 μm (a length chosen to approximate the limits of resolution imposed by microscopy of individual microtubules in the complex environment of living cells) between any two points are considered pause. However, if a change of less than 0.5 μm occurs between two points within a growth event, it is considered a part of the growth event. If a microtubule grows and shortens over several points and has an overall length change of 1.0 μm or less, the microtubule is also considered to be in a paused state. A catastrophe occurs when a microtubule switches from a pause or growth event to shortening; a rescue is when a shortening microtubule switches to either growth or pause. The frequency of catastrophe is the number of catastrophes divided by the total time in growth and attenuation. It can be calculated for each microtubule and then averaged for the population or it can be calculated as a single number for the entire population of microtubules analyzed. The frequency of rescue is the number of rescues divided by the total time in shortening. It can also be calculated on either a "per microtubule" basis or on a population basis.

III. Discussion and Summary

Individual microtubule dynamic instability parameters vary depending on the cell type (Shelden and Wadsworth, 1993). Data from mammalian cells in interphase either injected with rhodamine-tubulin or transfected with GFP-tagged α-tubulin indicate a range of parameter values. For example, the mean growth rates range from 6 to 20 μm/min in A549 human lung carcinoma cells and Chinese hamster ovary (CHO) fibroblasts, respectively. Mean shortening rates vary from 9 to 32 μm/min in A498 human kidney carcinoma and CHO cells, respectively. Mean catastrophe frequencies range from 1/min to 4/min in CaOV3 ovarian adenocarcinoma cells and MDCK kidney cells, respectively. Mean rescue frequencies range from 4/min to 7/min in A549 and A498 cells, respectively (Goncalves *et al.*, 2001; Shelden and Wadsworth, 1993; Wadsworth, 1999; Yvon *et al.*, 1999).

Overall the system described above has allowed the analysis of the mechanisms of action in cells of a large number of drugs and endogenous cellular regulators and has established the crucial importance of microtubule dynamic instability in numerous cellular processes and as a drug target.

Acknowledgments

We thank Dr. Nikki LaPointe and Ms. Jennifer Smith for critically reading the chapter. This study was supported by National Institute of Health CA57291.

References

Balasubramani, M., Nakao, C., Uechi, G., Cardamone, J., Kamath, K., Balogh, K., Balachandran, R., Wilson, L., Day, B., Jordan, M. (2010). Characterization and detection of cellular and proteomic alterations in stable stathmin-overexpressing and taxol-resistant BT549 cells using offgel IEF/PAGE difference gel electrophoresis. Mutation Research. In press.

Ballestrem, C., Wehrle-Haller, B., Hinz, B., Imhof, B. (2000). Actin-dependent lamellipodia formation and microtubule-dependent tail retraction control-directed cell migration. Mol Biol Cell **11**, 2999–3012.

Brust-Mascher, I., Scholey, J. (2002). Microtubule flux and sliding in mitotic spindles of Drosophila embryos. Mol Biol Cell **13**, 3967–75.

Bunker, J. M., Wilson, L., Jordan, M. A., Feinstein, S. C. (2004). Modulation of microtubule dynamics by tau in living cells: implications for development and neurodegeneration. Mol Biol Cell **15**, 2720–8.

Cassimeris, L. (1999). Accessory protein regulation of microtubule dynamics throughout the cell cycle. Curr. Opin. Cell Biol **11**, 134–141.

Desai, A., Mitchison, T. (1997). Microtubule polymerization dynamics. Annu Rev Cell Dev Biol **13**, 83–117.

Dhamodharan, R., Wadsworth, P. (1995). Modulation of microtubule dynamic instability in vivo by brain microtubule associated proteins. J Cell Sci **108** (Pt 4), 1679–89.

Dhamodharan, R. I., Jordan, M. A., Thrower, D., Wilson, L., Wadsworth, P. (1995). Vinblastine suppresses dynamics of individual microtubules in living cells. Mol. Biol. Cell **6**, 1215–1229.

Faire, K., Waterman-Storer, C. M., Gruber, D., Masson, D., Salmon, E.D., Bulinski, J. C. (1999). E-MAP-115 (ensconsin) associates dynamically with microtubules in vivo and is not a physiological modulator of microtubule dynamics. J Cell Sci **112** (Pt 23), 4243–55.

Faller, E., Brown, D. (2009). Modulation of microtubule dynamics by the microtubule-associated protein 1a. J Neurosci Res. **87**, 1080–9.

Galmarini, C. M., Kamath, K., Vanier-Viorney, A., Hervieu, V., Peiller, E., Puisieux, A., Jordan, M. A., Dumontet, C. (2003). Drug resistance associated with loss of p53 involves extensive alterations in microtubule composition and dynamics. Br. J. Cancer **88**, 1793–9.

Goncalves, A., Braguer, D., Kamath, K., Martello, L., Briand, C., Horwitz, S., Wilson, L., Jordan, M. A. (2001). Resistance to taxol in lung cancer cells associated with increased microtubule dynamics. Proc. Natl. Acad. Sci. USA **98**, 11737–11741.

Hyman, A., Drechsel, D., Kellogg, D., Salser, S., Sawin, K., Steffen, P., Wordeman, L., Mitchison, T. (1991). Preparation of modified tubulins. Methods Enzymol **196**, 478–85.

Jordan-Sciutto, K., Logan, T., Norton, P., Derfoul, A., Dodge, G., Hall, D. (1997). Reduction in fibronectin expression and alteration in cell morphology are coincident in NIH3T3 cells expressing a mutant E2F1 transcription factor. Exp Cell Res **236**, 527–36.

Jordan, M., Kamath, K. (2007). How do microtubule-targeted drugs work? An overview.. Curr Cancer Drug Targets **7**, 730–42.

Jordan, M. A. (2002). Mechanism of action of antitumor drugs that interact with microtubules and tubulin. Curr.Med Chem - Anti-Cancer Agents **2**, 1–17.

Jordan, M. A., Wilson, L.1999. , The use and action of drugs in analyzing mitosis. Methods in Cell Biology, **Vol. 61**. Academic Press, 1999, pp. 267–295.

Kamath, K., Okouneva, T., Larson, G., Panda, D., Wilson, L., Jordan, M.A. (2006). 2-Methoxyestradiol Suppresses Microtubule Dynamics and Arrests Mitosis without Depolymerizing Microtubules. Molecular Cancer Therapeutics **5**, 2225–33.

Landen, J., Lang, R., McMahon, S., Rusan, N., Yvon, A., Adams, A., Sorcinelli, M., Campbell, R., Bonaccorsi, P., Ansel, J., et al., (2002). Noscapine alters microtubule dynamics in living cells and inhibits the progression of melanoma. Cancer Res **62**, 4109–14.

Lee, A., Suter, D. (2008). Quantitative analysis of microtubule dynamics during adhesion-mediated growth cone guidance. Dev Neurobiol. **68**, 1363–77.

Lee, G. M. (1989). Measurement of volume injected into individual cells by quantitative fluorescence microscopy. J. Cell Sci. **94**, 443–447.

Mikhailov, A., Gundersen, G.G. (1998). Relationship between microtubule dynamic and lemellipodium formation revealed by direct imaging of microtubules in cells treated with nocodazole or taxol. Cel Motility and Cytoskeleton **41**, 325–340.

Miller, H., Wilson, L.2010. , Preparation of Microtubule Protein and Purified Tubulin from Bovine Brain by Cycles of Assembly and Disassembly and Phosphocellulose Chromatography. In: J. Correia, L. Wilson, Eds.), Methods in Cell Biology, **Vol. 95**. Elsevier, San Diego, 2010, pp. 1–13.

Minaschek, G., Bereiter-Hahn, J., Bertholdt, G. (1989). Quantitation of the volume of liquid injected into cells by means of pressure. Exp Cell Res **183**, 434–42.

Mitchison, T. J., Kirschner, M. (1984). Dynamic instability of microtubule growth. Nature **312**, 237–242.

Moore, A. T., Wordeman, L. (2004). The mechanism, function and regulation of depolymerizing kinesins during mitosis. Trends Cell Biol **14**, 537–546.

Oroudjev, E.2010. , Life history analysis procedures (LHAP) in Igor Pro software to analyze dynamic instability of microtubules in vitro. In, : J. Correia, L. Wilson, Eds.), Methods in Cell Biology, Microtubules in vitro, **Vol. 95**. Elsevier, San Diego, 2010, Appendix to Chapter 11, pages 203–206.

Pasquier, E., Honore, S., Pourroy, B., Jordan, M., Lehmann, M., Briand, C., Braguer, D. (2005). Antiangiogenic concentrations of paclitaxel induce an increase in microtubule dynamics in endothelial cells but not in cancer cells. Cancer Res **65**, 2433–40.

Ringhoff, D., Cassimeris, L. (2009). Stathmin regulates centrosomal nucleation of microtubules and tubulin dimer/polymer partitioning. Mol Biol Cell **20**, 3451–8.

Rose, G.G., Pomerat, C. M., Shindler, T., Trunnell, J. (1958). A cellophane strip technique for culturing tissue in multipurpose culture chambers. J. Biophys. Biochem. Cytol. **4**.

Rusan, N., Fagerstrom, C., Yvon, A., Wadsworth, P. (2001). Cell cycle-dependent changes in microtubule dynamics in living cells expressing green fluorescent protein-alpha tubulin. Mol Biol Cell **12**, 971–80.

Salaycik, K.J., Fagerstrom, C. J., Murthy, K., Tulu, U.S., Wadsworth, P. (2005). Quantification of microtubule nucleation, growth and dynamics in wound-edge cells. J. Cell Sci. **118**, 4113–22.

Saxton, W.M., Stemple, D. L., Leslie, R.J., Salmon, E.D., Zavortnik, M., McIntosh, J.R. (1984). Tubulin dynamics in cultured mammalian cells. J. Cell Biology **99**, 2175–2186.

Schaefer, A., Kabir, N., Forscher, P. (2002). Filopodia and actin arcs guide the assembly and transport of two populations of microtubules with unique dynamic parameters in neuronal growth cones. J Cell Biol. **158**, 139–52.

Shelden, E., Wadsworth, P. (1993). Observation and quantification of individual microtubule behavior in vivo: microtubule dynamics are cell-type specific. J. Cell Biol. **120**, 935–945.

Shelden, E., Wadsworth, P. (1996). Stimulation of microtubule dynamic turnover in living cells treated with okadaic acid. Cell Motility and the Cytoskeleton **35**, 24–34.

Tischfield, M., Baris, H., Wu, C., Rudolph, G., Van Maldergem, L., He, W., Chan, W. M., Andrews, C., Demer, J., Robertson, R., et al., (2010). Human TUBB3 mutations perturb microtubule dynamics, kinesin interactions, and axon guidance. Cell **140**, 74–87.

Toso, R. J., Jordan, M. A., Farrell, K. W., Matsumoto, B., Wilson, L. (1993). Kinetic stabilization of microtubule dynamic instability in vitro by vinblastine. Biochemistry **32**, 1285–93.

Wadsworth, P. (1999). Regional regulation of microtubule dynamics in polarized, motile cells. Cell Motil Cytoskeleton **42**, 48–59.

Wadsworth, P., Studying Mitosis in Cultured Mammalian Cells. Cold Spring Harb. Protoc., 2007 doi:10.1101/pdb.prot4674

Wadsworth, P., Bottaro, D.P. (1996). Microtubule dynamic turnover is suppressed during polarization and stimulated in hepatocyte growth factor scattered Madin-Darby canine kidney epithelial cells. Cell Motil. Cyt. **35**.

Walczak, C.E. (2000). Microtubule dynamics and tubulin interacting proteins. Curr Opin Cell Biol **12**, 52–6.

Walker, R. A., O'Brien, E. T., Pryer, N. K., Soboeiro, M. F., Voter, W. A., Erickson, H., Salmon, E. D. (1988). Dynamic instability of individual microtubules analyzed by video light microscopy: Rate constants and transition frequencies. J Cell Biol. **107**, 1437–1448.

Waterman Storer, C., Salmon, E. D. (1997). Actomyosin-based retrograde flow of microtubules in the lamella of migrating epithelial cells influences microtubule dynamic instability and turnover and is associated with microtubule breakage and treadmilling. J. Cell Biol. **139**, 417–34.

Waterman-Storer, C. M., Salmon, W. C., Salmon, E. D. (2000). Feedback interactions between cell-cell adherens junctions and cytoskeletal dynamics in newt lung epithelial cells. Mol Biol Cell **11**, 2471–83.

Yvon, A. -M., Wadsworth, P., Jordan, M. A. (1999). Taxol suppresses dynamics of individual microtubules in living human tumor cells. Mol Biol Cell **10**, 947–949.

Yvon, A. M., Wadsworth, P. (1997). Non-centrosomal microtubule formation and measurement of minus end microtubule dynamics in A498 cells. J Cell Sci **110** (**Pt 19**), 2391–401.

Zhang, D., Wadsworth, P., Hepler, P. K. (1990). Microtubule dynamics in living dividing plant cells: confocal imaging of microinjected fluorescent brain tubulin. Proc Natl Acad Sci U S A **87**, 8820–4.

CHAPTER 2

Analysis of Microtubule Polymerization Dynamics in Live Cells

Sarah Gierke, Praveen Kumar, *and* Torsten Wittmann
Department of Cell & Tissue Biology, University of California, San Francisco, California 94143-0512

Abstract

The spatiotemporal regulation of intracellular microtubule polymerization dynamics, by numerous microtubule-associated proteins and other mechanisms, is central to many cell processes. Here, we give an overview and practical guide on how to acquire and analyze time-lapse sequences of dynamic microtubules in live cells by either fluorescently labeling entire microtubules or by utilizing proteins that specifically associate only with growing microtubule ends and summarize the strengths and weaknesses of different approaches. We give practical recommendations for imaging conditions, and discuss important limitations of such analysis that are dictated by the maximum achievable spatial and temporal sampling frequencies.

978-0-12-381349-7
DOI: 10.1016/S0091-679X(10)97002-7

I. Introduction

Microtubules (MTs) are highly dynamic cytoskeletal polymers composed of α/β tubulin dimers. Precise regulation of intracellular MT dynamics is important for many biological processes ranging from proper attachment and segregation of chromosomes during mitosis (Wittmann *et al.*, 2001) to local stabilization of MTs toward the front of migrating cells (Wittmann and Waterman-Storer, 2001). MTs in cells and *in vitro* stochastically switch between phases of growth and shortening. This nonequilibrium polymerization behavior has been termed dynamic instability (Mitchison and Kirschner, 1984) and is driven by different structural states of the MT end, which are ultimately the result of polymerization-coupled GTP hydrolysis in the MT lattice (Fig. 1) (Nogales and Wang, 2006). At the plus end of a growing MT, GTP-loaded tubulin dimers initially polymerize as a relatively flat open sheet which subsequently closes into a tube. Shortly after polymerization GTP is hydrolyzed to GDP within the

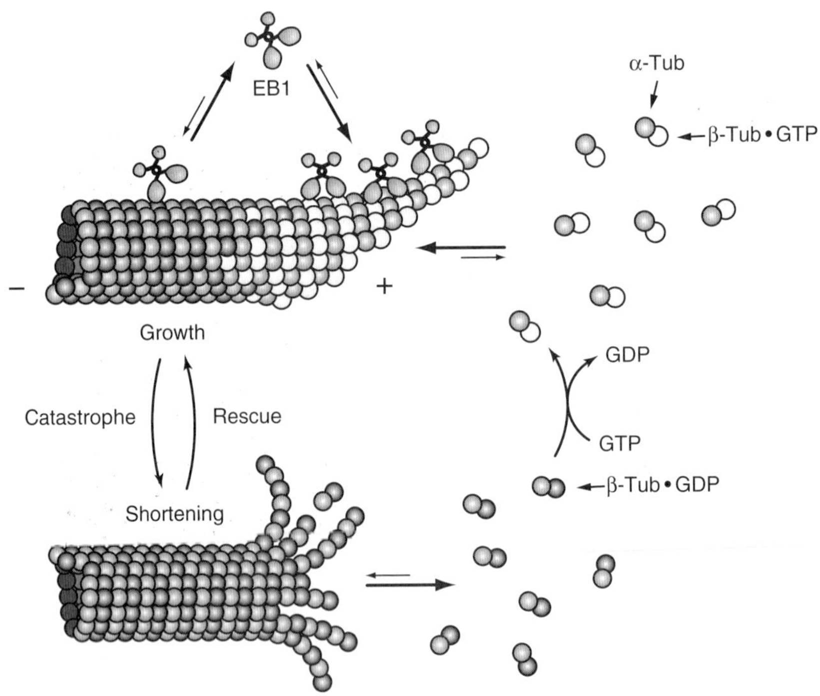

Fig. 1 Diagram of different phases of MT polymerization dynamics. MT growth is thought to be accompanied by a protective cap of GTP-loaded tubulin at growing MT ends. The structurally distinct GTP-tubulin cap also provides a platform for the binding of MT plus end tracking proteins, +TIPs (Akhmanova and Steinmetz, 2008), that can be used as indirect reporters of intracellular MT polymerization dynamics.

MT lattice, which is thought to leave a short GTP-tubulin cap at the tip of the MT. This remaining GTP-cap stabilizes the growing end and supports further addition of GTP-loaded tubulin subunits, resulting in a stable growth phase. In contrast, loss of the GTP-cap results in catastrophic depolymerization, and by electron microscopy highly curved protofilaments seem to peel away from the depolymerizing MT end. These frayed MT ends reflect the high intrinsic curvature of GDP-loaded tubulin dimers, and do not support addition of new GTP-loaded tubulin subunits. Thus, this large structural difference between polymerizing and depolymerizing MT ends is sufficient to explain the high persistency and abrupt switching between growing and shortening that characterizes MT dynamic instability *in vitro* (Kueh and Mitchison, 2009).

Four parameters are generally measured to describe MT polymerization dynamics: the rates of growth (polymerization), shortening (depolymerization), and the transition frequencies between these two states. The transition from growth to shortening is referred to as "catastrophe," and the transition from shortening to growth is referred to as "rescue" (Fig. 1). These parameters can be determined quite easily in *in vitro* polymerization reactions with purified components because MT growth and shortening rates are relatively constant and transitions occur infrequently. In cells, however, MT polymerization dynamics are spatiotemporally highly regulated by a large number of accessory proteins (van der Vaart *et al.*, 2009) as well as physical interactions with other intracellular structures (Dogterom *et al.*, 2005). As a result intracellular MT polymerization dynamics are significantly more complex and more difficult to quantify. *In vivo*, only MT plus ends exhibit dynamic instability. Free minus ends are either stabilized or depolymerize. Both growth and shortening rates are highly variable, and rates of individual MTs fluctuate significantly over relatively short time periods. In addition, MT polymerization dynamics in cells often include relatively long periods of pause during which MT ends do not appear to grow or shorten within the resolution limit of the light microscope. Furthermore, intracellular MTs are subject to pulling and pushing forces, which result in MT buckling, breakage, and movements of MT ends that are not due to polymerization dynamics (Brangwynne *et al.*, 2007; Waterman-Storer and Salmon, 1997; Wittmann *et al.*, 2003).

II. Rationale

The objective of this chapter is to give an overview of different techniques to observe and analyze intracellular MT dynamics either by continuous MT labeling or by the expression of fluorescently labeled proteins that specifically recognize growing MT ends. We aim to emphasize the strength and limitations of each approach, and discuss the theoretical boundaries of intracellular MT dynamics analysis that are imposed by the spatial and temporal resolution limits of light microscopy. Finally, it is important to note that these fundamental limitations similarly impact other intracellular tracking problems such as, for example, vesicular trafficking.

III. Imaging and Analysis of Homogeneously Labeled MTs

A. Probes to Visualize Dynamic MTs

Conventional analysis of intracellular MT dynamics involves homogenous labeling of the entire MT network. This was initially achieved by microinjection of tubulin in which surface amino groups that are exposed in polymerized MTs are chemically labeled using *N*-hydroxysuccinimide-derivatized fluorescent dyes (Sammak and Borisy, 1988; Shelden and Wadsworth, 1993). Several protocols are published describing tubulin labeling and microinjection procedures (Hyman *et al.*, 1991; Waterman-Storer, 2002; Wittmann *et al.*, 2004), and fluorescently labeled tubulin is also available commercially (e.g., from Cytoskeleton Inc.). Because microinjection is technically difficult, time-consuming, and only very few cells are available for analysis per experiment, this has mostly been replaced by the exogenous expression of tubulin tagged with fluorescent proteins (FPs). However, it should be noted that fluorescent dye-conjugated tubulin has some advantages over FP-tagged tubulin. Synthetic fluorescent dyes are generally brighter than FPs due to a higher extinction coefficient and better quantum yield, and because fluorescent dyes are small, dye-conjugated tubulin appears to be more efficiently incorporated into MTs. Together, this results in a higher MT to cytoplasm fluorescence ratio than FP-tagged tubulin. Finally, photobleaching of synthetic fluorophores is largely oxygen dependent and can be efficiently inhibited by oxygen depletion from the tissue culture medium (Waterman-Storer, 2002; Wittmann *et al.*, 2003, 2004).

Nevertheless, we and others have successfully used FP-tagged α or β tubulin to image and analyze intracellular MT polymerization dynamics (Kumar *et al.*, 2009; Rusan *et al.*, 2001). Although a growing toolbox of FPs is now available, EGFP-tagged tubulin still appears to be the brightest and most photostable variant. For dual-color imaging we have also successfully used mCherry-tagged tubulin (Fig. 2 A). In addition, recent advances in elucidating the mechanisms of FP photobleaching indicate that FP photostability can be significantly improved by using riboflavin-free media (Bogdanov *et al.*, 2009). Plasmid vectors suitable for transient transfections encoding different FP-tubulin fusion proteins are available from a variety of sources. We also routinely use adenovirus particles to transiently introduce FP-tagged tubulin and other cytoskeleton proteins into difficult-to-transfect cells (Kumar *et al.*, 2009). However, because the correct folding of α/β tubulin dimers relies on a complex pathway involving several specific chaperones (Szymanski, 2002), care should be taken not to overwhelm the cells biosynthetic machinery by using too much virus. Too rapid expression of tagged tubulin results in poor incorporation into the MT cytoskeleton and excessive cytoplasmic background.

While stable mammalian cell lines expressing FP-tubulins have been made and are viable (Rusan *et al.*, 2001), it has so far not been possible to generate whole animals expressing FP-tagged tubulin, indicating that the FP-tag does disrupt developmentally important tubulin functions. In an alternative approach, mice expressing the MT-binding domain of a MT-associated protein, ensconsin, display homogeneously labeled MTs and are viable (Lechler and Fuchs, 2007). In addition, the MT signal can be increased by attaching up to five GFP moieties to ensconsin (Bulinski *et al.*, 2001). Although GFP-ensconsin does not appear to modify intracellular MT dynamics, appropriate controls should be included when expressing exogenous MT-binding domains.

B. Preparation of Purified, Concentrated Adenovirus Particles

Although extensive adenovirus methods are published (Luo *et al.*, 2007), we include a short reference protocol for the preparation of concentrated, purified adenovirus particles that we routinely use to prepare virus stock to introduce FP-tagged proteins in many different cell types. AdEasy-based viral genomes for the expression of EGFP-tubulin and EB1-EGFP from our lab are available through AddGene. Although these viruses are replication deficient and new virus can only be produced in the packaging cell line, the experimentalist should be aware that these are infectious particles. At all times adhere to the required BSL-2 safety precautions, and sterilize and dispose infectious material according to the appropriate local regulations.

1. Required Materials

PacI-linearized and purified AdEasy viral plasmid containing the gene of interest
Transfection reagent (e.g., Lipofectamine 2000, Invitrogen Cat. No. 11668-027)
Adenovirus packaging cell line (AD-293, Stratagene)
Dulbecco's Modified Eagle Medium (DMEM, Invitrogen Cat. No. 10313) supplemented with 10% Fetal Bovine Serum (FBS, Invitrogen Cat. No. 26140), 10 mM $MgCl_2$, 2 mM L-glutamine (Invitrogen Cat. No. 25030), 1× Penicillin/Streptomycin (Invitrogen Cat. No. 15140)
10 mM Tris-Cl pH 8.0
Low-density CsCl buffer ($\rho = 1.2$ g/ml): Dissolve 35 g CsCl in a final volume of 100 ml 10 mM Tris-Cl pH 8.0
High-density CsCl buffer ($\rho = 1.45$ g/ml): Dissolve 53 g CsCl in a final volume of 100 ml 10 mM Tris-Cl pH 8.0
ARCA buffer: 10 mM Tris-Cl, pH 8.0, 1 mM $MgCl_2$, 5% sucrose, 1% glycine, 0.05% Tween-80
Beckman ultracentrifuge and tubes: 38.5 ml (Beckman Cat. No. 41103909 for SW 28 Ti rotor) and 13.2 ml (Beckman Cat. No. 41103909 for SW 41 Ti rotor)
Econo-Pac 10DG Desalting Columns (Bio-Rad Laboratories Cat. No. 732-2010)

2. Adenovirus Production and Amplification

1. Transfect ~50% confluent AD-293 cells in a 6 cm dish with the linearized AdEasy viral genome according to the manufacturer's instructions in antibiotic-free DMEM. We commonly use Lipofectamine 2000. AD-293 cells do not adhere well to tissue culture plastic and care should be taken not to disrupt the monolayer after transfection.

2. Grow transfected cells until the cytopathic effect (CPE) of virus production becomes visible. At this point, cells start to round up and lift off from the bottom of the plate. Depending on transfection efficiency, this will take approximately 1–2 weeks. When the medium becomes acidic, gently change medium without disrupting the cell monolayer. Transfection efficiency can also be estimated by fluorescence microscopy because transfected cells will express the EGFP-tagged fusion protein.
3. Harvest all cells and media by gently tapping the plate and pipetting up and down into a 15-ml centrifuge tube. Centrifuge at 4°C, 1200 rpm, 5 min. Discard supernatant and resuspend pellet in 1 ml sterile 10 mM Tris-Cl pH 8.0.
4. Lyse cells by freezing for 5–10 min in a dry ice/EtOH bath, then thaw completely in a 37°C water bath. Repeat freeze/thaw cycle three times. Centrifuge to remove cell debris at 4000 rpm for 20 min, 4°C. The supernatant now contains first generation virus, which is used for all subsequent amplifications. Remove 200 μl for the next step and freeze the rest at –80°C.
5. For subsequent virus amplification, infect a 10 cm dish of ~80% confluent AD-293 cells: Remove medium from cells and gently add 200 μl of the first generation virus diluted to 2 ml in DMEM. After 1 h add additional 8 ml DMEM. Harvest cells as in Step 3 as soon as CPE becomes evident.
6. Repeat the infection and amplification cycle (Step 3–5) until CPE is visible within 48 h postinfection. This should take a total of 2–3 cycles.

3. Large-Scale Adenovirus Production

7. Prepare 20 large plates (15 cm diameter or similar) of AD-293 cells. To ensure that cells are plated evenly, prepare cell suspension in a media bottle, mix, and plate 20 ml into each dish. At this point, DMEM with 5% FBS can be used to conserve serum.
8. Grow cells to ~90% confluency. Dilute 0.5 ml virus from Step 6 into 100 ml DMEM. Infect each plate with 5 ml of the diluted virus, mix gently, and return plates to incubator. After 48 h the virus-induced CPE should be clearly visible.
9. Harvest the cells and medium by pipetting up and down or by using a cell scraper. Centrifuge at 1200 rpm, 10 min, 4°C. Discard supernatant and resuspend pellet in 13 ml of 10 mM Tris-Cl pH 8.0.
10. Lyse cells with three freeze/thaw cycles (as in Step 4), and vortex for 30 s after each thaw. Centrifuge at 4000 rpm, 20 min, 4°C. Collect and keep supernatant on ice. It is very important for the subsequent purification step that all cell debris is removed.

4. Adenovirus Purification by Cesium Chloride Density Gradient Centrifugation

11. Prepare two CsCl step gradients in 38.5-ml Beckman Ultraclear centrifuge tubes: First, pipet 10 ml of low-density CsCl buffer into centrifuge tube. Then gently underlay 10 ml of high-density CsCl by carefully inserting the pipet to the bottom of the tube

and slowly dispelling the solution. In one of the tubes, gently overlay virus supernatant on top of the step gradient. Top off tube completely with 10 mM Tris-Cl pH 8.0 to avoid collapse during ultracentrifugation. Top off the second CsCl gradient tube with 10 mM Tris-Cl pH 8.0 to use as a balance. Make sure the tubes are well balanced and that everything is kept sterile in a BSL-2 laminar flow tissue culture hood.

12. Centrifuge at 20,000 rpm in an SW 28 Ti swinging bucket rotor with slow acceleration and no brakes, for 2 h at 4°C. Unload tubes in tissue culture hood and place tube in ring stand near eye level. A bluish band about halfway down the tube contains purified, concentrated virus. Just above there is usually a second, fainter band containing defective virus particles. During extraction of the virus band, care should be taken to avoid contamination from this upper band.
13. Remove some of the excess liquid from above the virus, being extremely careful not to disturb the band. Wipe the outside of the tube with ethanol to sterilize. Stick a piece of adhesive tape on the side of the tube where it is to be punctured to avoid leakage. Carefully pierce the tube with a hypodermic syringe with an 18-gauge needle slightly below the bluish virus band. Be sure to use as little force as possible in order not to pierce through both sides of the tube. Remember that the band contains highly concentrated virus particles. Thus, keep hands out of the way, and make sure to wear appropriate personal protective equipment. Tilt the needle upward into the virus band and draw approximately 3 ml of the virus slowly into the syringe. Leave the syringe with needle in the tube and pipet liquid from the tube until it is below the needle. Now remove the syringe and expel the virus into a sterile 15-ml polypropylene tube.
14. Dilute virus to 4 ml with 10 mM Tris-Cl pH 8.0. This is essential to reduce the density to below 1.2 g/ml. Prepare a smaller step gradient with 4 ml of each CsCl buffer as in Step 11, but in smaller 13.2 ml tubes. Overlay with the diluted virus and fill tube with 10 mM Tris-Cl pH 8.0. Centrifuge at 20,000 rpm for 2 h at 4°C in an SW 41 Ti rotor and isolate virus band as in Step 13.
15. Desalt virus on a 10DG column: Equilibrate column with 30 ml ARCA buffer. Load 3 ml of the virus and collect 0.5 ml fractions on ice. Concentrated virus should start eluting in the third fraction. Concentrated, opaque fractions can be pooled, aliquoted, and stored at –80°C. This procedure yields highly infectious adenovirus particles, and we generally use less than 1 μl adenovirus stock per 3.5 cm dish to infect cells for microscopy. Virus titer can be estimated by measuring optical density at 260 nm in PBS containing 1% SDS (1 OD = ~1.1 × 10^{12} virus particles per ml).

C. Imaging of Intracellullar MT Dynamics

Analysis of intracellular MT dynamics relies on time-lapse imaging of fluorescently labeled proteins. Different modalities of fluorescence microscopy including wide-field epifluorescence, confocal, and total internal reflection (TIRF) microscopy can and have been used to image MT dynamics, although each has its limitations. In wide-field images, out-of-focus blur severely limits the ability to observe MTs in thicker cell regions and only works reasonably well in flat, peripheral cell areas (Wittmann *et al.*, 2003).

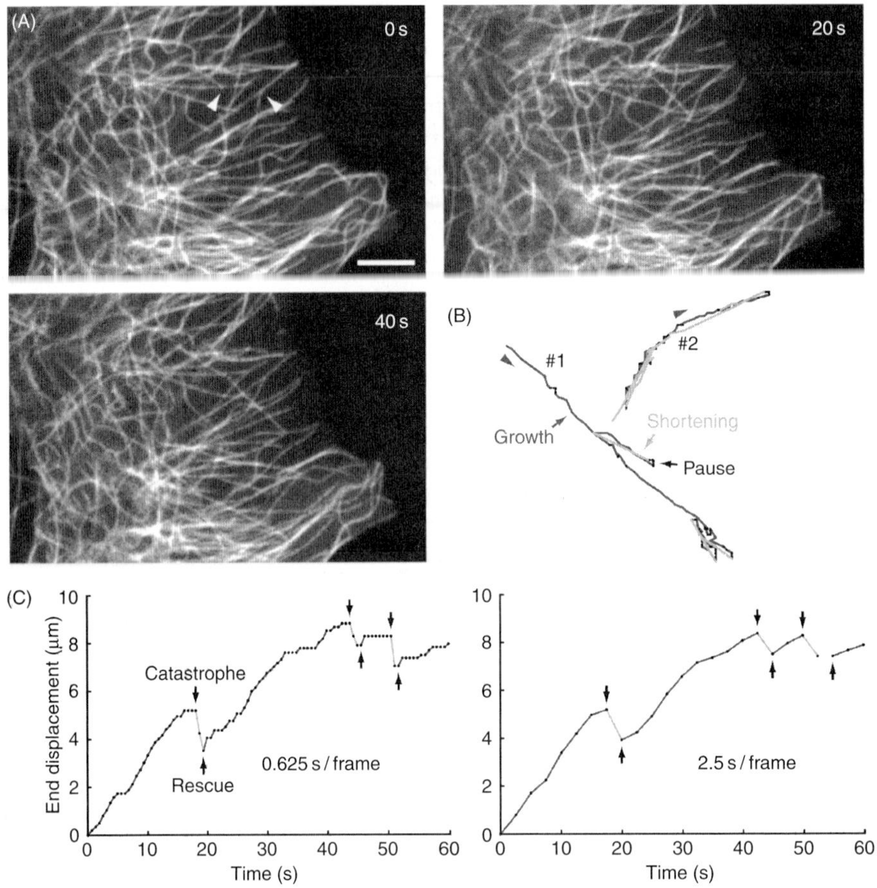

Fig. 2 Semi-manual analysis of continuously labeled MTs. (A) Images from a time-lapse sequence of mCherry-tagged tubulin. Images were acquired every 625 ms (1.6 frames per second) for 1 min. Scale bar, 5 μm. (B) Traces of the two MT ends indicated by arrowheads in A obtained by computer-assisted hand-clicking. (C) Life history plots of MT #1 at the original time resolution, and at a simulated frame rate of 0.4 frames per second by only analyzing every fourth image. This demonstrates the averaging of the high variability of intracellular MT polymerization dynamics at slower frame rates.

Although TIRF microscopy produces superior contrast, it only allows imaging of MTs in very close proximity of the bottom cell membrane (Krylyshkina *et al.*, 2003)(Chapter 6, this book). We mostly use spinning disk confocal microscopy as a versatile method to image dynamic MTs because it combines thin optical sectioning, which largely eliminates out-of-focus blur, with better sample penetration compared with TIRF.

Independent of the exact microscopy modality used, it is important to acquire images at the best possible spatial resolution. At the emission wavelength of EGFP of about 510 nm, the resolution limit of a fluorescence light microscope with a 1.4 NA objective lens under optimal conditions is $\sim$250 nm. Because tubulin dimers are about 8 nm long and are thus small compared to this approximate diameter of the microscope point spread function, any discernible MT length change represents the addition or removal of several hundred tubulin subunits. In order to be able to image changes as close to the molecular scale as possible, it is important that the digital camera used to acquire images is of sufficiently high resolution to oversample the optical resolution to fulfill the Nyquist sampling criterion (Wittmann *et al.*, 2004). This means that the effective pixel size of the captured image should be $\sim$two- to three-fold smaller than the resolution limit. On a microscope system with no intermediate magnification this can be achieved with a 60× or 100× objective and a high-resolution scientific grade charge-coupled device (CCD) camera. Such cameras from different manufacturers typically use a Sony Interline CCD sensor with 6.45 × 6.45 μm pixels resulting in an effective pixel size of $\sim$107 nm in object space at 60× magnification. Electron multiplying CCD (EMCCD) cameras have much larger pixels up to 16 × 16 μm (effective pixel size of 266 nm at 60×), which is insufficient for optimal sampling of the optical resolution without additional magnification (Up-to-date information and excellent tutorials about many aspects of light microscopy can also be found online, for example, at www.microscopyu.com). For similar reasons, the use of two-photon microscopy (which halves the optical resolution) or binning (which halves the detector resolution) is not optimal for MT dynamics imaging. Finally, it is important to remember that optimized imaging conditions are only achieved when cells are grown in two dimensions preferably on #1.5 cover glasses. We typically use glass-bottom tissue culture dishes (e.g., from MatTek Corporation) or custom-made sealed cover glass chambers (Wittmann *et al.*, 2004). The imaging system should also be environmentally controlled because, as any biochemical reaction, the rate of MT polymerization is temperature dependent. Imaging at high resolution becomes significantly more challenging for more physiological, three-dimensional samples.

Our lab uses a Yokogawa CSU-10 spinning disk confocal head with 200 mW 488 nm and 561 nm solid-state lasers in an LMM5 laser launch (Spectral Applied Research). In combination with 60× or 100× 1.49 NA TIRF objective lenses (Nikon) and a high-resolution CoolSNAP cooled CCD camera (Photometrics), we can achieve the highest spatial resolution possible with conventional light microscopy. Image acquisition is controlled by Nikon NIS-Elements software. With this setup we have been able to achieve frame rates of up to 4 frames per second at below 200 ms exposure times, although as outlined below manual tracking of continuously labeled MTs becomes highly error prone at time intervals shorter than 2–5 s between frames. Thus, slightly longer exposure times ($\sim$500 ms) combined with lower illumination intensities are usually acceptable.

Because MT polymerization dynamics analysis requires long time-lapse sequences, it is crucial to minimize sample exposure and photodamage. There are relatively simple aspects in the design of an imaging system that are often overlooked, but significantly improve the performance of fluorescent live cell microscopy. Excitation shutters should be hardware controlled and triggered directly from the camera so that shutters are only open when an image is actually acquired. In contrast, software shutter control typically adds several hundred milliseconds of sample exposure before and after image acquisition, which results in significantly faster photobleaching especially at low exposure times. In addition, it is important to use optimized emission filter sets. For example, for single-channel EGFP imaging, we use a long-pass dichroic mirror and emission filter (Semrock) with edge wavelengths of ~500 nm. This allows maximum detection over most of the EGFP emission spectrum and results in a greater than 50% signal increase at the same exposure settings compared to high-quality multiband pass filter sets.

D. Semi-manual Tracking and Analysis of Dynamic MTs

Accurate, automatic detection of the ends of continuously labeled MTs is a highly challenging computational problem. Although substantial progress is being made to develop computer vision algorithms to detect and track movement of continuously labeled MTs (Altinok *et al.*, 2007; Hadjidemetriou *et al.*, 2008) a robust solution is not currently available. However, computer-assisted hand-tracking in which the user manually selects MT ends in a time-lapse sequence can be done with a variety of available software packages (Fig. 2 B). We now mostly use the tracking function in NIS-Elements (Nikon), but similar functions are available in other image analysis software packages including MetaMorph (Molecular Dynamics) and ImageJ.

The output of such semi-manual tracking is a list of positions or displacements of the MT end as a function of time. It is important to remember that this list does not include information on whether an end displacement represents MT growth or shortening, and after the initial tracking we go through the list a second time to designate shortening events by negative numbers. MT end displacement can then be plotted (Fig. 2 C) and analyzed in different ways to calculate growth and shortening rates, transition frequencies, and the percent time MTs spend in growth, shortening, or pause phases. Because of the large variations of intracellular MT growth rates, we find it difficult to a priori designate phases of constant growth or shortening velocities in such life history plots (Walker *et al.*, 1988). Instead, we calculate instantaneous growth and shortening rates on a frame-to-frame basis. Because of the positional uncertainty introduced by image formation in the microscope and the error introduced by hand-clicking on the image, we then set a lower threshold of MT end displacement in the range of the optical resolution (Kumar *et al.*, 2009; Wittmann *et al.*, 2003). Growth or shortening events below this threshold are classified as a pause.

A drawback of this method is that the positional error becomes dominant if sequences are analyzed that are acquired at very short time intervals, which results in completely different measurements of MT dynamics parameters depending on the frame rate. In the example in Fig. 2, images were acquired at 1.6 frames per second (625 ms between images). Assuming a localization error of 1 pixel, which is about half the diameter of the point spread function and probably an overestimation of hand-clicking accuracy, any growth or shortening event below 10 μm/min is classified as a pause, overestimating the time MTs spend in a pause state (Fig. 3A). This also

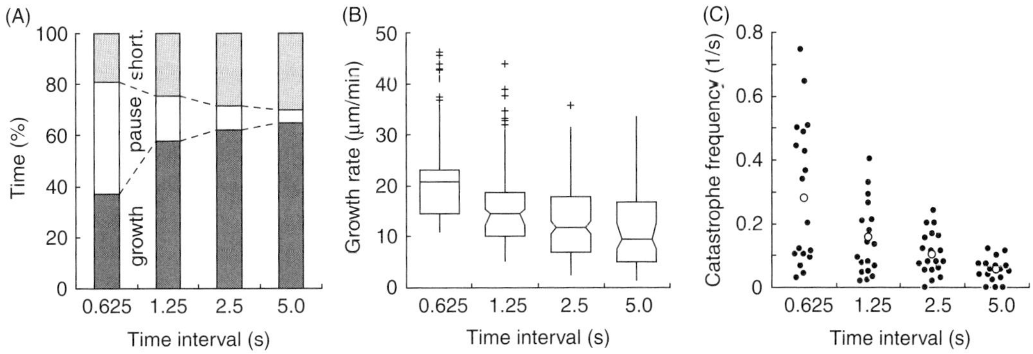

Fig. 3 Dependence of MT polymerization dynamics quantification on temporal resolution. (A) Time MTs spend in different phases, (B) growth rates, and (C) catastrophe frequencies as a function of frame rate. 19 MTs from the sequence in Fig. 2 were analyzed, and simulated frame rates were obtained by temporal subsampling.

truncates the lower end of the true growth and shortening rate populations and results in a gross overestimation of the average rates (Fig. 3B). At lower frame rates, this truncation of the rate populations becomes more reasonable, and the measured growth and shortening rates better approach average intracellular rates. However, at lower frame rates some of the heterogeneity of intracellular MT dynamics is lost (Fig. 2C). In addition, because of the positional detection limit, slow MT polymerization events cannot be distinguished from *bona fide* pauses during which no polymerization or depolymerization occurs. Thus, the threshold below which an event is considered a pause is relatively arbitrary and largely depends on the imaging conditions.

Transition frequencies are defined as follows: The catastrophe frequency is the number of transitions from growth to shortening divided by the time MTs spend growing. This is mathematically identical to the inverse of the average time MTs spend in an uninterrupted growth phase. Because true pauses cannot be clearly defined, we allow growth phases preceding a catastrophe to be interrupted by apparent pauses. Importantly, because events that occur between two acquired images are not observable, the imaging frame rate also defines an upper boundary for transition frequencies, and faster frame rates will result in the measurement of higher transition frequencies (Fig. 3C). For example, at a slower time resolution of 0.2 frames per second (i.e., an image acquired every 5 s), the shortest observable growth interval is 5 s. Thus, the maximal observable catastrophe frequency is identical to the frame rate. Similarly, pauses shorter than 5 s cannot be observed contributing to the apparent increase in the time MTs spend in growth at lower frame rates (Fig. 3A). Likewise, the rescue frequency is defined as the number of transitions from shortening to growth divided by the time MTs spend shortening, and the same limitations as for catastrophe frequencies apply.

In conclusion, because of the inevitable MT end localization error semi-manual analysis of MT dynamics is best performed on time-lapse sequences with images acquired every 2–5 s at high spatial resolution, which represents the range of frame rates used most frequently in published analysis of MT dynamics. These conditions are a good compromise to minimize the error introduced by positional inaccuracy, while still maintaining a relatively high time resolution for the quantification of transition frequencies. However, one should be aware of the inherent limitations of this analysis, and it is imperative to only compare quantifications obtained from time-lapse sequences acquired at identical magnification and frame rate (Shelden and Wadsworth, 1993). In addition, because semi-manual tracking is extremely time-consuming, the number of MTs analyzed is often low and it is challenging to obtain statistically sufficiently large data sets. Nevertheless, growth and shortening rates can be determined with good accuracy. Relatively large standard deviations on rate measurements reflect the variability of intracellular MT polymerization dynamics rather than measurement errors. In contrast, the number of observed catastrophes or rescues tends to be low, and care should be taken in interpreting differences in transition frequencies in different conditions. For any meaningful quantification, we would recommend to track 5–10 MTs per cell in at least five cells and an observation time of around 5–10 min per MT.

IV. MT Fluorescent Speckle Microscopy

In time-lapse sequences of continuously labeled MTs, it is not possible to distinguish between MT polymerization dynamics and MT translocation. However, fluorescent speckle microscopy (FSM) can be used to test whether MT end displacements are due to movements of the entire MT. The principle underlying MT FSM is simple and relies

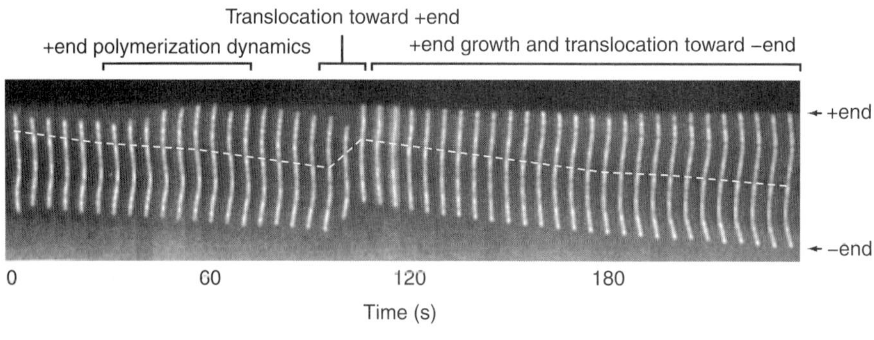

Fig. 4 MT fluorescent speckle microscopy. Montage of a time-lapse sequence of a MT fragment in a cell expressing constitutively active Rac1 (Wittmann *et al.*, 2003) injected with X-rhodamine-labeled tubulin. One bright speckle is followed over time and highlighted by the dashed line. The speckle pattern reveals different types of MT movement that would be misinterpreted as polymerization dynamics if the MT were homogenously labeled. For example, in the second half of the sequence, the MT plus end remains in close contact with the cell edge and is apparently stationary. However, the appearance of new speckles is evidence for polymerization at the plus end and translocation of the speckle pattern demonstrates movement of the MT polymer.

on the stochastic incorporation of fluorescently labeled tubulin subunits along the MT. At low intracellular ratios of fluorescently labeled to unlabeled tubulin, convolution with the microscope's point spread function results in intensity variations (fluorescent speckles) along the MT (Waterman-Storer and Salmon, 1998; Wittmann *et al.*, 2004). Although best speckle contrast is achieved at labeling ratios of below 1% labeled tubulin, intensity variations along MTs can be observed at much higher labeling ratios and are often evident in cells expressing low to moderate levels of FP-tagged tubulin. Simple image processing such as low-pass filtering to remove camera pixel noise, and sharpening with an unsharp mask filter can be used to greatly increase speckle contrast (Wittmann *et al.*, 2004). Because MTs only exchange subunits at the ends, the pattern of intensity variations along the MT is stable and can be used as a direct read-out for MT translocation (Fig. 4).

V. Imaging and Analysis of Growing MT Ends

A. Probes to Visualize Growing MT Ends

Analysis of MT dynamics in time-lapse sequences of continuously labeled MTs suffers from a fundamental limitation. Even in cell areas in which MTs are only moderately dense, it quickly becomes impossible to clearly observe growing ends. Thus, conventional analysis of intracellular MT dynamics is subjective and regionally biased because it is limited to a small subpopulation of MTs near the cell periphery, where MT ends can be observed clearly over sufficient periods of time (Wittmann *et al.*, 2003).

An alternative strategy to visualize intracellular MT dynamics utilizes fluorescently tagged proteins that specifically recognize growing MT plus ends (Akhmanova and Steinmetz, 2008; Salaycik *et al.*, 2005). Because these proteins, commonly referred to as +TIPs, bind only weakly along MTs, growing MT ends are clearly visible in more central cell areas in which MTs are too dense to visualize MT ends directly (compare images of the same cell in Figs. 2A and 5A).

End binding proteins (EBs) are the +TIP prototype and are thought to directly recognize the structurally distinct GTP-tubulin cap at growing MT ends (Fig. 1). EBs are small dimeric proteins containing an N-terminal MT-binding domain, and a C-terminal cargo-binding domain, and most if not all other +TIPs associate with growing MT ends through interactions with this C-terminal domain (Akhmanova and Steinmetz, 2008; Bieling *et al.*, 2008; Honnappa *et al.*, 2009). Because N-terminal FP-tags interfere with EB localization to MT ends (Skube *et al.*, 2009), C-terminally tagged EB1- or EB3-EGFP constructs have been predominantly used to highlight growing MT ends in cells. The exponential decay of available binding sites results in the characteristic comet-like fluorescence profiles of EGFP-tagged EBs on MT ends. In addition, rapid binding kinetics of EBs (Dragestein *et al.*, 2008) causes rapid loss of EB fluorescence from nonpolymerizing MT ends and rapid appearance of EB comets when MTs start growing. However, because EBs are central adaptor proteins that recruit many other +TIPs to growing MT ends, concerns have grown that EB-EGFP constructs may disrupt endogenous localization of other +TIPs and thus alter MT polymerization dynamics and cell behavior (Skube *et al.*, 2009). Nevertheless, low-level EB-EGFP expression appears relatively benign and we and others have made multiple stable EB1-EGFP-expressing cell lines that appear to behave normally. Similar to FP-tagged tubulin we also use adenovirus to introduce EB1-EGFP into difficult-to-transfect cells.

One way to increase EB-EGFP signal at growing MT ends without increasing the overall expression level and background along MTs is to add multiple EGFP tags. Alternatively, EB-binding domains of other +TIPs such as CLIP-170 or CLASPs can be used to visualize intracellular MT polymerization dynamics (Komarova *et al.*, 2009; Kumar *et al.*, 2009; Wittmann and Waterman-Storer, 2005). Thus, different types of +TIPs can be used to validate experimental results and help eliminate potential +TIP overexpression-induced artifacts. Most importantly, a minimal EB-binding motif has recently been identified that is sufficient for plus end localization (Honnappa *et al.*, 2009). It should therefore be possible to engineer artificial +TIPs that only minimally interfere with intracellular MT dynamics. The same considerations about high-resolution fluorescent live cell imaging as outlined in Section II. C. also apply for FP-tagged +TIPs.

B. Computational Tracking and Analysis of +TIP Dynamics

Because +TIPs specifically associate with growing MT ends, it is straight forward to determine growth rates from time-lapse sequences of FP-tagged +TIPs. For example, computer-assisted hand-tracking as described above for continuously labeled MTs can be used. Alternatively, maximum intensity projections of +TIP comet time-lapse sequences can be used to calculate growth rates directly from the comet-to-comet distance (Wittmann and Waterman-Storer, 2005).

However, one immediately obvious challenge in using +TIPs to analyze MT polymerization dynamics is that MT ends are not visible during pause and shortening phases. Thus, only growth rates can be measured directly. To extract additional parameters of MT dynamic instability, a computational framework has recently been developed that breaks down the analysis of +TIP time-lapse sequences into three steps (Matov *et al.*, 2010): First, a band-pass filter is used to detect objects of the size scale of +TIP comets. Second, these objects are tracked by single-particle tracking. Such automated tracking is highly accurate (Fig. 5 C and D) and yields statistically large populations of MT growth rates (Jaqaman *et al.*, 2008). For example, over 400 individual EB1-EGFP growth tracks were detected in the 1 min sequence shown in Fig. 5. To eliminate tracks that result from detection errors, we generally only consider tracks for further analysis that have a life-time of at least 4 frames. In order to achieve good EB1-EGFP comet correspondence between subsequent frames and minimize the number of tracking errors, computational tracking approaches generally require higher frame rates (~1–2 frames per second) as compared to manual analysis. We have

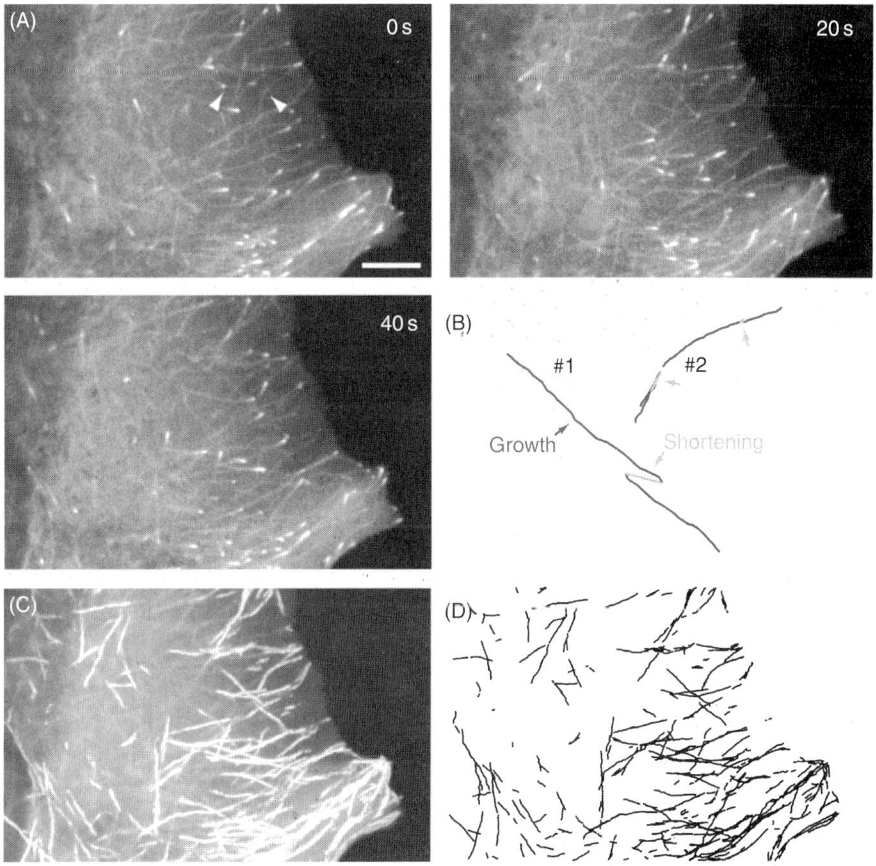

Fig. 5 +TIPs as reporters of MT polymerization dynamics. (A) Images of EGFP-tagged EB1 from the same dual-wavelengths time-lapse sequence as in Fig. 2. Scale bar, 5 um. (B) Computer-generated growth tracks of the same two MTs as in Fig. 2 with shortening events (light gray arrows) inferred by geometrical cluster analysis. (C) Maximum intensity projection of the entire image sequence directly showing EB1-EGFP growth tracks. (D) Computer-generated growth tracks with a minimum lifetime of 4 frames demonstrating high tracking fidelity.

successfully used this approach to demonstrate spatial gradients of MT polymerization dynamics in migrating cells (Kumar *et al.*, 2009).

Finally, +TIP growth tracks are linked by geometrical cluster analysis. This analysis relies on a priori knowledge of the physical characteristics and behavior of MTs. Intracellular MTs are laterally relatively immobile and are stiff and bend very little over the short time window used to record +TIP dynamics. As a result, MT shortening, rescues, and pauses predominantly occur in very close proximity to the path defined earlier by the growing end of the same MT (Fig. 2 B). A global combinatorial optimization algorithm that utilizes a cost function based on geometrical and temporal constraints between the beginning and the end of +TIP growth tracks can be used to determine those tracks that most likely belong to the same MT (Matov *et al.*, 2010). Because of the statistical nature of this algorithm, it will make errors and clearly has some fundamental limitations. For example, terminal shortening phases are hidden because a shortening phase has to be followed by a growth phase in order to be detected by the clustering algorithm. In addition, to minimize the number of clustering errors, the constraints for linking individual growth tracks have to be kept quite stringent only allowing the search for a rescue event ~10–20 frames in the future. Thus, long-shortening phases such as the one in MT #2 are frequently missed by the algorithm (compare Figs. 2B and 5B). Finally, growth phases are missed by the tracking algorithm if they are too slow and too short to produce a sufficient +TIP signal such as the short growth phases at the end of MT #1.

Thus, data derived by geometrical clustering of +TIP growth tracks cannot be directly compared to conventionally determined MT dynamics parameters. Nevertheless, because statistically large MT numbers can be analyzed relatively quickly using this approach, we expect this algorithm to be extremely useful to compare different experimental conditions on a relative basis. plusTipTracker, a Matlab-based open source software package enabling +TIP comet detection, growth track reconstruction, visualization, subcellular regional analysis, and MT subpopulation analysis will soon be available from http://lccb.hms.harvard.edu.

VI. Conclusion

Because spatiotemporal regulation of intracellular MT dynamics is important in many aspects of cell biology, the interest in generally applicable methods for quantitative analysis of MT dynamics is high. In this chapter, we give a brief overview and practical guide on how to acquire and analyze time-lapse sequences of dynamic MTs in cells by either fluorescently labeling entire MTs or by utilizing proteins that specifically associate only with growing MT ends. Because of the optical resolution limit it is not possible to measure the "true" position of a MT end with conventional light microscopy. Therefore, quantification of intracellular MT polymerization dynamics depends to a very large extent on the imaging conditions, sampling frequencies, and analysis methods used, which define theoretical limits of rates and transition frequencies that can be determined from a particular data set. Thus, absolute MT dynamics parameters are only of very limited value, and measurements should only be used to compare different experimental conditions for which time-lapse sequences were recorded identically. It will be exciting to see if and how modern super-resolution techniques will be used to more precisely observe MT polymerization dynamics in cells. In addition, because of the structural and biochemical processes, and nanoscale fluctuations of polymerization dynamics that occur on MT plus ends (Kerssemakers *et al.*, 2006; Schek *et al.*, 2007), the conventional definition of MT growth and shortening rates and transition frequencies is most likely not sufficient to accurately capture the complexity of MT polymerization dynamics. This problem is further aggravated in cells, where MT polymerization dynamics are even more variable as a result of regulation by associated proteins. Novel methods to analyze MT dynamics are needed, and we provide an outlook on one such method that utilizes computational clustering of MT growth tracks defined by the association of fluorescently labeled +TIPs with growing MT ends. It will be interesting to see how such new computational approaches will impact our future understanding of intracellular MT function and dynamics.

Acknowledgments

S.G. is the recipient of an NSF Graduate Research Fellowship T.W. is supported by National Institutes of Health grant R01 GM079139. This research was in part conducted in a facility constructed with support from the Research Facilities Improvement Program grant C06 RR16490 from the National Center for Research Resources of the National Institutes of Health.

References

Akhmanova, A., and Steinmetz, M. O. (2008). Tracking the ends: A dynamic protein network controls the fate of microtubule tips. *Nat. Rev. Mol. Cell Biol.* **9**, 309–322.

Altinok, A., Kiris, E., Peck, A. J., Feinstein, S. C., Wilson, L., Manjunath, B. S., and Rose, K. (2007). Model based dynamics analysis in live cell microtubule images. *BMC Cell Biol.* **8(Suppl. 1)**, S4.

Bieling, P., Kandels-Lewis, S., Telley, I. A., van, D. J., Janke, C., and Surrey, T. (2008). CLIP-170 tracks growing microtubule ends by dynamically recognizing composite EB1/tubulin-binding sites. *J. Cell Biol.* **183**, 1223–1233.

Bogdanov, A. M., Bogdanova, E. A., Chudakov, D. M., Gorodnicheva, T. V., Lukyanov, S., and Lukyanov, K. A. (2009). Cell culture medium affects GFP photostability: A solution. *Nat. Methods* **6**, 859–860.

Brangwynne, C. P., MacKintosh, F. C., and Weitz, D. A. (2007). Force fluctuations and polymerization dynamics of intracellular microtubules.. *Proc. Natl. Acad. Sci. U.S.A.* **104**, 16128–16133.

Bulinski, J. C., Odde, D. J., Howell, B. J., Salmon, E. D., and Waterman-Storer, C. M. (2001). Rapid dynamics of the microtubule binding of ensconsin in vivo. *J. Cell Sci.* **114**, 3885–3897.

Dogterom, M., Kerssemakers, J. W., Romet-Lemonne, G., and Janson, M. E. (2005). Force generation by dynamic microtubules. *Curr. Opin. Cell Biol.* **17**, 67–74.

Dragestein, K. A., van Cappellen, W. A., van, H. J., Tsibidis, G. D., Akhmanova, A., Knoch, T. A., Grosveld, F., and Galjart, N. (2008). Dynamic behavior of GFP-CLIP-170 reveals fast protein turnover on microtubule plus ends. *J. Cell Biol.* **180**, 729–737.

Hadjidemetriou, S., Toomre, D., and Duncan, J. (2008). Motion tracking of the outer tips of microtubules. *Med. Image Anal.* **v**, 689–702.

Honnappa, S., Gouveia, S. M., Weisbrich, A., Damberger, F. F., Bhavesh, N. S., Jawhari, H., Grigoriev, I., van Rijssel, F. J., Buey, R. M., Lawera, A., Jelesarov, I., Winkler, F. K., et al. (2009). An EB1-binding motif acts as a microtubule tip localization signal. *Cell* **138**, 366–376.

Hyman, A., Drechsel, D., Kellogg, D., Salser, S., Sawin, K., Steffen, P., Wordeman, L., and Mitchison, T. (1991). Preparation of modified tubulins. *Methods Enzymol.* **196**, 478–485.

Jaqaman, K., Loerke, D., Mettlen, M., Kuwata, H., Grinstein, S., Schmid, S. L., and Danuser, G. (2008). Robust single-particle tracking in live-cell time-lapse sequences. *Nat. Methods* **5**, 695–702.

Kerssemakers, J. W., Munteanu, E. L., Laan, L., Noetzel, T. L., Janson, M. E., and Dogterom, M. (2006). Assembly dynamics of microtubules at molecular resolution. *Nature* **442**, 709–712.

Komarova, Y., De Groot, C. O., Grigoriev, I., Gouveia, S. M., Munteanu, E. L., Schober, J. M., Honnappa, S., Buey, R. M., Hoogenraad, C. C., Dogterom, M., Borisy, G. G., Steinmetz, M. O., *et al.* (2009). Mammalian end binding proteins control persistent microtubule growth. *J. Cell Biol.* **184**, 691–706.

Krylyshkina, O., Anderson, K. I., Kaverina, I., Upmann, I., Manstein, D. J., Small, J. V., and Toomre, D. K. (2003). Nanometer targeting of microtubules to focal adhesions. *J. Cell Biol.* **161**, 853–859.

Kueh, H. Y., and Mitchison, T. J. (2009). Structural plasticity in actin and tubulin polymer dynamics. *Science* **325**, 960–963.

Kumar, P., Lyle, K. S., Gierke, S., Matov, A., Danuser, G., and Wittmann, T. (2009). GSK3beta phosphorylation modulates CLASP-microtubule association and lamella microtubule attachment. *J. Cell Biol.* **184**, 895–908.

Lechler, T., and Fuchs, E. (2007). Desmoplakin: An unexpected regulator of microtubule organization in the epidermis. *J. Cell Biol.* **176**, 147–154.

Luo, J., Dong, Z. L., Luo, X., Tang, N., Song, W. X., Chen, J., Sharff, K. A., Luu, H. H., Haydon, R. C., Kinzler, K. W., Vogelstein, B., and He, T. C. (2007). A protocol for rapid generation of recombinant adenoviruses using the AdEasy system. *Nat. Protoc.* **2**, 1236–1247.

Matov, A., Applegate, K. T., Kumar, P., Thoma, C. R., Krek, W., Danuser, G., and Wittmann, T. (2010). Analysis of microtubule dynamic instability using a plus end growth marker. *Nat. Methods.* in press.

Mitchison, T., and Kirschner, M. (1984). Dynamic instability of microtubule growth. *Nature* **312**, 237–242.

Nogales, E., and Wang, H. W. (2006). Structural intermediates in microtubule assembly and disassembly: How and why? *Curr. Opin. Cell Biol.* **18**, 179–184.

Rusan, N. M., Fagerstrom, C. J., Yvon, A. M., and Wadsworth, P. (2001). Cell cycle-dependent changes in microtubule dynamics in living cells expressing green fluorescent protein-alpha tubulin. *Mol. Biol. Cell* **12**, 971–980.

Salaycik, K. J., Fagerstrom, C. J., Murthy, K., Tulu, U. S., and Wadsworth, P. (2005). Quantification of microtubule nucleation, growth and dynamics in wound-edge cells. *J. Cell Sci.* **118**, 4113–4122.

Sammak, P. J., and Borisy, G. G. (1988). Direct observation of microtubule dynamics in living cells. *Nature* **332**, 724–726.

Schek, III, H. T., Gardner, M. K., Cheng, J., Odde, D. J., and Hunt, A. J. (2007). Microtubule assembly dynamics at the nanoscale. *Curr. Biol.* **17**, 1445–1455.

Shelden, E., and Wadsworth, P. (1993). Observation and quantification of individual microtubule behavior in vivo: Microtubule dynamics are cell-type specific. *J. Cell Biol.* **120**, 935–945.

Skube, S. B., Chaverri, J. M., and Goodson, H. V. (2009). Effect of GFP tags on the localization of EB1 and EB1 fragments in vivo. *Cell Motil. Cytoskeleton* **67**(1), 1–12.

Szymanski, D. (2002). Tubulin folding cofactors: Half a dozen for a dimer. *Curr. Biol.* **12**, R767–R769.

van der Vaart, B., Akhmanova, A., and Straube, A. (2009). Regulation of microtubule dynamic instability. *Biochem. Soc. Trans.* **37**, 1007–1013.

Walker, R. A., O'Brien, E. T., Pryer, N. K., Soboeiro, M. F., Voter, W. A., Erickson, H. P., and Salmon, E. D. (1988). Dynamic instability of individual microtubules analyzed by video light microscopy: Rate constants and transition frequencies. *J. Cell Biol.* **107**, 1437–1448.

Waterman-Storer, C. M. (2002). Fluorescent speckle microscopy (FSM) of microtubules and actin in living cells. *In* "Current Protocols in Cell Biology" (K. S. Morgan, ed.), 4.10.1–4.10.26John Wiley & Sons, Inc, New York.

Waterman-Storer, C. M., and Salmon, E. D. (1997). Actomyosin-based retrograde flow of microtubules in the lamella of migrating epithelial cells influences microtubule dynamic instability and turnover and is associated with microtubule breakage and treadmilling. *J. Cell Biol.* **139**, 417–434.

Waterman-Storer, C. M., and Salmon, E. D. (1998). How microtubules get fluorescent speckles. *Biophys. J.* **75**, 2059–2069.

Wittmann, T., Bokoch, G. M., and Waterman-Storer, C. M. (2003). Regulation of leading edge microtubule and actin dynamics downstream of rac1. *J. Cell Biol.* **161**, 845–851.

Wittmann, T., Hyman, A., and Desai, A. (2001). The spindle: A dynamic assembly of microtubules and motors. *Nat. Cell Biol.* **3**, E28–E34.

Wittmann, T., Littlefield, R., and Waterman-Storer, C. M. (2004). Fluorescent speckle microscopy of cytoskeletal dynamics in living cells. *In* "Live Cell Imaging: A Laboratory Manual, D" (L. Spector and R. D. Goldman, eds.), pp. 187–204. Cold Spring Harbor Press, New York.

Wittmann, T., and Waterman-Storer, C. M. (2001). Cell motility: Can rho GTPases and microtubules point the way? *J. Cell Sci.* **114**, 3795–3803.

Wittmann, T., and Waterman-Storer, C. M. (2005). Spatial regulation of CLASP affinity for microtubules by rac1 and GSK3beta in migrating epithelial cells. *J. Cell Biol.* **169**, 929–939.

CHAPTER 3

The Use of Fluorescence Redistribution After Photobleaching for Analysis of Cellular Microtubule Dynamics

Claire E. Walczak[*], **Rania S. Rizk**[†], *and* **Sidney L. Shaw**[‡]

[*]Medical Sciences, Indiana University, Bloomington, Indiana 47405

[†]Department of Molecular Genetics and Cell Biology, University of Chicago, Chicago, Illinois 60637

[‡]Department of Biology, Indiana University, Bloomington, Indiana 47405

METHODS IN CELL BIOLOGY, VOL. 97

978-0-12-381349-7
DOI: 10.1016/S0091-679X(10)97003-9

Abstract

Microtubules (MTs) are highly dynamic polymers that serve as tracks for vesicular movement during interphase and as structural components of the mitotic spindle, which is used to segregate the genetic material. MT dynamics are highly regulated wherein MTs turnover differentially between interphase and mitosis. Within the mitotic spindle, there are distinct classes of MTs with different dynamic properties. To understand how cellular proteins regulate the dynamics of MTs, it is necessary to have methods to assess their turnover properties. In this chapter we present approaches to assess MT dynamics in cultured mammalian cells using fluorescence redistribution after photobleaching. We include a discussion of cell culture and imaging conditions that maintain cell viability. We also provide an extensive discussion of both data collection and analysis that are utilized to estimate the turnover dynamics of MTs.

I. Introduction

Microtubules (MTs) are major structural components of the cell that are critical for a wide variety of tasks such as motility, cargo transport, spindle assembly, and chromosome segregation during mitosis. MTs are composed of α/β-tubulin heterodimers that associate longitudinally into protofilaments, 13 of which associate laterally into the hollow-tube polymer of the MT. MTs are highly dynamic polymers that exhibit a unique property known as dynamic instability where both polymerizing and depolymerizing MTs exist in the same population, interconverting between states of shrinkage and growth in a random fashion (Mitchison and Kirschner, 1984). While pure MTs can exhibit dynamic instability, the dynamics of MTs in the cell are typically more rapid due to their regulation by cellular proteins (Kline-Smith and Walczak, 2004).

MTs within many cell types are organized by the positioning of their dynamically distinct ends. In a typical interphase cell, the less dynamic minus-ends are at the MT organizing center, known as the centrosome in mammalian cells, while the more dynamic plus-ends extend toward the cell cortex. During mitosis, the duplicated centrosomes separate and focus the MT minus-ends at the two spindle poles. The MT plus-ends in mitotic cells extend either toward the cell cortex (astral MTs) or toward the spindle equator to form intraspindle cross-links (interpolar MTs) or attach to chromosomes at kinetochores (K-fibers).

The transition from interphase to mitosis involves a dramatic increase in MT dynamics (Saxton *et al.*, 1984) that has been correlated with an increase in catastrophe and a decrease in rescue frequency for astral MTs in cells (Rusan *et al.*, 2001). During mitosis, each MT class adopts unique dynamic properties. The K-fibers seldom show long growth or shortening excursions but exhibit rapid polymer turnover at the kinetochore attachment site (Mitchison and Salmon, 1992; McIntosh *et al.*, 2002). The nonkinetochore interpolar MTs are highly dynamic and turn over at a faster rate than the MTs that form the K-fibers (Cimini *et al.*, 2006; Saxton *et al.*, 1984). Astral

MTs turn over faster than MTs in interphase cells but slower than K-fiber MTs (Rusan *et al.*, 2001; Saxton *et al.*, 1984). In addition to MT turnover that occurs in the spindle due to dynamic instability, spindle MTs exhibit a specialized behavior known as MT flux wherein tubulin subunits undergo net addition at the plus-end and net loss at the minus-end creating a poleward translocation of tubulin subunits through the spindle (Kwok and Kapoor, 2007; Mitchison, 1989). Regulation of each MT subclass, via numerous MT-associated proteins, is essential for proper spindle assembly and/or chromosome segregation. Key to our understanding of how the cytoskeleton becomes organized is determining how these proteins modulate the dynamics of distinct MT populations.

A. Rationale

FRAP (fluorescence redistribution after photobleaching) is a tool that can be used to study dynamic cellular processes. The method has been used since the 1970s to understand mobility and dynamics of fluorescently labeled proteins in or associated with cells (Axelrod *et al.*, 1976). In an FRAP experiment, fluorescently labeled molecules are irreversibly photobleached in a specified area of the cell using a high-powered light source. Subsequent loss or diffusion of the bleached molecules out of the region and concomitant movement of nonphotobleached fluorescently labeled molecules into the bleached region results in the recovery of the fluorescence signal. Analysis of the rate of recovery and the percent of recovery at the photobleached region can reveal information regarding the dynamic properties of a protein or cytoskeletal structure, such as the MT.

In this chapter, we describe methods for using FRAP to estimate the general level of MT dynamics in both interphase and mitotic cells, with our focus being on mammalian cells in culture. FRAP experiments provide useful comparative information about MT turnover in living cells; however, the technique does not yet permit inference of individual MT dynamics parameters, such as assembly rates or transition frequencies. Unlike with conventional FRAP measurements, photobleached polymers undergoing dynamic instability will not exhibit simple diffusion limited recovery of fluorescent tubulin. The recovery profile is an estimate of the total turnover in polymer for a given cellular area dependent upon the composite dynamic profile of the MT array. Despite these limitations, FRAP experiments have provided significant insight into the complex regulation of MT dynamics within cells.

II. Choice and Preparation of Cells

A. Choice of Cell Type

A variety of cell lines have been successfully used for FRAP studies of the MT cytoskeleton. Cells must be either expressing GFP-tubulin, or they must be injected with fluorescently labeled tubulin. Because cells normally round up during mitosis, FRAP analysis is more readily carried out in flatter cells such as PtK-1 or 2, COS-7,

RPE-1, NIH/3T3, or LLCPK. In our studies, we routinely use PtK2 cells stably expressing GFP-alpha tubulin (PtK-T) (Khodjakov *et al.*, 2003).

B. General Tissue Culture Conditions

PtK-T cells are grown at 37°C at 5% CO_2. All media preparations, cell passaging, and cell plating are performed in a sterile laminar flow hood. Media is stored at 4°C in the dark. PtK-T cells are maintained in Opti-MEM Medium (Invitrogen; 31985-070) supplemented with 10% fetal bovine serum (Invitrogen; 16140-089), 1% penicillin/strepto mycin (Invitrogen; 15140-122), and 1% GlutaMAX (Invitrogen; 35050-061). This media formulation is referred to as Complete Opti-MEM. The general tissue culture maintenance techniques are as follows.

1. Using an inverted phase contrast microscope, check the confluency of the cells prior to passaging. Cells must be semiconfluent to passage.
2. Prepare coverslips for plating cells for FRAP experiments by placing the desired number of poly-L-lysine-coated coverslips in a plate [see (Stout, 2009) for preparing coverslips], while making sure that the coverslips do not overlap. Sterilize the coverslips by adding 100% ethanol so that the coverslips are submerged, then aspirate off the ethanol, and place the dish cover slanted over the plate to allow airflow for the coverslips to dry.
3. Trypsinize cells and prepare cellular dilutions. For plating cells for FRAP experiments, add the diluted cells in the dish containing the sterile poly-L-lysine-coated coverslips. We plate cells at concentrations ranging from 1/3 to 1/5 and allow growth time in the CO_2 incubator for 3–5 days, respectively. A 35 mm plate holds 2 ml of media and one 22 × 22 mm coverslip; a 60 mm plate holds 4 ml of media and two 22 × 22 mm coverslips; and a 100 mm plate holds 10 ml of media and up to four 22 × 22 mm coverslips.

III. Maintaining Cell Viability While Imaging

A. Maintenance and Imaging Media

For FRAP/imaging of PtK-T cells, we use complete Opti-MEM medium supplemented with 20 mM HEPES (titrated to pH 7.2 with potassium hydroxide) to maintain pH in the absence of CO_2. Additionally, an oxygen scavenging mix (Oxyrase Inc; EC-0050) is added to the media at a final concentration of 0.3 U/ml just prior to imaging chamber assembly to prevent photodamage.

B. Maintaining Constant Temperature

There are a variety of ways to maintain a constant temperature of the cells throughout the experiment. We use a device that flows air of a constant temperature onto the microscope stage, known as an air stream incubator (ASI) (Nevtek, Burnsville, VA).

The temperature is monitored using a digital thermometer with a remote probe (Acu-Rite; Model # 00890A1). The probe is taped to the specimen holder on the microscope stage and used to monitor the temperature throughout the experiment. The ASI can be replaced with any available temperature-controlled stage or chamber.

C. Choice of FRAP/Imaging Chamber

Imaging chambers for FRAP experiments need to be constructed such that the cells remain healthy while maximizing optical resolution. The major difference between chambers has to do with the amount of fluid that they can hold, and thus, the duration of the imaging experiment that can be carried out. For a detailed discussion of different viewing chambers, refer to Khodjakov and Rieder (2006).

1. Short-Term Imaging

For FRAP/imaging experiments lasting 20–30 min, we carefully mount the coverslip with cells on to a microscope slide and seal the chamber with VALAP. This chamber holds approximately 40 μl of media.

a. Prepare the sealant VALAP (vaseline, lanolin, and paraffin) by combining equal parts of vaseline, lanoline, and paraffin in a glass container, and melt on a hot plate set to low. Prepare the imaging media in a capped conical and warm by placing in a 37°C water bath. Warm the microscope stage and metal ring using the ASI and warm the microscope slide by placing it on a slide warmer set to 37°C.
b. Transfer the 22 × 22 mm coverslip on which cells are grown to a 35 mm culture plate containing 1 ml of prewarmed imaging media. Set the plate on a slide warmer.
c. Using forceps, carefully pick up the coverslip, overlay it on top of a microscope slide with the cells facing the microscope slide, and make sure that no air bubbles are formed as you overlay the coverslip.
d. Seal the coverslip on the slide using melted VALAP. Using a cotton swab, add a drop of VALAP to each corner of the coverslip, and let it dry. Next, connect the drops with additional VALAP to completely seal the coverslip to the microscope slide. This will prevent evaporation of the imaging media.
e. To clean the slide and coverslip, wet a kimwipe or lens paper with ddH_2O and gently draw the paper across the coverslip and slide. Repeat the cleaning with 70% EtOH.
f. Mount the slide on the microscope stage and record the time. Discard the chamber after 20–30 min have elapsed.

2. Long-Term Imaging

We routinely use a rose chamber for long-term live imaging (Khodjakov and Rieder, 2006; Stout *et al.*, 2009). This chamber is designed to hold up to 1 ml of media under mineral oil while still allowing high-resolution imaging through a cover glass. We find that this chamber allows us to image for approximately 5 h without any noticeable change in cell health or morphology.

a. Prewarm the slide warmer and imaging media to 37°C. Warm the microscope stage with the ASI.
b. Transfer a 22 × 22 mm coverslip with cells to a 35 mm dish containing 2 ml of prewarmed imaging media and place on the slide warmer set to 37°C.
c. Assemble the rose chamber as described in Khodjakov and Rieder (2006) and Stout *et al.* (2009) and add 1 ml of prewarmed media. The assembly needs to be done quickly to prevent the cells from cooling.
d. Clean the rose chamber with a kimwipe or lens paper wet with ddH_2O, then wipe with a kimwipe wet with 70% EtOH.
e. To prevent evaporation of media during imaging, cover the media with mineral oil and place the assembled chamber on the microscope stage.
f. Record the time at which chamber assembly is complete. Discard the coverslip after 5 h have elapsed.

IV. Imaging and Data Analysis

A. The Imaging and Photobleaching System

FRAP experiments of MTs have been successfully performed using both widefield and confocal microscope systems. An excellent discussion of live cell imaging parameters can be found in Goldman *et al.* (2010) to assist in determining what type of system might best suit your needs. The essential components are an ability to rapidly extinguish fluorescence in a user-defined area of the cell and to record high dynamic range images with enough temporal resolution to capture the fluorescence recovery. We use a spinning disk confocal microscope that is equipped with a photobleaching system. The microscope platform is a Nikon TE2000U inverted microscope equipped with a Yokogawa CSU-10 spinning disk confocal head. Illumination for imaging is provided by a 75 mW krypton/argon laser. Images are collected with a Photometrics Cascade II EM-CCD camera (EM-CCD chip cooled to –60°C, 512 × 512 pixels, 16 μm pixels). To increase the throughput of the system, we incorporated hard-coated (Semrock) wide-band emission filters. Attached to the device is a separate laser (300 mW, 488 nm Argon), coupled to a Photonics Instruments (St. Charles, IL) configurable mask system, used for photobleaching the specimen. The microscope, camera, filters, shutters, and imaging laser are controlled by via Metamorph software (Molecular Devices), and the photobleaching mask and laser are controlled by Mosaic software (Photonics Instruments).

B. Microscopy

The interpretability of the FRAP data depends upon balancing two often opposing goals. First, enough images need to be taken to accurately reconstruct the time evolution of fluorescence recovery within the bleached spot. Second, each image needs to have enough recorded signal to accurately estimate the relative level of fluorescence recovery at that time point. Since imaging the specimen inevitably results in photobleaching that decreases measurable signal, determining a reasonable protocol for image acquisition requires planning. Below we summarize the parameters that can be varied within the experiment to find an appropriate compromise between exposure duration and exposure interval. We provide the starting parameters for our typical FRAP experiments before adjustment for cell type or experimental condition.

1. Image Acquisition Settings

a. **Exposure Time.** We recommend that the image exposure time and illumination intensity be set to provide a signal-to-noise ratio of 5–10 before bleaching. This can be approximated by taking the average value of the region you wish to bleach, subtracting the average value of some region outside the cell, and dividing the result by the standard deviation of the region outside of the cell (Shaw *et al.*, 2003; Waters, 2009). Higher values will likely be associated with too much bleaching during the postbleach imaging of the cell. Lower values will result in low fidelity estimates of the fluorescence recovery. We further recommend, especially when using laser scanning or an EM-CCD camera, that the linearity of the detector be verified over the appropriate dynamic range using beads with calibrated relative fluorescence intensities (Invitrogen). Because GFP expression and microinjection will produce variable levels of fluorescently labeled tubulin, it is not possible to set a constant exposure time that can be used for all cells. We found that 500 ms (mitotic PtK-T cells) and 300 ms (interphase PtK-T cells) were useful in our setup.

b. **Time Interval Between Exposures.** The interval between images will be based on the dynamic properties of the sample being imaged. It is important to consider how quickly you expect the fluorescence recovery to occur so that a sufficient number of images are collected to provide an accurate estimate of the $t_{1/2}$ for recovery. For interphase cells with a $t_{1/2}$ of approximately 3–5 min, we collect images every 10–20 s for up to 1000 s total. For mitotic cells that have a $t_{1/2}$ of approximately 10 s for the fast phase, we collect images at a 2 s interval for $\geq$90 s.

2. The Photobleaching Pulse

Both the size of the bleach zone and its location are important factors when carrying out a FRAP experiment. Ideally, the bleach mark should be the minimum size that still allows you to collect the information you need. We find that a 3 × 10 μm box worked well for mitotic cells and a 3 × 18 μm box worked well for interphase cells when sampled to 100 or more pixels. For our studies, we determined that it was critical to bleach an equivalent position in each cell because the MT dynamics are different in the interior versus the edge of an interphase cell (Wadsworth, 1999) and near the pole or near the spindle equator of a mitotic cell (Buster *et al.*, 2007).

A second important variable is the duration of the photobleaching pulse. This will need to be determined empirically to identify the exposure time for the photobleaching laser that rapidly eliminates most of the prebleach fluorescence signal without killing the cell. While it is not critical to bleach 100% of the fluorescence, determining the recovery properties will be easier with more extensive bleaching. We find that a photobleaching pulse of approximately 500 ms works well for our setup.

3. The FRAP Experiment

Below is a step-by-step summary for how we carry out a FRAP experiment based on the parameters discussed above.

a. Turn on all the components of the FRAP/imaging setup, including the imaging laser and the bleaching laser, and prewarm the stage and all needed pieces.
b. Place the appropriate cell-containing chamber on the microscope stage and record the start time to keep track of the amount of time the cells are in the viewing chamber.
c. Through the eyepiece, scan the surface of the coverslip to find the cell of interest. It is critical that scanning is carried out as fast as possible to avoid inducing photodamage. It would be ideal if you have a nonfluorescence-based means to identify the sample.
d. Center the cell of interest in the eyepiece. For experiments analyzing MT dynamics in mitotic cells, we look for cells at late prometaphase to metaphase, which are easy to identify because they contain bipolar spindles.
e. Switch the view from the eyepiece to live view on the camera, focus on the cell, and adjust the location of the cell to be centered on the screen. Be sure to include areas of the cell that will not be photobleached as well as areas outside of the cell, as these regions will be important in data collection and analysis. This step should be carried out as quickly as possible and with the lowest amount of light to avoid excessive photobleaching.
f. Stop live view and set the acquisition settings for image collection as described above (*see section IV B Microscopy*). Determine the signal to noise for the region of interest and adjust the acquisition parameters accordingly. The same image exposure time must be used for the duration of a single experiment.
g. Use the software that controls the photobleaching laser to specify the region of interest (ROI). We use Mosaic software, which provides a means to draw an ROI as well as to specify the duration of photobleaching.
h. Acquire at least three image frames as a reference for the amount of fluorescence before bleaching.
i. Use the photobleaching software to bleach the sample.
j. Continue to acquire images postbleach at the same exposure time and interval as the prebleach settings. Figure 1 provides an example of a series of images from a FRAP experiment of an interphase and mitotic cell.

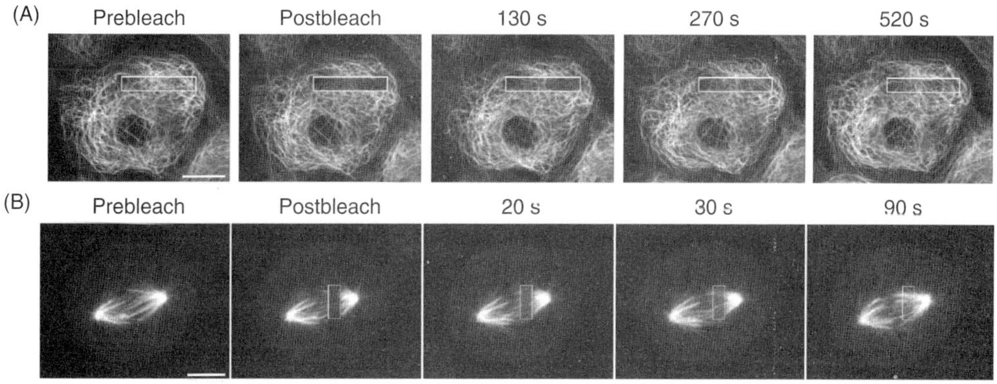

Fig. 1 FRAP analysis in an interphase and mitotic PtK-T cell. (A) Selected images before (prebleach) and then immediately after photobleaching (postbleach) of an interphase cell are shown. The cell was bleached with a laser at a specified 3 × 18 μm box on the MT array. Subsequent images were collected at 20 s intervals for ≥800 s, and three additional images are shown. Time is indicated above the figure. (B) Selected time-lapse images before (prebleach) and then immediately after (postbleach) of a mitotic cell. The cell was bleached with a laser at a specified 3 × 10 μm box on the MT array. Subsequent images were collected at 2 s intervals for ≥90 s. Time is indicated above the figure. Scale bar = 10 μm.

C. Measuring Fluorescence Recovery

We provide step-by-step instructions for estimating the quantitative changes in relative image intensity required to generate reliable $t_{1/2}$ and fluorescence recovery measurements. Image series showing obvious specimen drift or focal drift are rejected for analysis, as we have found no uniformly reliable way of compensating for these issues when analyzing our data sets. We use Metamorph software to create ROIs in the image series and tabulate the fluorescence intensity values for each region with their file-derived time stamps for when the image was captured. Once collected, the data are typically exported to a spreadsheet (e.g., Microsoft Excel), scientific graphing (e.g., Prism, Graph Pad), or command line-driven data analysis (e.g., MATLAB, The Mathworks) software.

We emphasize that there are several quantifiable properties from FRAP experiments that can be used to compare between experimental conditions. In general, the extent, the modal property (i.e., mono or multiphase), and the halftime(s) of fluorescence recovery relate to properties of the MT populations under study. The extraction of numerical data from the experiments is handled identically for determining each of these properties. Below are the regions that we use to estimate values for data analysis.

1. *Background (bg):* This measurement comes from a region outside of the cell. This mean value accounts for the offset or background in the imaging system and should remain nearly constant for each experiment. To collect the data, open the first image of a time-lapse series. Draw a fixed-size box in an area outside of the cell (Fig. 2A; boxes labeled bg), and log the average fluorescence intensity of that box for each frame of the time-lapse series. We have used fixed-size boxes of 2.8 and 2 μm^2 for interphase and mitotic cells, respectively.

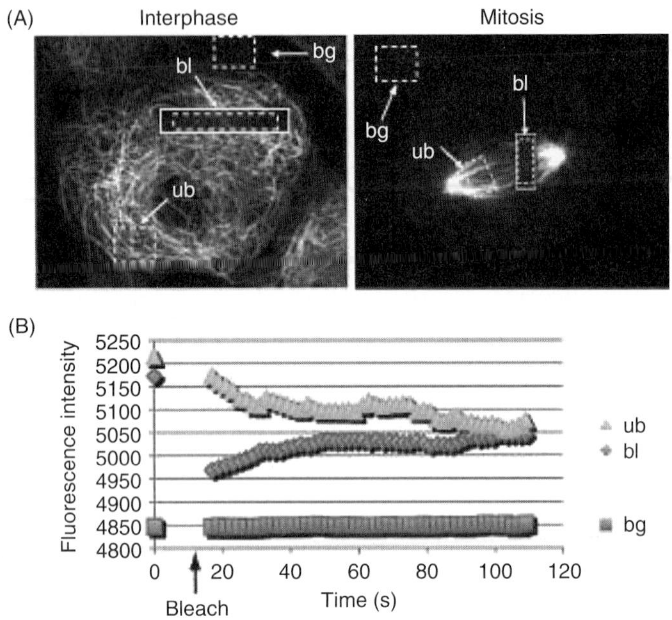

Fig. 2 Defining regions for data collection. (A) For data collection a series of boxes (dashed lines) are placed outside the cell periphery to collect average fluorescence intensity measurements of background noise (bg); within the bleached zone to collect the bleached date (bl), and on the unbleached areas on the MT array to collect the unbleached data (ub). In each case the bleached region is indicated by the solid white box on each image. To generate background-corrected data, the background value is subtracted from either the bleached (bl) or the unbleached (ub) region to create bl′ and ub′. (B) An example of a plot of the raw fluorescence intensity data collected from one time-lapse image series.

2. *Unbleached region (ub):* Unbleached portions of the specimen most closely resembling the bleached portion should be measured to determine the rate of photobleaching or other signal loss that occurs during imaging. For interphase cells, we draw a box or freehand ROI positioned on a part of the MT array that was not exposed to the bleaching laser (Fig. 2A, box labeled ub) and record the average fluorescence intensity for each frame of the time-lapse series. For mitotic cells, draw an ROI on the half of the spindle that was not exposed to the bleaching laser (Fig. 2A, region labeled ub). Collect the average fluorescence intensity from each frame of the time-lapse series.
3. *Bleached Region (bl):* This is the region in which we will measure fluorescence recovery after the bleaching. A single ROI is used inside the borders of the bleached region to measure relative average intensity values. Draw a box for interphase or mitotic cells within the bleached region being careful to exclude the unbleached area. Record the average fluorescence intensity for the ROI for each frame of the time-lapse series.

D. Data Analysis

1. Plot the Raw Data

For any experiment, we begin by plotting the raw data in excel or other graphing program (Fig. 2B). This allows for a rapid inspection for fluctuations in the intensity measurements of the bleached and unbleached regions as well as to assess whether the background measurement remains constant across the duration of imaging. Any fluctuations in the graphs should prompt you to go back to the original movie and determine whether the fluctuation is due to lamp flicker, a shift in the cell, or some other variable in the image capture. This is necessary to insure that the measured parameters of fluorescence recovery are due solely to biological changes in the MT cytoskeleton.

2. Background Subtraction

The general goal of the data analysis is to estimate the relative fraction of the original MT-generated fluorescence that is recovered at each time point after the photobleaching pulse. Since these are relative intensity values, the background (bg) value representing the camera or PMT offset for each image is directly subtracted from all other values associated with that image. Assuming use of spreadsheet-based software, we create a separate column for the background subtracted bleached (bl′) and unbleached (ub′) data.

3. Correcting for the Photobleaching that Occurs While Imaging the Specimen

The FRAP data must be corrected for the photobleaching that occurs due to the normal process of image capture in the fluorescence microscope. There are several methods to account for photobleaching, and selection of an appropriate method will depend upon the noise in the intensity measurements and the degree of photobleaching over the course of the experiment. The photobleaching estimate is created from the measurements of the unbleached region of the cell. When plotted as background-corrected intensity (ub′) over time, an ideal experiment would show a shallow and relatively smooth exponential decay owing to progressive bleaching over time. More commonly, the trace will trend from higher to lower values, will show point-to-point fluctuations, and will often exhibit short lived trends due to specimen movement or other nonbiological issues.

The first decision is whether to use the individual ub′ data points to correct the bl′ data or to use a model for fluorescence loss over time. If the data show relatively small point-to-point variation and few discernable trends, the background-corrected data (ub′) may be the easiest estimator of photobleaching and may better correct for nonbiological variation in the image acquisition. If the data show larger point-to-point fluctuations (i.e., appear noisy), then using the primary measurements for photobleach correction may introduce significant error into the recovery curve. In this case, a model can be fit to the unbleached measurements and used to correct for photobleaching. The choice of model will depend upon the character of the ub′ data and will require inspection of the original time-lapsed images to understand whether variations are coming chiefly from low signal, nonbiological acquisition problems (e.g., lamp flicker) or biological issues (e.g., specimen movement). For low signal samples or where small regions are measured for the ub′ estimate, we typically fit a single exponential decay curve to our data as developed below. For more complicated effects, it is permissible to use other models, such as a box-car or sliding average of the data, provided there is sufficient reason to believe that the entire specimen, including the photobleached spot, are properly corrected via the model. Since the method and measurement used for photobleach correction will have a critical effect on the interpretation of the experiment, the methods used should be stated explicitly in any manuscript reporting these measurements.

Once an estimate of the photodissipation is created, the next decision is by what means to use that estimate for correcting the bl′ data. The photodissipation estimate can either be added to the bleach recovery measurement (bl′) or, alternatively, the bl′ value can be divided by the ub′ value. As a practical matter, adding the photodissipation values to the ub′ data should only be performed if the prebleach bl′ intensity is approximately equal to the initial ub′ intensity. Otherwise, this real-valued correction will over or under correct when the initial intensity values are not equivalent. In our work, we typically generate a model for photobleaching from the unbleached data and correct via addition of the modeled values. This correction step is repeated for each individual FRAP experiment because photobleaching from cell to cell is typically different. Below is an outline of the steps we use to correct data for photodissipation.

a. Generate a new column in your file for an individual experiment and place the time data in this column. The first spreadsheet cell is zero time, which represents the first frame of the image series. Enter a formula to represent your imaging time interval. For example, during our imaging of mitosis, this formula would be +2 since we image at 2 s intervals so that the column would read 0, 2, 4, 6, etc.
b. The background-corrected ub data (ub′) will be pasted into a subsequent column where each cell in the column will represent a single time point.
c. Plot the data as a scatter plot where $X =$ time and $Y =$ ub′.
d. Using the fitting routines in your spreadsheet or scientific graphing software, determine the best mathematical model for the ub′ data. We generally fit the data to exponential, power, or logarithmic models and use either an r^2 value or other residuals test to find the best-fit model (Fig. 3A). Record values for the equation for use in analysis as the *p*hoto*d*issipation *m*odel (PDM).

$$\mathrm{PDM} = (Y = a \times \exp(b \times X)) \tag{1}$$

e. Take the PDM equation and calculate the theoretical Y for each time point by plugging in the value for X ($t = 0, 2, 4, 6\ldots.$). Paste the new theoretical Y values into a new column called theoretical fluorescence intensity (tfi), which represents the theoretical value for Y at any point along the curve.
f. The tfi is now used to calculate the actual *f*luorescence *l*oss (fl) at each point along the curve by subtracting the tfi at each time from the tfi at $t = 0$. If you elect not to fit

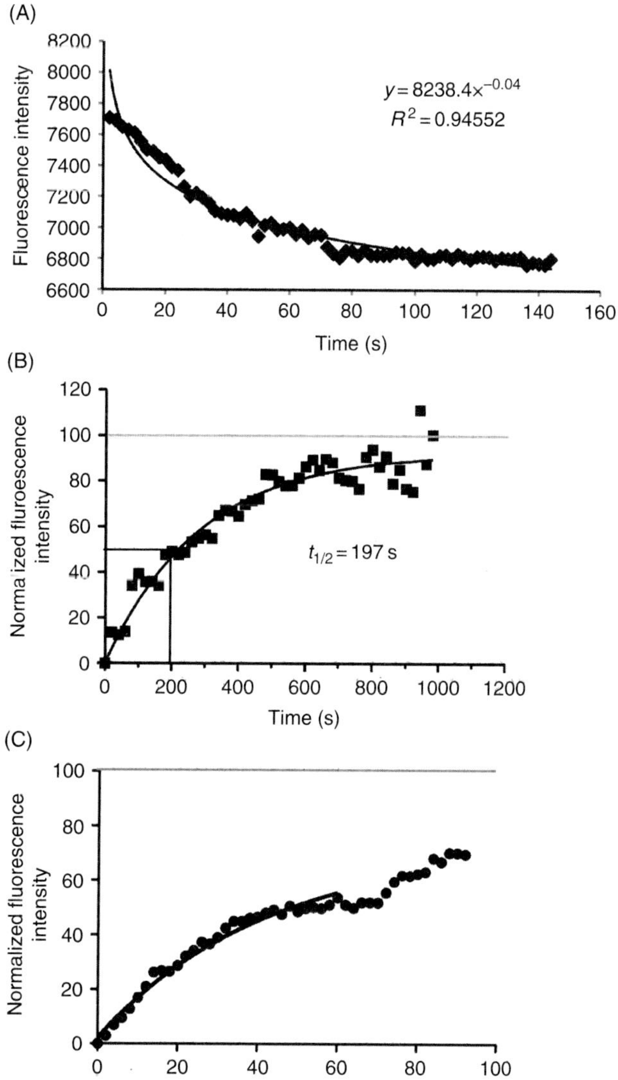

Fig. 3 Data analysis. (A) An example of the data from the unbleached part of the sample, which is used to create a PDM to account for the bleaching that occurs to the sample due to the imaging laser. This model is used to correct the fluorescence recovery data from each experiment. (B) An example of a fluorescence recovery curve for a single FRAP experiment in an interphase cell. The data fit a monophasic recovery model with a single $t_{1/2}$, reflecting the rate of recovery of fluorescence within the bleached region. (C) An example of a fluorescence recovery curve for a mitotic cell. In this example the data was normalized where 0 was the fluorescence intensity immediately postbleach, and 100% was set to the prebleach fluorescence intensity. The first part of the data clearly fit to a single-site exponential recovery model, indicated by the black line. After ~60 s, there is a clear difference in the shape of the recovery curve, indicating the possibility of a second recovery event. In this example, the fluorescence does not recover to the prebleach level (100% value) during the time course of imaging, indicating either that there is a hyperstabilized population of MTs or that the second recovery phase is very slow.

a model to the ub′ data, each ub′ value is subtracted from the first ub′ value (ub(1)′) to create fl.

g. The most straightforward method for using fl to correct for bleaching is to simply add the fl estimate to the background-corrected bleached data (bl′). If the initial bl′ prebleach and ub′ intensity values are not equal, then the data can be multiplied by the ratio of the initial values ((ub′(1)/bl′(1)) × corrected bl′) before the percent recovery of fluorescence is determined. The alternative method for correcting the data, as previously mentioned, is to divide the bleached (bl′) data by the unbleached (ub′) data for each data point. Again, these data will need to be renormalized for determining percent recovery to account for any discrepancy between the prebleached (bl(1)′) value and starting unbleached (ub(1)′) values as detailed below.

4. Developing Normalized Fluorescence Recovery Curves

a. Copy the columns for time and for corrected bl data into a program that allows you to perform curve fitting. We like Prism (GraphPad), for ease of use, or MATLAB for more advanced analyses.

b. We have two objectives in fitting a model to the recovery data. We first want to ask if the data are fit well by a single exponential recovery model or if the recovery data are more likely to be multicomponent. Once decided, we next want to determine the best estimator(s) to describe the recovery, which is typically specified as the halftime ($t_{1/2}$) value(s). For all cases, we first subtract off the intensity value of the first bleached measurement from all subsequent points and reset the time axis so that this value is time point 0. Further normalization of the data requires careful consideration of the primary data and the model-fitting software being used. Dividing all of the values by the prebleach intensity estimate is the easiest normalization because you can plot the data and visually estimate the percent of total recovery, the modal nature of the curve, and the approximate halftime to recovery. However, you must be very clear about whether the model (curve) fitting routine is assuming that the model goes from 0 to 1 (assumes 100% recovery) or from 0 to a value that is the best fit for the data. For cases where the software requires user input for the fitting range, and it is not clear that fluorescence would recover to 100%, it may be better to use the mean value of the last 3–10 intensities from the recovery data as an estimate for the data fitting range.

c. The next step is to examine the fit of the data to the various recovery models. A single exponential recovery model assumes that there is an equal probability of replacing a bleached tubulin molecule with an unbleached molecule at every point in time and that the decreasing number of available bleached sites at each time interval results in the exponential fluorescence recovery curve. A second MT population in the bleach zone, having different kinetic properties, would result in a recovery curve composed of two exponentially recovering populations. Many other phenomena could be occurring in perturbed cells that do not correspond to either of these hypotheses. What we look for is a model that fits all or a portion of the recovery curve consistently across all experiments for use in comparing experimental versus control specimens where the model and section of the recovery data are specified in the experimental protocol. We initially attempt to fit a single exponential model to all or a significant portion of the data. The first 1–2 points of the recovery curve can sometimes be obscured by repopulation of the bleach area with unincorporated fluorescent tubulin. In this case, it is possible to fit exponential models beginning at time points after 0 to ask if the fit better approximates the remainder of the recovery data. Determining what model to use for comparing control and experimental samples is ultimately up to the investigator and should be justified to both model-fitting criteria and extrapolation to the biological properties being explored. In our experience, the data generated from interphase cells were best described by a single (monophasic) exponential association model (Eq. 2, Fig. 3B), whereas the majority of data generated from the spindles of a control mitotic cell were best fit by a biphasic exponential association model (Eq. 3, Fig. 3C). The number of phases represented by each model is consistent with the number of dynamically different MT subclasses hypothesized for interphase and mitotic arrays.

$$Y = Y_{\max} \times (1 - \exp(-k \times X)) \tag{2}$$

$$Y = Y_{\max 1} \times (1 - \exp(-k1 \times X)) + Y_{\max 2} \times (1 - \exp(-k2 \times X)) \tag{3}$$

d. Determining $t_{1/2}$ values from the recovery curves: using Eqs. (2) and (3) above, the halftime to fluorescence recovery ($t_{1/2}$) can be calculated directly using the graphing software. Alternatively the $t_{1/2}$ values can be calculated manually using the values for the exponentials (k) provided from the equation of the best-fit models with the equation shown below (Eq. 4). For a monophasic model, there will be a single $t_{1/2}$, whereas for a biphasic model, there will be two $t_{1/2}$ values.

$$t_{1/2} = \mathrm{Ln}\, 2/k \tag{4}$$

5. Determining Percent of Fluorescence Recovery from a Monophasic Recovery Curve

If there is a stable underlying population of MTs in the specimen, the corrected fluorescence recovery values may not return to the prebleach intensity level. Determining the percent recovery of fluorescence requires renormalizing the corrected bleach data (bl′) to be relative to the prebleach fluorescence intensity. The simplest way to do this is to renormalize the data from 0 to 1 (or 0–100%) by setting the first recovery value to 0 intensity at 0 time and the prebleach background-corrected value as 100%. As above, Prism allows you to choose this type of data normalization and then replots the data as time (X axis) versus % fluorescence (Y axis). Alternatively, the percent recovery of fluorescence can be calculated by using the equation of the exponential function (Eq. 2) to extrapolate the recovery data to time $=X$ s postbleach after rescaling the data. Divide the calculated fluorescence intensity value at the desired time point by the prebleach fluorescence intensity value (should be 100%) and multiply by 100 to obtain the percent fluorescence recovery.

6. Determining Percent Fluorescence Represented in Each Phase of a Biphasic Recovery Curve

For a double exponential fit, the data will be used to extrapolate the fluorescence recovery for each phase using the values of the multiplier, $Y_{\max1}$ and $Y_{\max2}$, for each phase [Eq. (3)]. The values of both multipliers are added, and the percent contribution of each is calculated using the value for $Y_{\max1} + Y_{\max2}$ to represent the total fluorescence.

a. From Eq. (3), copy the values for multipliers ($Y_{\max1}$ and $Y_{\max2}$) and paste into the new Excel file so that the first column contains the movie name, the second column the $Y_{\max}$ value for the fast phase ($Y_{\max1}$), and the third column the $Y_{\max}$ value for the slow phase ($Y_{\max2}$).
b. In a new column (column 4), enter a formula to add the two values ($Y_{\max1} + Y_{\max2}$) for a given cell. The value from $Y_{\max1} + Y_{\max2}$ will represent the total fluorescence.
c. To calculate the percent fluorescence representing each phase, use the equations shown below [Eqs. (5) and (6)] to calculate the percent fluorescence for the first and the second phase, respectively.

$$Y_{\max1}/(Y_{\max1} + Y_{\max2}) \times 100 \tag{5}$$

$$Y_{\max2}/(Y_{\max1} + Y_{\max2}) \times 100 \tag{6}$$

7. Compiling Data from Multiple FRAP Experiments

To carry out statistical comparisons of different experimental conditions tested, we use spreadsheet software as outlined in the following steps. The sample sizes used to estimate reliable statistics are critical. First of all, we strongly advise pooling data from experiments done on different days and with different batches of cells to account for the real biological variation in the experiments. Our FRAP data are usually collected from over three different sets of experiments, and the compiled number of cells imaged in a single experimental condition varies from 10 to 26. However, the sample size that can be used to infer information about the whole population may vary considerably. For example, if a given treatment results in large variation in values across the different cells, a larger sample size needs to be collected before the results can be interpreted.

a. Calculating the $t_{1/2}$ values for a given experimental condition

   - Open the files for each individual movie for a given experimental condition and a new worksheet. Copy the $t_{1/2}$ value(s) for each movie and paste in a new worksheet. Organize the data within the worksheet so that the first column will contain the experiment's name, the second the fast $t_{1/2}$ values, and the third the slow $t_{1/2}$ values and label the columns accordingly. In adjacent columns, repeat entering the $t_{1/2}$ values for all the experimental conditions tested.

- Enter a formula to calculate the average values for all the slow or fast $t_{1/2}$ values for each experimental condition. Subsequently, you can enter formulas to calculate the standard deviation and standard error of the mean (SEM) for each experimental condition tested.

b. Calculating the percent fluorescence recovery for a given experimental condition

- Copy the individual percent fluorescence recovery values for each movie and paste into a new worksheet so that the first column will contain the experiment name and the second column, the percent fluorescence recovery value for this particular experiment. Repeat in adjacent columns for each of the different experimental treatments tested.
- In a new cell, enter the formula for calculating the average percent fluorescence recovery value and repeat for each given experimental condition.

c. Compiling the average percent fluorescence recovery representing each phase

- Copy the average percent fluorescence recovery representing each phase for each movie and paste into a new worksheet so that the first column will contain the experiment name and the second column the percent fluorescence recovery value for this particular movie. Repeat in adjacent columns for each of the different experimental treatments tested.
- For a given experimental condition, enter a formula to calculate the average value for percent fluorescence for each of the different conditions.

To compare whether there is a statistically significant difference between the different experiment conditions (control or perturbation), use a two-tailed student's *t*-test with a confidence interval of 95% or better.

V. Summary and Conclusions

FRAP of cellular MTs can be used to assess the overall dynamic state of MTs. The methods described above were developed to simplify acquisition and analysis of such data. Using this approach, it should be possible to compare both the $t_{1/2}$ and the fluorescence recovery values after experimental manipulations. This should provide a tool for understanding the impact of different proteins or treatments, such as MT-associated proteins and molecular motors, on the dynamics and function of MTs at different stages during the cell cycle.

Similar FRAP experiments and analyses have also been applied to yeast spindles (Maddox *et al.*, 2000). The methods presented here can also be applied to yeast and plant cells (Shaw *et al.*, 2003). However, because the yeast spindle is dramatically smaller than a mammalian cell, some alterations such as a decrease in the region sizes used to collect fluorescence data may be necessary.

Acknowledgments

We thank Jim Powers for many discussions about imaging acquisition and data analysis and Stephanie Ems-McClung for sharing expertise in statistics. Work in the PIs labs is supported by NIH Grant GM059618 to CEW and NSF Grant 0920555 to SLS. We also thank Lesley Weaver and Jessica Lucas for comments on early drafts of this chapter.

References

Axelrod, D., *et al.* (1976). Mobility measurement by analysis of fluorescence photobleaching recovery kinetics. *Biophys. J.* **16**, 1055–1069.

Buster, D. W., *et al.* (2007). Poleward tubulin flux in spindles: Regulation and function in mitotic cells. *Mol. Biol. Cell* **18**, 3094–3104.

Cimini, D., *et al.* (2006). Aurora kinase promotes turnover of kinetochore microtubules to reduce chromosome segregation errors. *Curr. Biol.* **16**, 1711–1718.

Goldman, R. D., *et al.* (2010). "Live Cell Imaging: A Laboratory Manual." Cold Spring Harbor Press, Cold Spring Harbor, NY.

Khodjakov, A., *et al.* (2003). Minus-end capture of preformed kinetochore fibers contributes to spindle morphogenesis. *J. Cell Biol.* **160**, 671–683.

Khodjakov, A., and Rieder, C. L. (2006). Imaging the division process in living tissue culture cells. *Methods* **38**, 2–16.

Kline-Smith, S. L., and Walczak, C. E. (2004). Mitotic spindle assembly and chromosome segregation: Refocusing on microtubule dynamics. *Mol. Cell* **15**, 317–327.

Kwok, B. H., and Kapoor, T. M. (2007). Microtubule flux: Drivers wanted. *Trends Cell Biol.* **19**, 1–7.

Maddox, P. S., *et al.* (2000). The polarity and dynamics of microtubule assembly in the budding yeast *Saccharomyces cerevisiae*. *Nat. Cell Biol.* **2**, 36–41.

McIntosh, J. R., *et al.* (2002). Chromosome-microtubule interactions during mitosis. *Annu. Rev. Cell Dev. Biol.* **18**, 193–219.

Mitchison, T. J. (1989). Polewards microtubule flux in the mitotic spindle: Evidence from photoactivation of fluorescence. *J. Cell Biol.* **109**, 637–652.

Mitchison, T., and Kirschner, M. (1984). Dynamic instability of microtubule growth. *Nature* **312**, 237–242.

Mitchison, T. J., and Salmon, E. D. (1992). Poleward kinetochore fiber movement occurs during both metaphase and anaphase-A in newt lung cell mitosis. *J. Cell Biol.* **119**, 569–582.

Rusan, N. M., *et al.* (2001). Cell cycle-dependent changes in microtubule dynamics in living cells expressing green fluorescent protein-alpha tubulin. *Mol. Biol. Cell* **12**, 971–980.

Saxton, W. M., *et al.* (1984). Tubulin dynamics in cultured mammalian cells. *J. Cell Biol.* **99**, 2175–2186.

Shaw, S. L., *et al.* (2003). Sustained microtubule treadmilling in Arabidopsis cortical arrays. *Science* **300**, 1715–1718.

Stout, J. R., *et al.* (2009). Protein inhibition by microinjection and RNA-mediated interference in tissue culture cells: Complementary approaches to study protein function. *Methods Mol. Biol.* **518**, 1–21.

Wadsworth, P. (1999). Regional regulation of microtubule dynamics in polarized, motile cells. *Cell Motil. Cytoskeleton* **42**, 48–59.

Waters, J. C. (2009). Accuracy and precision in quantitative fluorescence microscopy. *J. Cell Biol.* **185**, 1135–1148.

CHAPTER 4

Kinetochore–Microtubule Dynamics and Attachment Stability

Jennifer G. DeLuca

Department of Biochemistry and Molecular Biology, Colorado State University, Fort Collins, Colorado 80523

Abstract

Mitosis is the process by which a cell divides its genetic material equally into two daughter cells. Successful division requires that the two identical sister chromatids of a mitotic chromosome attach to the plus-ends of spindle microtubules (MTs) via their kinetochores, which are large protein structures built on centromeric DNA. Attachments between kinetochores and MTs must be persistent so that forces can be generated for chromosome movements, but at the same time they must be compliant,

978-0-12-381349-7
DOI: 10.1016/S0091-679X(10)97004-0

because attached MT plus-ends continuously polymerize and depolymerize to provide force for chromosome congression to the spindle equator. Both the attachment stability of kinetochore–MTs and the degree of dynamic instability exhibited by kinetochore–MTs must be precisely controlled to avoid errors in chromosome segregation. This chapter provides an overview of techniques used in cultured mammalian cells that measure stability and polymerization/depolymerization dynamics of kinetochore–MTs during mitosis.

I. Introduction

Eukaryotic cells orchestrate chromosome segregation on the mitotic spindle, a complex apparatus comprised of highly organized arrays of microtubules (MTs) and MT-associated proteins. During the process of mitotic chromosome segregation, chromosomes must become firmly tethered to the dynamic plus-ends of MTs of the mitotic spindle. They do this via their kinetochores, which are elaborate protein structures built atop centromeric DNA. Stable attachments between kinetochores and the plus-ends of spindle MTs are required to generate forces for chromosome movement and to generate kinetochore tension to silence the mitotic spindle assembly checkpoint. However, kinetochore–MTs must also be dynamic, since directed chromosome movements are driven, in large part, by forces derived from depolymerization and polymerization of MT plus-ends. Precise control of the attachment of kinetochores to spindle MTs is essential for successful mitosis and the preservation of genomic stability. Mitotic cells that lose control of attachment regulation and generate hyper- or hypostable kinetochore–MTs fail to execute proper mitotic cell division and produce daughter cells that are in many cases aneuploid, containing either too many or too few chromosomes. This dangerous condition is linked to both the initiation and the progression of human tumors and the formation of birth defects (Cimini, 2008; Kops *et al.*, 2005; Yuen *et al.*, 2005). Not surprisingly, understanding how cells form, stabilize, and regulate kinetochore–MTs is an active area of research. This chapter will describe experimental approaches that assess kinetochore–MT attachment stability and kinetochore–MT dynamics. Discussion will focus on fluorescence microscopy-based assays in mammalian tissue culture cells.

II. Materials

A. Cells

While any adherent cell line can be assayed using the protocols described here, several factors should be considered when choosing a cell line. These include ease of culture maintenance, tendency to remain flat during mitosis (which impacts quality of optical imaging), number of chromosomes, genome sequence availability (which affects the ability to carry out molecular manipulations), ease of transfection and/or microinjection, and presence of genetic abnormalities. This chapter will include protocols for both HeLa cells, immortalized human epithelial carcinoma cells derived from adult cervical tissue, and PtK1 cells, immortalized marsupial epithelial cells derived from the kidney of an adult female rat kangaroo. Advantages of HeLa cells for mitotic studies are that they are easily maintained in culture, they transfect well using lipid-based transfection or electroporation techniques, they can be easily synchronized to the same cell cycle stage, and their genome has been sequenced. This cell line also has several disadvantages: they round up during mitosis, which results in unfavorable conditions for microscopic imaging, and they are aneuploid carcinoma cells with accumulated genetic alterations. PtK1 cells have the advantages of remaining flat during mitosis, making them amenable to microscopic imaging, and they have only 12 large chromosomes, which allows for each kinetochore to be easily identified and assessed. Unfortunately, the rat kangaroo (*Potorous tridactylus*) genome has not been sequenced, which makes the identification and management of PtK1 genes more difficult, but still feasible (Stout *et al.*, 2006; Guimaraes *et al.*, 2008). Furthermore, PtK1 cells are not easily synchronized using standard methods. A thorough comparison of commonly used cell lines for mitosis experiments can be found in Wadsworth (2007).

B. Cell Culture, Treatments, Transfection, and Live-Cell Imaging

HeLa cell culture media Dulbecco's Modified Eagle Medium (DMEM) (Invitrogen, cat#11995) supplemented with 10% fetal bovine serum (FBS) (Atlanta Biologicals, cat# S11150) and antibiotic-antimycotic (Invitrogen, cat#15240-062)

PtK1 cell culture media Ham's F-12 Medium (Invitrogen, cat#11765) supplemented with 10% FBS and antibiotic–antimycotic

Sterile coverslips Prepare sterile coverslips using the following procedure: Empty the contents of 4 packages of # 1.5, 22 mm × 22 mm glass coverslips (VWR, cat#48366-227) into a 1 l glass beaker, taking care to gently separate the coverslips from each other. Add 500 ml of 1 M HCl, cover the beaker with a watch glass (150 mm diameter), and heat at 60°C for 12–16 h. Cool to room temperature and gently pour out the HCl and replace with double-distilled water. Pour out and refill with 500 ml fresh double-distilled water. Place in sonicator bath and sonicate for 30 min. Pour out liquid and refill with 500 ml fresh double-distilled water. Sonicate for 30 min, pour out liquid, and repeat once more. Pour out liquid and add 500 ml 50% ethanol diluted in double-distilled water. Sonicate 30 min. Pour out liquid and add 500 ml 70% ethanol. Sonicate 30 min. Pour out liquid and add 500 ml 95% ethanol. Sonicate 30 min. Pour out liquid, add 300 ml 95% ethanol, and carefully transfer coverslips and ethanol to a glass jar equipped with a screw-on lid for storage. To use, remove one coverslip at a time using forceps and flame prior to seeding cells.

Opti-MEM (Invitrogen, cat#31985) After opening, aliquot Opti-MEM into 10 ml aliquots and store at 4°C. For Opti-MEM + 10% serum, mix 9 ml Opti-MEM + 1 ml FBS and store at 4°C.

Oligofectamine (Invitrogen, cat#12252-011) Store at 4°C and keep lid tightly closed when not in use.
Fugene 6 (Roche, cat#11-814-443-001) Store at 4°C and keep lid tightly closed when not in use.
Monastrol (Tocris Biosciences, cat#1305) Prepare a 100 mM stock solution, diluted in DMSO.
Nocodazole (Sigma, cat#M1404) Prepare a 20 mM stock solution, diluted in DMSO.
Taxol (LCLabs, cat#P-9600) Prepare a 20 mM stock solution, diluted in DMSO.
MG132 (VWR, cat#100508-004) Prepare a 20 mM stock solution, diluted in DMSO.
PtK1 MT destabilizing buffer 60 mM PIPES, 25 mM HEPES, 1 mM $MgCl_2$, and 1 mM $CaCl_2$, pH 7.0
Filming media Leibovitz L-15 media (Invitrogen, cat#11415) containing L-glutamine without phenol red supplemented with 10% FBS and 7 mM HEPES, pH 7.2.
G418 (Invitrogen, cat#11811-031).
35 mm polystyrene culture dishes (Corning, cat#430165).
#1.5 22 mm × 22 mm glass coverslips (VWR, cat#16004-302).
Glass-bottom 35 mm culture dishes, #1.5 (MatTek, cat#P35G-1.5-20-C).
Modified Rose chambers (top and bottom plates; silicone gaskets; screws).
21 gauge needles (BD, cat#305165).
5 ml syringes (BD, cat#301603).

C. Immunofluorescence

PHEM 60 mM PIPES, 25 mM HEPES, 10 mM EGTA, 4 mM $MgSO_4$, pH 7.0.
2X PHEM 120 mM PIPES, 50 mM HEPES, 20 mM EGTA, 8 mM $MgSO_4$, pH 7.0.
PHEM-T PHEM + 0.1% Triton X-100 sonicated for 5 min in sonicator bath immediately prior to use.
Permeabilization buffer PHEM + 0.5% Triton X-100 sonicated for 5 min in sonicator bath immediately prior to use.
Fixative 4% paraformaldehyde diluted from a 16% stock solution (Ted Pella, cat#18505) into PHEM buffer.
10% Boiled donkey serum (BDS) To prepare mix 10 ml lyophilized normal donkey serum (Jackson ImmunoResearch, cat#017-000-121) with 25 ml double-distilled water and boil for 10 min. Add 2X PHEM buffer to bring the total volume of the solution to 50 ml. Add sodium azide to a final concentration of 0.05%. Spin solution using a JA-20 rotor (Beckman) or SS34 rotor (Sorvall) for 1 h at 40,000×*g*. Filter solution through a 0.22 μM filter and store at 4°C.
4′,6-diamidino-2-phenylindole (DAPI) (Invitrogen, cat#D3571) Diluted to 2 ng/ml from a 100 μg/ml stock solution in water.
Mounting media/antifade solution 0.5% *N*-propyl gallate (AlfaAesar, cat#A10877) in 90% glycerol diluted in PHEM buffer.

Primary antibodies Anti-α-tubulin antibody DM1-α (Sigma, cat#T9026); anti-Hec1 antibody 9G3 (Abcam, cat#3613); anti-centromere antibodies (Antibodies, Inc., cat#15-134); CENP-A antibody (Abcam, cat#ab13939).

Secondary antibodies Donkey anti-human IgG, Cy-5 conjugated (Jackson Immuno Research, cat#709-175-149); donkey anti-mouse IgG, rhodamine Red-X conjugated (Jackson Immuno Research, cat#715-296-150); donkey anti-mouse IgG, DyLight 488 conjugated (Jackson Immuno Research, cat#715-486-150).

Coverglass staining jars (Ted Pella, cat#21036).

Microscope slides (VWR, cat#16004-368).

III. Methods

A. General Methods: RNA Interference-Based Protein Depletion

RNAi (RNA interference)-based protein depletion is commonly used to study how various proteins contribute to kinetochore–MT attachment stability and dynamics in cultured cells. This technique is especially powerful when paired with rescue experiments in which wild-type or mutant versions of a protein of interest are expressed in cells depleted of endogenous protein. Described below are protocols for RNAi-based protein silence and rescue experiments using small, interfering RNAs (siRNAs) in both HeLa and PtK1 cells.

For HeLa cells, seed onto sterile #1.5, 22 mm × 22 mm coverslips contained in 35 mm-diameter polystyrene Petri dishes. Maintain cells in growth media at 37°C in 5% CO_2 until they reach 50% confluency. For one coverslip: in a microfuge tube add 48 μl of prewarmed (37°C) Opti-MEM media with no additional supplements. Pipette 6 μL Oligofectamine directly into the Opti-MEM and incubate for 5 min, flicking intermittently to mix. In a separate tube, mix 175 μl Opti-MEM with 8 μl of a 20 μM stock solution of siRNA. This will result in a final siRNA concentration of 80 nM. Combine contents of the two tubes and incubate for 30 min with occasional flicking to mix. In a separate tube, add 100 μl Opti-MEM (no supplements) prewarmed to 37°C. To this tube, pipette 3 μl Fugene 6 directly into the Opti-MEM. Flick to mix and incubate for 5 min. Add 1 μl of 1 μg/μl DNA that encodes the wild-type or mutant protein of interest. We routinely express the protein of interest as a fusion to green fluorescent protein (or an alternative fluorescent protein) so that both expression levels and protein localization can be monitored by fluorescence microscopy. Incubate for 30 min with occasional flicking to mix. Rinse cells with Opti-MEM and add 1.66 ml fresh Opti-MEM + 10% serum to the cells. Pipette the contents of both tubes (siRNA and DNA solutions) to the Petri dish containing the coverslip and media. Gently swirl to mix. Twenty-four hours posttransfection, add 1 ml fresh Opti-MEM + 10% serum to the existing solution.

For one coverslip of PtK1 cells, add 150 μl prewarmed Opti-MEM (no supplements) to a tube. Pipette 7 μl Oligofectamine into the Opti-MEM solution. Flick intermittently for 5 min to mix. To this tube, add the following: 8 μl siRNA from a 20 μM stock solution (for a final siRNA concentration of 80 nM), 1 μl of 1 μg/μl DNA, and an additional 150 μl Opti-MEM (no supplements). Incubate for 30 min and flick intermittently to mix. Rinse cells with Opti-MEM and add 1.68 ml fresh Opti-MEM + 10% serum. Add the siRNA/DNA solution to the coverslip. Twenty-four hours posttransfection, add 1 ml fresh Opti-MEM + 10% serum.

1. Points to Consider

Before embarking on silence and rescue experiments, it is important to first thoroughly characterize the phenotype of siRNA-transfected cells alone to determine the extent of the depletion (analyzed by both Western blotting and fluorescence intensity measurements) and timing of depletion to determine the optimal posttransfection time points for carrying out kinetochore–MT assays. In some cases, mitotic proteins are degraded after the completion of mitosis, and when dealing with such proteins, we assay cells 24–48 h posttransfection (DeLuca *et al.*, 2002; Guimaraes *et al.*, 2008). In other cases, proteins of interest may remain stable throughout multiple cell cycles, and assay times may extend out to 72, 96, or even 120 h (Goshima *et al.*, 2003; Kops *et al.*, 2005).

For silence and rescue experiments, it is also important to ensure that exogenously expressed protein (for the "rescue") is stable, and its coding sequence is not targeted for destruction by the RNAi machinery. One option is to design siRNAs such that they target a noncoding region of the sequence of interest, such as the 3′ or 5′ untranslated region. Alternatively, expression constructs can be mutated at the siRNA target site so that the sequence encoding for the protein of interest is no longer recognized by the siRNA. In this case, we generate silent mutations so that the translated protein sequence is not altered, and we make three nucleotide changes in the region of the sequence that is targeted by the siRNA.

Finally, for both PtK1 and HeLa cells, we often transfect with directly fluorescently labeled siRNAs, in which a Cy5 or Cy3 fluorescent dye is conjugated to the 3′ end of the siRNA sense strand, to confirm positive transfection on a cell-to-cell basis by fluorescence microscopy.

B. General Methods: Immunofluorescence

Using forceps, remove coverslips from Petri dishes and place in a coverslip staining jar containing PHEM buffer prewarmed to 37°C for an initial rinse. Transfer coverslips to staining jars containing freshly prepared fixative prewarmed to 37°C and incubate for 10 s. This initial incubation in fixative helps prevent loss of cell adherence from the coverslip during the subsequent permeabilization step. Transfer coverslips to a staining jar containing freshly prepared permeabilization solution prewarmed to 37°C and incubate for 5 min at 37°C. Following permeabilization, fix cells by placing coverslips in staining jars containing fixative pre-warmed to 37°C, and incubate at room temperature for 20 min. Carry out all subsequent steps at room temperature unless otherwise noted. Rinse cells by transferring coverslips to staining jars containing PHEM-T and incubate for 5 min. Pour off solution and rinse by incubation with fresh PHEM-T for 5 min. Repeat rinses for a total of 3 × 5 min. Rinse quickly with PHEM and place coverslips, cell-side up, in a humid chamber (we use 150 mm-diameter polystyrene Petri dishes lined with damp kimwipes covered with a layer of Parafilm). To block nonspecific antibody binding, pipette 150 μl of 10% BDS onto coverslips and incubate for 1 h. Following incubation, remove 10% BDS by aspiration and replace with primary antibody solution diluted in 5% BDS (in PHEM), and incubate for 1 h at 37°C, or alternatively, for 12 h at 4°C. For both PtK1 and HeLa cells, we use anti-α-tubulin antibodies at 1:200 and anti-centromere antibodies at 1:300. After primary antibody incubation, place cells in staining jars and wash with PHEM-T (3 × 5 min) followed by a single quick rinse in PHEM. Return coverslips, cell-side up, to the humid chamber lined with fresh Parafilm and incubate for 1 hr at room temperature, protected from light, with fluorescently conjugated secondary antibodies diluted 1:300 in 5% BDS (in PHEM). Rinse coverslips in staining jars containing PHEM-T (3 × 5 min), followed by a quick-rinse in PHEM. Transfer coverslips to staining jars containing DAPI freshly diluted to a final concentration of 2 ng/ml in PHEM and incubate for 1 min to stain DNA. Rinse coverslips in staining jars containing PHEM-T (3 × 5 min), followed by a quick-rinse in PHEM. Pipette approximately 10 μl of mounting media/antifade solution on a clean, 75 × 25 × 1 mm microscope slide and gently place the coverslip on top of the solution, cell-side down. Place a Kimwipe over the slide and with *very* light pressure, run a gloved finger along the top of the Kimwipe to soak up any excess liquid from the edges of the coverslip. Secure the coverslip to the slide by sealing edges of the coverslip with nail polish. Store slides in the dark at 4°C.

1. Points to Consider

For each new antibody used, the immunofluorescence protocol should be optimized. In our lab, when characterizing a new antibody, we initially try the protocol described above and a protocol with methanol as the fixative, where cells are first quickly rinsed in PHEM buffer and then incubated at –20°C in ice-cold methanol containing 5 mM EGTA for 5 min. The methanol both fixes and permeabilizes the cells, and after incubation, cells are immediately rinsed 3 × 5 min in PHEM, and the procedure is continued as described above.

C. Cold-induced MT Depolymerization Assay

1. Background

At the onset of mitosis in mammalian cells, kinetochore–MT attachment is usually initiated between a kinetochore and a MT emanating from the closest spindle pole. The newly attached chromosome is rapidly transported along the length of the MT toward the pole to which it has become attached. Lateral attachments are then converted into "end-on" attachments, where MT plus-ends directly embed into the kinetochore, and these end-on associated MTs form a bundle referred to as a kinetochore fiber. Eventually, the opposing sister captures MTs emanating from the opposite spindle pole to produce a bioriented chromosome in which both sister kinetochores are connected to a kinetochore fiber. This specific configuration of stable kinetochore–MT attachment is required for proper chromosome alignment at the spindle equator and for the initiation of anaphase (reviewed in Tanaka and Desai, 2008; Walczak and Heald, 2008). Several decades ago, it was observed that kinetochore–MTs are selectively stable against conditions that induce depolymerization of non-kinetochore–MTs including cold-treatment (Brinkley and Cartwright, 1975; Lambert and Bajer, 1977; Rieder, 1981), heat-treatment (Rieder and Bajer, 1977), hydrostatic pressure (Salmon *et al.*, 1976), and incubation with MT-depolymerizing drugs (Cassimeris *et al.*, 1990). To determine if stable kinetochore–MT attachments are present in mitotic cells, we carry out a cold-induced MT depolymerization assay in which the cells are subjected to cooling in order to depolymerize labile MTs. After cooling, cells are immunostained with anti-α-tubulin antibodies, and the amount of MT polymer is quantified. Described below is a cold-induced MT depolymerization protocol for HeLa cells, and modifications are noted for PtK1 cells. The major differences are (1) PtK1 cells require the addition of calcium for adequate non-kinetochore–MT destabilization and (2) PtK1 cells require a substantially longer cooling time.

2. Procedure

Seed HeLa cells onto sterile #1.5, 22 mm × 22 mm coverslips contained in 35 mm-diameter polystyrene Petri dishes. Maintain cells in growth media until they reach ~80% confluency. Aspirate culture media, replace with growth media precooled to 4°C, and place Petri dishes in a shallow ice bath for 15 min. Remove Petri dishes from bath and aspirate off ice-cold media. Carry out immunofluorescence as described in Section B, except use both the fixative and the permeabilization buffers at room temperature rather than 37°C. To induce MT depolymerization in PtK1 cells, replace growth media with ice-cold permeabilization buffer plus 1 mM ATP and incubate for 2 min. Following permeabilization, incubate PtK1 cells in ice-cold calcium-containing MT destabilization buffer for 30 min in a shallow ice bath. Continue following the immunofluorescence protocol described in Section B, except use the fixative at room temperature rather than 37°C.

3. Data Acquisition and Analysis

Image cells using a 60X or 100X objective on an inverted fluorescence microscope. For these experiments, we use an Olympus 60X/1.42NA Planapochromat DIC oil immersion lens on a DeltaVision PersonalDV Imaging System (Applied Precision). Collect a through-series ("Z-stack") of images for each cell from bottom to top using a 200 nm step-size. For data analysis, we convert collected images to TIFF (tagged image format file) files which are imported into the MetaMorph Imaging software program (Molecular Devices), where a stack file (STK) is generated. However, many image processing and analysis software packages are available and are appropriate for these analyses. To quantify the cold-stable MT polymer, first sum the total tubulin fluorescence of a 20-image Z-stack using

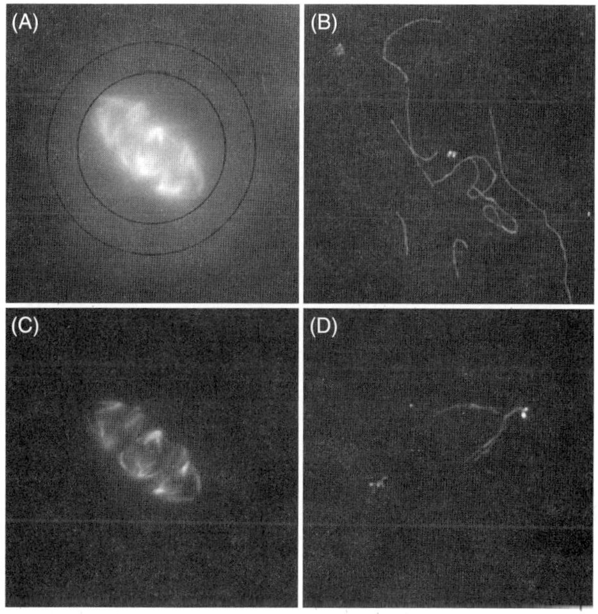

Fig. 1 Immunofluorescence images of HeLa cells subjected to a cold-induced microtubule depolymerization assay. (A) Cold-treated control mitotic HeLa cell. Shown is a composite image stack generated by summing 20 individual fluorescence images. For quantification, total integrated fluorescence intensities within both the small circle (spindle tubulin fluorescence) and the large circle (background fluorescence) are obtained. (B) Maximum projection image of a 20-image Z-stack obtained from a cold-treated, control interphase HeLa cell. Note the few thick, rope-like bundles that remain even after incubation in the cold. (C) Maximum projection image of a 20-image Z-stack obtained from a cold-treated, mock siRNA-transfected mitotic HeLa cell (treated with transfection reagent only). (D) Maximum projection image of a 20-image Z-stack obtained from a cold-treated, Nuf2 siRNA-transfected mitotic HeLa cell. All images were acquired on an Applied Precision DeltaVision PersonalDV Imaging System using a 60X/1.42NA Planapochromat DIC oil immersion objective. Postcold treatment, cells were fixed and immunostained using anti-α-tubulin antibodies.

the "Stack Arithmetic → Sum" function. Specifically, the 20 image planes should encompass the 10 planes above and below the mid-point of the cell. Create a circle that includes the entire mitotic spindle to be used as the constant area for quantification (small circle). Create a circle twice the area of the original circle to measure the background fluorescence (large circle) and place it around the small circle (Fig. 1A). Determine the total integrated fluorescence intensity for the tubulin fluorescence within the small and large circles using "Region Measurements" function. To calculate the portion of total fluorescence due to background fluorescence, we use a modified version of a procedure published for determining the background fluorescence from fluorescently labeled kinetochores (Hoffman *et al.*, 2001; King *et al.*, 2000). Specifically, subtract the total integrated fluorescence intensity value of the small circle from that of the large circle, and scale this value to the size of the small circle by multiplying the difference by the quotient of the area of the small circle divided by the difference between the area of the large circle and the area of the small circle. Calculate the background-subtracted tubulin fluorescence intensity by subtracting the background fluorescence value from the total fluorescence intensity value of the small circle. The following equation (modified from Hoffman *et al.*, 2001) describes these operations: $F_{\text{spindle}} = F_S - [(F_L - F_S)*(A_S/(A_L - A_S))]$, where F_S = total integrated fluorescence intensity with the small circle, F_L = total integrated fluorescence intensity with the large circle, A_S = area of the small circle, A_L = area of the large circle, and F_{spindle} is the final value for background-corrected spindle fluorescence intensity. For each experimental condition, measure the tubulin fluorescence intensity from at least 20 cells. Average spindle fluorescence intensities for control cells and normalize to 1 for comparison against other experimental conditions. Figure 1 shows an example of a cold-treated control mitotic HeLa cell (Fig. 1C) and a cold-treated mitotic HeLa cell depleted of the kinetochore protein Nuf2 by RNAi (Fig. 1D).

4. Points to Consider

The level of cold-stable MT polymer remaining in a cell is dependent on the length of time spent incubating in the cold. Ideally, cooling conditions are chosen such that the majority of non-kinetochore–MTs in control, untreated cells are depolymerized, and only the kinetochore fibers are retained. A good gauge of adequate depolymerization is to confirm that interphase cells contain little MT polymer. However, even with prolonged cooling, it is common to find one or two thick, rope-like cold-stable fibers in interphase cells (Fig. 1B).

It is also important to note that different cell types may require unique assay conditions for MT depolymerization; therefore, each new cell line considered for this assay should be tested for optimal MT depolymerization conditions. One modification we find works well for several types of human cultured cells is to incubate cells in ice-cold media in an ice bath for only 10 min, to precool both the permeabilization and the fixative solutions to 4°C, and to carry out both the permeabilization and the fixation steps at 4°C.

D. Inter-kinetochore Tension

1. Background

When both sister kinetochores of a mitotic chromosome accumulate stably bound MTs, tension is established across the centromere region, resulting in stretching of the centromeric chromatin. This stretching results in an increase in the distance between the two sister kinetochores from their "rest length," when kinetochore–MT attachments are absent (i.e., prior to nuclear envelope breakdown). Measuring inter-kinetochore distances is a useful assay to gauge inter-kinetochore tension generation (Waters *et al.*, 1996) and thus the presence of stable kinetochore–MT attachments.

2. Procedure

Permeabilize, fix, and immunostain cells using an anti-kinetochore primary antibody as described in Section B. We routinely use an antibody to the outer kinetochore protein Hec1 (Abcam, cat#3613); however, in some cases, we find it necessary to use an antibody to an inner kinetochore protein (i.e., if an experimental condition has resulted in the loss of outer kinetochore proteins). In these cases, we use an antibody to CENP-A (for HeLa cells) or an anti-centromere antibody that recognizes multiple inner kinetochore proteins (for HeLa and PtK1 cells). We also routinely stain with an anti-α-tubulin antibody, as MT staining can sometimes be helpful in the identification of sister kinetochore pairs. Using a fluorescence microscope, identify mitotic cells and collect a *Z*-stack of images for each cell using a 200 nm step-size. With appropriate imaging software, measure the distance between two kinetochores from a sister pair by identifying and marking the centroid of each fluorescent kinetochore puncta. Beginning at the bottom of a cell image stack, identify and measure sister pairs throughout the image series. If there is a question as to the certainty of two puncta belonging to a sister pair, do not measure. Also take care to measure only pairs whose sisters are in the same plane of focus; this can be checked by measuring the intensity maxima for each sister kinetochore, and only measuring those pairs whose maxima are located on the same plane. To determine the control "rest length," measure the inter-kinetochore distance between sister kinetochores in prophase cells (Fig. 2A). Alternatively, the rest-length can be measured from cells treated with 20 μM nocodazole for 1 h to depolymerize MTs. To determine the control "stretched length," measure sister kinetochore pairs in metaphase of untreated cells (Fig. 2B). We typically measure at least 250 kinetochore pairs from a total of 25 cells for each condition analyzed.

3. Points to Consider

Depending on the experimental condition, some cells may contain a uniform population of kinetochore pairs, while others may contain discrete subpopulations of pairs that have significantly different inter-kinetochore distances. For example, in cells depleted of any component of the NDC80 complex, in which chromosome biorientation is uniformly impaired, the inter-kinetochore distances of most sister kinetochore pairs are similar, as expected if kinetochore–MT attachment is perturbed homogenously throughout the kinetochore population (Fig. 2C). However, in HeLa cells depleted of CENP-E, most chromosomes align at the spindle equator, while a subset commonly remain at one or both of the spindle poles (McEwen *et al.*, 2001; Putkey *et al.*, 2002; Fig. 2D). In this case, the average inter-kinetochore distance between sister pairs that are able to congress is significantly higher than the average inter-kinetochore distance between the pairs that are stranded at the spindle poles (Putkey *et al.*, 2002; Fig. 2D). In such cases, it is helpful to keep distinct populations segregated when analyzing and interpreting datasets.

The measurements described here reflect *inter*-kinetochore tension, rather than *intra*-kinetochore tension. Recent reports have indicated that silencing of the spindle

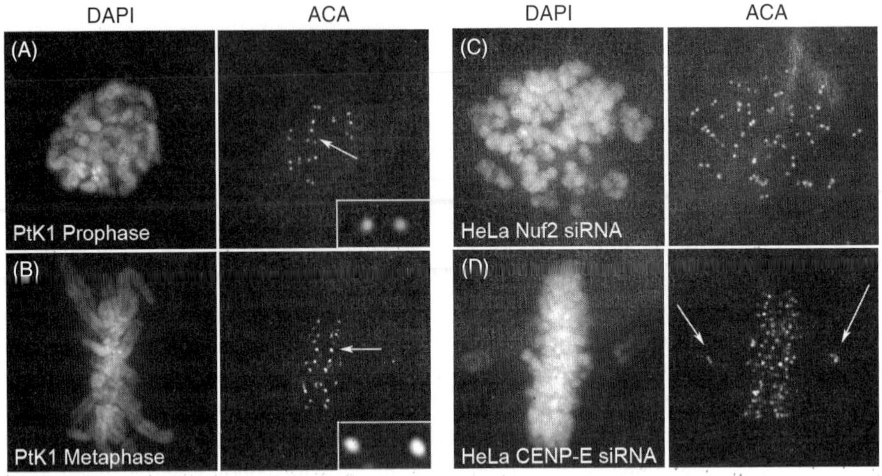

Fig. 2 Immunofluorescence images of HeLa and PtK1 cells illustrating variation in inter-kinetochore distances. (A) Kinetochores in a prophase PtK1 cell at "rest length," where kinetochore–MT attachments have not yet formed (arrow indicates the sister kinetochore pair shown in the inset) (B) Kinetochores in a metaphase PtK1 cell. Chromosomes have bioriented and kinetochore–MT attachments have been made, and sister kinetochore pairs exhibit inter-kinetochore stretching (arrow indicates the sister kinetochore pair shown in the inset). (C) HeLa cell depleted of Nuf2 by siRNA transfection. Kinetochore–MT attachments are uniformly impaired, therefore the variation in inter-kinetochore distances is low. (D) HeLa cell depleted of CENP-E by siRNA transfection. Most sister kinetochore pairs biorient and chromosomes align at the spindle equator; however, some chromosomes remain stranded at the spindle poles. In CENP-E-depleted cells, sister kinetochores on bioriented chromosomes establish inter-kinetochore tension, while those on chromosomes stranded at the poles do not (arrows). For all images in A–D, maximum intensity projections are shown so that multiple kinetochores can be displayed. For analysis, however, inter-kinetochore distances are measured from single images rather than maximum intensity projections, so that individual pairs can be resolved.

assembly checkpoint requires the generation of *intra*-kinetochore tension, where protein components within an individual kinetochore become stretched from each other. For these experiments, intra-kinetochore distance is measured between a protein in the kinetochore outer domain and a protein in the kinetochore inner domain (Maresca and Salmon, 2009; Uchida *et al.*, 2009).

Finally, be aware that certain situations exist where the inter-kinetochore distance is not necessarily indicative of kinetochore–MT attachment stability. For example, depletion of proteins that result in a weakening of sister kinetochore cohesion may result in an increase in inter-kinetochore distance that is not due to generation of centromere tension, but from a defect in sister chromatid cohesion (Kitajima *et al.*, 2005; Ritchie *et al.*, 2008).

E. Kinetochore–MT Turnover

The stability of kinetochore–MT attachments must be regulated throughout mitosis to ensure accurate chromosome segregation. For example, in early mitosis kinetochore–MTs are labile, so that erroneous kinetochore–MT attachments can be released. Conversely, in late mitosis kinetochore–MTs are stabilized so that forces can be generated for chromosome congression and to silence the spindle assembly checkpoint. The search for factors that contribute to this differential regulation is ongoing, and for many mitotic studies it is useful to measure the stability of kinetochore–MTs by quantifying the kinetochore–MT turnover rate in living cells. For this purpose, we utilize an assay in which fluorescent marks are generated on tubulin subunits within the mitotic spindle via fluorescence photoactivation. Kinetochore–MT turnover rates were originally measured in mitotic cells injected with caged fluorescein-labeled tubulin, which upon irradiation with a certain wavelength of light became uncaged and fluoresced green when excited with blue light (Mitchison, 1989; Zhai *et al.*, 1995). Kinetochore-MT half-lives have also been accurately measured by incubation of cultured cells with MT depolymerizing drugs for increasing amounts of time followed by fixation and quantification of kinetochore-MT numbers using serial-section electron microscopy (Cassimeris et al., 1990). More recently, photoactivatable (PA) and photoswitchable fluorescent proteins have been developed (Fernández-Suárez and Ting, 2008; Sample *et al.*, 2009), which allows cells to genetically encode tubulin fused to a fluorophore that can be specifically turned on at a given time by activation with a certain wavelength of light. For the study of kinetochore–MT turnover, a commonly used fusion is PA-GFP-tubulin, which becomes fluorescent when activated with light in the 405 nm range. When a small region within the mitotic spindle of a cell expressing PA-GFP-tubulin is activated, all PA-GFP-tubulin within that region becomes fluorescent. Fluorescent marks on free dimeric tubulin and non-kinetochore–MTs dissipate quickly due to the highly dynamic nature of these tubulin populations, while fluorescent marks persist on MTs within kinetochore fibers due to their selective stability. The turnover rate of spindle MTs can be calculated by determining tubulin fluorescence decay over time. To account for both kinetochore and non-kinetochore–MTs, fluorescence decay data are fit to a double exponential equation, and subsequent analysis allows the determination of the relative percentages of stable versus unstable MTs in the spindle (non-kinetochore–MTs vs kinetochore–MTs) and their respective half-lives (Zhai *et al.*, 1995).

1. Data Acquisition

Culture PA-GFP tubulin-expressing PtK1 cells in PtK1 growth media on sterile #1.5 22 mm × 22 mm coverslips (cells provided kindly by Dr. Alexey Khodjakov). For this cell line, we also include 1 mg/ml G418 in the culture media to continuously select for cells expressing PA-GFP-tubulin. Assemble modified Rose chambers (described in Rieder and Hard, 1990) with a top coverslip, and using a 21 gauge needle and 5 ml syringe, fill chamber with filming media prewarmed to 37°C. Alternatively, culture cells on #1.5 glass-bottom 35 mm culture dishes and prior to the experiment, remove growth media, and replace with filming media. Place the chamber or dish on the stage of an inverted fluorescence microscope in an incubation chamber warmed and stabilized to 37°C and equipped with the appropriate imaging light source and activation light source. For GFP imaging, we use a 473 nm laser and for activation we use a mercury arc lamp and a 420 nm short pass dichroic mirror, which transmits light in the 390–420 nm range and reflects light in the 420–700 nm range (note: lasers in the 405 nm range can also be used for PA-GFP activation). The microscope must be additionally equipped with an apparatus that permits a specific region of the field to be illuminated with activation light. For this purpose, we carry out photoactivation experiments on a spinning disc confocal microscope equipped with a Mosaic Laser Ablation/Photoactivation System (Photonics Instruments, Inc.), which allows for software-driven demarcation and activation of user-defined regions on the specimen field. Locate a mitotic cell by scanning the sample under transmitted light using a 60× DIC objective and capture both a DIC and fluorescence (473 nm) preactivation image (a phase-contrast objective can also be used to identify mitotic cells). Using the microscope software that drives the Mosaic system, create and save a narrow rectangular region (we routinely use a 1.2×21 μm region), and using the DIC image as a guide, drag the rectangle to the cell image on the computer monitor and rotate the rectangle so that it is positioned perpendicular to the spindle axis, adjacent to the chromosome mass. Through the software, activate the tubulin fluorescence within the rectangular region with the mercury light source. We activate for 100 ms; however, activation times will vary based on the intensity of the activation light source and the focus of the activating light source on the specimen plane. Acquire a single DIC and 473 nm excitation image immediately after activation, 10 s postactivation, and every 20 s thereafter for 10 min. After the final image is collected at 10 min, draw a new rectangular region that encompasses the entire cell, activate the tubulin fluorescence within this larger region, and acquire a final fluorescence image. This final step will allow for the visualization of the entire mitotic spindle to determine if spindle abnormalities are present (Fig. 3).

During the course of imaging the fluorescently labeled tubulin, some fluorescence will be lost due to photobleaching. In order to correct for this loss, a data set must be collected in which the decrease in tubulin fluorescence is due only to photobleaching and not MT turnover. For this purpose, we pretreat a population of cells with taxol to stabilize MTs and prevent tubulin exchange. Prior to assembling the imaging chambers, incubate cells with PtK1 growth media + 10 μM taxol for 1 h and fill the imaging chambers with filming media + 10 μM taxol. This concentration of taxol will induce assembly of the free tubulin into MT polymer and stabilize the MTs against depolymerization, preventing MT turnover (Schiff and Horwitz, 1980). Identify mitotic cells using transmitted light, capture preactivation images, activate a bar-shaped region close to the chromosome mass, and image as described above. The calculated fluorescence loss from MTs in the presence of taxol will be subtracted from the total loss of fluorescence measured in a control or experimental condition to correct for photobleaching through subsequent data analysis (see below).

2. Data Analysis

First determine the loss of tubulin fluorescence due to photobleaching. For taxol-treated cells, measure the fluorescence intensity of the activated tubulin by drawing a rectangular region, kept constant per cell, that encompasses the initial bleached region (Fig. 3B). The size of the rectangle will vary based on the objective used and the cell that is being analyzed. Record the total integrated fluorescence within that region using

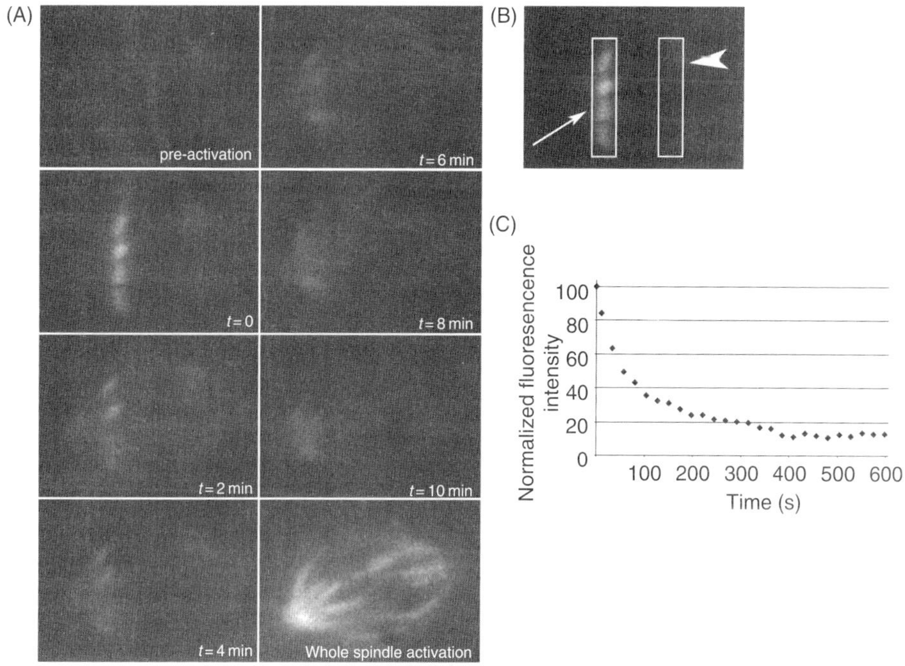

Fig. 3 Activation of PA-GFP-tubulin in a PtK1 cell. (A) Eight frames from a time-lapse sequence of a control PtK1 cell stably expressing PA-GFP-tubulin (cells kindly provided by Dr. Alexey Khodjakov). All images were acquired using a 473 nm laser for excitation. The image in the upper left was acquired prior to activation with ~405 nm light. A small bar-shaped region was photoactivated and images were acquired immediately after activation ($t = 0$), 10 s after activation, and then every 20 s for the duration of the 10 min time-lapse (selected images are shown). After 10 min, the entire cell was exposed to ~405 nm light to activate all the PA-GFP-tubulin in the cell. The post-whole cell activation image is shown in the bottom right panel. (B) Image demonstrating how the fluorescence was measured for each time point. The box on the left encompasses the fluorescent tubulin (arrow), and the box on the right encompasses a region in the opposite half-spindle to account for background fluorescence (arrowhead). (C) Graph showing the decrease in tubulin fluorescence intensity over time of the activated region in the cell shown in (A).

the "Region Measurements" function in MetaMorph. Move the rectangular region off of the activated tubulin region and place it at a similar distance from the chromosome mass within the opposite, non-activated half-spindle (Fig. 3B). Record the total integrated fluorescence within this region, which will serve as a background fluorescence measurement. At each time point, move the rectangle to encompass the activated region and record the total integrated fluorescence within that region, as well as within a region residing in the opposite half-spindle. In taxol-treated cells, the activated bar will not move significantly from its initial location, while in control cells, the activated bar will move from the chromosome mass to the spindle pole (Fig. 3A). For each time point, subtract the total fluorescence intensity measurement acquired from the background rectangle (taken from the half-spindle that was not activated) from the fluorescence intensity measurement acquired from the activated region ("background-subtracted value").

Calculate the percent loss of fluorescence at each time point from the initial postactivation time point ("photobleaching correction value"). Because MTs under these conditions are not dynamic, the percentage loss of fluorescence at each time point represents fluorescence loss due to photobleaching. To calculate the average fluorescence loss due to photobleaching, collect data from at least 10 cells and average the percent loss at each time point.

For an experimental data set, determine the fluorescence dissipation at each time point as described for the taxol-treated cells. In this case, the fluorescent region may significantly move away from its original location at the time of activation, so the bar encompassing the activated region and the bar used for the background measurement must be moved at each time point. Determine the background-subtracted value for each time point and average the background-subtracted fluorescence values at each time point for multiple cells (we average data from at least 10 cells per experimental condition). Determine the loss of fluorescence due to photobleaching by multiplying the average background-subtracted value by the average percent fluorescence loss due to photobleaching at each time point. Subtract this product from the background-subtracted value at each time point. Finally, normalize fluorescence values to the background-subtracted, photobleaching-corrected value at the first time point after activation to 100.

To determine the half-life of the kinetochore–MTs, subject the data to a non-linear regression analysis using appropriate data analysis software such as SigmaPlot (Systat Software), GraphPad (GraphPad Software), or Kaleidagraph (Synergy Software). Due to the presence of kinetochore- and non-kinetochore–MTs, the data are best fit to a double exponential curve represented by the equation: $y = A_1 * e^{(-k_1 * t)} + A_2 * e^{(-k_2 * t)}$. A_1 represents the percentage of non-kinetochore–MTs (less stable), A_2 represents the percentage of kinetochore–MTs (more stable), k_1 and k_2 represent the fluorescence dissipation rates of each population of MTs, and t is the time postphotoactivation. To calculate MT half-life values ($t_{1/2}$), use the equation $t_{1/2} = \ln 2/k$ (use k_1 to determine the non-kinetochore–MT half-life and k_2 to obtain the kinetochore–MT half-life). From these analyses, both the stability of kinetochore–MTs can be calculated ($t_{1/2}$) and the percentage of MTs that are incorporated into kinetochore fibers.

3. Points to Consider

An alternative to the Mosaic system for photoactivating a small, bar-shaped region on the specimen is to focus the activating light onto a slit that has been inserted into the light path, which results in activation of a very narrow region on the specimen (Mitchison, 1989; Salmon *et al.*, 2007). One disadvantage to this method is the lack of flexibility in regard to the orientation of the beam that has been focused through the slit. In this case, cells whose spindles are in the correct orientation in regard to the focused beam must be chosen, or alternatively, a rotating stage could be used (Mitchison, 1989).

Finally, useful comments regarding live-cell fluorescence imaging, including choice of microscope, imaging chambers, temperature considerations, and minimizing photobleaching, can be found in Waters (2007).

F. Kinetochore–MT Attachment Error Correction

For chromosomes to achieve biorientation, pairs of sister kinetochores must have one sister attached to MTs emanating from each of the two spindle poles (amphitelic attachment). During the course of chromosome biorientation, however, kinetochore–MT attachment errors are frequently made. One kinetochore can attach to MTs from both poles (merotelic attachment), both kinetochores of a sister pair can attach to MTs from one pole (syntelic attachment), or one kinetochore can attach to MTs emanating from one pole, while its sister remains unattached (monotelic attachment) (Biggins and Walczak, 2003; Cimini and Degrassi, 2005; Nicklas and Ward, 1994). All attachments except those that are amphitelic must be destabilized to allow for new, correct attachments to form. In order for error correction to occur, kinetochore–MTs must be capable of dynamic cycles of attachment and release, a process that is dependent on the highly conserved serine/threonine kinase, Aurora B (Biggins and Murray, 2001; Tanaka *et al.*, 2002). The ability of kinetochores to correct MT attachments can be assayed experimentally by carrying out an Eg5-inhibitor washout experiment. Eg5 is a plus-end directed mitotic kinesin-5 protein that is required for bipolar spindle assembly. This motor protein forms homotetramers that cross-link and translocate along spindle MTs to provide forces for centrosome separation (reviewed in Valentine *et al.*, 2006). Treatment of cell populations with the Eg5 inhibitor monastrol results in the accumulation of cells with monopolar spindles (Mayer *et al.*, 1999), and a consequence of this is the formation of many syntelically attached sister kinetochore pairs (Kapoor *et al.*, 2000). Upon removal of the drug by a thorough washout, monopolar spindles are quickly converted to bipolar spindles, and syntelic kinetochore attachments are replaced with amphitelic attachments (Kapoor *et al.*, 2000). Cells that have faulty kinetochore–MT error correction machinery fail to efficiently correct syntelic attachments and, as a result, bipolar spindle conversion is delayed and upon bipolar spindle assembly, chromosome alignment is perturbed (Lampson *et al.*, 2004; Vader *et al.*, 2007). This assay is therefore well suited to gauge the ability of cells to correct errors in kinetochore–MT attachment.

1. Procedure

Culture HeLa or PtK1 cells on multiple sterile #1.5 22 mm × 22 mm coverslips until they reach ~80% confluency. Remove growth media from cells, replace with growth media containing 100 μM monastrol, and incubate at 37°C in 5% CO_2 for 2 h. After incubation, prepare one coverslip for immunofluorescence (time point 0) as described in Section B. With the remaining coverslips, remove monastrol-containing media and rinse quickly 3X with fresh growth media prewarmed to 37°C. Immediately place cells in fresh growth media containing 5 μM MG132 prewarmed to 37°C to

Sample, V., Newman, R. H., and Zhang, J. (2009). The structure and function of fluorescent proteins. *Chem. Soc. Rev.* **38**, 2852–2864.

Schiff, P. B., and Horwitz, S. B. (1980). Taxol stabilizes microtubules in mouse fibroblast cells. *Proc. Natl. Acad. Sci. U.S.A.* **77**, 1561–1565.

Shelby, R. D., Sullivan, K., Wilson, L., and Jordan, M. A. (1996). Dynamic elastic behavior of alpha-satellite DNA domains visualized in situ in living human cells. *J. Cell Biol.* **135**, 545–557.

Skibbens, R. V., Skeen, V. P., and Salmon, E. D. (1993). Directional instability of kinetochore motility during chromosome congression and segregation in mitotic newt lung cells: A push-pull mechanism. *J. Cell Biol.* **122**, 859–875.

Skoufias, D. A., DeBonis, S., Saoudi, Y., Lebeau, L., Crevel, I., Cross, R., Wade, R. H., Hackney, D., and Kozielski, F. (2006). S-trityl-L-cysteine is a reversible, tight binding inhibitor of the human kinesin Eg5 that specifically blocks mitotic progression. *J. Biol. Chem.* **281**, 17559–175569.

Stout, J. R., Rizk, R. S., Kline, S. L., and Walczak, C. E. (2006). Deciphering protein function during mitosis in PtK cells using RNAi. *BMC Cell Biol.* **7**, 26.

Stumpff, J., von Dassow, G., Wagenbach, M., Asbury, C., and Wordeman, L. (2008). The kinesin-8 motor Kif18A suppresses kinetochore movements to control mitotic chromosome alignment. *Dev. Cell* **14**, 252–262.

Sunder-Plassmann, N., Sarli, V., Gartner, M., Utz, M., Seiler, J., Huemmer, S., Mayer, T. U., Surrey, T., and Giannis, A. (2005). Synthesis and biological evaluation of new tetrahydro-beta-carbolines as inhibitors of the mitotic kinesin Eg5. *Bioorg. Med. Chem.* **13**, 6094–6111.

Tanaka, T. U., and Desai, A. (2008). Kinetochore-microtubule interactions: The means to the end. *Curr. Opin. Cell Biol.* **20**, 53–63.

Tanaka, T. U., Rachidi, N., Janke, C., Pereira, G., Galova, M., Schiebel, E., Stark, M. J., and Nasmyth, K. (2002). Evidence that the Ipl1-Sli15 (Aurora kinase-INCENP) complex promotes chromosome bi-orientation by altering kinetochore-spindle pole connections. *Cell* **108**, 317–329.

Tirnauer, J. S., Canman, J. C., Salmon, E. D., and Mitchison, T. J. (2002). EB1 targets to kinetochores with attached, polymerizing microtubules. *Mol. Biol. Cell.* **13**, 4308–4316.

Uchida, K.S.K., Takagaki, K., Kumada, K., Hirayama, Y., Noda, T., and Hirota, T. (2009). Kinetochore stretching inactivates the spindle assembly checkpoint. *J. Cell Biol.* **184**, 383–390.

Vader, G., Cruijsen, C. W., van Harn, T., Vromans, M. J., Medema, R. H., and Lens, S. M. (2007). The chromosomal passenger complex controls spindle checkpoint function independent from its role in correcting microtubule kinetochore interactions. *Mol. Biol. Cell.* **18**, 4553–4564.

Valentine, M. T., Fordyce, P. M., and Block, S. M. (2006). Eg5 steps it up! *Cell Div.* **1**, 31–39.

Varma, D., Monzo, P., Stehman, S. A., and Vallee, R. B. (2008). Direct role of dynein motor in stable kinetochore-microtubule attachment, orientation, and alignment. *J. Cell Biol.* **182**, 1045–1054.

Wadsworth, P. (2007). Studying Mitosis in Cultured Mammalian Cells. In *Live Cell Imaging: A Laboratory Manual*, Second Edition. Cold Spring Harbor Laboratory Press. 2007. Chapter 27, pages 491-502.

Walczak, C. E., and Heald, R. (2008). Mechanisms of mitotic spindle assembly and function. *Int. Rev. Cytol.* **265**, 111–115.

Wan, X., O'Quinn, R. P., Pierce, H. L., Joglekar, A. P., Gall, W. E., DeLuca, J. G., Carroll, C. W., Liu, S. T., Yen, T. J., McEwen, B. F., Stukenberg, P. T., Desai, A., *et al.* (2009). Protein architecture of the human kinetochore microtubule attachment site. *Cell* **137**, 672–684.

Waterman-Storer, C. M., Desai, A., Bulinski, J. C., and Salmon, E. D. (1998). Fluorescent speckle microscopy, a method to visualize the dynamics of protein assemblies in living cells. *Curr. Biol.* **8**, 1227–1230.

Waters, J. (2007). Live-cell fluorescence imaging. *Methods Cell Biol.* **81**, 115–140.

Waters, J. C., Skibbens, R. V., and Salmon, E. D. (1996). Oscillating mitotic newt lung cell kinetochores are, on average, under tension and rarely push. *J. Cell Sci.* **109**, 2823–2831.

Wendell, K. L., Wilson, L., and Jordan, M. A. (1993). Mitotic block in HeLa cells by vinblastine: Ultrastructural changes in kinetochore-microtubule attachment and in centrosomes. *J. Cell Sci.* **104**, 261–274.

Wordeman, L., Wagenbach, M., and von Dassow, G. (2007). MCAK facilitates chromosome movement by promoting kinetochore microtubule turnover. *J. Cell Biol.* **179**, 869–879.

Yang, Z., Tulu, U. S., Wadsworth, P., and Rieder, C. L. (2007). Kinetochore dynein is required for chromosome motion and congression independent of the spindle checkpoint. *Curr Biol.* **5**, 973–80.

Yuen, K. W., Montpetit, B., and Hieter, P. (2005). The kinetochore and cancer: What's the connection? *Curr. Opin. Cell Biol.* **17**, 576–582.

Zhai, Y., Kronebusch, P. J., Borisy, G. G. (1995). Kinetochore microtubule dynamics and the metaphase-anaphase transition. *J. Cell Biol.* **131**, 721–734.

CHAPTER 5

Photoactivatable Green Fluorescent Protein-Tubulin

U. Serdar Tulu*, **Nick P. Ferenz**†, *and* **Patricia Wadsworth****

*Department of Biology, Duke University, Durham, North Carolina 27708

†Precept Medical Communications, Berkeley Heights, New Jersey 07922

**Department of Biology, University of Massachusetts, Amherst, Massachusetts 01003

Abstract

Direct observations of live cells expressing fluorescently tagged tubulin have led to important advances in our understanding of mitosis. A limitation of this approach is that all of the cells' microtubules are fluorescent and thus observation of the behavior of specific subsets of microtubules is precluded. To address this problem, we have tagged tubulin with a photoactivatable variant of green fluorescent protein (PA-GFP), thereby allowing one to follow the behavior of a subset of tagged molecules in the cell. Here, we describe methods to tag and express proteins with PA-GFP, locally photoactivate the recombinant protein and record the dynamic behavior of the photoactivated molecules in live cells. Use of photoactivatable proteins is a powerful approach to examine dynamic processes, including spindle formation, in diverse cells.

978-0-12-381349-7
DOI: 10.1016/S0091-679X(10)97005-2

I. Introduction

Understanding the behaviors of cells in an organism and molecules in a cell has been an essential motivation of many scientists in order to understand how organisms function and survive and also to find out possible ways to prevent malfunctions observed in various disease states. Live imaging in this context provides an effective and useful tool to elucidate both cellular and molecular behaviors in their natural environment. With the cloning of green fluorescent protein (GFP) from jellyfish and its expression in other organisms and cells, live imaging technology had a tremendous renaissance which allowed scientists to ask deeper and broader questions.

Vast usage of GFP in biological sciences over the last 2 decades has demonstrated that GFP tagging is a powerful technique that provides a new way to discover the workings of the microuniverse. In fact, it is widely agreed that one movie of a cell can tell more than a picture. GFP tagging provides us with a means to record cellular/subcellular dynamics in real time. With this approach we can observe subcellular architecture as well as the localization of the protein, and since molecules often localize where they function, following proteins tagged with GFP gives clues about their function while the cell/organism is still alive.

In our lab, we have used GFP-tagged tubulin to study microtubule behavior in live cells. We demonstrated that cell lines permanently expressing GFP-tubulin can be generated and that the presence of the GFP-tag does not deleteriously alter microtubule behavior (Rusan *et al.*, 2001). In the cell line that we generated, 17% of the total tubulin was GFP-tagged, and the level of unlabeled tubulin was reduced to 82% of that in the parental cell line. To determine if the GFP-tag on tubulin altered the parameters of microtubule dynamic instability, we compared dynamics in interphase cells expressing GFP-tubulin with parental cells injected with fluorescent tubulin. The results showed that the parameters of dynamic instability were not significantly different in the two cases, demonstrating that this cell line is a useful tool for measuring microtubule dynamic behavior. Moreover, the mitotic index and doubling time for these cells were not different from the parental cells. Using these cells, we showed for the first time in mammalian cells that microtubule dynamicity is increased as cells progress from interphase to mitosis and that the time microtubules spend in an attenuated state, or pause, is dramatically reduced (Rusan *et al.*, 2001).

Although these (and other similar cell lines) have been extremely useful for analysis of microtubule distribution and dynamics, the behavior of individual microtubules is difficult to follow because the microtubules are uniformly fluorescent [as viewed by wide-field or confocal fluorescence microscopy; but see (Bicek *et al.*, 2009)] and the density of microtubules can be high in many cell regions. One way to overcome these limitations is to microinject cells with lower levels of fluorescent tubulin, thereby creating microtubules with only a few fluorescent subunits, called "speckles," at any location along the microtubule (Waterman-Storer and Salmon, 1998). The speckles can then be followed over time to reveal additional features of microtubule dynamics. Although powerful, fluorescent speckle microscopy (FSM) usually requires that cells are individually microinjected with low levels of tubulin, a time-consuming procedure. Similarly, fluorescence recovery after photobleach (FRAP) can be used to locally reduce the fluorescence and record the kinetics of fluorescence recovery and reveal motion of the bleached regions (Bancaud *et al.*, 2010).

A. Rationale

An alternative to FSM and FRAP is the use of photoactivatable GFP. This variant of GFP has a threonine to histidine mutation at position 203 of the wild-type EGFP protein and is not fluorescent until it is exposed to ultraviolet light (Patterson and Lippincott-Schwartz, 2002). Because the UV illumination can be spatially controlled, a user-selected subset of photoactivatable variant PA-GFP-tagged proteins can be activated, while the remainder is left inactive and thus nonfluorescent. In the following sections, we describe the methods that we use to generate cells expressing PA-GFP-tubulin and to analyze the behavior of microtubules in these cells.

B. Expressing PA-GFP-Tubulin in Mammalian Cells

To generate PA-GFP tubulin, we replaced the EGFP sequence with the sequence encoding PA-GFP using either a pCMV or an IRES vector. We have had particularly good success when using the IRES vectors because the tagged gene, here PA-GFP-tubulin, and the selection marker are expressed from a single promoter, thus improving the selection step (Ferenz *et al.*, 2010). This is particularly important when generating PA-GFP cell lines because the tag is nonfluorescent until activated and so cloning is performed "blind." Using a pCMV vector expressing PA-GFP-tubulin, we screened ~72 colonies, each a potential cell line. These colonies were either nonphotoactivatable or not suitable for imaging. In contrast, using an IRES vector, we screened ~24 colonies and obtained one useful cell line expressing PA-GFP-tubulin.

Our cell line of choice for analysis of microtubule dynamics is LLC-Pk1 cells. These cells are derived from kidney epithelium and remain flat and spread throughout mitosis (Wadsworth, 2010). They are easy to culture, transfect, and image for long periods. To examine microtubule behavior in live cells we prepare clonal cell lines. Although generating such cell lines is time-consuming, it has several advantages. First, and most important, all the cells in a clonal line are identical. This is in contrast to transiently transfected cells, which show tremendous variation in expression level. Second, if transiently transfected cells are used for photoactivation, it is not possible to gauge the expression level until after the activation is performed, and thus it is difficult to perform experiments on cells with similar levels of expression. Third, several different clonal cell lines can be characterized (for example, by measuring doubling time, mitotic index, and the percentage of abnormal spindles) prior to performing experiments and the line most similar to the parental cells chosen for experimentation. Finally, stable lines are useful for live imaging of transient events. With transient transfection, catching these short moments in cells with appropriate expression levels is very difficult, if not impossible, and can always be questionable since different cells will have different amounts of expression. With stable cell lines, experiments can also be done much more efficiently because all the cells in the population are fluorescent.

To make a clonal cell line, we first transfect parental cells with PA-GFP-tubulin. For this step we find that nucleofection offers a comparatively higher level of transfection efficiency and requires a smaller amount of purified plasmid. Alternatively, cells can be transfected with any of a number of commercially available lipid-based transfection reagents.

Following transfection cells are grown for ~48 h to allow expression of the antibiotic resistance gene, and then subjected to ~2 weeks of positive selection in the presence of the appropriate antibiotic. During this period, the medium is changed frequently to remove dead and dying cells; surviving cells are subcloned as needed. Following the selection period, cells are trypsinized and seeded into 100 mm dishes at a very low density, so that individual cells generate colonies well separated from other colonies on the dish. Within 1–2 weeks of growth, these individual colonies will be visible by eye; each colony is a potential cell line. At the same time, aliquots of the antibiotic-selected cells are also frozen as a back-up in case a suitable cell line is not obtained. Cloning rings are used to harvest individual colonies, which are first transferred to individual wells of a 24 well plate, and then (after sufficient growth) transferred to two wells of a 6-well plate, one of which has a coverslip. As an alternative approach, cloning discs can be used to harvest colonies from the 100 mm plate. Cells on the coverslip can be examined for expression of the PA-GFP tubulin by photoactivation of full fields of view. Multiwell plates with a coverslip bottom are also convenient for screening large numbers of clones.

LLC-Pk1 cells are epithelial cells, which grow as colonies. For cells that do not form colonies, i.e., fibroblasts, the selected cells are counted and diluted so that approximately one cell is placed into each well of a 96-well plate. When growth of cells in each well becomes visible, they are transferred to each well of 24-well plates for further growth. Here, cells in each well can be treated as a potential cell line. They can be analyzed thereafter as we describe above.

For the most of the lines we generated, we were more comfortable using the first method with cloning rings or cloning discs partially because epithelial cells form colonies and do not crawl as much as other cells. In addition, when diluted and plated on a 100 mm dish, cells in a colony can be seen under a simple microscope and can be visually selected in terms of their morphological appearance, an indication of whether they are a clone. Finally, if more than one cell type is detected in a stable cell line (i.e., cells with different levels of fluorescence, an indication that the line is not actually clonal), these procedures above can be performed again to obtain a clonal cell line.

Western blotting of cell extracts is used to identify potentially useful clones and to determine the expression level of the tagged protein relative to the endogenous protein. In the cell lines that we have generated ~10–20% of the total tubulin is tagged with GFP or PA-GFP (Fig. 1). At this expression level, the protein is easily detected by

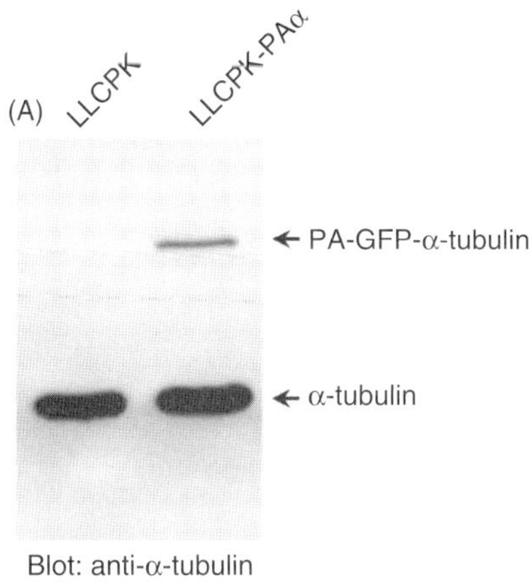

	LLCPK (control)	LLCPK-PAα
Mitotic index	2.6 + 0.80%	3.6 + 0.29%
Total tubulin	100%	103%
PA-GFP-α-tubulin	—	9.9%

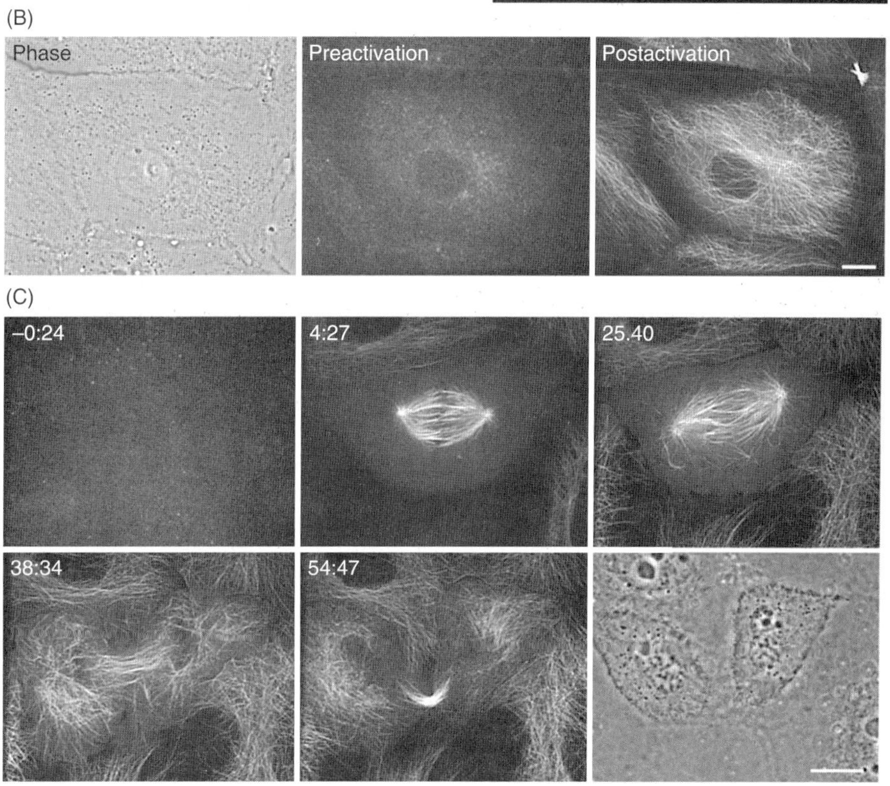

Fig. 1 Characterization of a PA-GFP-tubulin cell line. (A) Western blot of whole cell extract of parental cells (left) and a cell line expressing PA-GFP-tubulin (right); blot was stained for tubulin. (B) Phase contrast and fluorescence images of an interphase cell before and after activation. (C) Mitotic cell before and after activation; the cell progresses through mitosis without any detectable abnormality and all classes of mitotic microtubules are fluorescent. Time after activation is shown in the upper left. Bar = 10 μm.

fluorescence microscopy and the cells show no mitotic defects due to the expression of the tagged protein (Rusan *et al.*, 2001; Tulu *et al.*, 2003).

Once one or more cell lines are obtained that expresses a useful level of the tagged protein, additional assays are performed (measurement of doubling time, mitotic index, and % of abnormal spindles) to confirm that there is no detectable effect on cell division (Fig. 1). Depending on the particular cellular processes that are under investigation other assays can be performed to show that the presence of the tagged protein has no, or minimal, effect on these events.

C. Photoactivation

We perform photoactivation experiments on a Nikon Eclipse TE300 inverted microscope. To photoactivate, light from an X-cite 120 (EXFO America, Plano, TX) or 100W mercury arc lamp in the epi-illumination pathway is passed through a D405/20 filter cube (Chroma Technology, Rockingham, VT) to obtain light of the appropriate wavelength for activation (~413 nm) (Fig. 2). A laser of 405 nm can also be used. The activation time is determined by exposing cells to 413 nm light for various intervals,

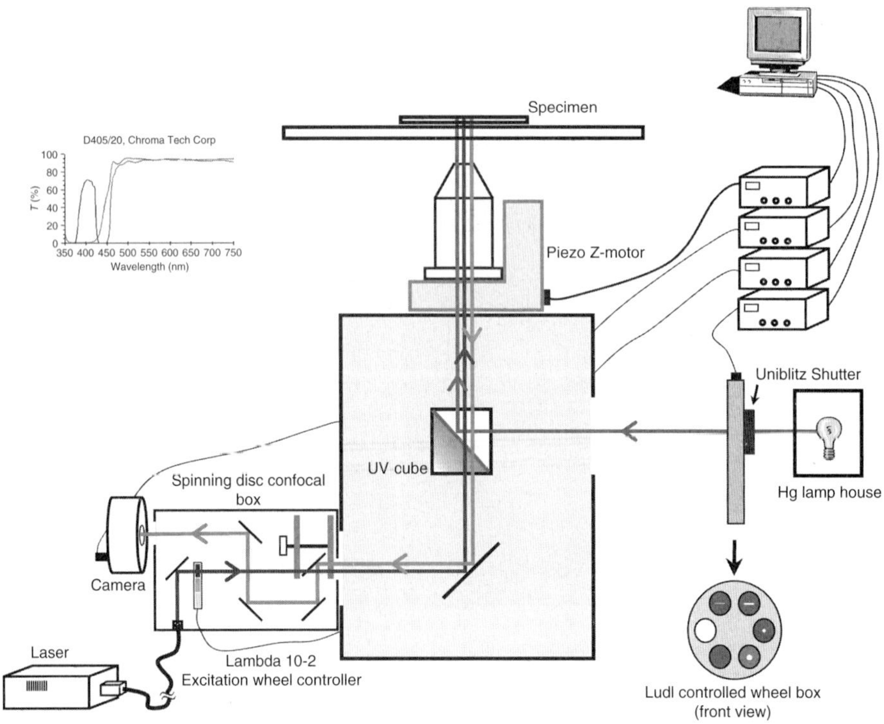

Fig. 2 Schematic diagram of the microscope used for photoactivation.

and quantifying the resulting fluorescence from images acquired using identical acquisition settings (Fig. 3). The minimal exposure that results in full fluorescence activation is then used for all subsequent experiments.

The region to be photoactivated is controlled by using a slit or a pinhole in a filter wheel placed in an image plane conjugate with the specimen image plane (Fig. 2). When photoactivation of a larger area is desired, the field diaphragm in the epi-illumination pathway can also be closed to a suitable diameter and used for this purpose. For activation of spindles, we find that a slit of dimensions 25 μm × 3 mm

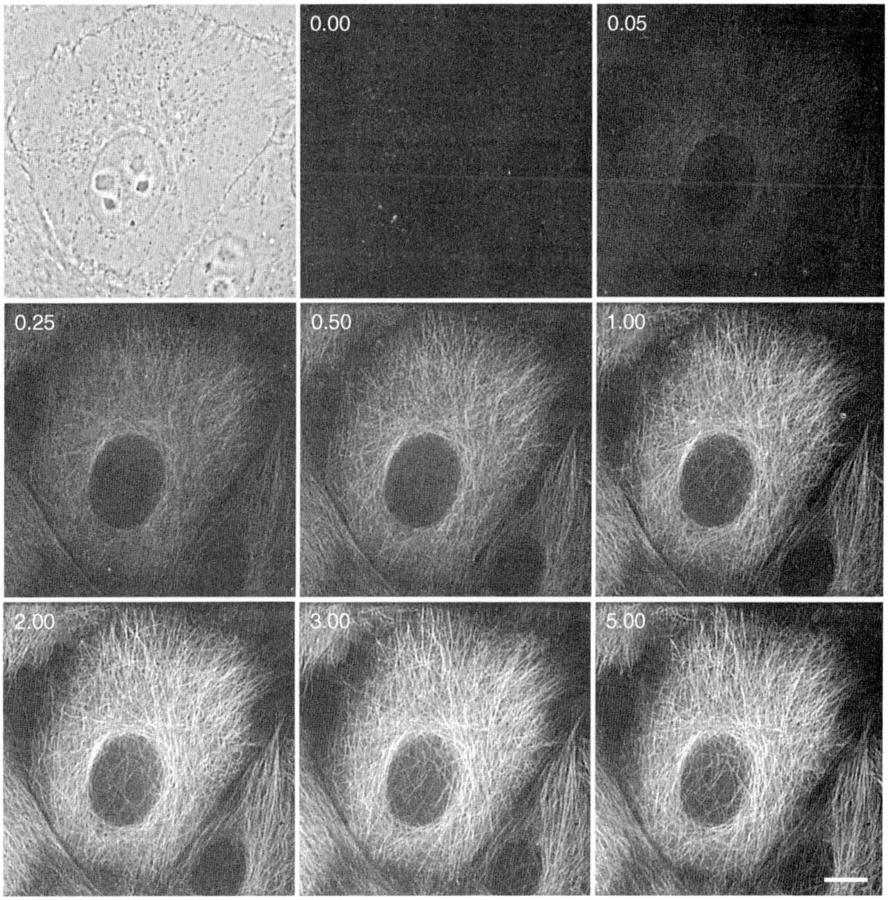

Fig. 3 Determination of the optimum time for activation of PA-GFP-tubulin. Phase contrast and fluorescence images of an interphase LLC-Pk1 cell expressing PA-GFP-tubulin. Exposure to 405 nm light is calculated cumulatively (i.e., exposure times on each panel show the total time that the cell was exposed to UV light). Images are acquired with the same settings after each exposure. Time of exposure (seconds) to activation light is shown in the upper left. Bar = 10 μm.

(Lennox Laser, Glen Arm, MD) generates an area of activation of ~2 μm width in the image plane. With this setup, the slit is well defined and sharp, facilitating tracking of the photoactivated area in time-lapse images. Because the slits and pinholes are in a fixed position, we rotate the sample (using a circular rotating stage) so that an appropriate cellular area is activated. For reference, the location of the slit can be noted on an eyepiece reticle or within the image acquisition software. We use a 100× objective lens for imaging.

We also need to take into account the fact that imaging wavelength and photoactivation wavelength are not the same; and they may not focus on the same plane (presumably due to chromatic aberration). We figured out in our system that laser light at 488 nm and epi-illumination at 405 nm have different focal distances, which we determined as ~3 μm. To be able to correct for this difference, we wrote a simple journal in MetaMorph that changes the position of objective using a piezo Z-motor (Physik Instrumente (PI), Auburn, MA) at the time of activation. Following activation, the objective is brought back to its original position and imaging is continued by the software. This procedure is essential since it is possible to image one plane and photoactivate another plane if enough care is not exercised.

Cells are imaged, prior to and following activation, using a Perkin Elmer spinning disk confocal scan head (Perkin Elmer, Waltham, MA) and a Hamamatsu Orca ER cooled charge-coupled device camera. Several models of spinning disc systems are now commercially available with lasers appropriate for activation (405 nm) and software controlled positioning of the activation light.

D. Analysis of PA-GFP Tagged Tubulin

We have used PA-GFP-tubulin to examine the dynamics of microtubules in live cells. In our experiments, cells were photoactivated using a rectangular slit or pinholes, and time-lapse images were acquired. For mitotic cells, the area of activation was perpendicular to the spindle long axis. Using this approach, we showed that a fraction of microtubules in prophase cells exhibit behavior consistent with flux, but other microtubules move much more rapidly, both toward and away from the spindle poles (Ferenz and Wadsworth, 2007). In prometaphase cells, microtubules are transported inward, toward the forming spindle, in a dynein-dependent manner (Tulu *et al.*, 2003). Microtubules in mitotic cells have also been activated in the centrosomal region, using a pinhole of appropriate diameter in the image plane. This experimental approach documented the release of microtubules from the centrosome in anaphase cells (Rusan and Wadsworth, 2005). Photoactivation of microtubules in interphase cells, using slits positioned perpendicular to the cell's long axis, has also been used to examine microtubule turnover (Fig. 4). Earlier work, using cells microinjected with tubulin chemically modified with caged fluorescein, showed that microtubules are transported in motile cells (Yvon and Wadsworth, 2000) and that antagonistic forces generated by dynein and myosin regulate microtubule turnover, organization, and movement (Yvon *et al.*, 2001). The use of cells expressing PA-GFP-tubulin to study microtubule behavior in interphase cells will greatly facilitate experimentation.

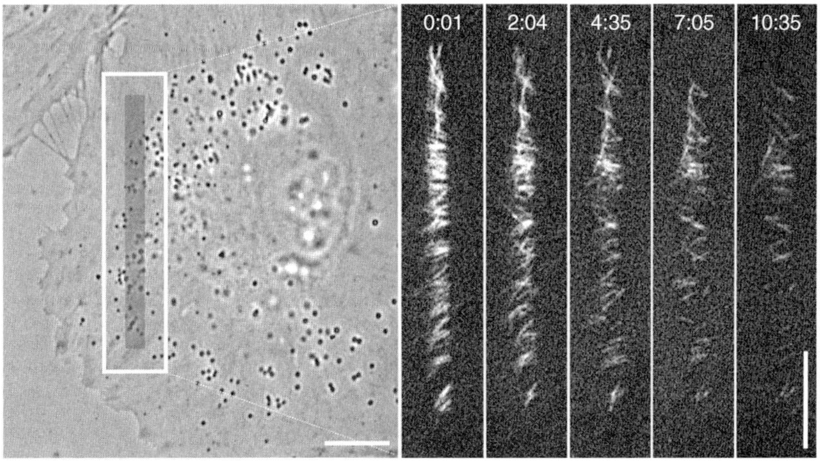

Fig. 4 Microtubule dynamics in an interphase LLC-Pk1 cell expressing PA-GFP-tubulin. The area of activation is shown by the magenta overlay in the phase image (left). Images from a time series acquired following activation; time following activation is shown at the top of each image (mins: seconds). Bars = 10 μm.

Following photoactivation, appropriate software (e.g., Metamorph and Image J) can be used to select the photoactivated area and examine its behavior. To quantify the motion of microtubules, rectangular boxes, typically 1–2 μm in height, were positioned parallel to the spindle's long axis were placed around a photoactivated mark of interest and a montage created. Rates of motion can be obtained from the slope within the montage. The turnover of microtubules can also be determined by measuring the dissipation of photoactivated fluorescence. Similar procedures can be used to analyze microtubule behavior in interphase cells.

II. Conclusions

In conclusion, we find that photoactivation of PA-GFP-tubulin is a powerful method to examine microtubule dynamics in live cells. Our work and that of others shows that the approach is suitable for diverse cell types and can be used to examine microtubule dynamics at all stages of the cell cycle. Moreover, photoactivation can be performed using standard laboratory equipment and data analysis is easily performed using image analysis software. Finally, a large number of proteins have already been shown to retain their native function after tagging with GFP, so substitution of PA-GFP, or other suitable photoactivatable FP, and performing photoactivation are experimentally straightforward and are likely to provide new information about protein dynamics in the physiological context of the live cell.

References

Bancaud, A., Huet, S., Rabut, G., and Ellenberg, J. (2010). Fluorescence perturbation techniques to study mobility and molecular dynamics of proteins in live cells: FRAP, photoactivation, photoconversion, and FLIP. *In* "Live Cell Imaging: A laboratory manual" (R. D. Goldman, J. R. Swedlow, and D. L. Spector, eds.), Cold Spring Harbor Press, Cold Spring Harbor, NY.

Bicek, A. D., Tuzel, E., Demtchouk, A., Uppalapati, M., Hancock, W. O., Kroll, D. M., and Odde, D. J. (2009). Anterograde microtubule transport drives microtubule bending in LLC-PK1 epithelial cells. *Mol. Bio. Cell.* **20**, 2943–2953.

Ferenz, N., Ma, N., Lee, W.-L., and Wadsworth, P. (2010). Imaging protein dynamics in live mitotic cells. *Methods* **51**(2), 193–196.

Ferenz, N. P., and Wadsworth, P. (2007). Prophase microtubule arrays undergo flux-like behavior in mammalian cells. *Molec. Biol. Cell.* **18**, 3993–4002.

Patterson, G. H., and Lippincott-Schwartz, J. (2002). A photoactivatable GFP for selective photolabeling of proteins and cells. *Science* **297**, 1873–1877.

Rusan, N. M., Fagerstrom, C., Yvon, A. C., and Wadsworth, P. (2001). Cell cycle dependent changes in microtubule dynamics in living cells expressing GFP-alpha tubulin. *Mol. Biol. Cell.* **12**, 971–980.

Rusan, N. M., and Wadsworth, P. (2005). Centrosome fragments and microtubules are released and transported asymmetrically Away From division plane in anaphase. *J. Cell Biol.* **168**, 21–28.

Tulu, U. S., Rusan, N., and Wadsworth, P. (2003). Peripheral, non-centrosome-associated microtubules contribute to spindle formation in centrosome containing cells. *Curr. Biol.* **13**, 1894–1899.

Wadsworth, P. (2010). Studying mitosis in cultured mammalian cells. *In* "Live Cell Imaging: A laboratory manual" (R. D. Goldman, J. R. Swedlow, and D. L. Spector, eds.), 571–582. Cold Spring Harbor Laboratory Press, Cold Spring Harbor, NY.

Waterman-Storer, C., and Salmon, E. D. (1998). How microtubules get fluorescent speckles. *Biophys. J.* **75**, 2059–2069.

Yvon, A. C., Gross, D. J., and Wadsworth, P. (2001). Antagonistic forces generated by myosin II and cytoplasmic dynein regulate microtubule turnover, movement, and organization in interphase cells. *Proc. Natl. Acad. Sci. U.S.A.* **15**, 8656–8661.

Yvon, A. C., and Wadsworth, P. (2000). Region specific microtubule transport in motile cells. *J. Cell Biol.* **151**, 1003–1012.

CHAPTER 6

Microtubule Dynamics at the Cell Cortex Probed by TIRF Microscopy

Ilya Grigoriev *and* Anna Akhmanova
Department of Cell Biology, Erasmus Medical Center, 3000 CA Rotterdam, The Netherlands

Abstract

Total internal reflection fluorescence (TIRF) microscopy is a technique that allows selective excitation of fluorescence at a liquid/solid interface within a short distance from the boundary. The penetration depth of TIRF microscopy depends on the angle of

METHODS IN CELL BIOLOGY, VOL. 97

978-0-12-381349-7
DOI: 10.1016/S0091-679X(10)97006-4

illumination resulting in a range of depths, which typically vary from ~70–200 nm up to ~500 nm. The advantages of TIRF microscopy include excellent signal-to-noise ratio, high sensitivity, low photobleaching, and low photodamage. TIRF microscopy is widely used for studying cell adhesion, exo- and endocytosis, and the dynamics of plasma membrane-associated molecules. TIRF microscopy can also be applied for selective visualization of any other cellular processes that occur near the basal membrane even if their localization is not restricted to this part of the cell. For example, microtubules are distributed throughout the cytoplasm, but the use of TIRF microscopy makes it possible to visualize specifically the microtubule subpopulation in the vicinity of the basal cortex and thus study cortical microtubule attachment and stabilization, interactions between microtubules and matrix adhesion structures, and the behavior of specific molecules involved in these processes. In this chapter we describe the application of a commercially available setup to analyze microtubule behavior in live mammalian cells using TIRF microscopy.

I. Introduction

Microtubules are hollow polymeric tubes present in all eukaryotic cells. They form tracks for intracellular transport and participate in essential cellular processes such as cell division and differentiation. They are 25 nm in diameter but can range in length from a few micrometers to tens of micrometers or even more. An important feature of microtubules is their asymmetry—the two ends, termed the plus and the minus end, are structurally and functionally different (Nogales and Wang, 2006). Minus ends (the slowly growing ends *in vitro*) never grow in cells and are typically attached to microtubule organizing centers such as the centrosome. Plus ends (the fast growing ends *in vitro*) are dynamically unstable and explore the cytoplasmic space (Desai and Mitchison, 1997; Howard and Hyman, 2003). Since microtubules are confined within the cell they inevitably come in contact with cell boundaries—the actin-rich cell cortex and/or the plasma membrane. These encounters can have important consequences for both the microtubule and the cortex. First, being a rigid barrier, the cortex can slow down microtubule growth and induce a catastrophe (a switch to depolymerization) (Janson *et al.*, 2003). Second, a microtubule might bend from its straight path and start to grow along the cortex. Third, a specific interaction might occur between the microtubule and some cortically associated molecules. This can either promote catastrophe or more frequently cause microtubule capture and stabilzation. As a result, a stable track will be formed from the cell interior to the cell periphery (Akhmanova and Steinmetz, 2008; Akhmanova *et al.*, 2009; Gundersen *et al.*, 2004). If cortical stabilization and destabilization of microtubules occur asymmetrically within the cell, this will cause polarization of the microtubule array. On the other hand, microtubules interacting with the cortex can deliver building blocks or signaling factors and thus promote remodeling of cortical structures such as the actin meshwork, cell matrix and cell–cell adhesion sites (Akhmanova *et al.*, 2009; Small and Kaverina, 2003).

In this way microtubules interacting with the cortex can guide polarized cell growth [as exemplified by budding yeast (Basu and Chang, 2007)], control polarized cell migration (as observed in some mammalian cell types such as fibroblasts and astrocytes) (Li and Gundersen, 2008; Siegrist and Doe, 2007), and regulated contractility (of which the positioning of the actomyosin contractile ring during cytokinesis is a striking example) (Oliferenko *et al.*, 2009).

Based on these considerations it is not surprising that specific and precise observation and quantification of microtubule–cortex interactions is of interest for understanding a wide range of cellular processes. Microtubules have been one of the favorite subjects of microscopic studies for many years because they are relatively easy to visualize both in live and in fixed cells by light microscopy and because such microscopic observations can be very revealing about microtubule organization and behavior (Semenova and Rodionov, 2007). The major part of our knowledge on microtubule dynamics comes from studying cells cultured on solid supports such as plastic or glass. Such cultured cells are usually relatively flat, and individual microtubules at the cell periphery, where the network is not too dense, can be visualized by wide-field or confocal microscopy. However, the difficulty with detecting specifically the events of microtubule–cortex interactions using these approaches is due to the poor resolution along the vertical (*Z*) axis: a typical optical slice that can be obtained with a confocal microscope is 0.6–1 μm, 25–40 times the width of a microtubule. Therefore, it is impossible to say whether microtubules undergo contacts with the underlying cell membrane or are separated from it by hundreds of nanometers. Better detection of microtubule–cortex interactions can be achieved at the outmost cell border, which is visible in the horizontal plane. However, in many cultured cell types this is a very thin actin-rich area that is hardly populated by microtubules while the relevant microtubule–cortex interactions occur in thicker, more proximal, and basally attached cell parts.

An alternative technique to visualize various events at the basal cell cortex [the region of contact between a cultured cell and the (glass) substrate] is total internal reflection fluorescence (TIRF) microscopy, an approach that will be discussed in detail in this chapter.

II. Rationale

A. TIRF Microscopy

TIRF microscopy, also known as evanescent wave excitation microscopy (Axelrod, 2001; Axelrod *et al.*, 1984; Schneckenburger, 2005; Toomre and Manstein, 2001), is based on the well-known optical phenomenon: when light strikes a boundary between two media with a different refractive index (n_1 and n_2), the angles of incidence (θ_1) and refraction (θ_2) are given by Snell's law:

$$n_1 \sin \theta_1 = n_2 \sin \theta_2 \qquad [1]$$

When the refractive index of the first medium is higher ($n_1 > n_2$), as it is the case when light passes from glass ($n_1 = 1.52$) into water ($n_2 = 1.33$) or a cell ($n_2 = 1.33$–1.38) (Beuthan *et al.*, 1996; Curl *et al.*, 2005; Farinas and Verkman, 1996; Rappaz *et al.*, 2005), there is a certain angle called the critical angle θ_c, where all light is reflected (total internal reflection).

$$\theta_c = \arcsin\,(n_2/n_1) \qquad [2]$$

Importantly, at angles of incidence that are equal to or exceed the critical angle, the electromagnetic field of the light still penetrates into the second medium, and the resulting "evanescent wave" can still excite fluorophores in the second medium [the nature of "evanescent wave" is described in Axelrod (2001), Axelrod *et al.* (1984), Schneckenburger (2005), and Toomre and Manstein (2001)]. The intensity of the evanescent wave penetrating the second medium decreases exponentially; this is the key point of TIRF microscopy. The penetration depth d_p (the distance where the light intensity is decreased by a factor of e) depends on the light wavelength, the angle of incidence, and the refraction index of the two media (in the microscope setup, a coverslip and an aqueous buffer):

$$d_p = \frac{\lambda}{4\pi}\left(n_1^2 \sin^2\theta_1 - n_2^2\right)^{-1/2} \qquad [3]$$

For light microscopy of cells, the penetration depth can be in the order of ~100 nm (Fig. 1A). In practical terms this means that one can use a TIRF microscope to illuminate specifically a coverslip-proximal optical slice with a thickness of ~70–200 nm. This provides superior z-resolution compared to a typical confocal microscope.

B. Objective-Type TIRF Setup

Two kinds of microscope setup can be used for TIRF microscopy: the prism-type and the objective-type [see Toomre and Manstein (2001) for a more detailed discussion and Manneville (2006) for a description of a prism-type setup applied for cell biological purposes]. The objective-type TIRF microscopes have gained popularity in the past years because they are commercially available from several major microscope manufacturers (Olympus, Nikon, Leica, Zeiss), can be easily combined with wide-field imaging, and allow performance of fluorescence recovery after photobleaching (FRAP) experiments. They are usually constructed in an "inverted" microscope configuration (with the objective beneath the coverslip) and are most convenient for maintenance and manipulation (e.g., microinjection) of cultured cells.

In the objective-type TIRF setup the numeric aperture of the objective is critical, because the laser beam is deflected to the required angle of incidence by the objective lens itself (Fig. 1B). The numeric aperture of an objective is given by

$$\text{NA} = n_1 \sin\theta \qquad [4]$$

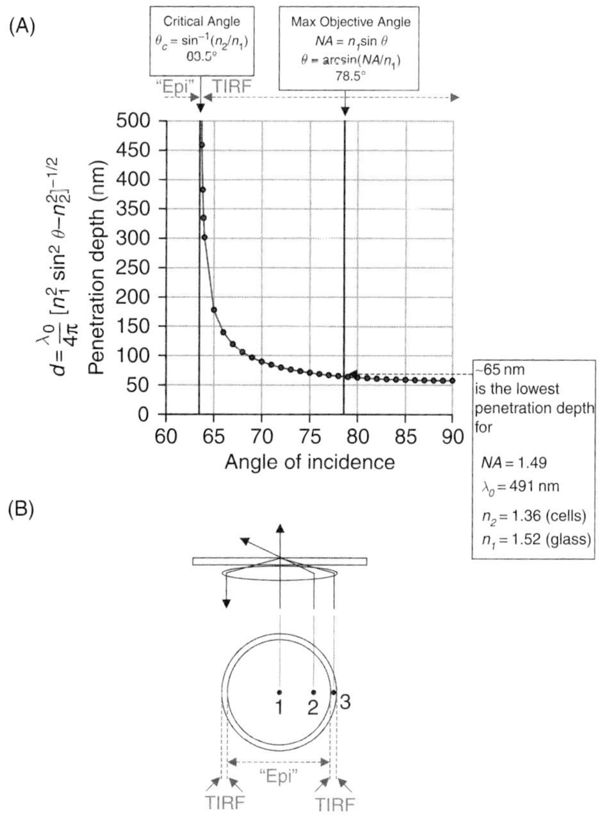

Fig. 1 The dependence of TIRF penetration depth on the angle of incidence. (A) The penetration depth of the evanescent wave is plotted against the angle of incidence of the laser beam. The critical angle calculated using Eq. (2) is 63.5°, if we assume that the refractive index of a cell is 1.36 and the refractive index of the cover glass equals 1.52. Images obtained with angles of incidence that are smaller than the critical angle do not differ significantly from the conventional epifluorescent images. Once the angle of incidence exceeds the critical angle, TIRF occurs. The dependence of the penetration depth on the angle of incidence is not linear. At the angles close to the critical angle the penetration depth will be in the order of 500–300 nm, while at high angles (~70°–80°) the penetration depth remains relatively constant at ~100–70 nm. The maximum objective angle determines the upper limit for the angles of incidence. Based on Eq. (4), it will be 78.5° for an objective with a numeric aperture of 1.49 (assuming the refractive index of a cover glass to be 1.52). (B) A schematic diagram of the objective lens. The upper part of the scheme shows a side view and the lower part of the scheme a view from the top. If the laser beam is positioned at the center of the lens (1), the beam passing through the lens will not bend from normal. When the beam is moved away from the center of the lens (2), it will bend from normal. TIRF will occur when this angle is higher than the critical angle. To achieve this, the laser beam should be moved close to the edge of the lens (3). The larger the numeric aperture of the lens, the further the beam can be moved, and a smaller penetration depth can be achieved. However, at very high angles of incidence the reduction in penetration depth is minimal.

where n_1 is the refractive index of the medium in which the lens is working (in this case, immersion oil/cover glass) and θ is the half-angle of the maximum cone of light that exits the lens. To achieve angles of incidence that are equal to or higher than the critical angle, the maximum objective angle, θ, must exceed the critical angle. Objectives with a high numeric aperture permit an increase in the angle of incidence and a decrease in penetration depth (Fig. 1). For many years a numeric aperture of 1.4 was common for high-power objectives ($\times$60 and $\times$100). Although TIRF can be achieved using such a lens, the angle of incidence will not surpass ~67° and the penetration depth will exceed 100 nm. Therefore, objectives with a high numeric aperture (1.45 and 1.49), which recently became commercially available, are preferable. Interestingly, the penetration depth does not change much at high angles (see Fig. 1A) and objectives with a very high numeric aperture (above 1.49) do not offer significant advantages.

III. Materials and Equipment

A. TIRF Setup

Our TIRF setup consists of a Nikon TIRF microscope together with Roper FRAP and laser launch (FRAP scanning system I-Las/I-Launch, Roper Scientific France/PICT-IBiSA, Institut Curie). We use a Nikon Eclipse Ti-E inverted microscope (Fig. 2) equipped with a perfect focus system that ensures stable maintenance of the z-position of the sample over time, a combined TIRF/EPI-motorized illuminator that allows us to use both TIRF and conventional epifluorescent illumination and a Nikon CFI Apo TIRF 100$\times$ 1.49 N.A. oil objective. This objective, when combined with a highly sensitive EM-CCD (electron-multiplying charge-coupled device) camera, which has a large pixel size (see below), is by itself insufficient to obtain the magnification of ~70 nm per pixel that is necessary for imaging microtubule end displacements; therefore, an additional lens of 2.5$\times$ (Nikon C mount adapter 2.5$\times$) is included in the setup. As a result, we obtain images at 0.065 μm/pixel. Since ~200 nm is the theoretical resolution limit, 65 nm per pixel obviously represents oversampling, but it is useful for performing measurements: when the structure of interest is represented in an image by a higher number of pixels the error of determining the position or the size of this structure is reduced.

For excitation, we use a mercury lamp HBO 100W/2 (Osram) for conventional epi-fluorescent illumination or lasers for TIRF. The system is equipped with two lasers: 491 nm 50 mW Calypso (Cobalt) and 561 nm 50 mW Jive (Cobalt). The 491 nm laser is used for the imaging of blue–green, green or green–yellow fluorophores (e.g., mTagBFP, CFP (cyan fluorescent protein), GFP (green fluorescent protein), or Venus/YFP (yellow fluorescent protein)), while the 561 nm laser is used for the imaging of red fluorophores (e.g., DsRed, mRFP (monomeric red fluorescent protein), mCherry, and mStrawberry).

The TIRF illuminator is motorized and there is also a manual option to control the angle of incidence. As explained above, the change of angle of incidence occurs simply by displacing the position of the laser beam: when projected at the edge of the lens it will bend from the normal allowing the critical angle to be achieved. The shift of the laser beam position in respect to the center of the lens occurs near the coupling of the

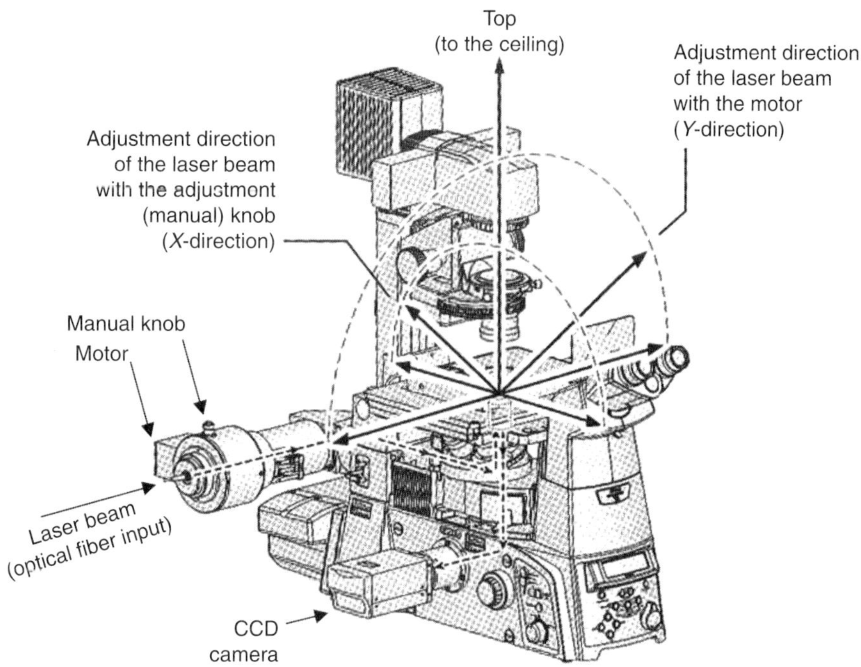

Fig. 2 TIRF microscope setup. The scheme shows key components of the Nikon TIRF microscope and the laser beam pathway. The laser beam is delivered into the microscope via an optical fiber. The position of the beam relative to the center of the lens can be adjusted with (i) manual knob or (ii) the motor. Mechanical displacement of the coupling of the optical fiber along the *X* and *Y* axes is sufficient to displace the laser beam (which is shown as dashed arcs) to the edge of the objective lens. It will decline from normal, and provided that the numeric aperture of the objective is sufficiently high, the required angle of incidence will be achieved. Both adjustment directions, *X* or *Y*, can be used to reach critical angle (manual control is possible for the *X*-direction and motorized control for the *Y*-direction). Manual control is faster but using the motor allows for better reproducibility. The figure is adapted from the instruction manual "Nikon Eclipse Ti series TI-TIRF TIRF/Epi-fl Illuminator Unit, TI-TIRF-E Motorized Illuminator Unit Instruction."

optical fiber to the TIRF illuminator (Fig. 2). There are two knobs for this in the non-motorized version (controlling *X* and *Y* displacement) or only one knob in the motorized version (controlling *X* displacement) (Fig. 2). *Y* displacement is controlled with the motor in the motorized version using the software in the microscope control pad. When controlling the angle of incidence manually the user "feels" what he/she is doing, but it is very hard to go back to exactly the same angle once it is changed. Using the motorized option, it is much easier to achieve reproducibility of the angle of incidence simply by choosing the same values in the software.

We use filter cubes (ET series from Chroma) but remove the excitation filters because they reduce the light intensity and are not necessary when lasers are used for excitation. However, since excitation filters are necessary for imaging with the mercury lamp (conventional epifluorescent mode) we keep them in front of the mercury lamp in a filter wheel (Sutter). For green and red fluorescent proteins (e.g., GFP and mCherry) we use Chroma ET-GFP (49002) and Chroma ET-mCherry (49008) filter sets along with Chroma ET-mCherry/GFP double dichroic (59022) for simultaneous green and red fluorophore imaging.

To obtain images of green and red fluorophores at the same time we use a beam splitter (DV2, Roper scientific), equipped with a dichroic 565dcxr (Chroma) and HQ530/30m emission filter (Chroma). The emission filter is required to minimize bleed-through between the two channels.

We acquire images either with a back-illuminated QuantEM 512SC EM-CCD camera (Roper scientific), installed on one microscope port (the beam splitter and 2.5× adapter are placed between the camera and the microscope), or with an interlined CoolSNAP HQ2 CCD camera (Roper scientific), installed on another microscope port. For both left and right ports the microscope is equipped with 100% mirrors. QuantEM is more sensitive and faster than CoolSNAP but has a smaller chip (512 × 512 pixels compared 1392 × 1040 pixels) and a larger pixel size (16 μm compared to 6.45 μm). Since we are using 2.5× adaptor for the QuantEM the final resolution is almost the same: 0.065 μm/pixel for the QuantEM and 0.063 μm/pixel for the CoolSNAP.

The system is also equipped with motorized stage (ASI, S21721010 closed loop XY stage with rotary encoders and MS two-axis stage controller for use with closed loop) and with two Smart-shutters (Sutter): one in front of mercury lamp (conventional epifluorescence) and one in front of the halogen lamp (transmitted light).

Image acquisition and control of the whole system is performed using MetaMorph 7.5 software (Molecular Devices).

To keep cells at 37° C we use a stage top incubator and an objective heater (model INUG2E-ZILCS, Tokai Hit). This incubator is also equipped with a water bath, to keep cells at a certain humidity, and a 5% CO_2 input. Besides the incubator, the objective is also heated by the same device, so there is no temperature gradient at the spot of imaging.

We plate cells onto 25 mm diameter round coverslips (N1, Menzel-Glaser). The coverslip is then assembled into a Attofluor cell chamber (A-7816, Molecular Probes) before the experiment.

B. Cell Lines

The choice of cell line depends on the experimental question, especially as microtubule organization and the complement of microtubule regulators vary greatly between different cell types. Another important consideration is the ease of cell maintenance and transfection. We routinely use HeLa cells, which are very easy to transfect with plasmids as well as synthetic siRNAs; they have a very dense microtubule array due to extensive stabilization at the cortex (Lansbergen *et al.*, 2006; Mimori-Kiyosue *et al.*, 2005). HeLa cells are thus an excellent model for studying microtubule–cortex interactions but are less convenient for imaging microtubule dynamics in internal cell regions. Human lung fibroblasts MRC5-SV are an excellent choice of cell type for distinguishing long stretches of individual microtubules since in these cells the microtubule network is exceptionally sparse. CHO cells might also be a good choice because their microtubule dynamics has been characterized extensively (Komarova *et al.*, 2002a, b; Komarova *et al.*, 2005). Cultured astrocytes (Manneville, 2006), Swiss 3T3 cells (Cook *et al.*, 1998), or B16 melanoma cells (Schober *et al.*, 2007) are well suited for studying microtubule–cortex and microtubule–actin cross talk during cell migration because these cells readily migrate in culture.

C. Fluorescent Markers

Microtubules in live cells can be visualized either by microinjection of purified tubulin chemically labeled with a fluorochrome, such as Cy3 or Rhodamine, or by expressing a tubulin cDNA fused to fluorescent tag, such as GFP (Semenova and Rodionov, 2007). The second method is generally less laborious and produces the best results when cells stably expressing fluorescently tagged tubulin are used. Alternatively, transient transfection can also be used, but it is better to image cells at least 48–120 h after transfection, as at early time points after transfection the incorporation of fluorescently tagged tubulin into microtubules might be poor and the cytoplasmic background is high. Our preferred marker is mCherry-α-tubulin (Shaner *et al.*, 2004); GFP or YFP-α-tubulin (Clontech) can also be used but appear to inhibit cell division, and cell lines established with these markers are often less stable and bright than those with mCherry-α-tubulin.

An alternative strategy aimed at specific visualization of growing microtubule ends is the use of microtubule plus end tracking (+TIP) markers (Akhmanova and Steinmetz, 2008; Morrison, 2007; Schuyler and Pellman, 2001). The most popular ones are EB1, EB3, and CLIP-170. We find EB3 to be the most convenient marker, because it is widely expressed in most standard cell lines as well as differentiated cells such as neurons (Stepanova *et al.*, 2003). Being a small protein (~30 kDa), it is easy to express, it displays an excellent microtubule tip-to-lattice ratio, brightly labels microtubule ends without causing microtubule bundling, and can be tagged with GFP at both the N- and the C-terminus (Komarova *et al.*, 2009). In contrast, EB1 does not tolerate N-terminal tags (Zhu *et al.*, 2009; our unpublished observations), while the widely used C-terminally tagged EB fusions can exhibit dominant-negative effects even at the levels when they are localized exclusively at the microtubule plus ends because of their inability to interact with CAP-Gly domain proteins such as CLIP-170 (Lomakin *et al.*, 2009). It should be noted that even low expression levels of EB3 and other +TIPs can affect microtubule dynamics and thus must be combined with imaging microtubules if the goal of the experiment is to determine the parameters of microtubule dynamic instability.

D. Cell Culture

Cells are maintained in humidified incubator at 37°C, with 5% CO_2. Typically, we culture cells in a medium with 45% DMEM and 45% Ham's F10 (Invitrogen), 10% fetal bovine serum, 100 U/ml penicillin, and 100 μg/ml streptomycin (Invitrogen).

IV. Methods

A. Generation of the Stable Cell Lines with Fluorescent Microtubule Markers

This protocol is written for HeLa cells but can be adapted to other cell lines that are easy to transfect. It has been successfully used to obtain stable cell lines expressing fluorescent microtubule markers. The concentration of G418 should be tested for each cell type.

1. Split the cells into 6-well plate to achieve ~40% confluence by the next day.
2. Prepare DNA for transfection. DNA should be linearized in such a way that neither the fluorescent protein expression cassette nor the selective marker expression cassette is disrupted. After restriction enzyme digestion, DNA should be purified by phenol extraction and ethanol precipitation or by using a DNA purification kit. Linearized DNA gives much lower initial transfection efficiency but works better for establishing stably expressing cells. Alternatively, circular plasmid DNA can also be used.
3. Transfect 40% confluent cells using Fugene 6 (Roche). For 1 well of a 6-well plate use ~0.5 μg of linearized DNA.
4. Next day split the cells and transfer them to 10 cm dish. Add 10 ml of the culture medium supplemented with 0.5–0.6 mg/ml G418 (dissolved at 100 mg/ml in phosphate-buffered saline and filter sterilized).
5. After the cells become confluent, perform fluorescence-activated cell sorting (FACS) to select fluorescent cells. We typically use FACS Aria II cell sorter equipped with 405, 488, and 633 nm lasers (BD Biosciences). Positive cells (~50,000–100,000) are collected and plated in the selection medium in one well of a 6-well plate.
6. After the cells become confluent they can be frozen down, used for another round of selection or for imaging.

B. Sample Preparation

1. Plate cells onto coverslips at the appropriate density. Different coverslips can be used depending on the observation chamber, including glass bottom Petri dishes. The thickness of the coverslips is important. The microscope objective is designed to be used with coverslips of 0.17 μm thickness. N1 coverslips have a thickness between 0.13 and 0.17 μm while N1.5 coverslips between 0.16 and 0.19 μm. N1.5 coverslips are preferable for imaging. Theoretically, to achieve perfect results, the thickness of each coverslip should be measured before cell plating. To overcome this problem some objectives (such as Nikon CFI Apo TIRF 100× 1.49 N.A. oil objective) have a correction collar, which can be adjusted to the coverslips of different thickness.
2. When not imaging a stable cell line transfect cells 1 day after plating. We use FuGene 6 (Roche) transfection reagent and follow the manufacturer's transfection protocol.

3. Change the culture medium before imaging. Mount the coverslip with cells into the observation chamber. Never allow the cells to dry during any preparation steps.

C. Imaging

1. Add immersion oil on top of the lens (we use Nikon type A immersion oil).
2. Insert the chamber with cells into the heating incubator or onto the stage (this depends on the microscope configuration).
3. Raise the objective until it touches the coverslip.
4. Use transmitted light or conventional epifluorescent illumination if transmitted light is not available to focus on the cells.
5. After finding a cell, acquire an image using conventional epifluorescent mode.
6. Switch into TIRF mode. Position the laser through the center of the objective. Start live mode acquisition. The acquired image should be very similar to the one acquired previously using conventional epifluorescent illumination.
7. Change the angle of incidence until there is a dramatic loss of background inside the cell (Fig. 3). This loss of background indicates TIRF mode. Continue to increase the angle of incidence until the signal is lost. Now reduce the angle of incidence until the signal reappears. This is TIRF with the smallest penetration depth (Fig. 3). As you reduce the angle of incidence, the penetration depth will increase until a certain point where you will observe an abrupt change in the picture, which indicates a switch from TIRF to epifluorescent imaging.
8. In order to maintain constant intensities and penetration depths, the angle of incidence must be determined. For example, this would be useful to estimate the z-dimensions of a given structure. Manual control of the angle of incidence is problematic because it can be difficult to return the control knobs to the same position as before. Motorized control is superior because a certain angle (in either absolute or arbitrary units) can be saved. To determine which penetration depth corresponds to a given angle, Eq. (3) can be used. Since the exact refractive index at the spot of observation is necessary for this equation, and penetration depth can vary depending on the sample and other properties of the setup, a calibration should be considered (Fiolka *et al.*, 2008; Gell *et al.*, 2009; Saffarian and Kirchhausen, 2008).
9. Choose the angle of incidence corresponding to the penetration depth that suits your needs and acquire an image or a movie. When collecting data on microtubule dynamics, keep in mind that in mammalian cells cultured at 37° C microtubules grow with a velocity of 0.1–0.5 μm/s. We typically collect 1–2 frames per second to reliably detect events of microtubule growth and shortening.

D. Image Analysis

In some cases counting the number or measuring the length of microtubule segments visible by TIRF microscopy might be sufficient for analysis (Lansbergen *et al.*, 2006; Webb *et al.*, 2009). If you wish to determine the parameters of microtubule dynamics, direct tracking of the microtubule ends (Shelden and Wadsworth, 1993) or kymograph

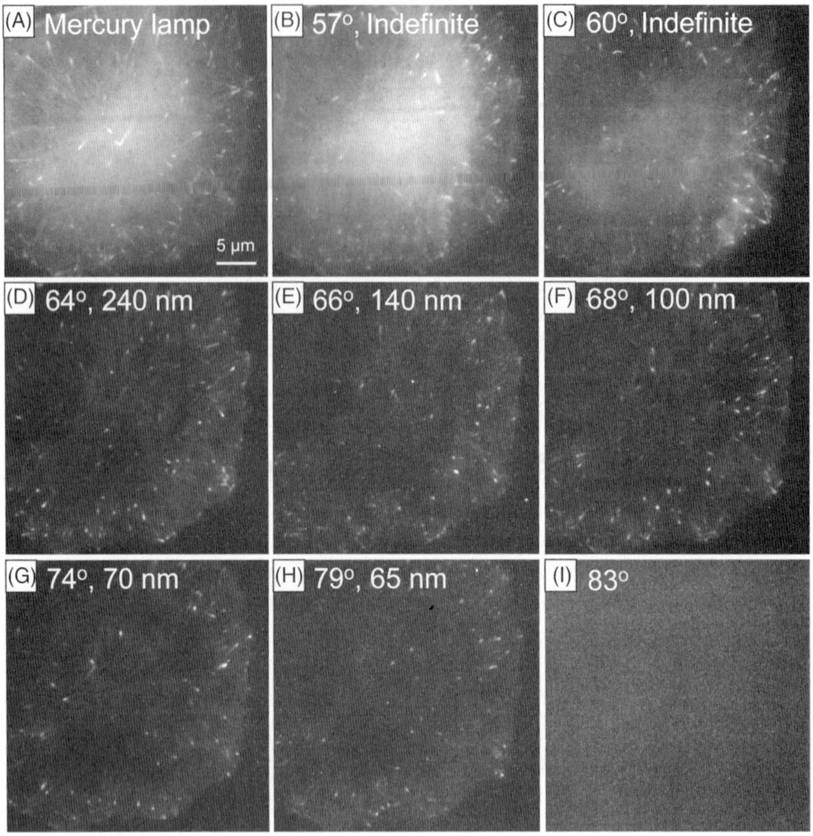

Fig. 3 Live cell imaging using a TIRF setup with different angles of incidence. Images of a HeLa cell stably expressing the microtubule plus end marker EB3-GFP were obtained in the same focal plane but with different angles of incidence. Images were collected with 500 ms exposure, using the mercury lamp (A) or the 491 nm laser line (B–I). (A) Conventional epifluorescent image. The whole cell is visible resulting in a high background within the cytoplasm. (B, C) The angle of incidence of the laser beam is less than the critical angle, and the resulting image does not significantly differ from the conventional epifluorescent image. (D) The angle of incidence has reached the critical angle. This is an example of a TIRF image. (E, F, G, H) If the numeric aperture of the objective allows displacing the laser beam further to the edge, increasing the angle of incidence, the resulting penetration depth will become even smaller [Eq. (3)]. (I) If the laser beam is displaced too far from the center of the lens, no light will penetrate into the objective and no signal above the CCD camera noise will be detected. The angle of incidence and calculated penetration depth is indicated in each image (calculations are based on the following assumptions: an objective numeric aperture of 1.49, laser wavelength 491 nm, refractive index of the glass 1.52, refractive index of the cell 1.36). Note that penetration depth is the distance over which the intensity of the evanescent field decreases to ~37% (1/*e*) of the value at the surface; therefore penetration depth does not represent an absolute penetration limit in the *Z*-direction.

analysis can be used. Kymographs illustrate the changes of fluorescent intensity relative to time. They can be plotted using different software (ImageJ, MetaMorph, etc.). Typically, a line is drawn along the selected microtubule and the kymograph option in the software is used to obtain a plot, which allows visualization of displacements along this line over time. The advantage of using kymograph is the reduced error of determining the position of the structure of interest. The use of MetaMorph software for plotting kymographs has been described in detail by Manneville (2006). All kinetic parameters such as velocity, track duration and length, frequencies of events, and the fluorescent intensities can be measured from the kymograph. An automated version of kymograph analysis specifically adapted for studying microtubule dynamics is available (Smal *et al.*, 2010).

V. Discussion

A. TIRF Microscopy in Studies of Cortical Microtubules

Since TIRF microscopy allows selective illumination of a relatively thin cell slice near the coverslip, it can be used to prove that certain structures, such as microtubules, come into close vicinity of the cell cortex. Krylyshkina *et al.* used this approach to show that polymerizing microtubules labeled with GFP-tubulin or a microtubule plus end marker GFP-CLIP-170 come within 50 nm of the substrate and target focal adhesions (Krylyshkina *et al.*, 2003). TIRF microscopy in combination with RNA interference and dominant-negative approaches was used to show that microtubule plus ends can be stabilized by protein complexes containing microtubule plus end bound and plasma membrane-associated components, such as APC-Dlg1 in migrating astrocytes (Etienne-Manneville *et al.*, 2005) or CLASP-LL5β in HeLa cells and fibroblasts (Lansbergen *et al.*, 2006; Mimori-Kiyosue *et al.*, 2005) (Fig. 4). TIRF microscopy has also been used to support the idea that gap junction hemichannels can be targeted directly by microtubules to adherens junctions (Shaw *et al.*, 2007). Interestingly, in cultured mammalian cells, growing microtubule ends in mitotic spindles are often completely "invisible" by TIRF microscopy, indicating that they do not come into close contact with the basal cortex (Fig. 5).

TIRF microscopy can also be applied to distinguish the cortical microtubule population and to describe its specific properties. For example, this approach has been applied to *Drosophila* embryos at the syncytial blastoderm stage (Webb *et al.*, 2009). As could be expected for mitotic spindles that have a three dimensional organization, microtubules were visible by TIRF as spots in cases where they approached the cortex vertically, or as lines, when they bent and started to extend along the cortex; spot to line transitions were frequently observed. Using this approach, the authors were able to describe the effect of mutations in regulatory factors APC2 and RhoGEF2 on the cortical microtubule organization (Webb *et al.*, 2009). TIRF microscopy was also used to probe spatial relationships between microtubules and filopodia (Schober *et al.*, 2007) and to explore the localization of myosin filaments, microtubules, and kinesin-6 during

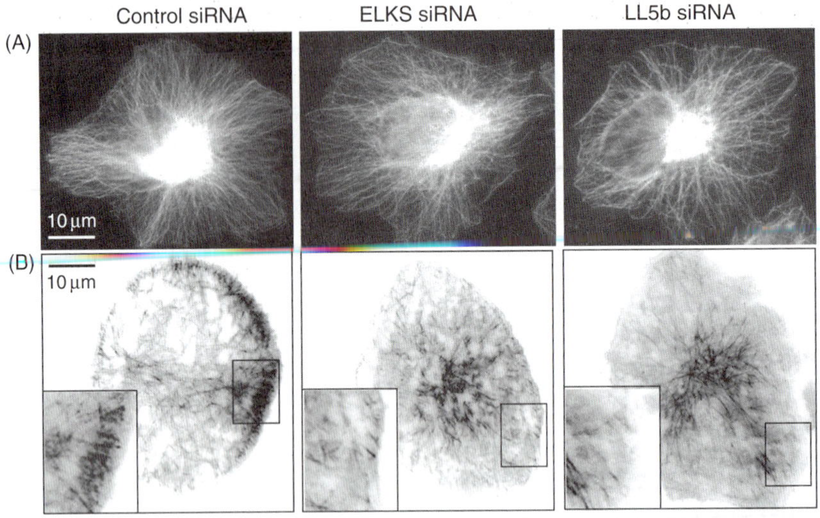

Fig. 4 Use of TIRF microscopy to demonstrate cortical microtubule attachment by LL5β–ELKS protein complex. (A) HeLa cells were cultured for 72 h after transfection with the control, LL5β or ELKS siRNAs, fixed with methanol and stained for α-tubulin. There is a mild decrease in microtubule density in LL5β or ELKS-depleted cells, but microtubule organization appears normal with this type of imaging. (B) HeLa cells, stably expressing GFP-α-tubulin, were imaged by TIRF microscopy 72 h after transfection with the control, LL5β or ELKS siRNAs. Insets show enlargements of the boxed areas. Note that there is a strong enrichment of TIRF-visible microtubule segments at the peripheral cortex in the control but not in the LL5β or ELKS-depleted cells, indicating that both proteins participate in cortical microtubule attachment. Images are reproduced with modification from Lansbergen *et al.* (2006).

cytokinesis in *Drosophila* S2 cells (Vale *et al.*, 2009). TIRF microscopy can also be applied to fixed cells (Manneville, 2006); for example, it was used to enhance the detection of microtubule plus end binding proteins EB1 and APC and the formin mDia at the tips of stable, detyrosinated microtubules (Wen *et al.*, 2004).

Since the evanescent electromagnetic field decays exponentially with the distance from the coverslip, quantitative information about microtubule position along the *Z*-axis can be obtained if the microtubules are uniformly labeled (Hadjidemetriou *et al.*, 2005). This approach has not yet been systematically applied to study microtubule organization, but interestingly, fluorescent microtubules assembled *in vitro* and embedded in agarose with some degree of tilt relative to the surface can be used to calibrate the evanescent field of TIRF microscope (Gell *et al.*, 2009).

B. TIRF Microscopy as a General Tool for Highly Sensitive Imaging

The possibility of obtaining a very thin optical section strongly improves the signal-to-noise ratio. This property makes TIRF microscopes the instruments of choice for single molecule *in vitro* experiments, which are widely used in the microtubule field to reconstitute and study the behavior of motors and microtubule-associated proteins. In

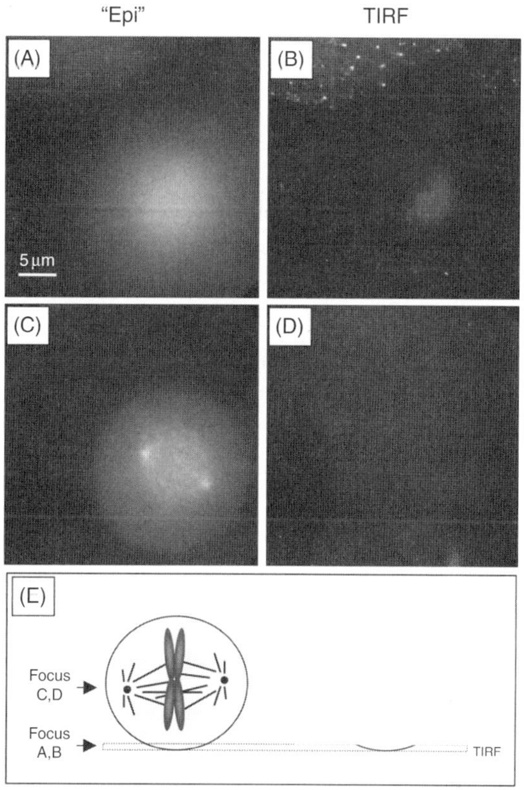

Fig. 5 Imaging of microtubule organization in mitotic cells using conventional epifluorescence and TIRF microscopy. Images of a metaphase HeLa cell stably expressing microtubule plus end marker EB3-GFP were obtained in two different focal planes (schematically shown in E) and with two different angles of illumination (A, C—conventional epifluorescent illumination with a mercury lamp and B, D—TIRF mode). Images were collected with 500 ms exposure, 491 nm laser line was used for TIRF excitation. (E) A schematic diagram showing that if the object is thick and the structure of interest, such as the mitotic spindle, is distant from the solid/liquid boundary, TIRF microscopy cannot be used to image it (as shown in D).

such experiments the samples are typically quite thick, but microtubules are attached to the glass surface, making them ideally suited for TIRF microscopy imaging. The extensive applications of this approach are beyond the scope of this chapter, but we would like to note that it is possible to implement it on the same setup as TIRF-based live cell imaging experiments.

Furthermore, because only a very thin part of the specimen is illuminated, photobleaching and cell photodamage are strongly reduced, and the contrast is dramatically improved, especially when compared to conventional wide-field epifluorescence or point-scanning confocal microscopes (Fig. 6). Use of TIRF microscopy in cells might thus be valuable not only when the research is focused on cortical events, but also when microtubules in the vicinity of the cortex are suitable for sampling the overall

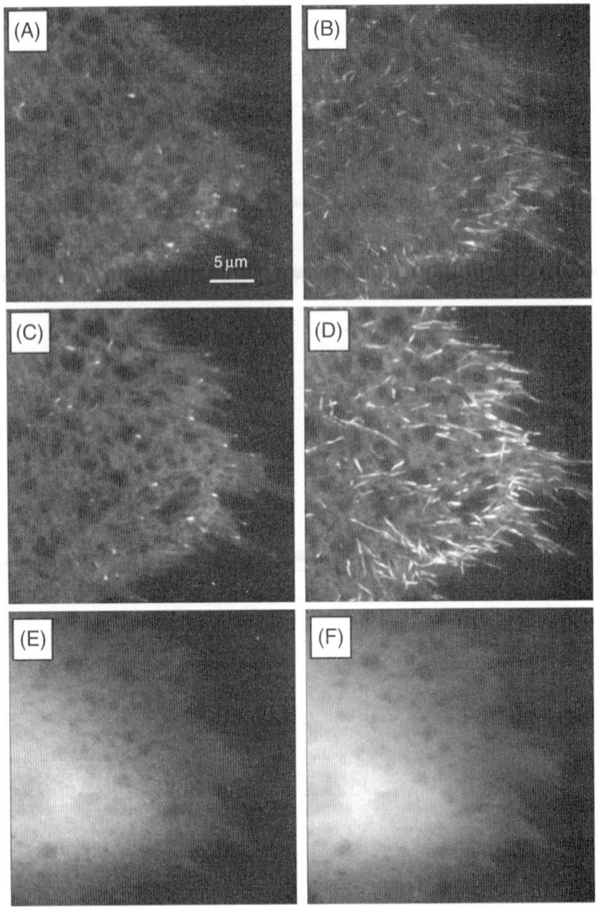

Fig. 6 Use of TIRF microscopy to enhance signal-to-noise ratio. Images of a CHO cell transiently expressing microtubule plus end marker EB1-YFP were obtained in the same focal plane but with different types of illumination. Images were collected with 500 ms exposure; 491 nm laser line was used for excitation. (A, B) Images were obtained in a TIRF mode with a high angle of incidence and low penetration depth. (A) Representative image from the movie and (B) maximum intensity projection of 100 planes of the same movie. (C, D) Images were obtained in TIRF mode with a lower angle of incidence and higher penetration depth, compared to (A, B). (C) Representative image from the movie and (D) maximum intensity projection of 100 planes of the same movie. (E, F) Images were obtained in the epifluorescence mode. (E) Representative image from the movie and (F) maximum intensity projection of 100 planes of the same movie. If the signal-to-noise ratio is poor (E, F), switching to the TIRF illumination (A–D) improves the detection of the structures of interest: in this case, growing microtubule ends. However, when the angle of incidence is high, only a relatively narrow optical plane (~100–200 nm) is visualized, and moving structures quickly move out of focus (microtubule growth tracks are short in B). If this is undesirable, decreasing the angle of incidence helps to obtain thicker optical sections (C, D); note that microtubule growth tracks are longer in D than in (B).

microtubule organization and dynamics. For example, using dual-color TIRF microscopy it was possible to detect single kinesin molecules moving on individual microtubule tracks and thus determine the preference of different types of kinesins for certain posttranslational tubulin modifications (Cai *et al.*, 2009). For this approach, optimal visualization of the structure of interest in combination with low photobleaching is more important than obtaining a very thin optical section. Therefore, it is often an advantage to decrease the angle of incidence and thus increase the penetration depth to ~500 nm (Figs. 1 and 6). This type of imaging is sometimes called semi- or near-TIRF and can be used to improve imaging contrast for the measurement of microtubule dynamics. We have also used it to observe simultaneous extension of microtubules and ER tubules in the ventral part of HeLa cells (Grigoriev *et al.*, 2008). Low photobleaching and photodamage make it also well suited for imaging microtubules in thin and vulnerable cells, such as hippocampal neurons (Jaworski *et al.*, 2009).

VI. Summary

The availability of commercial TIRF microscopes makes this technique increasingly popular among cell biologists. It can be used to prove that cytoskeletal elements localize in the close vicinity of the basal cortex and study the details of their behavior with different markers. It is also useful for generation of high-quality data suitable for quantitative analysis of processes that are not restricted to the basal cortical region but do occur in its vicinity. Both properties have been extensively used to study microtubule organization and dynamics, and we can expect that TIRF microscopy, in combination with FRAP and FRET approaches, will be one of the methods of choice in future studies.

Acknowledgments

This work was supported by the Netherlands Organisation for Scientific Research grants ALW-VICI and ZonMW-TOP to A.A. We thank Kris Leslie, Ihor Smal, and Ivan Maly for critically reading this chapter.

References

Akhmanova, A., Stehbens, S. J., and Yap, A. S. (2009). Touch, grasp, deliver and control: Functional cross-talk between microtubules and cell adhesions. *Traffic* **10**, 268–274.

Akhmanova, A., and Steinmetz, M.O. (2008). Tracking the ends: A dynamic protein network controls the fate of microtubule tips. *Nat. Rev. Mol. Cell Biol.* **9**, 309–322.

Axelrod, D. (2001). Total internal reflection fluorescence microscopy in cell biology. *Traffic* **2**, 764–774.

Axelrod, D., Burghardt, T.P., and Thompson, N.L. (1984). Total internal reflection fluorescence. *Annu. Rev. Biophys. Bioeng.* **13**, 247–268.

Basu, R., and Chang, F. (2007). Shaping the actin cytoskeleton using microtubule tips. *Curr. Opin. Cell Biol.* **19**, 88–94.

Beuthan, J., Minet, O., Helfmann, J., Herrig, M., and Muller, G. (1996). The spatial variation of the refractive index in biological cells. *Phys. Med. Biol.* **41**, 369–382.

Cai, D., McEwen, D. P., Martens, J.R., Meyhofer, E., and Verhey, K. J. (2009). Single molecule imaging reveals differences in microtubule track selection between kinesin motors. *PLoS Biol.* **7**, e1000216.

Cook, T. A., Nagasaki, T., and Gundersen, G.G. (1998). Rho guanosine triphosphatase mediates the selective stabilization of microtubules induced by lysophosphatidic acid. *J. Cell Biol.* **141**, 175–185.

Curl, C. L., Bellair, C. J., Harris, T., Allman, B.E., Harris, P.J., Stewart, A.G., Roberts, A., Nugent, K. A., and Delbridge, L. M. (2005). Refractive index measurement in viable cells using quantitative phase-amplitude microscopy and confocal microscopy. *Cytometry A* **65**, 88–92.

Desai, A., and Mitchison, T. J. (1997). Microtubule polymerization dynamics. *Annu. Rev. Cell Dev. Biol.* **13**, 83–117.

Etienne-Manneville, S., Manneville, J. B., Nicholls, S., Ferenczi, M. A., and Hall, A. (2005). Cdc42 and par6-PKCzeta regulate the spatially localized association of dlg1 and APC to control cell polarization. *J. Cell Biol.* **170**, 895–901.

Farinas, J., and Verkman, A. S. (1996). Cell volume and plasma membrane osmotic water permeability in epithelial cell layers measured by interferometry. *Biophys. J.* **71**, 3511–3522.

Fiolka, R., Belyaev, Y., Ewers, H., and Stemmer, A. (2008). Even illumination in total internal reflection fluorescence microscopy using laser light. *Microsc. Res. Tech.* **71**, 45–50.

Gell, C., Berndt, M., Enderlein, J., and Diez, S. (2009). TIRF microscopy evanescent field calibration using tilted fluorescent microtubules. *J. Microsc.* **234**, 38–46.

Grigoriev, I., Gouveia, S. M., van der Vaart, B., Demmers, J., Smyth, J. T., Honnappa, S., Splinter, D., Steinmetz, M. O., Putney, J. W. Jr., Hoogenraad, C.C., and Akhmanova, A. (2008). STIM1 is a MT-plus-end-tracking protein involved in remodeling of the ER. *Curr. Biol.* **18**, 177–182.

Gundersen, G. G., Gomes, E. R., and Wen, Y. (2004). Cortical control of microtubule stability and polarization. *Curr. Opin. Cell Biol.* **16**, 106–112.

Hadjidemetriou, S., Toomre, D., and Duncan, J.S. (2005). Segmentation and 3d reconstruction of microtubules in total internal reflection fluorescence microscopy (TIRFM). *Med. Image Comput. Comput. Assist. Interv.* **8**, 761–769.

Howard, J., and Hyman, A. A. (2003). Dynamics and mechanics of the microtubule plus end. *Nature* **422**, 753–758.

Janson, M.E., de Dood, M.E., and Dogterom, M. (2003). Dynamic instability of microtubules is regulated by force. *J. Cell Biol.* **161**, 1029–1034.

Jaworski, J., Kapitein, L. C., Gouveia, S. M., Dortland, B.R., Wulf, P.S., Grigoriev, I., Camera, P., Spangler, S. A., Di Stefano, P., Demmers, J., Krugers, H., Defilippi, P., *et al.* (2009). Dynamic microtubules regulate dendritic spine morphology and synaptic plasticity. *Neuron* **61**, 85–100.

Komarova, Y. A., Akhmanova, A. S., Kojima, S., Galjart, N., and Borisy, G.G. (2002a). Cytoplasmic linker proteins promote microtubule rescue *in vivo*. *J. Cell Biol.* **159**, 589–599.

Komarova, Y., De Groot, C. O., Grigoriev, I., Gouveia, S. M., Munteanu, E. L., Schober, J. M., Honnappa, S., Buey, R. M., Hoogenraad, C. C., Dogterom, M., Borisy, G. G., Steinmetz, M. O., *et al.* (2009). Mammalian end binding proteins control persistent microtubule growth. *J. Cell Biol.* **184**, 691–706.

Komarova, Y., Lansbergen, G., Galjart, N., Grosveld, F., Borisy, G.G., and Akhmanova, A. (2005). EB1 and EB3 control CLIP dissociation from the ends of growing microtubules. *Mol. Biol. Cell* **16**, 5334–5345.

Komarova, Y. A., Vorobjev, I. A., and Borisy, G.G. (2002b). Life cycle of MTs: Persistent growth in the cell interior, asymmetric transition frequencies and effects of the cell boundary. *J. Cell Sci.* **115**, 3527–3539.

Krylyshkina, O., Anderson, K. I., Kaverina, I., Upmann, I., Manstein, D. J., Small, J. V., and Toomre, D. K. (2003). Nanometer targeting of microtubules to focal adhesions. *J. Cell Biol.* **161**, 853–859.

Lansbergen, G., Grigoriev, I., Mimori-Kiyosue, Y., Ohtsuka, T., Higa, S., Kitajima, I., Demmers, J., Galjart, N., Houtsmuller, A. B., Grosveld, F., and Akhmanova, A. (2006). CLASPs attach microtubule plus ends to the cell cortex through a complex with LL5beta. *Dev. Cell* **11**, 21–32.

Li, R., and Gundersen, G. G. (2008). Beyond polymer polarity: How the cytoskeleton builds a polarized cell. *Nat. Rev. Mol. Cell Biol.* **9**, 860–873.

Lomakin, A. J., Semenova, I., Zaliapin, I., Kraikivski, P., Nadezhdina, E., Slepchenko, B. M., Akhmanova, A., and Rodionov, V. (2009). CLIP-170-dependent capture of membrane organelles by microtubules initiates minus-end directed transport. *Dev. Cell* **17**, 323–333.

Manneville, J. B. (2006). Use of TIRF microscopy to visualize actin and microtubules in migrating cells. *Meth. Enzymol.* **406**, 520–532.

Mimori-Kiyosue, Y., Grigoriev, I., Lansbergen, G., Sasaki, H., Matsui, C., Severin, F., Galjart, N., Grosveld, F., Vorobjev, I., Tsukita, S., and Akhmanova, A. (2005). CLASP1 and CLASP2 bind to EB1 and regulate microtubule plus-end dynamics at the cell cortex. *J. Cell Biol.* **168**, 141–153.

Morrison, E.E. (2007). Action and interactions at microtubule ends. *Cell Mol. Life Sci.* **64**, 307–317.

Nogales, E., and Wang, H. W. (2006). Structural mechanisms underlying nucleotide-dependent self-assembly of tubulin and its relatives. *Curr. Opin. Struct. Biol.* **16**, 221–229.

Oliferenko, S., Chew, T.G., and Balasubramanian, M. K. (2009). Positioning cytokinesis. *Genes Dev.* **23**, 660–674.

Rappaz, B., Marquet, P., Cuche, E., Emery, Y., Depeursinge, C., and Magistretti, P. (2005). Measurement of the integral refractive index and dynamic cell morphometry of living cells with digital holographic microscopy. *Opt. Express* **13**, 9361–9373.

Saffarian, S., and Kirchhausen, T. (2008). Differential evanescence nanometry: Live-cell fluorescence measurements with 10-nm axial resolution on the plasma membrane. *Biophys. J.* **94**, 2333–2342.

Schneckenburger, H. (2005). Total internal reflection fluorescence microscopy: Technical innovations and novel applications. *Curr. Opin. Biotechnol.* **16**, 13–18.

Schober, J. M., Komarova, Y. A., Chaga, O. Y., Akhmanova, A., and Borisy, G. G. (2007). Microtubule-targeting-dependent reorganization of filopodia. *J. Cell Sci.* **120**, 1235–1244.

Schuyler, S. C., and Pellman, D. (2001). Microtubule "plus-end-tracking proteins": The end is just the beginning. *Cell* **105**, 421–424.

Semenova, I., and Rodionov, V. (2007). Fluorescence microscopy of microtubules in cultured cells. *Methods Mol. Med.* **137**, 93–102.

Shaner, N. C., Campbell, R. E., Steinbach, P. A., Giepmans, B. N., Palmer, A. E., and Tsien, R.Y. (2004). Improved monomeric red, orange and yellow fluorescent proteins derived from discosoma sp. Red fluorescent protein. *Nat. Biotechnol.* **22**, 1567–1572.

Shaw, R. M., Fay, A. J., Puthenveedu, M. A., von Zastrow, M., Jan, Y. N., and Jan, L. Y. (2007). Microtubule plus-end-tracking proteins target gap junctions directly from the cell interior to adherens junctions. *Cell* **128**, 547–560.

Shelden, E., and Wadsworth, P. (1993). Observation and quantification of individual microtubule behavior *in vivo*: Microtubule dynamics are cell-type specific. *J. Cell Biol.* **120**, 935–945.

Siegrist, S. E., and Doe, C. Q. (2007). Microtubule-induced cortical cell polarity. *Genes Dev.* **21**, 483–496.

Smal, I., Grigoriev, I., Akhmanova, A., Niessen, W.J., and Meijering, E. (2010). Microtubule dynamics analysis using kymographs and variable-rate particle filters. *IEEE Trans. Image Process.* **19**, 1861–1876.

Small, J. V., and Kaverina, I. (2003). Microtubules meet substrate adhesions to arrange cell polarity. *Curr. Opin. Cell Biol.* **15**, 40–47.

Stepanova, T., Slemmer, J., Hoogenraad, C. C., Lansbergen, G., Dortland, B., De Zeeuw, C. I., Grosveld, F., van Cappellen, G., Akhmanova, A., and Galjart, N. (2003). Visualization of microtubule growth in cultured neurons via the use of EB3-GFP (end-binding protein 3-green fluorescent protein. *J. Neurosci.* **23**, 2655–2664.

Toomre, D., and Manstein, D.J. (2001). Lighting up the cell surface with evanescent wave microscopy. *Trends Cell Biol.* **11**, 298–303.

Vale, R. D., Spudich, J. A., and Griffis, E. R. (2009). Dynamics of myosin, microtubules, and kinesin-6 at the cortex during cytokinesis in *Drosophila* S2 cells. *J. Cell Biol.* **186**, 727–738.

Webb, R. L., Rozov, O., Watkins, S. C., and McCartney, B. M. (2009). Using total internal reflection fluorescence (TIRF) microscopy to visualize cortical actin and microtubules in the *Drosophila* syncytial embryo. *Dev. Dyn.* **238**, 2622–2632.

Wen, Y., Eng, C.H., Schmoranzer, J., Cabrera-Poch, N., Morris, E. J., Chen, M., Wallar, B. J., Alberts, A.S., and Gundersen, G.G. (2004). EB1 and APC bind to mDia to stabilize microtubules downstream of rho and promote cell migration. *Nat. Cell Biol.* **6**, 820–830.

Zhu, Z. C., Gupta, K. K., Slabbekoorn, A. R., Paulson, B. A., Folker, E. S., and Goodson, H. V. (2009). Interactions between EB1 and microtubules: Dramatic effect of affinity tags and evidence for cooperative behavior. *J. Biol. Chem.* **284**, 32651–32661.

CHAPTER 7

Microtubule Dynamics in Dendritic Spines

Lukas C. Kapitein, Kah Wai Yau, ***and*** **Casper C. Hoogenraad**

Department of Neuroscience, Erasmus Medical Center, 3015 GE, Rotterdam, The Netherlands

Abstract

Neuronal microtubules recently emerged as temporal and spatial regulators of dendritic spines, the major sites of excitatory synaptic input. By imaging microtubules in cultured mature primary hippocampal neurons using fluorescently tagged tubulin and microtubule plus-end binding (EB) protein EB3, dynamic microtubules were found to regularly depart from the dendritic shaft and enter dendritic spines.

METHODS IN CELL BIOLOGY, VOL. 97

978-0-12-381349-7
DOI: 10.1016/S0091-679X(10)97007-6

Evidence indicates that microtubule invasions into spines regulate spine actin dynamics and induce transient morphological changes, such as the formation of spine head protrusion and spine growth. Because alterations in spine morphology play an important role in synaptic plasticity and have been linked to learning and memory formation, it is possible that dynamic microtubules are engaged in adaptive processes in the adult brain. This chapter provides detailed methods for live imaging of dynamic microtubules in mature hippocampal neurons in culture. We describe protocols for culturing and transfecting mature hippocampal neurons and visualizing microtubules and microtubule plus-EB proteins by total internal reflection fluorescence microscopy and spinning disk confocal microscopy.

I. Introduction

Neurons are electrically excitable polarized cells and typically composed of a cell body, one axon, and multiple highly branched dendrites. Signal transmission from the axon of one neuron toward dendrites of other neurons occurs at specialized junctions called synapses. Neuronal differentiation and maintenance of neuronal function in a brain network requires a well-organized interplay of many cellular processes, in many of which the microtubule cytoskeleton plays an important role (Conde and Caceres, 2009; Hoogenraad and Bradke, 2009; Lowery and Van Vactor, 2009). Our understanding of neuronal structure and function has increased tremendously through the use of primary culture techniques that allow neurons to develop axons and dendrites *in vitro*, as well as specialized subdomains such as growth cones, axon initial segment, dendritic spines, and inhibitory and excitatory synapses. Culturing primary neurons has become an important tool in neuronal cell biology, especially when combined with the ability to express exogenous genes, stain endogenous molecules using immunocytochemistry, and perform live cell imaging to address the function of specific proteins in their native cellular context. For instance, most of our knowledge of regulated exocytosis (Sudhof and Rothman, 2009), long-distance microtubule-based transport (Hirokawa and Takemura, 2004), receptor dynamics (Newpher and Ehlers, 2008; Sheng and Hoogenraad, 2007), and local mRNA translation (Bramham, 2008; Lin and Holt, 2008) comes from experiments performed in cultured neuronal cells. In general, insight into the basic cellular mechanisms of neurons in culture will help to better understand how the brain functions in an entire animal.

Most cellular studies of mammalian microtubule behavior and function have been focused on cultured fibroblasts (Cheeseman and Desai, 2008; Howard and Hyman, 2009; Steinmetz and Akhmanova, 2008). However, recent experimental evidence shows that microtubules and their plus-end binding (EB) proteins, also named +TIPs, play essential roles in the process of neuronal differentiation as well as in diverse aspects of mature neuronal functioning (Conde and Caceres, 2009; Hoogenraad and Bradke, 2009). Several plus-EB proteins, including cytoplasmic linker proteins (CLIPs), CLIP-associated proteins, EB family members, navigator family proteins, and Lissencephaly 1 (LIS1), have been shown to be important during several stages of neuronal development (Jaworski *et al.*, 2008), such as in the formation and growth of axons (Lee *et al.*, 2004; Martinez-Lopez *et al.*, 2005; Tsai *et al.*, 2005; Zhou *et al.*, 2004) and proper signaling in more mature neurons (Gu *et al.*, 2006). Recently, dynamic microtubules and the microtubule plus-EB protein EB3 were shown to be important for dynamic changes in structure and function of dendritic spines (Jaworski *et al.*, 2009).

Microtubule arrays within neuronal processes appear highly organized with respect to their intrinsic polarity (Baas *et al.*, 1988, 1989; Dombeck *et al.*, 2003). Ultrastructural studies show that in axons, microtubules are generally long and uniformly oriented with their plus-ends distal to the cell body, whereas in proximal dendrites microtubules are much shorter and exhibit mixed polarity. More distal thinner dendrites of higher order, however, contain unipolar microtubules oriented the same way as the axonal ones (Baas *et al.*, 1989). Since a major role of microtubules in mature neurons is to act as transport routes (Hirokawa and Takemura, 2005), distinct patterns of microtubule polarity orientation can generate asymmetries in the composition of each neuronal compartment by promoting specific motor protein motility (Kapitein *et al.*, 2010). The specialized microtubule organization has recently been captured in action by visualizing fluorescently labeled plus-EB proteins in living neuronal cells (Jaworski *et al.*, 2009; Morrison *et al.*, 2002; Stepanova *et al.*, 2003). In particular, it was found that the microtubule plus-end signal of EB3-GFP in cultured neurons is brighter than that detected with the other plus-EB proteins (Stepanova *et al.*, 2003). Thus, fluorescently labeled EB3 can be used as tools for visualizing the dynamic behavior of microtubules in developing and mature neurons.

This chapter describes detailed protocols for imaging microtubules and microtubule plus-EB proteins in primary hippocampal neurons in culture. We provide protocols detailed enough to set up the primary hippocampal culture system and discuss the important principles of visualizing microtubules and microtubule plus-ends in neurons by total internal reflection fluorescence microscopy (TIRFM) and spinning disk confocal microscopy. We will specify the reagents and equipment necessary to conduct each subprocedure. First, we will describe how to prepare hippocampal neurons in culture (Section III). Second, we will explain how to express fluorescently labeled tubulin and microtubule plus-EB proteins in cultured neurons using plasmid DNA transfection (Lipofectamine 2000) (Section IV) and Semliki Forest virus (SFV)-mediated gene delivery (Section V). Third, we will describe how to image dynamic microtubules in mature hippocampal neurons (Section VI). Finally, we will briefly discuss the data analysis (Section VII).

II. Rationale

The brain is a network of electrically active neurons that communicate with each other through synapses. Chemical synapses are asymmetric contacts formed between axons of the presynaptic neuron and dendritic specializations of the postsynaptic cell. In excitatory synapses of the hippocampus, the presynaptic terminal typically

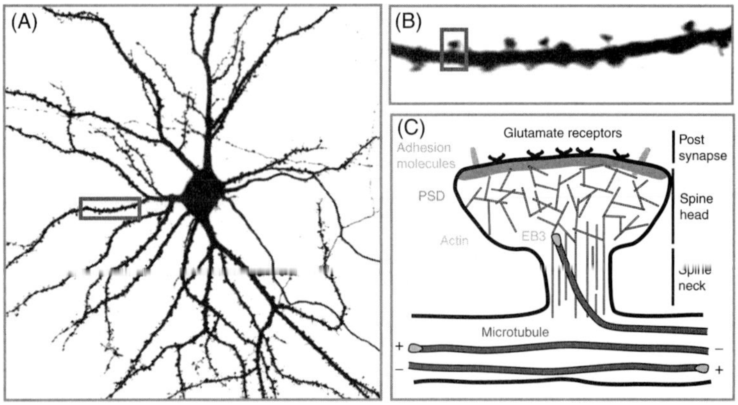

Fig. 1 Microanatomy of dendritic spines. (A) Image of hippocampal neurons transfected at 13 days in vitro (DIV13) with green fluorescent protein (GFP) as a marker to visualize neuronal morphology. Two days after transfection, neurons were fixed for 10 min with 4% formaldehyde/4% sucrose in PBS. Confocal images were acquired using a LSM510 confocal microscope (Zeiss) with a 40× oil objective and sequential acquisition settings at the maximal resolution of the microscope (1024 × 1024 pixels). Each image was a z-series of 6–10 images each averaged two times. The resulting z-stack was "flattened" into a single image using maximum projection. (B) Inset of the boxed area in (A) showing an enlarged dendritic segment containing several dendritic spines. (C) Schematic diagram of mature mushroom-shaped spine, showing the actin (brown lines) and microtubule (red) cytoskeleton and the postsynaptic membrane containing the PSD, adhesion molecules, and glutamate receptors. The actin cytoskeleton is connected to the PSD and determines spine structure and motility. Microtubule plus-end binding protein EB3 is symbolized as a green oval. Microtubules depart from the dendritic shaft, curve, and transiently enter dendritic spines. Some microtubules move all the way up in the spine head and even appear to touch the synaptic membrane. (See Plate no. 1 in the Color Plate Section.)

releases the neurotransmitter glutamate, which diffuses across the synaptic cleft to bind to and activate glutamate receptors in the postsynaptic membrane on top of dendritic spines, small membrane protrusions on neuronal dendrites. In addition to glutamate receptors, dendritic spines contain the postsynaptic machinery, including postsynaptic density (PSD), actin cytoskeleton, and a wide variety of membrane-bound organelles, such as smooth endoplasmic reticulum, mitochondria, and endosomes (Sheng and Hoogenraad, 2007) (Fig.1). Live imaging studies showed that spines are remarkably dynamic, changing size and shape over timescales of seconds to minutes and of hours to days (Holtmaat and Svoboda, 2009). Such dynamic changes in spine morphology are closely linked to changes in strength of synaptic connections and believed to be associated with learning and memory formation in the brain (Yuste and Bonhoeffer, 2001). Furthermore, the loss of synaptic stability or alterations of spine morphology are linked to many psychiatric and neurological diseases, including addiction, mental retardation, such as Fragile X syndrome, autism spectrum disorders, and neurodegenerative diseases such as Alzheimer's disease (Kauer and Malenka, 2007; Selkoe, 2002; Sudhof, 2008). Therefore, knowledge about molecular mechanisms underlying spine morphology and synaptic stability in primary neurons in culture will be important for understanding the basic molecular mechanisms underlying learning and memory formation, as well as synaptic disorders and traumatic injury (Blanpied and Ehlers, 2004; Hoogenraad and Bradke, 2009). Recent work from our laboratory revealed an important contribution of dynamic microtubules to the morphology and maintainance of dendritic spines (Fig. 1). These unexpected findings further increase the long list of important cellular functions mediated by dynamic microtubules and neccesitate future studies that closely examine microtubule dynamics in hippocampal neurons. The required procedures are described in detail in this chapter.

III. Culturing Primary Hippocampal Neurons

Over the last century, basic neuroscience research has taught us a great deal about the molecular and cellular mechanisms underlying neuronal development and plasticity. Pioneering work by the Banker lab using *in vitro* dissociated hippocampal neuron cultures provided an experimental system to study neuronal cell biology (Banker and Cowan, 1977). Using this culture method, hippocampal neurons from embryonic rats are cultured on glass coverslips in serum-free medium and then inverted and maintained above an astrocyte feeder layer (Kaech and Banker, 2006). The use of the neuron–glia cell coculture system allows hippocampal neurons to become appropriately polarized, develop extensive axonal and dendritic arbors, and form functional synaptic connections within a week after plating the cells (Dotti *et al.*, 1988). However, for detailed analysis of synaptic plasticity mechanisms and dendritic spine morphology we prefer to use fully developed, mature hippocampal neurons, which are maintained for a longer time in culture (over $>$2–3 weeks) (Fig. 1). The major differences with the original Banker protocol is that we plate neurons at a medium density (375–500 cells/mm^2), eliminate the need for astrocyte cocultures, and use serum-free Neurobasal medium supplemented with B27 (Brewer *et al.*, 1993). For the last decade, the medium-density neuronal cultures without cocultured astrocytes have been used by many investigators to address fundamental questions in neuronal cell biology (Ehlers, 2003; Lise *et al.*, 2006; Luscher *et al.*, 1999; Sala *et al.*, 2001; Shi *et al.*, 2003), including the role of microtubules in spine morphology (Jaworski *et al.*, 2009). In this section we will describe how to culture medium-density primary rat hippocampal neurons.

A. Buffers, Solutions, and Equipment

The following materials are needed for medium-density serum-free cultures.

Preparing Coverslips and Neurobasal/B27 Medium

- Glass coverslips [VWR 406-0189-32 (19 mm), 406-0189-50 (24 mm); Gallard Schlesinger]
- Porcelain racks (85422-E40; Thomas Scientific)

- Oven gloves (32885-804; VWR)
- 1 l glass beakers, pack of 6 (13912-284; VWR)
- Nitric acid, 65% solution (84382; Sigma)
- 0.1 M borate buffer, pH 8.5 (500 ml): Dissolve 1.24 g boric acid (B-0252; Sigma), 1.90 g borax (B-9876; Sigma) in 500 ml H_2O, and adjust pH to 8.5. Filter sterilize and store at 4°C.
- Poly-L-lysine (PLL) and laminin coating solution (40 ml): The PLL stock (P2636, solid powder; Sigma) is kept at 4°C. Dissolve 10 mg/ml PLL in H_2O and store in 150 µl aliquots at –80°C. The laminin stock (1243217, 1 mg/ml solution; Roche) is kept in 150 µl aliquots at –20°C. Before making the coating solution, slowly thaw PLL and laminin aliquots on ice. Add 450 µl PLL and 300 µl laminin in 120 ml to 0.1 M borate buffer to make fresh coating solution.
- 12-well plates (353043; Falcon) and 6-well plates (3506; Corning Costar).
- Neurobasal/B27 medium (100 ml): 97.5 ml Neurobasal medium (21103-049; Invitrogen), 2 ml B27 supplement (17504-044; Invitrogen), 1 ml penicillin/streptomycin (15140-148; Invitrogen), 250 µl glutamine (200 mM stock) (25030-081; Invitrogen), and 125 µl glutamic acid (glutamate; 1.84 mg/ml stock) (RBI G-100). Note that for feeding cultured neurons, B27 medium is used without glutamic acid.
- Tissue culture incubator (37°C, 5% CO_2).

Hippocampal Dissection and Plating Cells

- Timed pregnant Wister rat (E19) with about 10–15 embryos (from Harlan Laboratories).
- 100% carbon dioxide (CO_2) gas.
- Dissection tools: Medium forceps (11002-12; FST) and straight scissor (14001-12; FST) to dissect out embryos from pregnant rat. Two forceps #5 (11252-30; FST), two forceps, #4 (11242-40; FST), and small curved scissor (14061-09; FST) for brain and hippocampus dissection.
- Dissection M75 zoom stereomicroscope (Wild Heerbrugg).
- 0.3 M HEPES (100 ml): Dissolve 7.15 g HEPES (H-9136; Sigma) in 100 ml in H_2O (pH to 7.3). Filter sterilize and store at 4°C.
- Hanks Balanced Salt Solution (HBSS) (500 ml): 50 ml 10× HBSS (14185-052; Invitrogen), 16.5 ml 0.3 M HEPES (pH 7.3), 5 ml Pen/Strep (15140-148; Invitrogen), and 435 ml H_2O. Filter sterilize and store at 4°C.
- 100 ml glass beakers (13912-182; VWR).
- 10 cm dishes, TC dish 100 × 20 mm style TC-Treated BD Falcon (353003; VWR).
- 14 ml polypropylene round-bottom tube 17 × 100 mm style BD Falcon (352059; VWR).
- Trypsin (2.5%) (15090-046; Invitrogen). Aliquot in 200 µl and store at –20°C.
- Inverted phase contrast microscope (CKX31; Olympus) for cell counting and checking cell viability.

- Burker bright-line counting chamber (0642030; Marienfeld-superior) for cell counting.
- Cell hand tally counter (23609-102; VWR).

B. Preparing Coverslips for Neuronal Cultures

- Acid washing: Place coverslips (24 mm for live cell imaging and 19 mm for immunocytochemistry) in the porcelain racks and incubate in nitric acid for 2 days at room temperature.
- Water washing: Carefully transfer the racks into 1 l beakers (3 racks/beaker) containing about 600 ml of tap-distilled water. Allow the racks to sit undisturbed for 1 h at room temperature. Repeat washing for at least three more times for a total of 4 × 1 h of washing.
- Dry and bake coverslips: Remove racks from last wash and gently blot the base of each rack on a paper towel and allow the coverslips to air-dry on the bench for a while. Use vacuum suction to remove moisture caught between coverslips. Transfer the racks to a clean, dry beaker and cover the top with foil. If the coverslips are still wet, leave one edge of the foil slightly lifted. Bake coverslips overnight at 200°C.
- After baking, transfer the cooled coverslips to 12-well plates (19 mm) and 6-well plates (24 mm) in a sterile laminar flow hood. The coverslips can be directly used for coating or stored for later use.
- Coat coverslips overnight at room temperature (wrapped in aluminum foil in the tissue culture hood or in a drawer) with 1 ml per well of PLL and laminin-coating solution.
- Wash coated coverslips 4× with 2 ml per well of autoclaved MilliQ water. After the last wash, aspirate all the water and add 1 ml of Neurobasal/B27 medium to each well in a 12-well plate with 19 mm coverslips (2 ml of Neurobasal/B27 in a 6-well plate with 24 mm coverslips) and place in the incubator (37°C and 5% CO_2) until needed. This step can be performed the day before the dissection.

C. Hippocampal Dissection

- Before starting the dissection: Sterilize the dissecting instruments (all medium and small forceps and scissors) dipping them in 70% alcohol and flaming them. When the flames go out the forceps are sterile. Put on ice 2 × 100 ml beakers with about 25 ml HBSS each with sterile petri dish covers, 1 × 14 ml Falcon snap-top with 10 ml of HBSS, 2 × 50 ml Falcon tubes with about 35 ml HBSS (place one on ice, the other in 37°C water bath), and 1× petri dishes with about 12 ml HBSS on ice.
- Anesthetize a time-pregnant rat (E19; Wistar) by CO_2 and kill it by cervical dislocation. In the flow hood spray the abdomen and instruments with 70% ethanol. To minimize contamination, cut first through the skin using medium forceps and scissor and lay it back away from the abdomen. Rinse again the instruments with 70% ethanol and then cut through the abdomen wall.

- Lift out the uterus and cut away the connection tissues. Remove embryos one at a time from the uterus and decapitate them, placing the heads immediately into the 100 ml beaker with 25 ml HBSS on ice.
- Take one embryo head and remove the skin covering the skull. Place forceps under the skull and peel it to the side. Scoop the embryo brain out into the second 100 ml beaker with 25 ml HBSS on ice. Repeat for all embryos.
- Transfer one embryo brain into the petri dishes with 12 ml HBSS and place on the dissection microscope. Turn the brain so that the bottom faces upward. Cut along the area marked out by the blood vessels to separate the hemispheres from the brainstem.
- The hippocampus is located in the thicker end of the hemisphere. Place forceps in the narrow part of the hemisphere to hold it still and peel away the meninges from the medial part of the hemisphere. Usually it is possible to grasp the meninges and pull it away as a single sheet without tearing the underlying hippocampus.
- The inner edge of the hippocampus is now free. Using scissors cut away the adjoining tissue from the outer edge and the ends of the hippocampus to remove it from the hemisphere.
- Transfer the hippocampus into the 14 ml Falcon snap-top with 10 ml of HBSS on ice. Dissect out the hippocampi from the other hemisphere and repeat for all brains.
- After collecting all hippocampi, gently wash five times with 10 ml ice-cold HBSS (from the 50 ml Falcon tube); wait for the tissue to sink to the bottom, then aspirate HBSS as much as possible without sucking up the cells. After the last wash, remove all but 4 ml of HBSS and add 10 μl 2.5% trypsin/hippocampus (in general we add 200 μl trypsin for ~10 brains).
- Incubate for 15 min in the 37°C water bath and gently shake the tube every 5 min.
- Gently wash five times with 10 ml HBSS (from the 50 ml Falcon tube in the 37°C water bath), wait for tissue to sink to the bottom, and then aspirate HBSS as much as possible without sucking up the cells.
- After the last wash, remove all HBSS but leave 4 ml per ~12 brains and gently pipette the suspension up and down (about 10 times) with a 10 ml pipette to homogenize the tissue. This should result in virtually complete dissociation of the tissue into a homogenous single-cell suspension. If any debris is left, allow it to settle to the bottom of the tube and transfer the cell suspension to a new 14 ml Falcon tube.
- Next count the number of viable cells under an inverted phase contrast microscope using a Burker counting chamber (0.1 mm depth). Counting 16 small square zones will give the number of cells (c) per 0.1 μl suspension. For plating 75,000 cells per 19 mm coverslip (375 cells per mm^2), use the equation 75,000 cells/($c \times 10$) to calculate the volume of cell suspension (in μl) to pipette into one well.
- Incubate the cells in a 37°C and 5% CO_2 incubator for 2–3 weeks before DNA transfection or virus infection (Sections IV and V). Optional: after ~1 week in culture, exchange half of the medium with fresh Neurobasal/B27 medium. Never exchange all medium.

IV. Expression of EB3-GFP in Hippocampal Neurons Using Lipophilic Transfection

Expression of specific genes in mature neurons in culture has become an invaluable approach to study the subcellular localization of neuronal proteins (e.g., by visualizing fluorescently tagged recombinant proteins) and their functions (e.g., by introducing specific mutations, dominant-negative constructs, small hairpin RNA expression). Typically, the gene of interest is cloned into mammalian expression vectors and transfected into dissociated neurons in culture to obtain efficient neuronal expression. For optimal expression in mature hippocampal neurons in culture, we normally use either the cytomegalovirus-based pGW1-CMV expression vector for high-expression levels (Hoogenraad *et al.*, 2005) or the chicken β-actin promoter vector pβactin-16-pl for low expression of the transgene (Kaech *et al.*, 1996) (Fig. 2A, B). Several methods have been described to transfect hippocampal neurons in culture, including calcium phosphate transfection, microinjection, and electroporation (Jiang and Chen, 2006; Lappe-Siefke *et al.*, 2008; Leclere *et al.*, 2005), but we prefer to use Lipofectamine 2000 transfection (Dalby *et al.*, 2004) (Fig. 2C). Lipofectamine 2000 is a cationic liposome-based reagent that provides moderate transfection efficiency in primary cultured neurons (0.5–5%) using a relatively simple protocol. The positive surface charge on the liposome allows association of negatively charged groups such as DNA. The liposome–DNA complexes are subsequently carried through the positively charged and hydrophobic cell membrane into the cell, where it is released. Optimum transfection efficiency and subsequent neuronal viability depend on a number of experimental variables such as cell density, liposome and DNA concentrations, liposome–DNA complexing time, and presence or absence of media components.

The Lipofectamine 2000 transfection method is ideal for imaging subcellular structures in fully polarized neurons, so that axons and dendrites of a single neuron can be visualized and directionality of transport (anterograde vs retrograde) or microtubule orientations (plus-end movement outward vs plus-end movement inward) can be analyzed. We obtain optimal low expression of EB3-GFP and other +Tip markers in cultured neurons by using the pβactin vector in combination with Lipofectamine 2000 transfection reagent.

A. Buffers, Solutions, and Equipment

- Tissue culture incubator (37°C, 5% CO_2)
- 1.5 ml Eppendorf tubes
- Tabletop microcentrifuge (5415D; Eppendorf)
- 50 ml tube (227261; Greiner)
- Lipofectamine 2000 (11668-027; Invitrogen)
- Neurobasal medium (21103-049; Invitrogen)
- 12-well plates (353043; Falcon) and 6-well plates (3506; Corning Costar)
- 250 μl glutamine (200 mM stock) (25030-081; Invitrogen)
- Sterile forceps #5 (11252-30; FST)

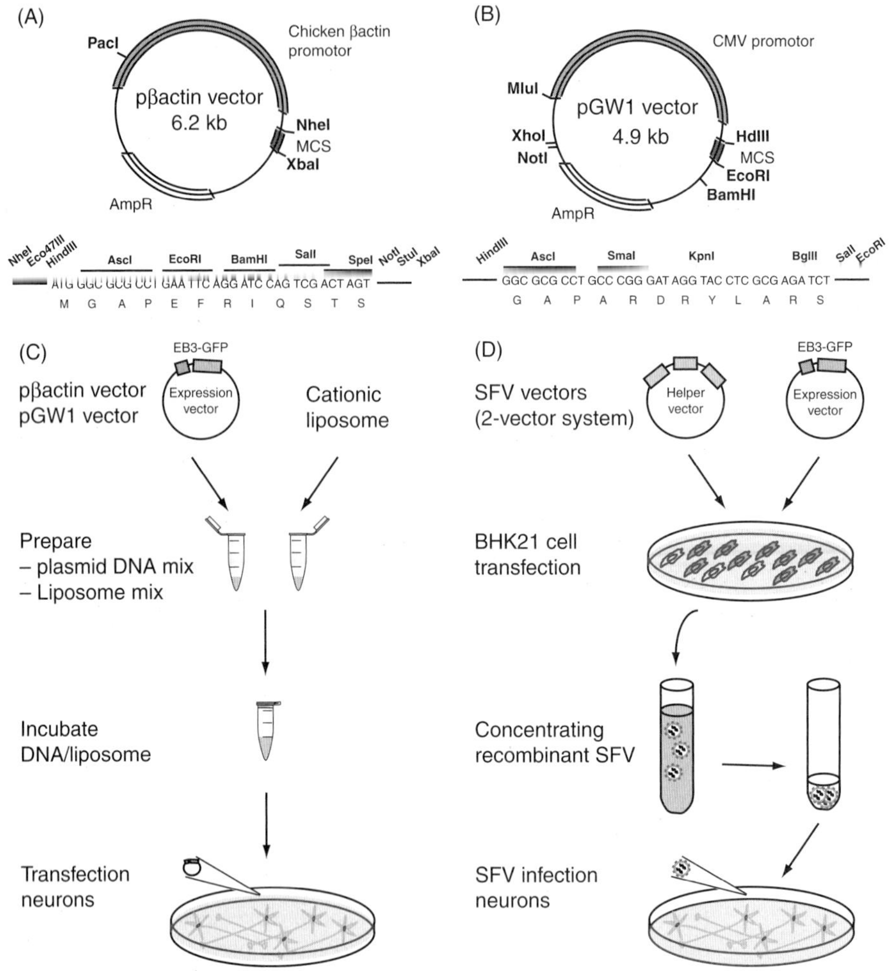

Fig. 2 Expressing microtubule plus-end tracking proteins (+Tips) in cultured neurons. (A–B) Mammalian expression vectors pβactin (A) and pGW1 (B) are used for optimal expression in primary hippocampal neurons. The original pβactin-16-pl vector was modified by inserting a multiple cloning site (NheI-Eco47III-HdIII-AscI-EcoRI-BamHI-SalI-SpeI-NotI-StuI-XbaI) and the pGW1-CMV vector contains the following multiple cloning site (HdIII-AscI-SmaI-KpnI-BglII-SalI-EcoRI). (C–D) Lipofectamine transfection procedure (C) and SFV flow diagram (D). See text for further details.

B. Transfection of Neurons Using Lipofectamine 2000

– Prepare the following before starting the transfection. (1) Plasmid DNA using Qiagen plasmid midi kit; (2) Incubation medium: for one 12-well plate containing 12 × 19 mm coverslips prepare 15 ml Neurobasal medium with 37.5 μl glutamine and put in 37°C water bath; (3) Fill 50 ml tube with Neurobasal medium and put in 37°C water bath.

- Thaw plasmid DNA (in this case βactin-EB3-GFP or βactin-cherry-α-tubulin) and spin in microcentrifuge for 5 min at 13,000 rpm to remove any microaggregates. For each Lipofectamine 2000 (L2K) transfection start with two 1.5 ml Eppendorf tubes, one containing plasmid DNA (DNA mix) and one containing the Lipofectamine 2000 reagent (L2K mix). All the incubation steps can proceed at room temperature.
- Prepare DNA mix in the first 1.5 ml tube: 1.8 μg plasmid DNA per 19 mm coverslip and add 100 μl Neurobasal medium. Gently mix. Use 3.6 μg plasmid DNA per 24 mm coverslip.
- Prepare L2K mix in the second 1.5 ml tube: 3.3 μl Lipofectamine 2000 per 19 mm coverslip and add 100 μl Neurobasal medium. Gently mix. Use 6.6 μg Lipofectamine 2000 per 24 mm coverslip. To preserve the quality of the Lipofectamine 2000 stock reagent, it should be removed from the refrigerator for as short a time as possible and placed on ice when not in use.
- Add 100 μl L2K mix to the DNA mixture and incubate for 30 min at room temperature.
- Note that no adjustments for cotransfections of two DNA constructs are required even though the Lipofectamine 2000 : DNA ratio will be halved in such cases. It is best to start testing 0.8 μg for each plasmid DNA for optimal expression.
- Approximately 10 min before the end of the incubation time pipette 1 ml of conditioned medium from the culture plate to the new 12-well plate and add 1 ml incubation medium to the original plate. Do this per 4 wells at the time and repeat for the next 4 wells. For the 6-well plate pipette 2 ml. Store the plate with conditioned medium in the 37°C and 5% CO_2 incubator.
- Gently add dropwise 200 μl of DNA/L2K mix to each well and incubate for 45 min in the 37°C and 5% CO_2 incubator.
- Rinse the coverslips containing the neurons in fresh warm Neurobasal medium by dipping them in a full 50 ml tube. Return the coverslip to the plate containing conditioned medium.
- Incubate the transfected neurons in the incubator at 37°C and 5% CO_2. Allow expression to proceed for the desired time before using them for immunocytochemistry or live cell imaging. We observed optimal microtubule labeling of cherry-α-tubulin and microtubule plus-end movement of EB3-GFP comets in cultured neurons in low-expressing neurons after 1–2 days after transfection by using the pβactin-16-pl vector in combination with Lipofectamine 2000 transfection reagent.

V. Expression of EB3-GFP in Hippocampal Neurons Using SFV

The SFV is an enveloped type alphavirus virus and has a single-stranded positive-sense RNA genome which functions directly as an mRNA after infection. Within host cells, translation can take place within a few hours after infection and thousands of exogenous protein copies can be made instantly. Compared to other viruses used to infect brain cells, SFVs exhibit a preference for neurons rather than glial cells

(Ehrengruber, 2002). SFV vectors have been modified to be defective in their production of infectious viral particles (in this case pSFV2), and cotransfection with a packaging-deficient helper vector (in this case pSFV-Helper2) allows viral propagation (Fig. 2D). By using this SFV vector system, SFV vector-mediated gene expression has been established in dissociated hippocampal neurons (de Hoop *et al.*, 1994) and slice cultures (Ehrengruber *et al.*, 1999). The advantages of SFV system is that it requires biosafety level 1, is relatively simple, and takes less than 2 days to make a high-titer virus. Moreover, cultured neurons can be infected with recombinant SFV with an efficiency of 60–80% within 3–4 h postinfection (de Hoop *et al.*, 1994). On the other hand, SFV vector infections are cytotoxic which permits only short-term transgene expression (4–24 h). The SFV vector system we use to produce SFV EB3-GFP particles is previously described by the Ehrengruber laboratory (Ehrengruber and Lundstrom, 2007; Ehrengruber *et al.*, 1999).

A. Buffers, Solutions, and Equipment

Culturing BHK-21 cells

- Baby hamster kidney-21 (BHK-21) cell line (c-13), American Type Culture Collection (ATCC; CCL-10)
- Glasgow MEM (G-MEM) BHK-21 (1×) liquid 500 ml (21710-025; Invitrogen)
- Streptomycin/penicillin (15140-114; Invitrogen)
- Phosphate-buffered saline (PBS) medium w/o Ca^{2+}/Mg^{2+} (14190-094; Invitrogen)
- Trypsin/EDTA (25300-054; Invitrogen)
- Fetal bovine serum (FBS) (PET10106169; Invitrogen)
- 5% FBS Media complete: 500 ml G-MEM, 5 ml streptomycin/penicillin, and 25 ml FBS
- 10 cm dishes, TC dish 100 × 20 mm style TC Treated BD Falcon (353003; VWR)
- 175 cm^2 flasks

Transfection BHK-21 cells

- SFV plasmid system: pSFV2 and pSFV-Helper2
- 1.5 ml sterile and RNAse-free Eppendorf tubes
- Phenol/chloroform/isoamylalcohol 25:24:1 (P-2069; Sigma)
- Roche Sp6 transcription kit containing Sp6 RNA polymerase and 10× transcription buffer, rNTP mix, 10–50 U/μl Rnase inhibitor (999644), 10 mM m^7G (5′)ppp(5′)G (904988; Roche), 50 mM dithiothreitol (DTT) (197777; Roche)
- Opti-MEM I reduced serum medium (1×) with L-glutamine (31985047; Invitrogen)
- DMRIE-C reagent 2 mg/ml (10459-014; Invitrogen)
- 40–60 units α-chymotrypsin (C4129; Sigma)
- 100 mg Aprotinin (4511388; Amersham Pharmacia Biotech)
- 0.22 μm 33 mm Millipore filters (SLGSO 3355; Millipore)

- Ultracentrifuge with SW41Ti rotor and ultra clear SW41Ti centrifuge tubes 13.2 ml 14 × 89 mm (344059; Beckman Coulter)
- 40 mM HNE buffer (pH 7.4): 40 mM HEPES, 138.5 mM NaCl, 0.1 mM EGTA, and adjust pH with 4 M NaOH. Measure osmolarity (should be 300–310 mosm/l), autoclave, and store at 4°C

B. Preparation of Packaged SFV EB3-GFP Replicons in BHK-21 Cells

- Clone the gene of interest into the multiple cloning site of the pSFV2 vector. In this case, the cDNA encoding for the microtubule plus-EB protein EB3 was fused to GFP (EB3-GFP) and subcloned in pSFV2.
- Digest ~10 μg of recombinant pSFV2-EB3-GFP vector with NruI, and pSFV-Helper2 with SpeI to linearize plasmid DNA. Clean up linear DNA by phenol/chloroform extraction, precipitate, add 15 μl RNAse-free H_2O, and use for *in vitro* transcription reaction.
- The Roche Sp6 transcription kit is used to synthesize single-stranded RNA for transfection in BHK-21 cells. The *in vitro* transcription mix contains 15 μl linear DNA, 15 μl 10× transcription buffer, 15 μl 10 mM $m^7G(5')ppp(5')G$, 15 μl 50 mM DTT, 15 μl rNTP mix, 9 μl SP6 RNA polymerase, and 5.6 μl 40 U/μl RNasin in a total volume of 150 μl dH_2O. Mix and incubate for 2 h at 37°C.
- Prepare BHK-21 cells. Grow BHK-21 stock cells in 10 cm^2 dishes with 5% FBS media complete in a tissue culture incubator (37°C + 5% CO_2). When cells form a confluent monolayer, cells are split using 1 ml of prewarmed trypsin/EDTA. One or two weeks before the transfection, prepare BHK-21 cells in 175 cm^2 flasks.
- One day before transfection split a confluent 175 cm^2 flask of BHK-21 cells, seed 2.5 ml cells in a new 175 cm^2 flask (50 ml medium), and incubate cells overnight at 37°C in a 5% CO_2 incubator. Next day discard medium and wash cells two times with ~30 ml of prewarmed Opti-MEM I-reduced serum.
- Add in a sterile tube 17.5 ml Opti-MEM I, 175 μl DMRIE-C, 450 μl SFV-Helper2 RNA, 225 μl SFV2-EB3-GFP RNA. Immediately add the transfection mix to the washed BHK-21 cells and mix gently. Incubate for 4 h at 37°C and 5% CO_2 in the incubator.
- Replace transfection media with 36 ml 10% FBS media complete and allow the BHK-21 cells to express and release virus particles into the media.
- After 24 h, harvest 36 ml of medium and filter through a 0.22 μm sterile Millipore filter.
- Virus was activated with 0.5 mg/ml α-chymotrypsin at room temperature for 30 min and aprotinin (900 μl, 250 μg/ml) added to stop protease activity.
- Concentrate the virus by centrifugation of the activated viral stock in a SW41 rotor for 2.5 h at 35,000 rpm. Remove supernatant very carefully; the virus pellet is hardly visible and very small. Leave the last ~50 μl of the supernatant on the virus pellet. Dissolve the pellet by carefully pipetting up and down. Avoid air bubbles while dissolving the virus pellet.

- At equal volume of 40 mM HNE buffer pH 7.4 (sterile) to the collected virus. Make 5–10 μl virus aliquots and store aliquots at –80°C. Aliquots can be used a few times but every freeze/thaw step will decrease the virus titer.
- Determine the virus titer by infecting BHK-21 cells (in a dilution series) and count the number of infected cells per total number of cells. The virus titer should be 10^7–10^9 replicons/ml after concentrating.
- Primary hippocampal neurons at 14–21 days *in vitro* (DIV) were infected with (1–3 × 10^3) SFV-EB3-GFP replicons. Expression of GFP-EB3 in neurons was visible after 4–6 h at 37°C and 5% CO_2. Longer incubation times or higher virus titer results in high levels of fluorescence in the cytoplasm masking EB3-GFP signals at microtubule plus-ends.

VI. Imaging EB3-GFP by TIRF and Spinning Disk Microscopy

A. Maintaining Neuronal Health

Observing and characterizing dynamic cellular processes, such as microtubule dynamics, often yields important information about cellular activity unavailable from static images. Although TIRF microscopy (Axelrod, 2008) and spinning disk confocal microscopy (Nakano, 2002) are now established tools to image living cells, experiments on living primary neurons pose some additional challenges. The most important challenge is to keep primary neurons alive under the microscope for a long time (hours to days).

Many high-resolution live cell experiments are typically performed using simple defined buffered solutions during imaging (e.g., Ringer's solution). Such defined solutions can be optimized to have minimal background fluorescence, are compatible with most chemical treatments, and do not require a special atmosphere for buffering (i.e., 5% CO_2). However, because such simple solutions lack most components required for survival, they are not compatible with long-term imaging. Furthermore, even in short-term experiments, we frequently observed differences in organelle dynamics between buffered solutions and the full, conditioned medium used to promote neuron survival over several weeks (Section III A). We therefore routinely use full, conditioned medium as the standard medium for all our live cell-imaging experiments. This medium is bicarbonate buffered and, hence, requires the atmosphere surrounding cells on the microscope to comprise 5% CO_2. In addition, for optimal survival, the temperature of the medium should be stable and uniform at 37° C. Finally, because the medium immersing the cells needs to be accessible for the CO_2, the sample cannot be sealed and, hence, evaporation of the medium needs to be prevented by maintaining high humidity of the surrounding atmosphere.

To meet all these requirements, we use a small incubator system that fits our motorized stage insert space (Tokai Hit; INUG2-ZILCS-H2) and keeps the cells in a temperature, humidity, and CO_2 concentration-controlled environment. A 24 mm cover glass with cultured neurons are removed from the 6-well plate and mounted on a metal ring (Invitrogen, Attofluor cell chamber A-7816) that is placed in the incubator. Environmental control is achieved by heating several parts of the microscope and the incubator (objective heater, 37°C; incubator water bath, 38°C; stage heater, 37°C; and a transparent top heater at 40.5°C to prevent condensation of water on the lid) and by connecting to an automated digital gas mixer that mixes pure CO_2 and air. In addition, to compensate the evaporation from the water bath that occurs over tens of hours, we use an automatic water supply system (Tokai Hit, IMF-I-W). Altogether, the stage incubator conditions closely mimic those of a regular cell culture incubator and facilitate cell survival for many days.

The use of local heating and atmosphere control, rather than enclosing the whole microscope with an environment chamber, facilitates rapid and accurate equilibration of the chamber conditions. However, the focal drift induced by the temperature gradient in the objective renders this approach incompatible with long-term experiments on microscopes without focus feedback. We use Nikon's Perfect Focus System on both the TIRF and the spinning disk confocal microscopy systems.

B. Total Internal Reflection Fluorescence Microscopy

Light traveling from one medium to another with different index of refraction (n_i) will be refracted according to Snell's law. When traveling from glass ($n_{i1} = 1.5$) to water ($n_{i2} = 1.33$) the exit angle will exceed the entry angle, and a critical angle θ_c exists above which light will not propagate through the water but instead be totally reflected on the internal surfaces of the interface (i.e., $\theta_c = \arcsin \left(n_{i2}/n_{i1}\right) \approx 62.5$). In this situation, a small exponentially decaying electromagnetic standing wave will emerge at the interface in the second medium. The penetration depth of this field depends on the wavelength and precise angle of incidence, but is of order 150 nm. This light can be used to selectively excite fluorophores present within this zone and therefore allows the specific high-contrast detection of a subset of fluorescent signals (down to single-molecule fluorescence) in situations where the overall particle density is high (Axelrod, 2008). For example, numerous studies have used TIRFM to detect single fluorophores in *in vitro* assays, while many others applied TIRFM to the study of living cells to examine the dynamics of structures close to the cell membrane, such as focal adhesions. When applying TIRFM to the study of cells, it should be noted that the precise properties of evanescent wave are hard to predict because cells do not have a uniform index of refraction. In general, cellular index of refraction is higher than that of water, necessitating a higher angle of incidence to achieve total internal reflection and affecting penetration depth. Furthermore, it is important to note that we and others frequently use the TIRFM setup in a semi-TIRF mode that allows excitation deeper into the sample, while still providing far better contrast than epiillumination (Nakata and Hirokawa, 2003). Since in this case there is no evanescent wave, this would be more appropriately called oblique illumination microscopy rather than TIRFM.

Our TIRF microscope is based on the inverted research microscope Nikon Eclipse TE2000E (Nikon). The microscope is equipped with the Perfect Focus System (Nikon, T-PFS) and a motorized stage (Prior, Proscan II). Metamorph 7.1 (Universal Imaging) is used to control all motorized parts and cameras. A HBO 103 W/2 Mercury Short Arc Lamp (Osram) is used for regular episcopic wide-field illumination, whereas evanescent wave excitation is achieved by a commercially obtained multicolor TIRF arm (Nikon). Lasers (113 mW 488 nm laser line of an argon laser (Spectra-Physics Lasers) and 11 mW 561 nm diode-pumped solid-state laser (Melles Griot), mounted on a Nikon laser combiner (Nikon, C-LU3EX 3) are coupled into the TIRF arm using an optical fiber whose exit point can be shifted in a plane conjugate to the objective back focal plane. Shifting the fiber exit will thus shift its image (the focused laser) in the objective back focal plane and induce a tilt in the (near-parallel) laser beam that exits the objective. Precise focusing of the laser in the objective back focal plane can be obtained by axially sliding one of the lenses in the optical path between fiber entry and objective back pupil.

The maximal (theoretical) exit angle θ that an objective can achieve is related to its numerical aperture (N.A.) by N.A. $= n_{i1}\sin\theta$. This means that N.A. $> n_{i2}$ is the minimal requirement for objectives in order to achieve total internal reflection of the entering laser beam on the interface between cover glass and medium. In practice, N.A. should be at least 1.4 to achieve internal reflection of the whole, spatially extending, laser beam. In our setup, we use a CFI Apo TIRF 100× 1.49 N.A. oil objective (Nikon). Laser light enters the objective after being reflected on a dichroic mirror. Correct laser entry depends critically on the precise positioning of the dichroic mirror and laser alignment needs to be adjusted for each dichroic separately to prevent oblique laser entry. If only a single laser color is used, we use single-band dichroics (Chroma, T495lp for 488 nm and T570lp for 561 nm). However, for simultaneous dual-color TIRF microscopy a multispectral dichroic is required (Chroma, 59022 bs). These dichroics are mounted in metal filter cubes (Nikon, C-FL-HQ) that allow fine-tuning of their tilt and also contain single-band emission filters (Chroma, ET525/25 and ET620/60, respectively) or a double-band emission filter (Chroma, 59022 m).

Fluorescence emission can be detected using two different cameras mounted to different exit ports of the microscope. A Coolsnap HQ2 camera (Photometrics) is attached to a high-speed filter wheel (LB10-NWE, Sutter Instruments) that is mounted to the right exit port of the microscope. This camera has a large detection area (1392 × 1040 pixels) and small pixel size (6.45 μm) and we use it to image large field of views (~90 × 67 μm or 224 × 167 μm with 100× or 40× magnification, respectively) of reasonably bright samples. Near-simultaneous dual-color TIRFM is achieved by alternating the emission filters in the filter wheel mounted to the exit port (either ET525/25 or ET630/75) in combination with the corresponding excitation laser.

At the left port of the microscope, we use a QuantEM:512SC EMCCD (Electron Multiplying CCD technology) camera (Photometrics). This camera is perfectly suited for low-light level microscopy, because the back-illuminated CCD detector of this camera has a high-quantum efficiency (generated electrons per photon >90% at 575 nm wavelength) and features on-chip multiplication to minimize readout noise. In addition, the camera can achieve high acquisition rates by using frame transfer, which means that before readout the charge distribution accumulated during exposure is rapidly shifted to the unexposed part of the CCD chip and read out while the exposed part is acquiring again. The camera can thus be used for continuous (unshuttered) exposure, with the minimum exposure time of frames being limited by the readout time, which for the full chip at 10 MHz is about 35 ms (~512 × 512 pixels × 10^{-7} s). The QuantEM:512SC camera has an exposed CCD area of 512 × 512 pixels with a pixel size of 16 μm. At 100× magnification this results in an effective pixel size of 160 nm and ~82 × 82 μm field of view. However, to prevent undersampling of the point spread function (width ~λ/(2N.A.)~200 nm, with emission wavelength λ~600 nm) we typically use an additional 2.5× magnification lens (Nikon, VM Lens C-2.5×) to achieve effective pixel sizes similar to the Coolsnap camera (i.e., 64 nm at 100×). Finally, to achieve true simultaneous dual-color TIRFM we used a DualView (MAG Biosystems, DV2) with beam splitter (Chroma, 565DCXR) and additional emitter (Chroma, ET525/25) in the GFP light path. This optical device features two relay lenses to project the native image plane of the microscope onto the displaced camera. In the infinity space between these two lenses, a dichroic mirror separates the emission from the two emitters (e.g., GFP and RFP) and adjustable mirrors redirect the two emission light paths onto the second relay lens at slightly different angles to create two spatially (and spectrally) separated images on the same CCD chip.

C. Spinning Disk Confocal Microscopy

TIRF microscopy is an excellent tool to study processes that occur close to the surface of the cover glass. However, neurons do not entirely attach to the surface and large parts of dendrites and axons are therefore out of reach for the evanescent wave. Confocal microscopes, in contrast, can acquire high-contrast images throughout the sample. Excitation in confocal microscopy occurs through a focused laser that rapidly scans the image plane. The power of confocal microscopy lies in the pinhole that is used to exclude out-of-focal-plane fluorescence emission from the (biological) specimen, allowing the high-contrast imaging of an optically sectioned slice. In combination with z-scanning of the sample by use of a motorized stage or objective, this allows complete sectioning of the specimen to create a three-dimensional reconstruction. In conventional confocal laser scanning microscopes, scanning of the excitation lasers is achieved through mechanical movement of mirrors. This method is rather slow and requires seconds to scan just one plane of a specimen, which makes this approach incompatible with fast live cell microscopy. This disadvantage can be overcome by using spinning disk confocal microscopy. In this technique, scanning of confocal excitation light is achieved through two spinning disks in which one contains thousands of pinholes and the other contains an equal number of microlenses to focus the laser beam into the pinholes. This enables fast, multicolor, three-dimensional live cell imaging.

To perform live cell spinning disk confocal microscopy, we use a Nikon Eclipse-Ti (Nikon) microscope with a CFI Apo TIRF 100× 1.49 N.A. oil objective (Nikon). The microscope is equipped with a motorized stage (ASI, PZ-2000) and Perfect Focus System (Nikon) and uses MetaMorph 7.6.4 software (Molecular Devices) to control the cameras and all motorized parts. Confocal excitation and detection is achieved using a 50 mW 491 nm laser (Cobolt Calypso) and a Yokogawa spinning disk confocal scanning unit (CSU-X1-A1N-E, Roper Scientific) equipped with a triple-band dichroic mirror (z405/488/568trans-pc, Chroma) and a filter wheel (CSU-X1-FW-06P-01, Roper Scientific) containing a GFP emission filter (ET525/50m, Chroma). Confocal images were acquired with a QuantEM:512 SC EMCCD camera (Photometrics) at a final magnification of 64 nm/pixel, including the additional 2.5× magnification introduced by an additional lens mounted between scanning unit and camera (VM Lens C 2.5×, Nikon). The small incubator system (Tokai Hit, INUG2-ZILCS-H2) and Nikon's Perfect Focus System are also used on this system.

VII. Data Analysis

Using TIRFM or spinning disk confocal microscopy, fluorescent EB3 comets highlighting the growing microtubule plus tip can be readily imaged and followed over time. Speeds and directionality of microtubule growth can then be analyzed by manual or automatic tracking of comet trajectories. We recently became particularly interested in using SFV vector-mediated expression of EB3-GFP to identify potential events of microtubules growth into dendritic spines (Jaworski *et al.*, 2009). To rapidly identify potential spine-entering events, we create maximum (Fig. 3A) and average (Fig. 3B) projections of entire time-lapse recordings. This results in a single image, in which each pixel value corresponds to the maximum or average value of that pixel position across the entire time series, respectively. Because comets are short-lived and disappear when microtubules undergo catastrophe and start shrinking, the average projection is a close approximation to the distribution of the diffuse pool of nonmicrotubule-bound EB3-GFP. This provides a direct marker for neuron morphology, without requiring coexpression of another fluorescent protein (such as mRFP). In many cases, the maximum projection of the time series reveals locations of microtubule-associated comets and can be used to identify potential spine-entering events. In other cases, however, the large pool of diffuse EB3-GFP not associated with microtubules obscures the unambiguous identification of microtubule-bound EB3-GFP comets. We therefore generally subtract the average projection from each frame of the time series before creating a maximum projection (Fig. 3C). This procedure dramatically enhances the appearance of EB3-GFP comets (Fig. 3C) and also permits creating color-coded merges of background and comets (Fig. 3D–E). If spine entry events are identified (marked by arrows in Fig. 3C), kymography can be used to display comet dynamics along a line over time (Fig. 3F).

We found that EB3 comet entry was often followed by an increase in spine size (Jaworski *et al.*, 2009). To quantify spine size change upon EB3-GFP comet entry, we compare the spine area before the first recorded entry event to the area 5 min after the last entry event that occurred within a window of 3 min after the first event, provided no new entries occurred in those 5 min. To obtain spine area, image regions including head and neck, but not dendritic shaft, are first low-pass filtered, then binarized by

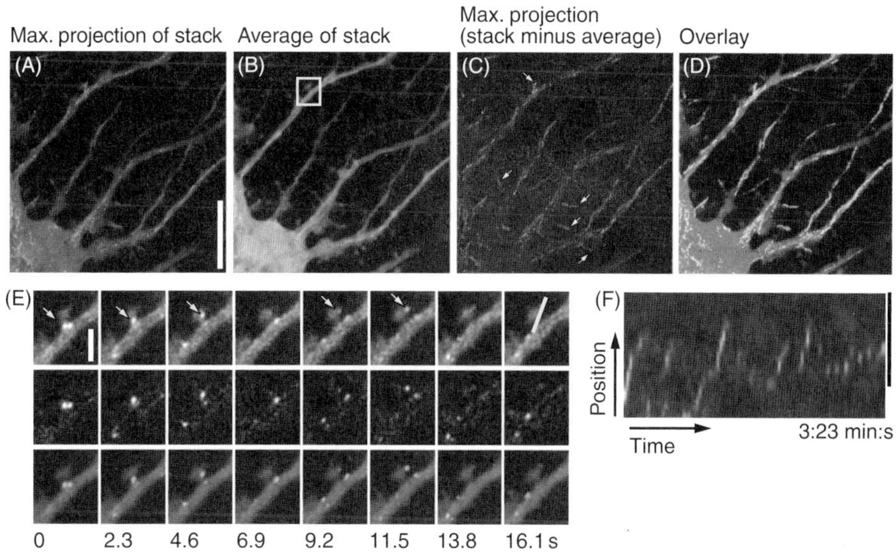

Fig. 3 EB3-GFP imaging using spinning disk and TIRF. (A–D) DIV15 Hippocampal neurons infected by SFV and expressing EB3-GFP were imaged using a spinning disk confocal microscope (Nikon). Each 2.3 s a series of 11 z-slices (spaced 0.5 μm) was recorded and merged into a single image by maximum projection. The resulting time series was then low-pass filtered before subsequent operations to produce (A–D). Scale bar represents 10 μm. (A) Maximum projection of the low-pass-filtered time-lapse recording. (B) Average projection of the time-lapse recording used to obtain an approximate background (noncomet) fluorescence image. Box indicates the region used in (E). (C) Maximum projection of the time series obtained by subtracting the average projection [shown in (B)] from the original, low-pass filtered, time series. EB3 comet spine entry events are now readily detectable. (D) Merge of (B) red) and (C) (green). (E) Video frames of the region shown in (B). Upper row shows the original, low-pass filtered data. Second row displays frames obtained by subtracting the average fluorescence shown in (B). The third row shows the merge of average and average-subtracted stills. Arrows indicate comet entering spine. Yellow line marks region used for kymography in (F). Scale bar represents 2 μm. (F) Kymograph over the total time-lapse recording for the line shown in (E). Scale bar represents 2 μm. (See Plate no. 2 in the Color Plate Section.)

thresholding at ~30% of maximum intensity above background, and finally closed by morphological filtering (dilation followed by erosion). Area is then measured as the number of nonzero pixels and multiplied by the calibration factor (1 pixel$^2 = 64^2$ nm^2). To analyze control spine growth, spines that exist throughout the imaging but do not show EB3 entry events can be selected.

VIII. Conclusion

Primary neuron cultures have become an important tool for addressing fundamental questions in molecular and cellular neurobiology, especially when combined with the ability to express specific genes and perform quantitative high-resolution live cell imaging experiments to address the subcellular localization and function of neuronal proteins in their native cellular context. Although the study of the neuronal cytoskeleton has a long and rich history that spans several decades, it seems that a new era has started now that state-of-the-art imaging techniques are readily available from commercial sources and can easily be combined with the rapidly expanding repertoire of techniques that allow expressing, tagging, or silencing of specific proteins and/or modulation of their functions using chemical biology toolboxes. We hope that the detailed protocols provided in this chapter will contribute to a better understanding of the dynamic cytoskeleton in neuronal cells.

Acknowledgments

We thank Samantha Spangler and Nanda Keijzer for preparing neuron cultures and establishing the dissection protocol. L.C.K. is supported by the Erasmus Medical Center (EMC fellowship) and the Netherlands Organization for Scientific Research (NWO-VENI). C.C.H. is supported by the Netherlands Organization for Scientific Research (NWO-ALW and NWO-CW), the Netherlands Organization for Health Research and Development (ZonMW-VIDI and ZonMW-TOP), the European Science Foundation (EURYI), EMBO Young Investigators Program (YIP), and the Human Frontier Science Program (HFSP-CDA).

References

Axelrod, D. (2008). Chapter 7: Total internal reflection fluorescence microscopy. *Methods Cell Biol.* **89**, 169–221.

Baas, P. W., et al., (1988). Polarity orientation of microtubules in hippocampal neurons: Uniformity in the axon and nonuniformity in the dendrite. *Proc. Natl. Acad. Sci. U. S.A.* **85**, 8335–8339.

Baas, P. W., et al., (1989). Changes in microtubule polarity orientation during the development of hippocampal neurons in culture. *J. Cell Biol.* **109**, 3085–3094.

Banker, G. A., and Cowan, W. M. (1977). Rat hippocampal neurons in dispersed cell culture. *Brain Res.* **126**, 397–42.

Blanpied, T. A., and Ehlers, M. D. (2004). Microanatomy of dendritic spines: Emerging principles of synaptic pathology in psychiatric and neurological disease. *Biol. Psychiatry* **55**, 1121–1127.

Bramham, C. R. (2008). Local protein synthesis, actin dynamics, and LTP consolidation. *Curr. Opin. Neurobiol.* **18**, 524–531.

Brewer, G. J., et al., (1993). Optimized survival of hippocampal neurons in B27-supplemented neurobasal, a new serum-free medium combination. *J. Neurosci. Res.* **35**, 567–576.

Cheeseman, I. M., and Desai, A. (2008). Molecular architecture of the kinetochore-microtubule interface. *Nat. Rev. Mol. Cell Biol.* **9**, 33–46.

Conde, C., and Caceres, A. (2009). Microtubule assembly, organization and dynamics in axons and dendrites. *Nat. Rev. Neurosci.* **10**, 319–332.

Dalby, B., et al., (2004). Advanced transfection with Lipofectamine 2000 reagent: primary neurons, siRNA, and high-throughput applications. *Methods* **33**, 95–103.

de Hoop, M. J., et al., (1994). Semliki forest virus as a tool for protein expression in cultured rat hippocampal neurons. *Gene Ther.* **1**(Suppl 1), S28–S31.

Dombeck, D. A., et al., (2003). Uniform polarity microtubule assemblies imaged in native brain tissue by second-harmonic generation microscopy. *Proc. Natl. Acad. Sci. U. S.A.* **100**, 7081–7086.

Dotti, C. G., et al., (1988). The establishment of polarity by hippocampal neurons in culture. *J. Neurosci.* **8**, 1454–1468.

Ehlers, M. D. (2003). Activity level controls postsynaptic composition and signaling via the ubiquitin-proteasome system. *Nat. Neurosci.* **6**, 231–242.

Ehrengruber, M. U., et al., (1999). Recombinant Semliki Forest virus and Sindbis virus efficiently infect neurons in hippocampal slice cultures. *Proc. Natl. Acad. Sci. U.S.A.* **96**, 7041–7046.

Ehrengruber, M. U. (2002). Alphaviral vectors for gene transfer into neurons. *Mol. Neurobiol.* **26**, 183–201.

Ehrengruber, M. U., and Lundstrom, K. (2007). Alphaviruses: Semliki Forest virus and Sindbis virus vectors for gene transfer into neurons. *Curr. Protoc. Neurosci.* Chapter 4, Unit 4 22.

Gu, C., et al., (2006). The microtubule plus-end tracking protein EB1 is required for kv1 voltage-gated K+ channel axonal targeting. *Neuron* **52**, 803–816.

Hirokawa, N., and Takemura, R. (2004). Molecular motors in neuronal development, intracellular transport and diseases. *Curr. Opin. Neurobiol.* **14**, 564–573.

Hirokawa, N., and Takemura, R. (2005). Molecular motors and mechanisms of directional transport in neurons. *Nat. Rev. Neurosci.* **6**, 201–214.

Holtmaat, A., and Svoboda, K. (2009). Experience-dependent structural synaptic plasticity in the mammalian brain. *Nat. Rev. Neurosci.* **10**, 647–658.

Hoogenraad, C. C., et al., (2005). GRIP1 controls dendrite morphogenesis by regulating EphB receptor trafficking. *Nat. Neurosci.* **8**, 906–915.

Hoogenraad, C. C., and Bradke, F. (2009). Control of neuronal polarity and plasticity—a renaissance for microtubules?. *Trends Cell Biol.* **19**, 669–676.

Howard, J., and Hyman, A. A. (2009). Growth, fluctuation and switching at microtubule plus ends. *Nat. Rev. Mol. Cell Biol.* **10**, 569–574.

Jaworski, J., et al., (2008). Microtubule plus-end tracking proteins in differentiated mammalian cells. *Int. J. Biochem. Cell Biol.* **40**, 619–637.

Jaworski, J., et al., (2009). Dynamic microtubules regulate dendritic spine morphology and synaptic plasticity. *Neuron* **61**, 85–100.

Jiang, M., and Chen, G. (2006). High ca2+-phosphate transfection efficiency in low-density neuronal cultures. *Nat. Protoc.* **1**, 695–700.

Kaech, S., et al., (1996). Cytoskeletal plasticity in cells expressing neuronal microtubule-associated proteins. *Neuron* **17**, 1189–1199.

Kaech, S., and Banker, G. (2006). Culturing hippocampal neurons. *Nat. Protoc.* **1**, 2406–2415.

Kapitein, L. C., et al., (2010). Mixed microtubules steer dynein-driven cargo transport into dendrites. Curr Biol. **20**(4): 290–9.

Kauer, J. A., and Malenka, R. C. (2007). Synaptic plasticity and addiction. *Nat. Rev. Neurosci.* **8**, 844–858.

Lappe-Siefke, C., et al., (2008). Microinjection into cultured hippocampal neurons: A straightforward approach for controlled cellular delivery of nucleic acids, peptides and antibodies. *J. Neurosci. Methods* **175**, 88–95.

Leclere, P. G., et al., (2005). Effective gene delivery to adult neurons by a modified form of electroporation. *J. Neurosci. Methods* **142**, 137–143.

Lee, H., et al., (2004). The microtubule plus end tracking protein orbit/MAST/CLASP acts downstream of the tyrosine kinase abl in mediating axon guidance. *Neuron* **42**, 913–926.

Lin, A. C., and Holt, C. E. (2008). Function and regulation of local axonal translation. *Curr. Opin. Neurobiol.* **18**, 60–68.

Lise, M. F., et al., (2006). Involvement of myosin Vb in glutamate receptor trafficking. *J. Biol. Chem.* **281**, 3669–3678.

Lowery, L. A., and Van Vactor, D. (2009). The trip of the tip: understanding the growth cone machinery. *Nat. Rev. Mol. Cell Biol.* **10**, 332–343.

Luscher, C., et al., (1999). Role of AMPA receptor cycling in synaptic transmission and plasticity. *Neuron* **24**, 649–658.

Martinez-Lopez, M. J., et al., (2005). Mouse neuron navigator 1, a novel microtubule-associated protein involved in neuronal migration. *Mol. Cell. Neurosci.* **28**, 599–612.

Morrison, E. E., et al., (2002). EB1 identifies sites of microtubule polymerisation during neurite development. *Brain Res. Mol. Brain Res.* **98**, 145–152.

Nakano, A. (2002). Spinning-disk confocal microscopy—a cutting-edge tool for imaging of membrane traffic. *Cell Struct. Funct.* **27**, 349–355.

Nakata, T., and Hirokawa, N. (2003). Microtubules provide directional cues for polarized axonal transport through interaction with kinesin motor head. *J. Cell Biol.* **162**, 1045–1055.

Newpher, T. M., and Ehlers, M. D. (2008). Glutamate receptor dynamics in dendritic microdomains. *Neuron* **58**, 472–497.

Sala, C., et al., (2001). Regulation of dendritic spine morphology and synaptic function by shank and homer. *Neuron* **31**, 115–130.

Selkoe, D. J. (2002). Alzheimer's disease is a synaptic failure. *Science* **298**, 789–791.

Sheng, M., and Hoogenraad, C. C. (2007). The postsynaptic architecture of excitatory synapses: A more quantitative view. *Annu. Rev. Biochem.* **76**, 823–847.

Shi, S. H., et al., (2003). Hippocampal neuronal polarity specified by spatially localized mPar3/mPar6 and PI 3 kinase activity. *Cell* **112**, 63–75.

Steinmetz, M. O., and Akhmanova, A. (2008). Capturing protein tails by CAP-gly domains. *Trends Biochem. Sci.* **33**, 535–545.

Stepanova, T., et al., (2003). Visualization of microtubule growth in cultured neurons via the use of EB3-GFP (end-binding protein 3-green fluorescent protein. *J. Neurosci.* **23**, 2655–2664.

Sudhof, T. C. (2008). Neuroligins and neurexins link synaptic function to cognitive disease. *Nature* **455**, 903–911.

Sudhof, T. C., and Rothman, J. E. (2009). Membrane fusion: Grappling with SNARE and SM proteins. *Science* **323**, 474–477.

Tsai, J. W., et al., (2005). LIS1 RNA interference blocks neural stem cell division, morphogenesis, and motility at multiple stages. *J. Cell Biol.* **170**, 935–945.

Yuste, R., and Bonhoeffer, T. (2001). Morphological changes in dendritic spines associated with long-term synaptic plasticity. *Annu. Rev. Neurosci.* **24**, 1071–1089.

Zhou, F. Q., et al., (2004). NGF-induced axon growth is mediated by localized inactivation of GSK-3beta and functions of the microtubule plus end binding protein APC. *Neuron* **42**, 897–912.

CHAPTER 8

Protein Micropatterns: A Direct Printing Protocol Using Deep UVs

Ammar Azioune*, Nicolas Carpi*, Qingzong Tseng†, Manuel Théry†, *and* Matthieu Piel*

*Systems Cell Biology of Cell Division and Cell Polarity, UMR144, Institut Curie, CNRS, Paris 75248, France

†Laboratoire de Physiologie Cellulaire et Végétale, iRTSV, CEA/CNRS/UJF/INRA, 38054 Grenoble, France

Abstract

The described protocol is a simple method to make protein micropatterns with a micron size resolution. It can be applied to control cell shape and adhesive geometry, and also for any other assay requiring protein patterning. It is based on the use of

978-0-12-381349-7
DOI: 10.1016/S0091-679X(10)97008-8

a photomask with microfeatures to locally irradiate with deep UV light (below 200 nm) an antifouling substrate, making it locally adsorbing for proteins. The entire process can be subdivided into three main parts. The first part describes the design of a photomask. The second part describes the passivation (antifouling treatment) of the substrate, its irradiation, and the binding of proteins. The entire process can be completed in a couple of hours. It requires no expensive equipment and can be performed in any biology lab. The last part describes cell deposition on the micropatterned substrate. We also provide a discussion with pitfalls and alternative techniques adapted to various substrates, including silicone elastomers.

I. Introduction

Microfabrication techniques applied to cell biology already have a rather long history [see Folch and Toner (2000) and Whitesides *et al.* (2001) for reviews of many micropatterning techniques developed from the 1970s to the 1990s]. The recent development of biological applications (from cell biology, tissue engineering, cell cocultures, bio-assays, bio-sensors, etc.) led to a huge burst of technical papers in the last 10 years, providing adaptation of micropatterning techniques to various substrates (glass, plastics, hydrogels, elastomers, etc.), molecules, and cell types, in two dimensions (2D) and in three dimensions. This wealth of information is often difficult to deal with when trying to choose the right method, as there is a multitude of alternative techniques. Four main processes are dominating the field: (1) photolithography and liftoff (and other stencil types of methods), (2) microcontact printing, (3) UV-based chemistry, and (4) laser/electron beam etching (as well as other micro/nanoprinting techniques).

Each method has drawbacks and advantages, and choosing one strongly depends on the application. There is unfortunately no universal solution. When working in a biology lab, being independent of specialized microfabrication facility is an advantage to consider. Two methods are easy to implement for biologists wanting to do simple micropatterning: microcontact printing and UV-based chemistry. The main advantage of microcontact printing is that, once a mold is available to produce the stamps, no special equipment and no special chemistry is needed to produce patterns [see Thery and Piel (2009) and Ostuni *et al.* (2009)]. Here we propose a deep UV (185 nm)-based protocol as an example of a technique well adapted to control cell adhesion geometry and cell shape (patterns of minimal dimensions of a few microns) and easy to implement in a cell biology lab. We present the simplest protocol which worked for most cell types we have tested so far, keeping them confined for several days (but not weeks). We chose PLL-g-PEG as a cell/protein-repellent molecule as it readily binds with strong affinity on glass and is commercially available at low cost [a technique first proposed by Csucs *et al.* (2003)]. For a recent contribution to UV-based technique for 2D surface micropatterning see Azioune *et al.* (2009), with an introduction reviewing the field and referring to papers previously describing similar methods.

II. Designing a Photomask

This section provides a few tips in the design and ordering of a proper photomask.

A. Materials

Software. Several software programs will allow design of mask features, depending on the requirements of the mask producing company. In the simplest case, drawing software with indication of sizes can be used, and the mask manufacturer will convert it into a proper file format (this will of course have a cost). A common file format used by manufacturers is GDS II (other formats are CIF and DXF). Any software which can produce such a file will work. Some software are specifically meant to design masks, they are not only very convenient but also often expansive, and will require a short learning phase (e.g., L-Edit, Clewin, AutoCAD).

Photomask. For deep UV irradiation, it is important to have a proper type of photomask, transparent to wavelengths below 200 nm. The material used for such masks is usually called fused silica, or synthetic quartz. It has fewer defects than natural quartz and has a better transparency to short wavelength. It is also more expensive and is not the basic material proposed by photomask producers, so it has to be specified. Many companies produce photomasks for the microelectronics industry, not all of them propose fused silica photomasks. Another important parameter to check is the resolution provided by the company. Examples of companies to which we have ordered photomasks which worked with deep UV are Delta Mask (The Netherlands), Toppan photomasks (present in many countries), and Microtronics Photomasks (USA).

B. Designing Features

Size limitations. Size of features is limited by two factors: the resolution of the photomask, which can go down to a fraction of microns for the most expensive ones and will be around 1 μm for regular ones. The second factor is the quality of the contact between the substrate and the photomask (see part II for more details). Features of 1 μm are possible to obtain with care, and features of a few microns are easy to obtain.

Single-cell micropatterns (see an example in figure 1). Designing patterns for single cells will depend on cell type: cells need enough space to spread [some would die if they do not have enough space, see Chen *et al.* (1997)], and if they have too much space, they will move around and lose their stereotyped morphology. Single-cell patterns usually range between 300 and 2000 μm^2 for mammalian cells (but some cells might need larger patterns, a good estimate can be found by looking at the spreading area of cells on regular nonpatterned adhesive substrates). A second important feature is the distance between patterns. Here again, it will depend on cell types. Some cells are highly mobile and protrusive and able to bridge large gaps (for example, fibroblasts), other will not, allowing higher density of individual micropatterns on the substrate. A distance of 100 μm will prevent most cells from going from one pattern to the next, but it can be lowered down to 50 μm, for example, for HeLa cells. Allowing cells to jump from one pattern to another can also produce interesting results (if the pattern size is well set, cells will jump only when they are big enough, like in G2, or when there are two cells on the pattern, in telophase).

Lines of various widths are interesting to study cell migration (Doyle *et al.*, 2009; Pouthas *et al.*, 2008), they can also be used to impose an axis for cell division without constraining cells on single-cell patterns, on which only a single round of division can be studied. Line width has been reported to affect cell speed and cell morphology: polarization effects occur for lines up to 15 μm in width, strong morphological and speed transition have been reported around 4 μm. Lines are easy to use as cells can be platted on the patterns and kept for several days. The distance between lines will allow or prevent cells from passing from one line to the other (spacing above 50 μm should avoid most cells from binding on two lines).

Larger features can be used simply to keep motile cells from leaving the field of observation during long-term time-lapse recording. This will allow tracking of multiple sequential divisions. The size of the feature has to be adapted to the field of view (for example, about 500 μm diameter disks will keep cells within the field of a 10× objective when using a camera with about one million pixels of around 6 μm, a standard size for camera chips).

It can be useful to include grids around patterns. Patterns can be grouped in squares of size equivalent to the most common field of view used in the experiments planned (for example, again using a 10× objective). This will help when scanning to select fields to be recorded. If no grid is present, it can be difficult to navigate through the coverslip, due to the repetition of identical patterns on a regular array. The grid width will of course bind cells; these cells will not be patterned, but they can serve as controls. We found that it is better to draw discontinuous lines if the mask is to be used for the protocol described in this article, as closed grids might trap air bubbles, introducing defects. Numbers and letters can be added on the grid for easier localization of cells on coverslips, for correlative microscopy.

General organization of the photomask. A photomask is usually much larger than a single coverslip. A regular size is 5 × 5 in. This allows production of several coverslips at once, or to irradiate a large surface with a single type of patterns. But it can also be used to test several types of micropattern. It is then useful to leave a large line, visible by eye, between the regions with different features, and a mark to orient the mask, so finding the region of interest can be performed easily, without a microscope. When starting with micropatterns, it is often useful to first order a mask with many small regions containing all the ideas and parameters to test. Then, when it is clear which set of patterns is most useful, order a second mask with only a few types of patterns to produce several patterned coverslips at once.

III. Micropatterned Substrate Fabrication

Here we present patterning on glass coverslips. The exact same protocol can be applied to cell culture polystyrene (PS) substrates. Such substrates usually work better but are less suited for fluorescence live cell imaging at high magnification.

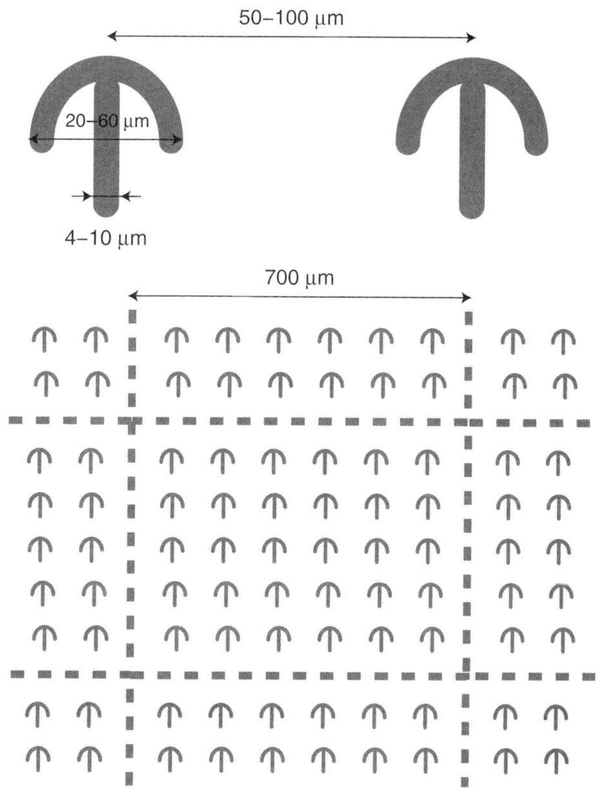

Fig. 1 Designing features. This scheme illustrates the basic geometrical parameters that has to be satisfied to ensure proper cell spreading on micropatterns (line width and micropattern size) and absence of cell spreading over two adjacent patterns (array step).

A compromise can be found by coating glass coverslips with a thin layer of PS (for a detailed method, refer to Thery and Piel (2009).

A. Materials

- Glass coverslip.
- MilliQ water.
- Phosphate-buffered saline (PBS).
- 10 mM HEPES buffer, pH 7.4. We noticed that problems with our protocol often came from bad buffers. Making new buffers is the first thing to try if the protocol works poorly.
- 100 mM $NaHCO_3$ buffer, pH 8.5.
- Ethanol 96%.
- PLL-g-PEG (Surface Solutions, Switzerland, ref: PLL(20)-g[3.6]-PEG(2), stock solution at 1 mg/ml in 10 mM HEPES buffer, pH 7.4, stored at + 4°C for several months).

- Fibronectin (Sigma, F1141) or other protein to be patterned. It can be useful to have a fluorescently labeled version of the protein (some are commercial, other can be produced using labeling kits like Invitrogen ref A-10239).
- PS (facultative) (Acros Organics, 178890250).
- Toluene (Sigma Aldrich, 32249).
- TI Prime (MicroChemicals).

B. Equipments

- Plasma cleaner (facultative, only needed if micropatterns are transferred to PS-coated glass coverslips, for example: PDC-32G, Harrick).
- UV ozone oven. Such ovens can be found, usually in large formats, in clean rooms. Smaller ones are often used to clean AFM tips. A small benchtop version, perfectly fit for 5 in photomasks, can be found at Jelight (UVO cleaner, ref. 342-220). It is important to also order the ozone killer, or the oven will have to be placed under a chemical hood. We also recommend buying a fan to avoid overheating. Alternatively, it is possible for a very modest cost to build a homemade deep UV oven. Bulbs are available at Heraeus Noblelight GmH (NIQ 60/35 XL longlife lamp, $\lambda = 185$ and 254 nm, quartz tube, 60 W). Four bulbs are enough. Be careful to order controllers allowing frequent switches of the bulbs (EVG 65-80W). A closed box has to be made containing bulb holders, allowing irradiation at a distance of about 10 cm. It is also recommended to add fans to the box to avoid overheating. Particular care has to be taken to avoid any direct exposition to deep UV light and to get rid of the ozone it produces.
- Vacuum mask holder (facultative). A mask holder can help ensure a better contact between the coverslip and the photomask (see Fig. 2). A homemade design may be obtained from the authors of this article.
- Spin coater (facultative, only needed for polystyrene-coated glass coverslips, for example, Laurell Technologies Corporation, WS-400-6NPP-LITE).

C. Method

1. Surface Preparation

1. The glass coverslip is washed with ethanol. Optionally it can be sonicated in ethanol to optimize dust removal. It is dried with filtered airflow or let dry under the hood.
2. Dried coverslips are exposed to air plasma for 1 min, or oxygen plasma for 10 s at 30 W, or to deep UV at 5 cm of the lamps, for 5 min.
3. Incubate clean coverslips with 0.1 mg/ml of PLL-g-PEG in 10 mM HEPES, pH 7.4, at room temperature (RT) for 1 h. It is not necessary to rinse, in fact not rinsing gives better results, just slowly lift off the coverslip to ensure complete PLL-PEG solution dewetting; if necessary remove the last drop of PLL-g-PEG with kimwipes, dry with airflow, and store at room temperature.

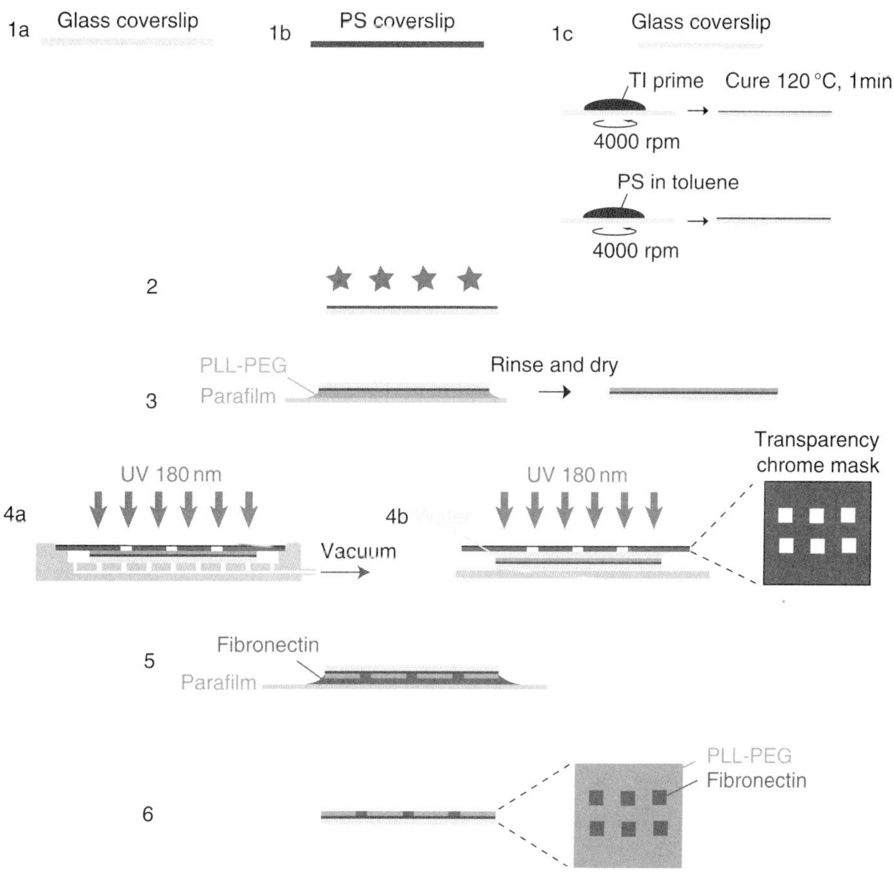

Fig. 2 Micropattern fabrication. This scheme summarizes the sequential steps to proceed through micropattern fabrication. Micropatterning can be performed on various substrates: generally on glass coverslip (1), on PS coverslip (2), or on PS-coated glass coverslip (3).Step 1: 1a. Take a clean glass coverslip and go to step 2,or 1 b: take a clean PS coverslip and go to step 2,or 1c: take a clean glass coverslip. Spin-coat "TI prime" for 30 s at 4000 rpm and cure1 min at 120°C, then spin-coat 0.5% PS in toluene for 30 s at 4000 rpm.Step 2. Oxidize with a plasma cleaner (30 W, 10 s).Step 3. Incubate with PLL-PEG (0.1 mg/ml in HEPES pH = 7.4, 30 min) and wash with MilliQ water.Step 4. 4a Place the coverslip and the chrome mask on a mask holder.or 4b put the coverslip in contact with the chrome mask using a water drop.Then place the sandwich under UV 180 nm (3 min) to oxidize the PLL-PEG under transparent areas. Step 5. Incubate with protein (Fibronectin in $NaHCO_3$ pH = 8.5, 20 µg/ml, 30 min).Step 6. Rinse in $NaHCO_3$ buffer and dry.

At that step the coverslip can be kept for several days, but the best quality is obtained if they are used in the following 24 h.

To ensure a strongest protein and cell adhesion to the substrate, the glass coverslip can be coated with a thin (less than 50 nm) layer of polystyrene (PS).

1. The glass coverslip is washed with ethanol and dried with filtered airflow.
2. Place the glass coverslip on a spin coater, cover with TI PRIME, and spin-coat 30 s at 3000 rpm.
3. Cure the coverslip 1 min at 120°C on a hot plate.
4. Place the glass coverslip on a spin coater, cover with 0.5% of PS in toluene, and spin-coat 30 s at 3000 rpm.

2. Surface Patterning

The following step can be performed in two ways: with a mask holder or a water drop. The mask holder uses vacuum to ensure a better contact between the coverslip and the mask. The contact is dry and coverslips are easier to remove (follow steps 1a–4a). The water drop is efficient without dedicated equipment (follow steps 1b and 4b).

With vacuum and mask holder

1a. Cautiously clean the photomask with acetone to remove organic residues (such as the one that sometimes result from contact with PS-coated slides) and then with isopropanol extensively to remove inorganic residues and acetone traces. If there is no organic residues on the mask, cleaning by isopropanol only will be sufficient and better. Dry with filtered airflow.

2a. Place the coverslip on the vacuum holder with the pegylated side in contact with the chrome-coated side of the photomask on the mask holder. Open vacuum to ensure intimate contact between the coverslip and the mask.

3a. Expose the mask-covered substrate to deep UV light for 3 min, at about 5 cm from the lamp.

4a. To remove the coverslip from the photomask after step 2a, use a 1 ml plastic micropipette tip and a plastic tweezer or better the vacuum suction to lift up the coverslip. The coverslip should detach the mask very easily if the mask has been cleaned with isopropanol only.

With water drop

1b. Place the photomask under deep UV for 5 min to make it more hydrophilic.

2b. Place the pegylated side of the coverslip in contact with the chrome-coated side of the photomask with a drop of water (the volume has to be adapted to the substrate size to allow complete coverage of the surface by water, but still keep close contact. A volume calculated to provide a spacing of about 5 μm is recommended). For 25 mm coverslips, use a drop of 1.5 μl of water on the mask. For 12 mm coverslips, use a drop of 0.5 μl of water on the mask. The formation of air bubbles between the mask and the substrate must be prevented.

3b. Expose the mask-covered substrate to deep UV light for 3 min, at about 5 cm from the lamp.

4b. To remove the coverslip from the photomask add water around it and wait until it is lifted by water. The lifetime of your mask depends of the way you are removing your coverslips. Use plastic tweezers, never a metallic one, or put a plastic micropipette tip on the tweezers when moving the coverslip over the surface of the mask. The best is to add enough water so the coverslip floats on it and can be taken without touching the mask with the tweezers.

Dried UV-patterned substrate can be kept on the bench in the lab atmosphere for a few months, but best results are obtained if used within the next few days. If they were kept dried, before using them, rehydrate them for 30 min in water or in PBS, so the PEG chains are well swollen before you incubate cells or proteins.

5. Incubate patterned substrates with a mixture of fibronectin and fibronectin-Alexa fluor 488 nm (Invitrogen) in 100 mM $NaHCO_3$, pH 8.5, at RT for 1 h, at a concentration of 25 and 10 μg/ml, respectively (note that if the surface used was PS or PDMS, the patterns are visible in phase contrast). You can use any protein, just note that for some proteins that tend to form films (like collagen or fibrinogen) not more than 10 μg/ml should be used or films could form over the PEG surface. It is important to incubate the proteins in a basic buffer, because proteins will then covalently bind to the patterned areas, which contain carboxyl groups. This reaction can by activated using a mix of 1-Ethyl-3-[3-dimethylaminopropyl]carbodiimide hydrochloride (EDC or EDAC) and Sulfo-NHS (N-hydroxysulfosuccinimide) (see Section V). To avoid using too much protein, the protein solution can be placed on a piece of parafilm and then covered by the coverslip.
6. Wash twice with PBS. At this step the substrate should not be dried. It can be kept overnight at 4°C in PBS but better results are obtained if used readily to plate cells.

IV. Cell Deposition

This part is identical to the similar part in Thery and Piel (2009). The described micropatterned substrates have been used successfully with the following cells: HeLa, RPE1, MCF10A, MCF7, NIH3T3, HepaRG, MDCK, and human mesenchymal stem cells, as well as mice bone marrow-derived dendritic cells.

A. Materials

1. Reagents

PBS (Invitrogen/Gibco, 14040-091)
Trypsin (0.5 g/l)-EDTA(0.2 g/l) (Invitrogen/Gibco, 25300-054)
DMEM or DMEM-F12 (Invitrogen/Gibco, 31331-028)
Fetal bovine serum (Dominique Dutscher, ref 500105)
Penicillin-streptomycin (Invitrogen/Gibco, 15140-122)
Flasks for cell culture (Sigma, 75 cm^2, 430641)
Pipettes

2. Equipments

Laminar Flow hood (Fisher Bioblock Scientific, ref B90649)
Incubator (Heracell 150, Thermo electron corp, ref 51022392)

Microscope (Olympus, CKX41)
Centrifuge (Eppendorf 5702, Dominique Dutscher, ref 033716)

B. Method

1. Adherent cells are washed in PBS and detached from their flask with trypsin-EDTA (for 5–10 min, depending on the dilution used) or Versen EDTA.
2. Complete culture medium (DMEM or DMEM-F12 + 10% SVF + 1% penicillin and streptomycin) is added to the flask and collected cells are centrifugated 3 min at 1500 rpm.
3. Supernatant is removed and cells are resuspended in culture medium at 150,000 cells/ml.
4. Cell solution is added on the micropatterned substrate (glass slide or TCPS dish). The final density should be about 10,000 cells/cm^2. The whole is placed in the incubator.
5. After a given time that varies from one cell line to the other (10–20 min for RPE1 and 20–60 min for HeLa-B) the coverslip is checked under the microscope to confirm that a sufficiently large proportion of cells have attached to the micropatterns.
6. Nonattached cells are removed by gently aspirating the medium with a 1 ml pipetman while simultaneously adding some warm new medium.

   Note: Pay attention not to aspirate all the medium, otherwise the dewetting of the solution due to PEG physicochemical properties could dry the attached cells.
7. Attached cells are placed back in the incubator to let them spread fully (1–5 h depending on cell type).
8. 1 h later cells can be fixed or video recorded.

V. Discussion

The protocol presented in this article is the simplest and yet robust protocol to directly pattern glass substrates with cell adhesion proteins like fibronectin. The success rate is high and the limited number of chemicals and steps involved reduces the potential sources of problems. Moreover, it does not involve any toxic reagent or expensive device.

This protocol can also be modulated to adapt to specific applications. There are mainly four reasons to adapt the protocol:

1. Cells. Some cells are easier to maintain on micropatterns than others. The main problems are cells which exert strong forces on the substrate and can tear off the proteins, and cells which are strongly motile and adhesive and will easily escape to invade the nonadhesive regions [see Fink *et al.* (2007) for a discussion of these problems].

2. Patterned proteins. Some proteins or peptides might react differently with the deep UV-patterned PLL-g-PEG-covered substrate. It is particularly true for small peptides, which might not be repelled properly by PLL-g-PEG, as it leaves small "holes" on the surface, and for charged peptides or proteins and proteins which tend to adsorb on hydrophobic surfaces. Such proteins might show either a poor contrast between UV irradiated and nonirradiated regions or even display an inverse contrast.
3. Timescale of the experiment. For experiments which require keeping cells on the patterns for more than 48 h, it might be important to adapt the protocol to prevent cells from escaping the patterns.
4. Substrate. The method presented can be applied on any substrate which can be made cell repulsive or is naturally cell repulsive, like glass, plastic, and "soft" substrates like silicone rubber or hydrogels.

To successfully adapt the protocol to various cells, proteins, or substrates, there are two parameters to modulate:

1. Antiadhesive surface coating. In general, the success of the method presented here mostly depends on the quality of the passivation of the substrate. PLL-g-PEG is the easiest molecule to use, but it does not provide the best passivation when just adsorbed on bare glass, especially for long-term experiments.
2. Protein binding to the substrate. Just adsorbing proteins on the irradiated substrate is often enough, but in some cases (cells which pull strongly or substrates on which proteins do not adsorb properly), it can be important to covalently bind proteins in a well-controlled way.

At this stage, many options are open: which surface treatment should be used in the cell/protein-repellent regions and which one should be used to optimally bind cell adhesion molecules? Various options include simple adsorption, covalent binding, electrostatic interactions, silanization, and hydrophobic/hydrophilic interactions. Some very simple techniques (direct patterning on bare glass without any backfilling with repellent molecule, for example) will work well with cells which do not bind on bare glass and do not pull too strongly on their adhesion molecules. For other cells, very good repellent molecules should be used and very strong binding of the adhesion molecules to the substrate is required.

Many technical papers have been published in the past few years, proposing various protocols for fabrication of adhesive micropatterns on different substrates and using a variety of methodologies. Most of these studies are performed by bioengineering teams and surface chemists who have both expertise and tools that are often hardly accessible to biologists. As there is no universal solution to produce micropatterns, one has to find a compromise between easiness, reproducibility, and quality of patterning, together with good optical quality of the substrate. Moreover each cell type or culture condition will impose precise constraints. In some difficult cases, it will be necessary to have the patterned substrates produced by physicists or chemists specializing in complicated surface chemistry and microfabrication methods.

A. Discussion of Alternative Methods for Passivation

A discussion of some methods can also be found in Fink *et al.* (2007).

- Short-term (<48 h) confinement of cells on glass substrate: use PLL-g-PEG as in this article.
- Longer term (>48 h) on glass substrate. The easiest improvement of passivation consists of coating the glass coverslip with PS.
- Covalent binding of PLL-g-PEG. Many substrates (PS, silicon rubber, polyacrylamide, PVA, etc.) can be activated by deep UVs to produce oxidized groups on their surfaces (like hydroxyl or carboxyl groups). These can be made reactive with amines with a mix of EDC and NHS (see below for a protocol for PDMS silicon rubber). PLL-g-PEG will then be bound covalently. A similar method can enable binding of proteins covalently, as regions irradiated with deep UV will contain such groups [see Azioune *et al.* (2009)].
- A more dense and efficient coating of glass with PEG molecules can be obtained using a silanization process [see Cuvelier *et al.* (2003), Blümmel *et al.* (2007), and Thery *et al.* (2005)). Such processes usually involve a first step of silanization, leading to glass coating with reactive groups (amines, sulphahydryl, etc.) and then binding of PEG chains with a fitting reactive group (NHS, maleimide, etc.). To get a denser coating, it is possible to use two PEGs with different lengths, a long one (2 kD) and a short one (PEG 4 or 8).
- On hydrophobic substrates like PS, direct deep UV irradiation will create hydrophilic regions. Substrates can then be incubated with Pluronic to prevent cell/protein binding to nonirradiated regions, followed by incubation with proteins which will then bind only in the irradiated regions. Pluronic is a very efficient antifouling molecule and will allow cell confinements for days if not weeks (Tan *et al.*, 2004).

B. Discussion of Alternative Methods for Protein Adsorption and Binding

As mentioned above, it is possible to obtain a covalent binding of proteins using incubation with EDC/NHS, prior to protein incubation.

It is also possible to invert the protocol proposed in this article: binding proteins on the substrate by any covalent method of choice (for example, silanization with amino silane, then activation with glutaraldehyde and reaction with the protein) and then destroy the protein with deep UVs through a photomask. Then backfilling the substrate with a passivation molecule like PLL-g-PEG. This can allow more flexibility in the protein binding process.

It is possible to pattern multiple proteins sequentially. After the first protein is patterned as described in this protocol, a second irradiation with deep UV can be performed, and a second protein can then be bound to the second set of patterns. Contamination of the first patterns with the second protein cannot be completely avoided but can be reduced by saturating the first pattern with bovine serum albumin before starting the second patterning process.

C. Example of an Alternative Protocol for Micropatterning of Silicon Elastomer with Deep UVs

Plating cells on thin silicon elastomer films allows controlled cell stretching. Combined with micropatterns, it is a good tool to study cell response to mechanical stress.

- Wash the PDMS in EtOH 70% (and sonicate if necessary), 10 min RT.
- Dry.
- Activate the PDMS with deep UV for 5 min (5–10 cm from bulb).
- Prepare fresh EDC/sulfo-NHS solution.

For 1 ml of solution (enough to cover about 12 cm^2 of PDMS)

- Weight 11.5 mg of Sulfo-NHS.
- Weight 19.2 mg of EDC.
- Dissolve in buffer 0.05 M MES + 0.5 M NaCl pH 6.0.

This solution cannot be stored; it has to be made freshly each time you prepare passivated PDMS.

- Wash PDMS with H_2O.
- Incubate 15 min at RT with EDC/Sulfo-NHS solution.
- Wash with PBS and H_2O.
- Incubate 3 h at RT (or O/N at RT) with a solution of PLL-g-PEG at 0.5 mg/ml in HEPES 10 mM, pH 8.6.
- Wash with PBS and then H_2O—can be stored at that stage at 4°C up to 1 week.
- Dry the PDMS well. No water should be left between PDMS and photomask.
- Place in close contact with the photomask.
- Illuminate with deep UV through the photomask for 5 min.
- Add H_2O and remove gently from the mask.
- Incubate the PDMS with a solution of 25 μg/ml fibronectin in $NaHCO_3$ 100 mM, pH 8.6, 1 h at RT.
- Rinse with H_2O and PBS.
- Place in cell culture medium
- You can plate the cells on the PDMS.

1. Material

- PDMS can either be made from scratch and casted in order to make a thin flat film, or you can buy it in ready to use form from Gel Pak (PF-60-X4; thickness 150 μm).
- EDC (*N*-(3-Dimethylaminopropyl)-*N*′-ethylcarbodiimide hydrochloride) can be purchased from Sigma (ref 03450).
- Sulfo-NHS (*N*-Hydroxysulfosuccinimide sodium salt) can also be purchased from Sigma (Ref 56485).
- MES (2-(*N*-morpholino)ethanesulfonic acid, 4-morpholineethanesulfonic acid) is bought from Sigma (M3671).

VI. General Conclusions

Successful micropatterning implies that cells show both a good adhesion in the patterns and a long-term confinement. We found this to be more difficult to achieve on single-cell patterns than on large areas and that some cell types are more demanding than others. Nevertheless optimization based only on these parameters leads to techniques which are often too heavy to handle on a daily basis in a biology lab. One has thus to find compromises to minimize specialized devices and techniques and keep a good micropattern quality.

References

Azioune, A., Storch, M., Bornens, M., Théry, M., and Piel, M. (2009). Manuel Théry and Matthieu Piel. Simple and rapid process for single cell micro-patterning. Lab Chip **9**, 1640–1642. DOI: 10.1039/B821581M.

Blümmel, J., Perschmann, N., Aydin, D., Drinjakovic, J., Surrey, T., Lopez-Garcia, M., Kessler, H., and Spatz, J. P. (2007). Protein repellent properties of covalently attached PEG coatings on nanostructured SiO (2)-based interfaces. Biomaterials **28**(32), 4739–4747.

Chen, C. S., Mrksich, M., Huang, S., Whitesides, G. M., and Ingber, D. E. (1997). Geometric control of cell life and death. Science **276**(5317), 1425–1428.

Csucs, G., Michel, R., Lussi, J. W., Textor, M., and Danuser, G. (2003). Microcontact printing of novel co-polymers in combination with proteins for cell-biological applications. Biomaterials **24**(10), 1713–1720.

Cuvelier, D., Rossier, O., Bassereau, P., and Nassoy, P. (2003). Micropatterned "adherent/repellent" glass surfaces for studying the spreading kinetics of individual red blood cells onto protein-decorated substrates. Eur. Biophys. J. **32**(4), 342–354.

Doyle, A. D., Wang, F. W., Matsumoto, K., and Yamada, K. M. (2009). One-dimensional topography underlies three-dimensional fibrillar cell migration. J. Cell Biol. **184**(4), 481–490.

Fink, J., Théry, M., Azioune, A., Dupont, R., Chatelain, F., Bornens, M., and Piel, M. (2007). Comparative study and improvement of current cell micro-patterning techniques. Lab Chip **7**(6), 672–680.

Folch, A., and Toner, M. (2000). Microengineering of cellular interactions. Annu. Rev. Biomed. Eng. **2**, 227–256.

Ostuni, E., Whitesides, G.M., Ingber, D.E., and Chen, C.S. (2009). Using self-assembled monolayers to pattern ECM proteins and cells on substrates. Methods Mol. Biol. **522**, 183–194.

Pouthas, F., Girard, P., Lecaudey, V., Ly, T.B., Gilmour, D., Boulin, C., Pepperkok, R., and Reynaud, E. G. (2008). In migrating cells, the Golgi complex and the position of the centrosome depend on geometrical constraints of the substratum. J. Cell Sci. **121**(Pt 14), 2406–2414.

Tan, J. L., Liu, W., Nelson, C. M., Raghavan, S., and Chen, C. S. (2004). Simple approach to micropattern cells on common culture substrates by tuning substrate wettability. Tissue Eng. **10**(5–6), 865–872.

Théry, M., Piel, M. (2009). Adhesive micropatterns for cells: a microcontact printing protocol. Cold Spring Harb Protoc. 2009(7):pdb.prot5255.

Théry, M., Racine, V., Pépin, A., Piel, M., Chen, Y., Sibarita, J.B., Bornens, M. (2005). The extracellular matrix guides the orientation of the cell division axis. Nat Cell Biol. **7**(10), 947–953.

Whitesides, G. M., Ostuni, E., Takayama, S., Jiang, X., and Ingber, D. E. (2001). Soft lithography in biology and biochemistry. Annu. Rev. Biomed. Eng. **3**, 335–373.

CHAPTER 9

New and Old Reagents for Fluorescent Protein Tagging of Microtubules in Fission Yeast: Experimental and Critical Evaluation

Hilary A. Snaith, Andreas Anders, Itaru Samejima, *and* Kenneth E. Sawin

Wellcome Trust Centre for Cell Biology, School of Biological Sciences, University of Edinburgh, Edinburgh EH9 3JR, United Kingdom

Abstract

The green fluorescent protein (GFP) has become a mainstay of *in vivo* imaging in many experimental systems. In this chapter, we first discuss and evaluate reagents currently available to image GFP-labeled microtubules in the fission yeast *Schizosaccharomyces pombe*, with particular reference to time-lapse applications. We then describe recent progress in the development of robust monomeric and tandem dimer red fluorescent proteins (RFPs), including mCherry, TagRFP-T, mOrange2, mKate, and tdTomato, and we present data assessing their suitability as tags in *S. pombe*. As part of this analysis, we introduce new PCR tagging cassettes for several RFPs, new pDUAL-based plasmids for RFP-tagging, and new RFP-tubulin strains. These reagents should improve and extend the study of microtubules and microtubule-associated proteins in *S. pombe*.

METHODS IN CELL BIOLOGY, VOL. 97

978-0-12-381349-7
DOI: 10.1016/S0091-679X(10)97009-X

I. Introduction

The green fluorescent protein (GFP) of the jellyfish *Aequorea victoria* was purified in the early 1960s (Shimomura *et al.*, 1962). After the gene encoding GFP was cloned (Prasher *et al.*, 1992), GFP was used as a marker for gene expression (Chalfie *et al.*, 1994) and was quickly adopted as a protein tag by cell biologists, including fission yeast researchers (Nabeshima *et al.*, 1995; Sawin and Nurse, 1996). In this chapter, we discuss and evaluate reagents currently available to image GFP-labeled microtubules in fission yeast, with particular reference to time-lapse applications. We also introduce new tagging cassettes for several novel red fluorescent proteins (RFPs) as well as new RFP-tubulin strains that can be added to the set of tools available for *in vivo* imaging of microtubules and microtubule-associated proteins (MAPs). We will not address basic fission yeast tagging and growth protocols as these subjects are well-covered elsewhere (e.g., Bähler *et al.*, 1998; Moreno *et al.*, 1991; Sato *et al.*, 2009).

GFP is a 26.9 kD protein consisting of an 11-stranded β-barrel surrounding a coaxial α-helix, with the chromophore, a cyclic derivative of the tripeptide sequence serine-dehydrotyrosine-glycine contained within the α-helix (Cody *et al.*, 1993; Örmo *et al.*, 1996). GFP has become the mainstay of *in vivo* imaging for several reasons, including its bright and photostable fluorescence, low phototoxicity upon prolonged illumination of the fluorophore, and (in most cases) relatively minimal impact on the function of proteins to which it is fused. GFP has been subject to multiple rounds of mutagenesis to improve its spectral characteristics and folding efficiency. The first major improvement in spectral characteristics of GFP was a single point mutation in the chromophore (S65T), which generated GFP with a fluorescence signal six-fold brighter than the original GFP (Heim *et al.*, 1995; Patterson *et al.*, 1997). This mutation also shifted the excitation maximum from 396 to 488 nm, making the fluorophore much more amenable to imaging with standard fluorescein filters. Most of the mutations affecting the spectral properties of GFP are contained within the central α-helix and the contacting β-strands, with mutations affecting folding more widely distributed through the protein (Shaner *et al.*, 2007). Versions of GFP with blue, cyan, and yellow fluorescence are available, and on-going studies continue to further improve the brightness, photostability, and brightness of these and other variants.

Microtubules are highly dynamic, intracellular polymers composed of dimers of α and β tubulin (Mandelkow and Mandelkow, 1985; Nogales *et al.*, 1998, 1999). The fission yeast *Schizosaccharomyces pombe* contains a single β-tubulin isoform encoded by *nda3+* (Hiraoka *et al.*, 1984) and two isoforms of α-tubulin, Nda2 and Atb2 (Toda *et al.*, 1984). The *nda2+* gene is essential, and its level of expression is tightly regulated by the total cellular α-tubulin concentration; the *atb2+ gene* is nonessential and is constitutively expressed (Adachi *et al.*, 1986). Microtubules are organized into relatively simple arrays in *S. pombe*, making it an attractive model for studying microtubule dynamics (Sawin and Tran, 2006). Fission yeast cells contain three to five bundles of antiparallel microtubules, which align with the long axis of the cell (Drummond and Cross, 2000; Marks *et al.*, 1986; Tran *et al.*, 2001). Two or three independently regulated microtubules are present within each bundle (Hoog *et al.*, 2007; Sagolla *et al.*, 2003).

Microtubules are nucleated from specific sites in the cell called microtubule organizing centers and, once nucleated, rapidly become bundled at their slow-growing, minus ends by the microtubule bundling protein Ase1 (Piel and Tran, 2009). The rapidly growing, plus ends exhibit behavior known as "dynamic instability," in which individual microtubules stochastically switch between periods of growth and shrinkage (Mitchison and Kirschner, 1984). Regulation of microtubule dynamics is complex and involves many factors, including the local concentration of tubulin dimers, cell cycle position, and the concerted activity of a host of MAPs. Much attention has focused on the group of proteins associated with the microtubules plus ends, the +TIPs (reviewed by Akhmanova and Steinmetz, 2008) including EB1, CLIP170, and the fission yeast protein Tea1.

The dynamic growth pattern of microtubules allows cells to respond to constantly changing cellular requirements by remodeling microtubule arrays. The most significant alteration in microtubule organization occurs at cell division, when the cytoplasmic microtubules depolymerize and an intranuclear mitotic spindle is formed. In fission yeast the spindle is nucleated from the nucleoplasmic face of the spindle pole bodies (SPB) as cells enter prophase. It remains at constant length while the chromosome become bioriented on the metaphase plate, and then rapidly elongates once the cells enter anaphase (Mallavarapu *et al.*, 1999; Nabeshima *et al.*, 1998; Tatebe *et al.*, 2001). While cells are in prometaphase, the SPB also nucleates highly dynamic short nuclear microtubules in addition to kinetochore microtubules (Sagolla *et al.*, 2003; Zimmerman *et al.*, 2004). Once the cell initiates spindle elongation in anaphase, astral microtubules are nucleated from the cytoplasmic side of the SPB. As the cell completes anaphase, microtubule organization changes again and a postanaphase array (PAA) of microtubules is nucleated from a novel equatorial microtubule organizing centre located on the division plane (Hagan, 1998).

II. Which GFP-Tubulin Should I Use?

This is what everyone reading this article *really* wants to know. However, before discussing in detail what versions of GFP-tubulin may be most appropriate for physiological imaging, it is important to consider some general criteria for what makes a "good" GFP-tubulin, as these have been satisfied by different GFP-tubulin expression systems with varying degrees of success. These criteria apply equally to tubulin fused to RFPs (discussed further below):

1. GFP-tubulin expression should not strongly alter the total amount of tubulin in the cell.
2. GFP-tubulin expression should not significantly perturb microtubule nucleation, dynamics, or function.
3. GFP-tubulin expression must be sufficiently high to allow useful imaging.
4. GFP-tubulin expression should be uniform in cells wherever possible.

Although some early studies of *in vivo* microtubule dynamics in fission yeast used GFP-tagged versions of Nda2 (Drummond and Cross, 2000), essentially all investigators currently use Atb2 tagged at its amino-terminus with GFP, and we will therefore restrict discussion to the different versions of GFP-Atb2 available. Initial experiments used a plasmid-expressing GFP-Atb2 (Ding *et al.*, 1998; Mallavarapu *et al.*, 1999) under control of the repressed form of the thiamine-regulated *nmt1* promoter (Maundrell, 1990). Plasmids were also constructed to allow expression of CFP-Atb2 under *nmt1* regulation (Glynn *et al.*, 2001). However, episomal expression is far from ideal, due to cell-to-cell variation in plasmid copy number, with concomitant effects on microtubule dynamics: it should now be considered obsolete, replaced by use of integrated forms of GFP-Atb2.

Two different approaches have been taken to generate integrated GFP-Atb2 strains. In one approach, PCR-based gene targeting (Bähler *et al.*, 1998) has been used to introduce a GFP cassette driven by the *nmt1* promoter (or its weaker variants *nmt41* and *nmt81*) at the amino-terminus of Atb2 at the endogenous *atb2* locus, creating a strain in which all of the Atb2 in the cell is GFP-Atb2 (Garcia *et al.*, 2001). Early imaging experiments used the repressed *nmt1* promoter or the induced *nmt41* promoter (Garcia *et al.*, 2002; Snaith and Sawin, 2003). However, with most current imaging systems, GFP-Atb2 expression driven by the weakest *nmt* promoter, *nmt81*, is sufficient for imaging microtubules (Grallert *et al.*, 2006; Sawin *et al.*, 2004). In the other approach, plasmids containing GFP-Atb2 driven by a promoter of choice have been integrated at extragenic (i.e., non-*atb2*) loci, preserving expression of endogenous untagged Atb2. Examples of GFP-Atb2 expression at extragenic loci include integration at the *ars1* locus, driven by the *nmt81* promoter or by a weakened (but uncharacterized) *nmt1* promoter (Anders *et al.*, 2006; Sawin *et al.*, 2004), and integration at the *leu1* locus, driven by the *SV40* early promoter (Bratman and Chang, 2007; Jones *et al.*, 1988; Pardo and Nurse, 2005).

Are there reasons to prefer one GFP-Atb2 strain over another? Because the coding sequences of the various versions of GFP-Atb2 are essentially identical (see discussion concerning linker sequences, below), the two main features distinguishing different GFP-Atb2 strains are the expression of GFP-Atb2 relative to Nda2 and whether or not untagged Atb2 is also present. We used quantitative fluorescence-based immunoblotting to measure the levels of GFP-Atb2 expression in several commonly used strains, both in the presence and in the absence of thiamine (using YE5S and EMM2 media, respectively; Fig. 1A). Cells were grown at both 25°C and 36°C to identify any possible effects of temperature on promoter activity, which is often ignored in the analysis of temperature-sensitive mutants. The graphs in Fig. 1B and C show levels of GFP- or mCherry-Atb2, untagged Atb2, and Nda2, as a percentage of the total α-tubulin in each strain.

In cells expressing *SV40:*GFP-Atb2 at the *leu1* locus in EMM2 at 25°C, the tagged Atb2 was 1.3% of the total cellular α-tubulin. At 36°C the levels of GFP-Atb2 increased four-fold, demonstrating an unexpected temperature-dependent activity for the *SV40* promoter. A similar increase in *SV40* promoter activity upon temperature-shift was detected in YE5S as well as in EMM2 plus thiamine (data not shown). Unlike the *SV40* promoter, the *nmt1* promoter and its variants did not exhibit temperature dependence. When induced in EMM2 at either 25 or 36°C, levels of *nmt81*:GFP-Atb2 at the *atb2* locus were approximately 20% of total α-tubulin and levels of *nmt41:* GFP-Atb2 approximately 64% of total α-tubulin. The amount of GFP-Atb2 produced

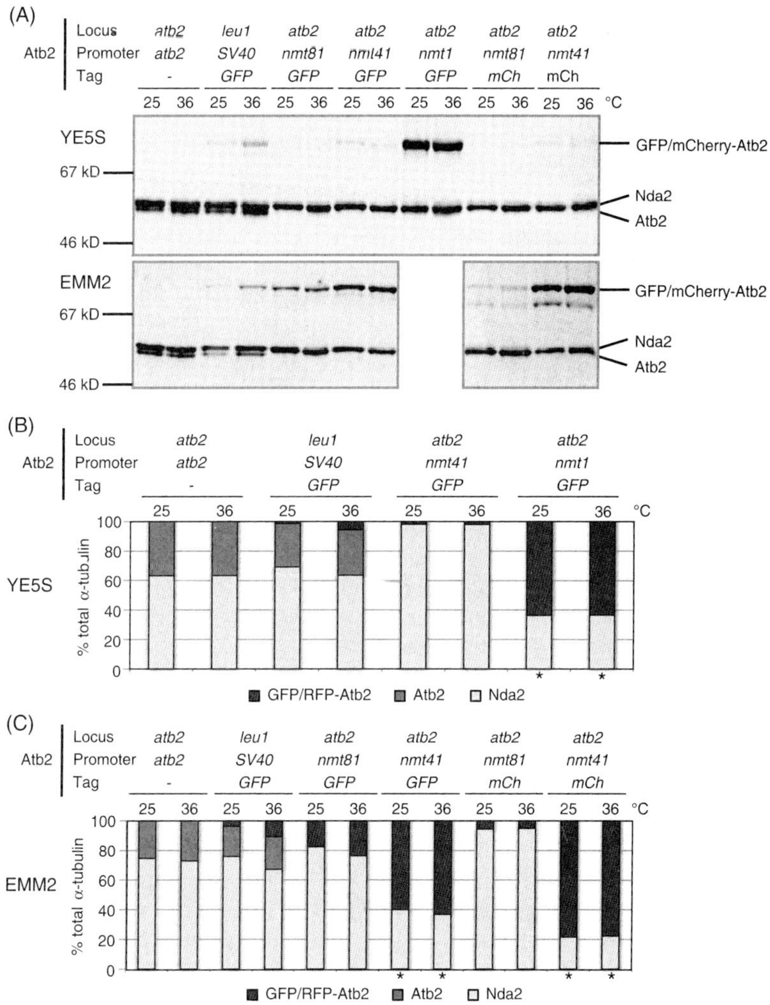

Fig. 1 Expression levels of different GFP-Atb2 and RFP-Atb2 strains. In all yeast strains discussed here, GFP contains the (S65T) mutation (Heim *et al.*, 1995). (A) Wild-type cells (strain KS515) and strains expressing either GFP-Atb2 at the *leu1* locus (*SV40:*GFP-Atb2, KS4956) or GFP-Atb2 or RFP-Atb2 at the *atb2* locus (*nmt81:*GFP-Atb2, KS1235; *nmt41:*GFP-Atb2, KS1231; *nmt1:*GFP-Atb2, KS261; *nmt81:*mCherry-Atb2, KS2789; *nmt41:* mCherry-Atb2, KS2790) were grown in the presence (YE5S, rich medium) or absence of thiamine (EMM2, minimal medium) at either 25°C for 48 h or at 25°C for 44 h followed by temperature shift to 36°C for 4 h. Total protein extracts were prepared by boiling cell pellets for 5 min and vortexing with glass beads (Moreno *et al.*, 1991). Samples were separated by SDS-PAGE, and α-tubulin (Nda2 and Atb2) levels were detected by a mouse α-TAT1 immunoblot (Woods *et al.*, 1989) with IRDye800 donkey antimouse secondary antibody (*LI-COR* Biosciences). Signals were quantitated with *Odyssey* Infrared Imaging System and software (*LI-COR* Biosciences). (B and C) Proportion of GFP- or mCherry-Atb2, untagged Atb2, and Nda2 present in each strain grown in YE5S (B) or EMM2 (C) as a percentage of total α-tubulin. Asterisks indicate samples in which the total levels of α-tubulin were increased approximately two-fold over wild type; the total α-tubulin concentration in all other samples remained similar to wild type. Full genotypes of all strains are given in Table III.

by the induced *nmt1* promoter was lethal (data not shown), and even when repressed in YE5S, the proportion of *nmt1*:GFP-Atb2 was nevertheless about 60% of total α-tubulin at both 25°C and 36°C. Figure 1B and C also emphasize the retention of untagged Atb2 when GFP-Atb2 was expressed at the *leu1* locus, and the loss of untagged Atb2 in strains expressing GFP-Atb2 at the *atb2* locus.

The overexpression of GFP-Atb2 from the repressed *nmt1* promoter or the induced *nmt41* promoter resulted in an approximately two-fold increase in the overall levels of total α-tubulin (Fig. 1A, data not shown). The increase in GFP-Atb2 was accompanied by a decrease in Nda2 as the cells attempted to maintain α-tubulin homeostasis. The same down-regulation of Nda2 was also observed when *nmt41*:mCherry-Atb2 was expressed at the *atb2* locus (Fig. 1C, 6H. 6I).

How do these differences in expression and/or the presence or absence of untagged Atb2 affect usability? Evidence from GFP-Atb2 expressed at the *atb2* locus suggests that too much GFP-Atb2 adversely affects microtubule behavior even when not lethal to cells. For example, astral microtubules in mitosis are not very common with *nmt41*:GFP-Atb2 but are readily observed with *nmt81*:GFP-Atb2 (Fig. 2A; Samejima *et al.*, 2005). Similar differences between *nmt41*:GFP-atb2 and *nmt81*:GFP-Atb2 have been observed in interphase microtubules in mutant strains such as *mto1*Δ and *mto2*Δ, both of which have defects in cytoplasmic microtubule nucleation (Samejima *et al.*, 2005; Sawin *et al.*, 2004). Thus, among GFP-Atb2s expressed from the *atb2* locus, *nmt81*:GFP-Atb2 is clearly to be preferred. We and others have used it extensively (Alvarez-Tabares *et al.*, 2007; Grallert *et al.*, 2006; Kerres *et al.*, 2007; Robertson and Hagan, 2008; Rosenberg *et al.*, 2006; Sawin *et al.*, 2004), and further, we have not found differences between it and strains expressing *nmt81*:GFP-Atb2 from the *ars1* locus (in which untagged Atb2 is also present; see below; Anders *et al.*, 2006). Accordingly, Grallert *et al.* (2006) used fixed-cell immunofluorescence to compare microtubules in wild-type cells with those expressing different GFP-Atb2 constructs and found that microtubules in *nmt81*:GFP-Atb2 strains were indistinguishable from wild type.

However, even though *nmt81*:GFP-Atb2 expressed at the *atb2* locus is similar (or slightly lower) in levels to untagged Atb2, there is some evidence that it may not fully substitute for untagged Atb2. Bratman and Chang (2007) reported that anaphase spindles sometimes bend severely in *nmt81:GFP*-Atb2 expressing cells, resulting in premature spindle disassembly. Moreover, Garcia *et al.* (2001) found that *nmt81*:GFP-Atb2 expressed at the *atb2* locus worsened the phenotype of a temperature-sensitive mutant (*alp14-1270*) of TOG/XMAP215 homologue Alp14, while Kerres *et al.* (2007) found that it partially rescued the phenotype of temperature-sensitive mutants in the Spc105/KNL-1 family kinetochore protein Spc7. These effects could result alternatively from the absence of untagged Atb2 (it is still unknown whether nonessential Atb2 nevertheless has a unique biological function, not shared by Nda2) or from the ratio of GFP-Atb2 to total untagged α-tubulin or from a combination of the two.

Given the possibility that *nmt81*:GFP-Atb2 expressed at the *atb2* locus may not be completely appropriate in all contexts, Bratman and Chang (2007) suggested that GFP-Atb2 driven by the *SV40* promoter at the *leu1* locus may be a better alternative, as

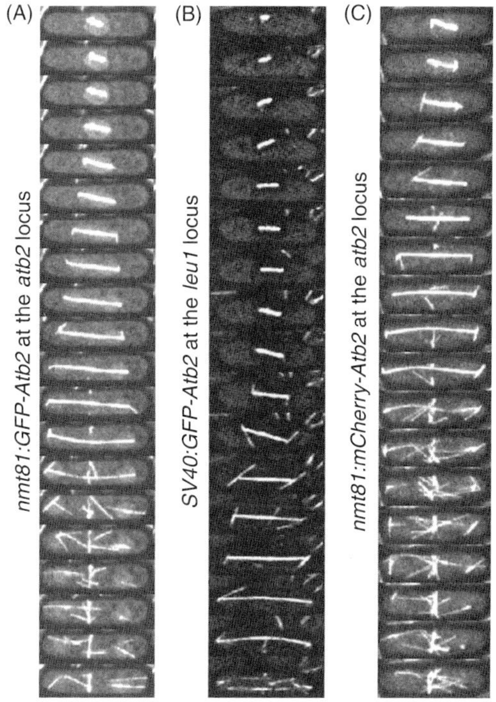

Fig. 2 Mitotic spindles in cells expressing GFP-Atb2 or mCherry-Atb2. Cells expressing *nmt81:*GFP-Atb2 at the *atb2* locus (A), *SV40:*GFP-Atb2 at the *leu1* locus (B), and *nmt81:*mCherry-Atb2 at the *atb2* locus (C). Note the long astral microtubules present in (B). Images in (A) and (C) were acquired using spinning disc confocal illumination. Images in (B) were acquired using wide-field illumination and analyzed by blind deconvolution (AutoQuant Imaging).

untagged Atb2 is present, and the phenotype of *alp14-1270* is not exacerbated in this strain. In many respects this would seem to be the ideal strain for imaging, as the total amount of expressed GFP-Atb2 is very low, and any intrinsic functional defects would be "diluted out" by endogenous Nda2 and Atb2 to the maximum possible extent. In our own experiments we have observed that cytoplasmic astral microtubules during anaphase are particularly long in these cells (recapitulating what is seen by immunofluorescence in untagged cells) compared to *nmt81*:GFP-Atb2 expressed at the *atb2* locus (Fig. 2A and B). However, it should be noted that the low expression (and thus brightness) of the *SV40*:GFP-Atb2 strain may place undesired constraints on imaging protocols, especially for long-term time-lapse imaging, and thus this strain may not be suitable for all applications.

In summary, there is no "perfect" GFP-tubulin, and the choice of which one to use should ultimately be based on practical considerations. Based on first principles, the ideal GFP-Atb2 strain would be one in which an essentially normal complement of untagged Nda2 and Atb2 is "spiked" with a low level of GFP-Atb2 expression, and

SV40:GFP-Atb2 expression at the *leu1* locus comes the closest to satisfying this. However, the very low level of GFP-Atb2 expression and/or issues concerning genetic markers may preclude use in some circumstances. In such instances, three alternatives would be nearly as good. The first two of these, *nmt81*:GFP-Atb2 expressed at the *atb2* locus, and the more recently introduced *atb2*:GFP-Atb2 expressed at the *atb2* locus (Sato *et al.*, 2009), perform well in most cases, although in exceptional cases (and depending on the needs of the investigator and/or referees) it may be important to validate function by comparison with *SV40*:GFP-Atb2 expression at the *leu1* locus. The third alternative, although it has not been as widely used, is *nmt81*:GFP-Atb2 expression at the *ars1* locus, as this produces higher levels of GFP-Atb2 than the *SV40*: GFP-Atb2 strain but leaves endogenous untagged Atb2 similarly undisturbed (Fig. 6).

III. Searching for the "GFP" of RFPs

It is becoming increasingly important to be able to image microtubules with other proteins of interest, using different fluorescent tags. While labeling microtubules with GFP has become commonplace, it has been much more difficult to identify a red fluorescent protein that is the equal of GFP in terms of brightness, photostability/phototoxicity, chromophore maturation time, and effects on protein function. The tetrameric dsRed protein from the mushroom coral *Discosoma striata* was subjected to several rounds of mutagenesis to generate a monomeric form, mRFP1 (Campbell *et al.*, 2002), and further rounds of mutagenesis of mRFP produced several new proteins, including mOrange and mCherry (Shaner *et al.*, 2004). Among these variant RFPs, mCherry has been used the most successfully to label a wide variety of structures, including microtubules, in many different organisms (Shaner *et al.*, 2005). In addition, a different mutant form of dsRed produced a very bright variant termed tdTomato, which is an intramolecular tandem dimer (Shaner *et al.*, 2004). Both mCherry and tdTomato have been used extensively for gene tagging in fission and budding yeast (Snaith *et al.*, 2005). In separate efforts, the RFP eq578, from the sea anemone *Entacmaea quadricolor*, has been subjected to rounds of mutagenesis to generate the monomeric proteins TagRFP and mKate (Merzlyak *et al.*, 2007; Shcherbo *et al.*, 2007). mKate has an emission wavelength maximum significantly longer than many other RFPs (50% of its fluorescence emission is beyond 650 nm), and it has been reported to be extremely photostable. Further mutagenesis on TagRFP has produced another new variant, TagRFP-T, with the same excitation and emission wavelengths as the parent, and with greatly improved photostability (Shaner *et al.*, 2008).

IV. Generation and Evaluation of New RFPs in Fission Yeast

Armed with several of the new RFPs we undertook a critical study to evaluate their properties in fission yeast and determine which might be most suitable for general use and for microtubules in particular. We previously generated tagging cassettes for tdTomato and mCherry (Snaith *et al.*, 2005). As part of the present study we created additional tagging cassettes for mKate, mOrange2, and TagRFP-T (Table I), as their properties suggested that they may be the most useful for *in vivo* time-lapse microscopy. We subcloned cDNAs encoding each of the RFPs into standard *pFA6a* plasmid backbones to allow either C-terminal gene tagging or N-terminal gene tagging, the latter under the control of the strong *nmt1*, or weaker *nmt41* and *81* promoters (Bähler *et al.*, 1998; see Table I). Each of the RFPs contains a seven-amino acid linker at the N- and C-termini identical to the corresponding sequences in GFP, as these were found to improve the function of fusions made with the RFPs (Shaner *et al.*, 2004). To test the expression level, function, brightness, and photostability of the RFPs we attempted to fuse each of them to three cytoskeletal proteins of interest in our laboratory: the cell polarity regulator and microtubule plus-tip-associated protein Tea1 (Mata and Nurse, 1997), the γ-TuRC component Alp4 (Vardy

Table I
Plasmid Templates for PCR Amplification of Integration Cassettes

Plasmid[a]	Lab reference	Tag	Tag position	Promoter	Promoter strength
pFA6a-GFP(S65T)-kanMX6[b,c,g]	–	GFP	C	Endogenous	NA
pFA6a-nmt1-GFP(S65T)-kanMX6[b,c,g]	–	GFP	N	*nmt1*	Strong
pFA6a-nmt41-GFP(S65T)-kanMX6[b,c,g]	–	GFP	N	*nmt41*	Medium
pFA6a-nmt81-GFP(S65T)-kanMX6[b,c,g]	–	GFP	N	*nmt81*	Weak
pFA6a-tdT-kanMX6[b,d]	pKS392/393[e]	tdTomato	C	Endogenous	NA
pFA6a-nmt1-tdT-kanMX6[d]	pKS397	tdTomato	N	*nmt1*	Strong
pFA6a-nmt41-tdT-kanMX6[d]	pKS398	tdTomato	N	*nmt41*	Medium
pFA6a-nmt81-tdT-kanMX6[d]	pKS399	tdTomato	N	*nmt81*	Weak
pFA6a-mCh-kanMX6[b,d]	pKS390/391[e]	mCherry	C	Endogenous	NA
pFA6a-nmt1-mCh-kanMX6[d]	pKS394	mCherry	N	*nmt1*	Strong
pFA6a-nmt41-mCh-kanMX[d]	pKS395	mCherry	N	*nmt41*	Medium
pFA6a-nmt81-mCh-kanMX6[d]	pKS396	mCherry	N	*nmt81*	Weak
pFA6a-TRT-kanMX6[b,f]	pKS965/969[e]	TagRFP-T	C	Endogenous	NA
pFA6a-nmt1-TRT-kanMX6[b,f]	pKS966/970[e]	TagRFP-T	N	*nmt1*	Strong
pFA6a-nmt41-TRT-kanMX6[b,f]	pKS967/971[e]	TagRFP-T	N	*nmt41*	Medium
pFA6a-nmt81-TRT-kanMX6[b,f]	pKS968/972[e]	TagRFP-T	N	*nmt81*	Weak
pFA6a-mO2-kanMX6[b,f]	pKS957/961[e]	mOrange2	C	Endogenous	NA
pFA6a-nmt1-mO2-kanMX6[b,f]	pKS958/962[e]	mOrange2	N	*nmt1*	Strong
pFA6a-nmt41-mO2-kanMX6[b,f]	pKS959/963[e]	mOrange2	N	*nmt41*	Medium
pFA6a-nmt81-mO2-kanMX6[b,f]	pKS960/964[e]	mOrange2	N	*nmt81*	Weak
pFA6a-mK-kanMX6[f]	pKS826	mKate	C	Endogenous	NA
pFA6a-nmt1-mK-kanMX6[f]	pKS827	mKate	N	*nmt1*	Strong
pFA6a-nmt41-mK-kanMX6[f]	pKS828	mKate	N	*nmt41*	Medium
pFA6a-nmt81-mK-kanMX6[f]	pKS829	mKate	N	*nmt81*	Weak

[a] All plasmids allow integration of PCR-amplified cassette at gene-specific locus.
[b] Plasmid also available with *natMX6* cassette to confer resistant to nourseothricin.
[c] Bähler *et al.* (1998).
[d] Snaith *et al.* (2005).
[e] pFA6a plasmid with *natMX6* cassette.
[f] This study.
[g] Van Driessche *et al.* (2005), Sato *et al.* (2005).

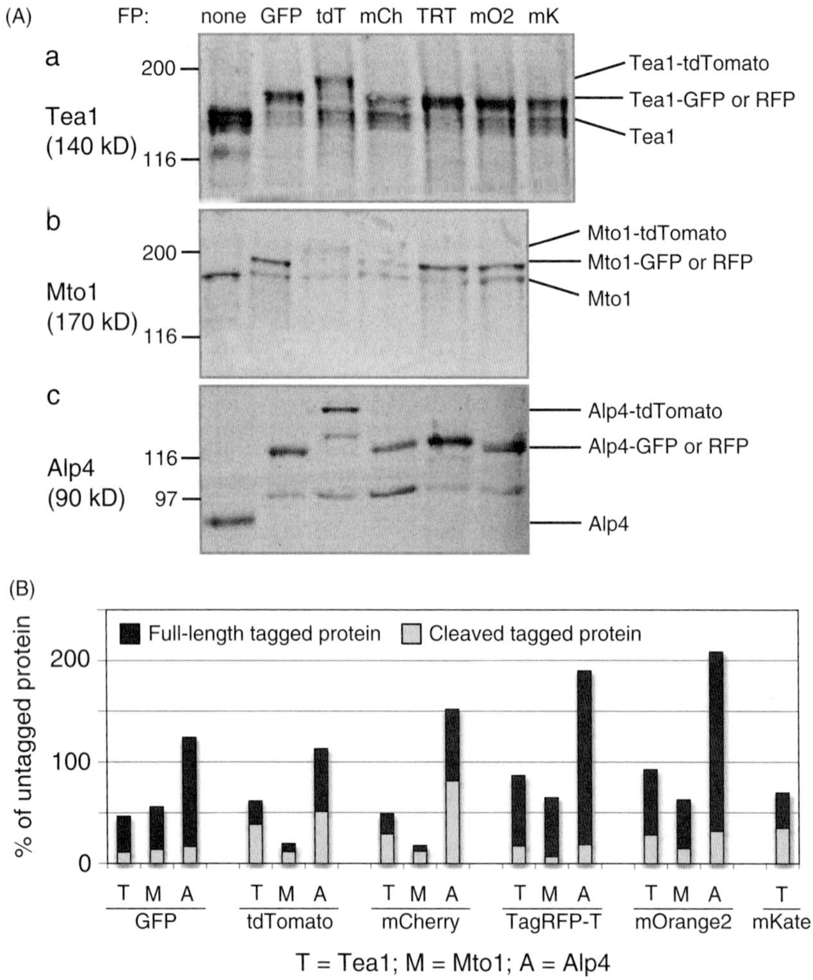

(C)

Activity	Tag / Protein	GFP	tdT	mCh	TRT	mO2	mK
Level	Tea1	47%	62%	50%	87%	92%	70%
Function		+++	+++	+++	+++	+++	+++
Level	Mto1	56%	19%	18%	64%	63%	NA
Function		+++	+/-	+/-	+++	+++	
Level	Alp4	124%	113%	152%	189%	208%	ND
Function		+++	+	ND	+++	+++	-

Fig. 3 (continued)

et al., 2002), and the γ-TuRC-associated protein Mto1 (Sawin *et al.*, 2004). It was relatively straightforward to obtain strains fusing mCherry, tdTomato, TagRFP-T, and mOrange2 to Tea1, Mto1, and Alp4, indicating that these RFPs probably do not have gross adverse effects on the function of the proteins to which they are fused. However, a different situation emerged with mKate; although cells expressing Tea1-Kate were apparently normal, cells expressing Alp4-mKate grew extremely poorly, and it was not possible to isolate any transformants expressing Mto1-mKate. This suggests that mKate could have a perturbing effect on the function of some proteins.

When we analyzed stability of the tagged proteins by Western blotting (Fig. 3), we found that all tagged proteins produced smaller cleavage products that were only marginally larger than the corresponding untagged proteins (Fig. 3A). This suggests that a proportion of each protein may in fact be "untagged" even when the tagged gene replaces the endogenous copy at the genomic locus, and this may contribute to the overall function of the protein. In this context, investigators should be cautious in any interpretations that tagged protein "fully substitutes" for untagged protein. The levels of both full-length tagged protein and total expressed protein (i.e., tagged protein plus cleavage products) were quantified relative to untagged protein in wild-type cells (Fig. 3B and C). The function of the fusions generated was also assessed (Fig. 3C). We found that the different tags had different effects on protein stability, depending on the protein to which they were fused. As might have been predicted from the relatively unperturbing nature of GFP, it had minor effects on the stability of all the proteins we analyzed, with levels of the full-length proteins decreasing to between 30 and 100% of the untagged proteins. Of all the RFPs examined, mOrange2 and TagRFP-T had the smallest negative effects on protein stability, with levels of full-length protein varying between 47 and 176% of the untagged protein. Both TagRFP-T and mOrange2 had minimal effects on the function of the proteins to which they were fused, with all strains appearing phenotypically wild type. Both tdTomato and mCherry had variable

Fig. 3 Stability and function of RFP-fusion proteins with Tea1, Mto1, and Alp4. A new suite of plasmids allowing chromosomal N- and C-terminal tagging with TagRFP-T, mOrange2, and mKate was constructed by replacing the *PacI/AscI* GFP sequence in the *pFA6a-GFP-kanMX6* or *-natMX6*, *pFA6a-nmt1-GFP-kanMX6* or *-natMX6*, *pFA6a-nmt41-GFP-kanMX6* or *-natMX6*, and *pFA6a-nmt81-GFP-kanMX6* or *-natMX6* plasmids with the corresponding TagRFP-T, mOrange2, or mKate sequences. mKate cDNA was PCR-amplified as a *PacI/AscI* fragment from a plasmid template kindly provided by Dmitriy Chudakov (Russian Academy of Sciences, Moscow). TagRFP-T and mOrange2 sequences were codon optimized for expression in *S. pombe* and synthesized by Geneart (Germany). (A) Whole protein extracts were prepared from wild-type cells and from cells in which Tea1 (a), Mto1 (b), or Alp4 (c) were tagged with GFP, tdTomato (tdT), mCherry (mCh), TagRFP-T (TRT), mOrange2 (mO2), and mKate (mK, for Tea1 only). Western blots were probed with antibodies to Tea1 (a), Mto1 (b), or Alp4 (c), and IRDye800 donkey anti-sheep secondary antibody (*LI-COR* Biosciences). (B) Total levels of each protein and levels of uncleaved full-length protein were quantitated from immunoblots in (A) using *Odyssey* Infrared Imaging System and software (*LI-COR* Biosciences). Levels of full-length tagged protein and cleaved tagged protein are presented as a percentage of the endogenous untagged protein present in a wild-type strain. (C) Level and function of each of the Tea1-, Mto1-, and Alp4-RFP fusion strains. Level of combined full-length tagged and cleaved tagged protein is presented as a percentage of the endogenous untagged protein present in wild type. Function of each fusion is shown with +++ for wild type, ± for likely functional but significantly reduced in expression, and − for nonfunctional. NA = not available and ND = not determined. Full genotypes of all strains are given in Table III.

effects, depending on the protein to which they were fused. Whereas Alp4 was relatively stable with either tag, Mto1 and Tea1 were both destabilized by tagging with tdTomato and mCherry, with levels of the full-length tagged proteins decreasing to approximately 10% of the untagged proteins. Cells expressing Mto1-mCherry and Mto1-tdTomato also displayed phenotypes characteristic of *mto1*Δ mutants—curved

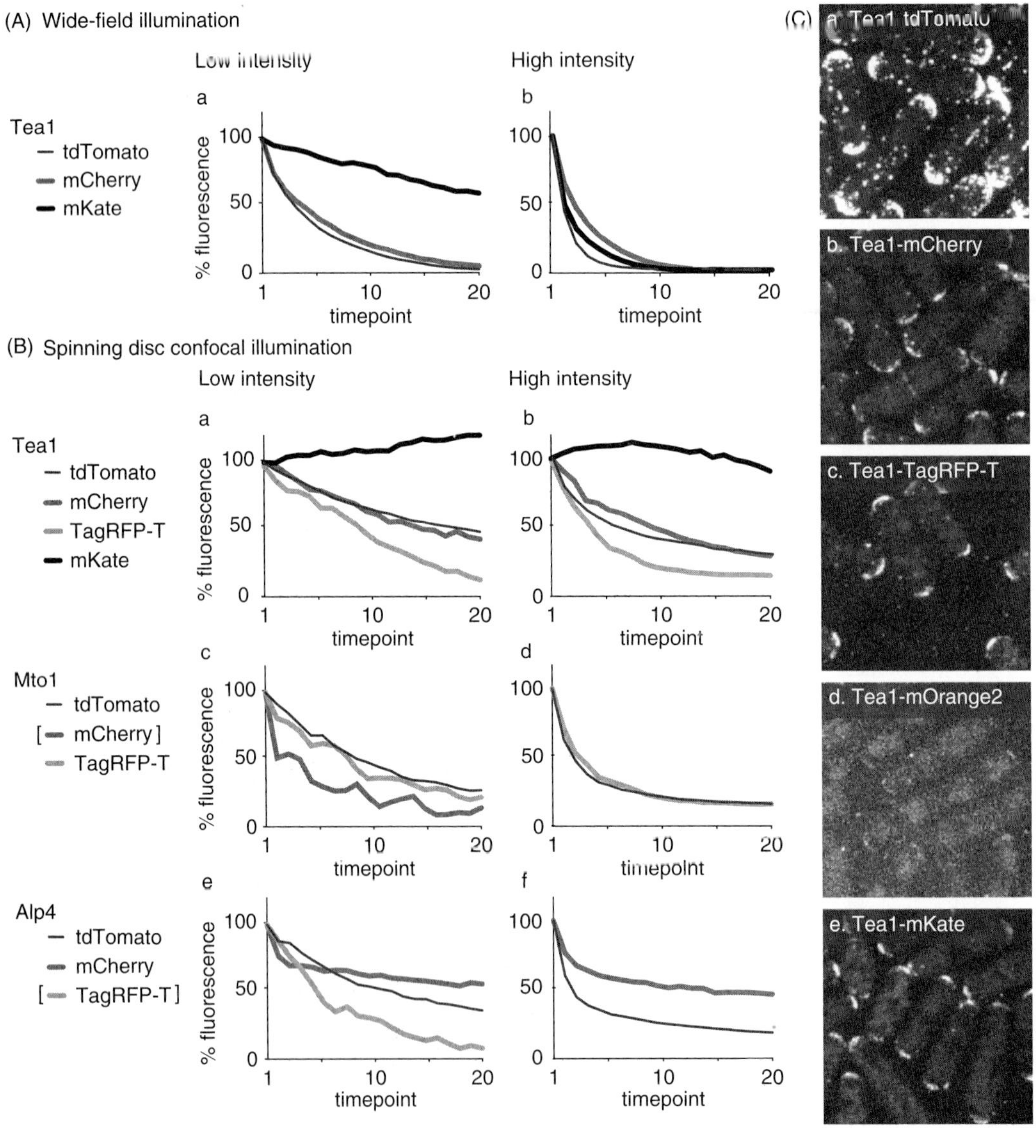

Fig. 4 (continued)

cells, and a reduction in microtubule bundles. This is likely due to the low expression levels rather than impairment of intrinsic function, as we found that increasing expression by heterologous promoters could restore function in related strains (data not shown). Despite the modest effect of tdTomato on Alp4 stability, it is likely that tdTomato-tagging of Alp4 affects function to a small extent, as strains that carry Alp4-tdTomato together with a deletion of the g-TuRC protein Alp16 are temperature sensitive for germination (Samejima *et al.*, 2008). In contrast, although stability was affected by fusion with some of the RFPs, Tea1 function was normal irrespective of the tag used. These data illustrate the difficulties in predicting the effect of a particular RFP on the stability and function of a fusion protein. In addition, they underscore the importance of raising antibodies to a protein of interest, so that effects of tagging on protein stability, and hence function, can be measured properly before use.

To assess the usefulness of each RFP for *in vivo* imaging in fission yeast, we analyzed the photostability of RFP-fusions using wide-field and spinning disc confocal microscopy. We chose low and high levels of illumination intensity and measured the decay in fluorescence signal over a defined time course of imaging (Fig. 4A and B). To make direct comparisons between each RFP we used the same excitation conditions for all the RFPs tested (530–560 nm on the wide-field system and 561 nm on the spinning disc confocal system). In our experimental conditions, each particular RFP displayed similar fluorescent properties independent of the protein to which it was fused. Our data show that where it was possible to obtain a functional fusion with mKate (Tea1-mKate), the resulting protein was the most photostable of the RFPs examined, exhibiting negligible photobleaching even under high-intensity illumination. mCherry also displayed good photostability, exhibiting strong fluorescence even after prolonged illumination. Under our

Fig. 4 Photostability of RFP-tagged proteins. (A) Photobleaching profiles of Tea1-tdTomato (thin dark gray line), Tea1-mCherry (thick medium gray line), and Tea1-mKate (thick black line) at low (a, HBO100 fluorescence excitation attenuated by ND 1.0 filter) and high (b, excitation attenuated by ND 0.6 filter) intensity wide-field illumination with HQ545/30× excitation filter, Q570LP dichroic mirror, and HQ610/75m emission filter (Chroma). Images were acquired using a Nikon TE300 inverted microscope system as described previously (Snaith and Sawin, 2003; Snaith *et al.*, 2005). The complete cell volume was imaged every 10 s for 20 time points in 7 Z-sections at 0.6 μm spacing. Photobleaching was calculated using Metamorph software (Molecular Devices) by measuring the total signal above a defined lower threshold value in average projections of each time point. Each value in a curve was normalized to the total signal present at the first time point. (B) Photobleaching profiles of Tea1- (a, b) Mto1- (c, d), and Alp4-RFPs (e, f) using low (a, c, e)- or high (b, d, f)-intensity spinning disc confocal illumination. Images were acquired with Yokogawa spinning disc confocal system mounted on Nikon TE2000 inverted microscope with an Andor Ixon+ DU888 EMCCD camera, controlled by Metamorph software. The complete cell volume was imaged as in (A). Low-intensity illumination was achieved with a Coherent 15 mW 561 nm laser operated at using 30% power and a 1000 ms exposure time; high-intensity illumination was achieved with the same laser at 100% power and a 300 ms exposure time. Photobleaching profiles were calculated as in (A). RFP fusions are indicated in thin dark gray line (tdTomato), thick medium gray line (mCherry), thick light gray line (TagRFP-T), and thick black line (mKate). It was not possible to collect accurately quantifiable data for Mto1-mCherry or Alp4-TagRFP-T at high-intensity illumination; absent samples are indicated by []. (C) Average projections of spinning disc confocal images of (a) Tea1-tdTomato, (b) Tea1-mCherry, (c) Tea1-TagRFP-T, (d) Tea1-mOrange2, and (e) Tea1-mKate taken under low-intensity illumination, displayed at identical contrast levels.

experimental conditions tdTomato appeared slightly less stable than mCherry, particularly under high-intensity illumination. However, if the intensity of illumination were reduced such that the apparent brightness of tdTomato fusions was comparable to those of mCherry, the photostability of tdTomato would almost certainly be improved. Somewhat disappointingly, TagRFP-T fusions exhibited lower photostability than either mCherry or tdTomato. However, since Tag-RFP-T is relatively nonperturbing to function, it may still be a valuable tool, particularly for proteins that are destabilized by mCherry or tdTomato, such as Mto1. tdTomato was the brightest of all the fluorophores, followed by mCherry, mKate, and TagRFP-T, with mOrange2 being particularly dim when expressed in fission yeast (Fig. 4C). This last observation was surprising, as mOrange2 was reported to be as bright as eGFP (Shaner *et al.*, 2008). Clearly it is important to test the properties of each RFP in different experimental systems.

These results suggest that no single RFP has all the positive attributes of GFP. Although some of the RFPs are very photostable, they can also affect the function of proteins to which they are fused. Conversely, some RFPs that have little effect on protein function can have much poorer fluorescence properties. This means that in many cases it will be necessary to evaluate several different RFP-fusions to determine the optimal tag for one's protein of interest. However, nevertheless some general rules emerge from these experiments. For many proteins, mCherry and tdTomato appear to be the best options. tdTomato is very bright, has low levels of photobleaching, and in many cases has minimal effects on functionality. The brighter signal obtained with tdTomato will be particularly useful for proteins with low expression levels, where detecting a signal above cellular autofluorescence is challenging. However, despite the brightness of tdTomato, its large size means that in many cases it will not be appropriate. In these cases mCherry provides a satisfactory alternative as it is monomeric, reasonably bright and displays robust photostability. It should be remembered that fusion with both mCherry and tdTomato can reduce the stability of some proteins, and in such cases an alternative tag such as TagRFP-T could be considered. Alternatively, a heterologous promoter could be used to boost expression back to wild-type levels. Our experience suggests that mOrange2 is really too faint to be of practical value for *in vivo* imaging in fission yeast. mKate is extremely photostable, making it a very valuable tool in experiments requiring high-intensity or prolonged sample illumination. However, it is the most functionally perturbing of all the RFPs we tested and therefore should be used with care. Very recently an mKate2 protein has been developed, with reportedly improved fluorescent properties and low toxicity in transgenic *Xenopus laevis* embryos. This may be another useful RFP to add to the armory (Shcherbo *et al.*, 2009).

V. The Hunt for Red Tubulin

In light of our experience in tagging a variety of cytoskeletal proteins with the different RFPs, we wanted to investigate which RFP would be best-suited to tagging tubulin. Plasmids expressing *nmt1*:mRFP1-Atb2 (Yamashita *et al.*, 2005), *nmt1*:mCherry-Atb2 (Terenna *et al.*, 2008), or *nmt81*:mCherry-Atb2 (Grallert *et al.*, 2006; Hauf *et al.*, 2007) have all been described, and these have been used primarily to image mitotic or meiotic spindles. However, episomal expression of RFP-tubulin is subject to copy number variation (see above). Therefore we wanted to examine microtubules in cells expressing integrated versions of some of the new RFPs fused to Atb2. Since mOrange-2 and TagRFP-T did not perform well in our initial studies, we analyzed Atb2 fusions with mCherry, tdTomato, and mKate. We generated strains in which RFP-Atb2 fusions were expressed under the control of either the medium-strength *nmt41* promoter or the weak *nmt81* promoter. Integration was either at the *atb2* locus, replacing the endogenous protein, or at the *leu1* locus, leaving the endogenous *atb2+* gene intact (Fig. 5). To generate plasmids for RFP-Atb2 integration at the *leu1* locus we modified plasmids

Expression construct	Integration locus	Untagged Atb2?	Tagged Atb2 as % of total α-tubulin	Microtubules	Reference
nmt41 | *GFP* | *Atb2*	*atb2*	no	61%	Near wild-type	Snaith and Sawin (2003)
nmt81 | *GFP* | *Atb2*	*atb2*	no	27%	Wild type	Sawin *et al.* (2004)
nmt81 | *GFP* | *Atb2*	*ars1*	yes	25%	Wild type	Sawin *et al.* (2004)
SV40 | *GFP* | *Atb2*	*leu1*	yes	10%	Wild type	Pardo and Nurse (2005); Bratman and Chang (2007)
atb2 | *GFP* | *Atb2*	*atb2*	no	NA	Wild type	Sato *et al.* (2009)
nmt41 | *mCherry* | *Atb2*	*atb2*	no	78%	Aberrant	This study
nmt81 | *mCherry* | *Atb2*	*atb2*	no	30%	Wild type	This study
pADH15 | *mCherry* | *Atb2*	*Z*	yes	NA	Wild type	Kawashima (2010)
nda3 | *mCherry* | *Atb2*	*aur1*	yes	NA	Wild type	Unsworth *et al.* (2008)
nmt41 | *tdTomato* | *Atb2*	*leu1*	yes	3.5%	Wild type	This study
nmt81 | *tdTomato* | *Atb2*	*leu1*	yes	ND	Wild type	This study
nmt81 | *mKate* | *Atb2*	*atb2*	no	46%	Aberrant	This study
nmt41 | *mKate* | *Atb2*	*leu1*	yes	17%	Wild type	This study

Fig. 5 Selected available GFP-Atb2 and RFP-Atb2 expressing strains. Description of the structure of the expression construct, the genomic locus at which the construct is integrated, presence or absence of endogenous untagged Atb2, level of tagged Atb2 expressed as a percentage of the total α-tubulin present (in cells grown at 32°C), and the phenotype of microtubules in selected strains is given. Strains with Atb2 expression under control of the *nmt1* promoter are grown in the absence of thiamine; all other strains are insensitive to thiamine. New pDUAL Gateway vectors allowing integration of 6xHis-FLAG-tdTomato- and 6xHis-FLAG-mKate-Atb2 at the *leu1* locus under the *nmt41* or *nmt81* promoters were constructed as follows: mKate and tdTomato cDNAs were cloned as *NcoI/BglII* fragments into pHFF1 (Matsuyama *et al.*, 2004) to create pHFmK1 and pHFtdT1. The *nmt1* promoter in pHFmK1 and pHFtdT1 was replaced with an *SphI/BamI* fragment containing the *nmt41* or *nmt81* promoter (Basi *et al.*, 1993; Maundrell, 1990) to generate pHFmK41/81 and pHFtdT41/81. The *ccd*B-Rfa cassette (Invitrogen) was cloned into the *EcoRV* site of pHFmK1/41/81 and pHFtdT1/41/81 to generate pHFmK1c/41c/81c and pHFtdT1c/41c/81c. Finally, the *SphI/XhoI* fragment from pHFmK1c/41c/81c and pHFtdT1c/41c/81c (containing *nmt*-HFmK/tdT-*ccd*B) was cloned into pDUAL (Matsuyama *et al.*, 2004) digested with *SphI/SalI* to create pDUAL-pHFmK1c/41c/81c and pDUAL-pHFtdT1c/41c/81c.

Table II
Selected Useful pDUAL Gateway-Compatible GFP and RFP Plasmids

Plasmid[a]	Lab reference	Tag	Tag position	Promoter	Promoter strength
pDUAL-GFH1c[b]	–	GFP-FLAG-6xHis	C	*nmt1*	Strong
pDUAL-GFH41c[b]	–	GFP-FLAG-6xHis	C	*nmt41*	Medium
pDUAL-GFH81c[b]	–	GFP-FLAG-6xHis	C	*nmt81*	Weak
pDUAL-HFG1c[b]	–	6xHis-FLAG-GFP	N	*nmt1*	Strong
pDUAL-HFG41c[b]	–	6xHis-FLAG-GFP	N	*nmt41*	Medium
pDUAL-HFG81c[b]	–	6xHis-FLAG-GFP	N	*nmt81*	Weak
pDUAL-GFH31c[c]	–	GFP-FLAG-6xHis	C	*cam1*	Weak
pDUAL-GFH51c[c]	–	GFP-FLAG-6xHis	C	*tif51*	Strong
pDUAL-HFG31c[c]	–	6xHis-FLAG-GFP	N	*cam1*	Weak
pDUAL-HFG51c[c]	–	6xHis-FLAG-GFP	N	*tif51*	Strong
pDUAL-HFtdT1c[d]	pKS903	6xHis-FLAG-tdTomato	N	*nmt1*	Strong
pDUAL-HFtdT41c[d]	pKS905	6xHis-FLAG-tdTomato	N	*nmt41*	Medium
pDUAL-HFtdT81c[d]	pKS907	6xHis-FLAG-tdTomato	N	*nmt81*	Weak
pDUAL-HFmK1c[d]	pKS902	6xHis-FLAG-mKate	N	*nmt1*	Strong
pDUAL-HFmK41c[d]	pKS904	6xHis-FLAG-mKate	N	*nmt41*	Medium
pDUAL-HFmK81c[d]	pKS906	6xHis-FLAG-mKate	N	*nmt81*	Weak

[a] All plasmids allow integration at the *leu1* locus after digestion with *NotI*.
[b] Matsuyama *et al.* (2004).
[c] Matsuyama *et al.* (2008).
[d] This study.

derived from pDUAL-HFG Gateway-compatible vectors (Matsuyama *et al.*, 2004), replacing the GFP coding sequence with either tdTomato or mKate (see Table II).

Cells expressing *nmt41:*mCherry-Atb2 at the *atb2* locus were abnormally shaped and had very short microtubule bundles compared to cells expressing *nmt41:*GFP-Atb2 at the same locus (Fig. 6C; Snaith and Sawin, 2003). However, cells expressing *nmt81:*mCherry-Atb2 at the *atb2* locus had wild-type shape and normal microtubule distribution, similar to that observed with *nmt81:*GFP-Atb2 (Fig. 6A and B). In addition, astral microtubules and spindle elongation were normal (Fig. 2A and C; additional data not shown). There was some cell-to-cell variation in microtubule brightness with *nmt81:*mCherry-Atb2 expressed at the *atb2* locus, and wild-type microtubule arrays were present only if the cells were grown for 72–96 h at 25°C prior to analysis, rather than the usual 48 h that we use in the lab. The reasons for this are still not understood. By contrast, cells expressing *nmt81:*mKate-Atb2 at the *atb2* locus displayed short, aberrant microtubules (Fig. 6D).

Cells expressing either *nmt81:*tdTomato-Atb2, *nmt41:*tdTomato-Atb2, or *nmt41:* mKate-Atb2 at the *leu1* locus (which allows unperturbed expression of endogenous untagged Atb2 at the *atb2* locus) displayed apparently wild-type microtubules (Fig. 6E–G). Even though the level of *nmt81:*tdTomato-Atb2 integrated at *leu1* was too low for quantitation by western blotting, it was sufficient to generate fluorescent

Table III
List of Strains Described in this Study

Genotype	Expression locus	Our lab reference	Reference
ade6-M216 leu1-32 ura4-D18 h+	N/A	KS515	Lab stock
kanMX6:nmt81:GFP-atb2 ade6-216 leu1-32 ura4-D18 h+	*atb2*	KS1235	Sawin *et al.* (2004)
kanMX6:nmt41:GFP-atb2 ade6-210 leu1-32 ura4-D18 h+	*atb2*	KS1231	Snaith and Sawin (2003)
kanMX6:nmt1:GFP-atb2 h-	*atb2*	KS261	This study
lntp[nmt81:GFP-atb2 LEU2] ade6-M210 leu1-32 ura4-D18 h+	*ars1*	KS1225	Anders *et al.* (2006)
lntp[SV40:GFP-atb2 leu1+] ade6-216 leu1-32 ura4-D18 h-	*leu1*	KS4956[a]	Bratman and Chang (2007)
hphMX6:nmt81:mCh-atb2 ade6-210 leu1-32 ura4-D18 h-	*atb2*	KS2789	This study
nda3:mCh-atb2 Klp6-GFP:aur ade6-M216 h-	*aur1*	AR616[b]	Unsworth *et al.* (2008)
PAdh15:mCh-atb2:natMX6 Sgo2-GFP:kanMX6 leu1-32 ade6	*Z*	PM26[c]	Kawashima *et al.* (2010)
hphMX6:nmt41:mCh-atb2 ade6-210 leu1-32 ura4-D18 h-	*atb2*	KS2790	This study
lntp[nmt81:tdT-atb2 leu1+] ade6 leu1-32 ura4-D18	*leu1*	KS5080	This study
lntp[nmt41:tdT-atb2 leu1+] ade6 leu1-32 ura4-D18	*leu1*	KS5077	This study
kanMX6:nmt81:mK-atb2 ade6-M216 leu1-32 ura4-D18 h+	*atb2*	KS4408	This study
lntp[lnmt41:mK-atb2 leu1+] ade6 leu1-32 ura4-D18	*leu1*	KS5075	This study
tea1-GFP:kanMX6 ade6-M210 leu1-32 ura4-D18 h-	*tea1*	KS1259	Snaith and Sawin (2003)
tea1-tdT:natMX6 ade6-M210 leu1-32 ura4-D18 h-	*tea1*	KS3138	This study
tea1-mCh:natMX6 ade6-M210 leu1-32 ura4-D18 h-	*tea1*	KS3137	This study
tea1-TRT:kanMX6 ade6-M210 leu1-32 ura4-D18 h-	*tea1*	KS5380	This study
tea1-mO2:kanMX6 ade6-M210 leu1-32 ura4-D18 h-	*tea1*	KS5379	This study
tea1-mK:kanMX6 ade6-M210 leu1-32 ura4-D18 h-	*tea1*	KS4310	This study
mto1-GFP:kanMX6 ade6-M216 leu1-32 ura4-D18 h+	*mto1*	KS819	Sawin *et al.* (2004)
mto1-tdT:natMX6 ade6-M210 leu1-32 ura4-D18 h-	*mto1*	KS5445	This study
mto1-mCh:natMX6 ade6-M210 leu1-32 ura4-D18 h-	*mto1*	KS5446	This study
mto1-TRT:kanMX6 ade6-M210 leu1-32 ura4-D18 h-	*mto1*	KS5378	This study
mto1-mO2:kanMX6 ade6-M210 leu1-32 ura4-D18 h-	*mto1*	KS5377	This study
alp4-GFP:kanMX6 ade6-M216 leu1-32 ura4-D18 h+	*alp4*	KS1368	Anders *et al.* (2006)
alp4-tdT:natMX6 ade6-M210 leu1-32 ura4-D18 h-	*alp4*	KS3104	Samejima *et al.* (2008)
alp4-mCh:natMX6 ade6-M216 leu1-32 ura4-D18 h+	*alp4*	KS5253	This study
alp4-mCh:kanMX6 ade6-M216 leu1-32 ura4-D18 h+	*alp4*	KS5253	This study
alp4-TRT:kanMX6 ade6-M210 leu1-32 ura4-D18 h-	*alp4*	KS5376	This study
alp4-mO2:kanMX6 ade6-M210 leu1-32 ura4-D18 h-	*alp4*	KS5375	This study

[a] Strain FC1234, Fred Chang lab, Columbia University, New York.
[b] Takashi Toda lab, CRUK, London.
[c] Yoshinori Watanabe lab, Tokyo University, Tokyo

microtubules. However, the incorporation of tdTomato-Atb2 was somewhat uneven, producing speckled microtubules, which may be useful for speckle microscopy applications. These results suggest that tdTomato-Atb2 and mKate-Atb2 are probably not fully functional but can incorporate into wild-type microtubules if untagged Atb2 is also expressed (see above for discussion concerning GFP-Atb2). It should be noted that the level of tagged tubulin in the strain expressing *nmt41:*mKate-Atb2 at the *leu1* locus (by a *pDUAL* plasmid) was lower than in the strain expressing mKate-Atb2 under the nominally weaker *nmt81* promoter at the *atb2* locus (Fig. 6I and J). Similarly low expression was seen with *nmt81*:tdTomato-Atb2 and *nmt41*:tdTomato-Atb2 at the *leu1* locus (Fig. 6I and J; additional data not shown). This could indicate a general reduction in activity of *nmt* promoters when integrated at *leu1*; alternatively, the 6xHis

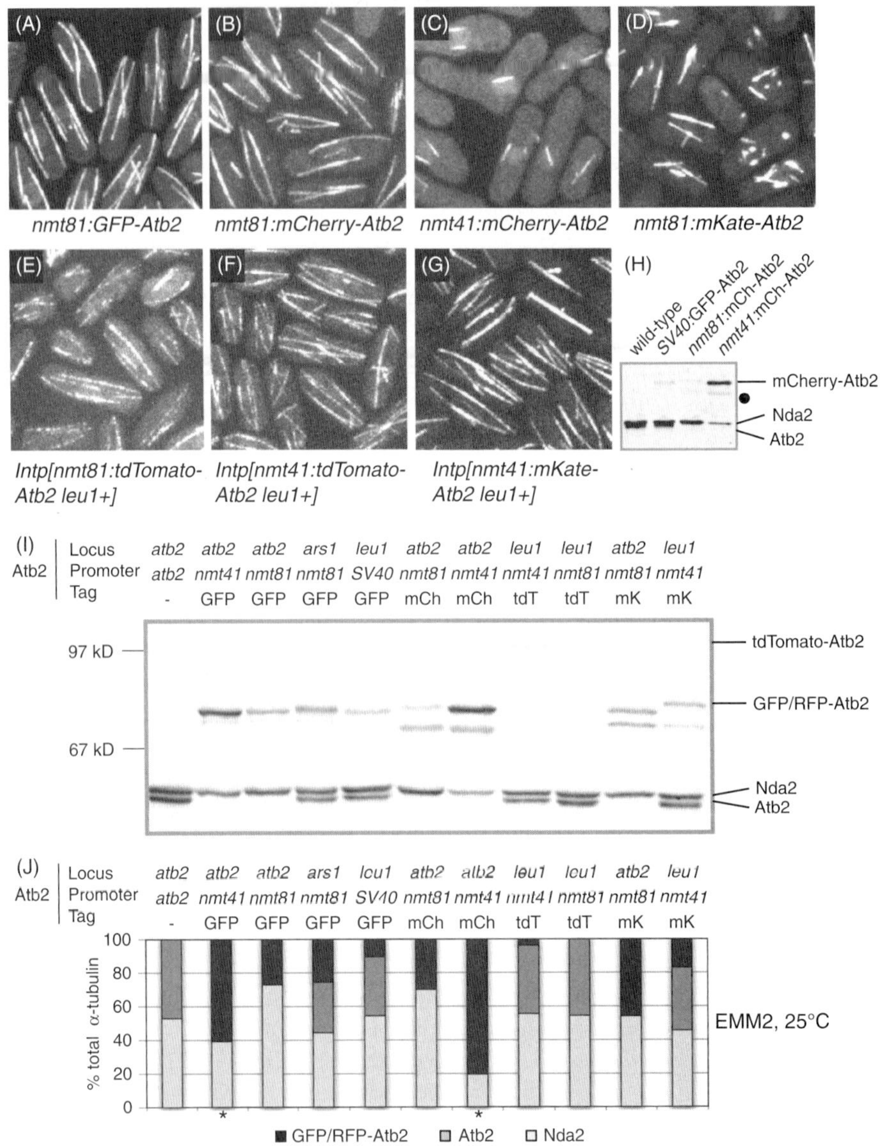

Fig. 6 (continued)

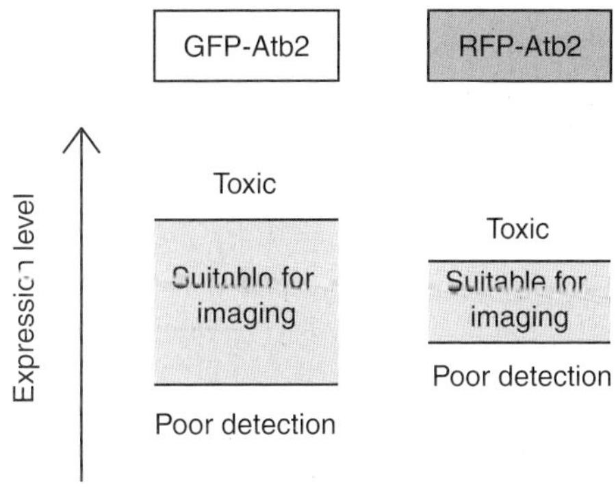

Fig. 7 Useable range of expression for GFP-Atb2 and RFP-Atb2 constructs. GFP-Atb2 can be used over a greater range of expression levels than RFP-Atb2.

and FLAG tags on the *leu1*-integrated tubulins could affect protein stability. Overall, these data support the general notion that GFP-Atb2 produces "wild-type" microtubule arrays suitable for imaging over a broader range of expression than RFP-Atb2 (Fig. 7).

In summary, if used carefully, *nmt81*:mCherry-Atb2 integrated at the *atb2* locus produces fluorescently robust and wild-type microtubules and currently can be considered the construct of choice for many applications. In experiments requiring simultaneous imaging of microtubules with MAPs, an obvious question is whether to use *nmt81*:mCherry-Atb2 with MAP-GFP, or *nmt81*:GFP-Atb2 with MAP-RFP. Since all the evidence shows that GFP is still the preferred fluorophore for optimal imaging and protein function, if microtubules are the focus of the study it would be better to use GFP-Atb2 with the second

---

Fig. 6 Microtubules in cell expressing RFP-Atb2. (A–G) Images of cells expressing different versions of GFP-Atb2 or RFP-Atb2 at the *atb2* locus (A–D) or at the *leu1* locus (E–G). (A) *nmt81*:GFP-Atb2 at *atb2* (strain KS1235). (B) *nmt81*:mCherry-Atb2 at *atb2* (KS2789). (C) *nmt41*:mCherry-Atb2 at *atb2* (KS2790). (D) *nmt81*:mKate-Atb2 at *atb2* (KS4408). (E) *nmt81*:tdTomato-Atb2 at *leu1* (KS5080). (F) *nmt41*:tdTomato-Atb2 at *leu1* (KS5077). (G) *nmt41*:mKate-Atb2 at *leu1* (KS5075). Because different strains were imaged by spinning disc confocal microscopy at different illumination intensities, relative expression, and brightness of each RFP cannot be judged from the images. (H) Nda2 levels are reduced upon overexpression of mCherry-Atb2 from the *nmt41* promoter. Total protein extracts were prepared from wild-type cells (KS515), cells expressing *SV40*:GFP-Atb2 at the *leu1* locus (KS4956), and cells expressing either *nmt81*:mCherry-Atb2 (KS2789) or *nmt41*:mCherry-Atb2 (KS2790) at the *atb2* locus, after 4 days of growth in EMM2 at 25°C. Western blots detecting α-tubulin were performed as described in Fig. 1. The dot shows the position of 67 kD marker. (I) Western blot of total α-tubulin in GFP- and RFP-Atb2-expressing strains. Total protein extracts were prepared from wild-type cells (KS515) and from cells expressing *nmt41*:GFP-Atb2 at *atb2* (KS1231), *nmt81*:GFP-Atb2 at *atb2* (KS1235), *nmt81*:GFP-Atb2 at *ars1* (KS1225), *SV40*GFP-Atb2 at *leu1* (KS4956), *nmt81*:mCherry-Atb2 at *atb2* (KS2789), *nmt41*:mCherry-Atb2 at *atb2* (KS2790), *nmt41*:tdTomato-Atb2 at *leu1* (KS5077), *nmt81*:tdTomato-Atb2 at *leu1* (KS5080), *nmt81*:mKate-Atb2 at *atb2* (KS4408), and *nmt41*: mKate-Atb2 at the *leu1* locus (KS5075). Cells were grown for 2 days at 32°C in EMM2 prior to harvesting. (J) Levels of α-tubulin in the western blot in (I) were quantitated as described in Fig. 1 and the proportions of tagged Atb2, untagged Atb2, and Nda2 in each strain are presented. Asterisks are as in Fig. 1. Full genotypes of all strains are given in Table III.

protein labeled with RFP to ensure wild-type microtubule distribution. Alternatively, if the MAP is of greater interest, MAP-GFP should be used with *nmt81:*mCherry-Atb2. If highly photostable microtubules are required for long time-lapse imaging, the mKate-Atb2 expressed under the *nmt41* promoter integrated at *leu1* can be used successfully. Recently, strains expressing mCherry-Atb2 expressed under the control of the *nda3* or *Adh15* promoters have been reported (Kawashima *et al.*, 2010; Unsworth *et al.*, 2008). Development of a *SV40:*mCherry-Atb2 construct and *pDUAL* plasmids allowing expression of *nmt41:* or *nmt81:*mCherry-Atb2 would be additional valuable tools.

VI. Successful Fluorescent Imaging of Fission Yeast Microtubules and Associated Proteins

Several factors contribute to high-quality fluorescence imaging of fission yeast. Some of these are hardware-related (e.g., the quality of the imaging system) and may be expensive to optimize. However, there are many simple, less expensive ways to maximize image quality.

1. Emission filters should be matched to fluorophores to ensure that the maximum signal emitted by one's protein of interest is collected. Widely available fluorescein filters match the spectral properties of GFP(S65T) very well and should cause no difficulties. However, the excitation and emission maxima of some RFP or GFP variants may fall outside the useful range of one's filters, and this may compromise results.
2. Many of the currently available GFP- and RFP-tagging vectors contain only minimal linker sequences. The C-terminal Bähler-style tagging vectors (Bähler *et al.*, 1998) add a seven-amino acid linker (RIPGLIN) between the C terminus of the tagged protein and the downstream GFP or RFP tag. The N-terminal tagging vectors do not add any extra amino acids at all; these must be provided by the PCR primers. In the case of Atb2, tagging has generally used the two-amino acid linker Gly-Ser (derived from a BamHI site) that was used in the original *nmt1*:GFP-Atb2 plasmid (Ding *et al.*, 1998). The presence or absence of linker sequences between the fluorescent protein tag and one's protein of interest may greatly affect the behavior of the fusion protein. When the *S. pombe* EB1 homologue Mal3 was originally tagged with GFP, only the plasmid-derived seven-amino acid linker sequence was present between the C-terminus of Mal3 and GFP, and fluorescence of the Mal3-GFP fusion protein was very low (Beinhauer *et al.*, 1997). Addition of a 24-amino acid linker sequence between Mal3 and GFP greatly improved the fluorescence, making it readily amenable to *in vivo* imaging (Sandblad *et al.*, 2006). If the linkers required are relatively short, they can be added easily into the primers used to amplify the integration cassettes. For longer sequences it could be necessary to introduce them into the template vector before PCR amplification.
3. The choice of media in which the cells are grown is important. Rich media (YE5S) has an associated autofluorescence that can mask the specific fluorescence of one's

protein of interest, especially if it is a low-abundance protein. Even EMM minimal medium is somewhat yellow after standard preparation, mainly due to caramelization of glucose during autoclaving. Unwanted coloration of growth medium can be almost completely avoided if the nitrogen source and any necessary supplements are added to the medium after autoclaving.

4. Cells should be grown as near as possible to the final imaging temperature. For example, although wild-type fission yeast is commonly grown at 30–32°C, it also grows well at 25°C. Since many microscope rooms are maintained at about this temperature, pregrowth at 25°C means that no complicated temperature controlled equipment is required, and any temperature variation to the cells upon transfer from incubator to microscope is minimized.
5. Special care should be taken to ensure that the cells are in early to mid-log phase during imaging, as cells taken from overgrown cultures do not represent actively growing cells.
6. To obtain the best data, cells must be imaged in physiological conditions with a nutrient source. Commonly this is done either with agarose pads or with tissue culture dishes. Cells incubated in such conditions are able to divide several times, indicating near physiological growth conditions; in comparison, cells placed directly between a slide and coverslip usually fail in cytokinesis (our unpublished data).

A. Agarose pads (Sawin, 1999) allow cells to be mounted in high-density monolayers. Agarose pads are also useful if you want to examine the effects of drug treatment on cells, as the drugs can be added directly to the agarose prior to slide preparation.

- Dissolve 2% agarose in the same growth medium as the cell culture and keep at 70°C for the day (aliquots can prepared in bulk in advance, allowed to solidify, and stored in sealed microfuge tubes before use; they can be remelted in a 95°C hot block on the day of use. Discard such tubes at the end of a day of imaging, as the medium will begin to yellow if kept at 70°C for long periods of time). Place a piece of 3 M "Scotch" adhesive tape at either end of a microscope slide to act as spacers between the slides. Pipette 16 μl of molten agarose onto the microscope slide, warmed to 37°C on a hot block. Immediately cover with second microscope slide and transfer slides to lab bench to cool.
- While the agarose pad is setting, concentrate 200–300 μl of cell culture to 4 μl by centrifugation. Thoroughly resuspend the cell pellet.
- Within 10 min of preparing the agarose pad, gently slide the top slide away from the bottom slide and pipette 1 μl of cell suspension onto one side of the pad.
- Place a coverslip on top, and seal with a 1:1:1 mixture of vaseline, lanolin, and paraffin (commonly known as VALAP). Since GFP is often best imaged under relatively low-oxygen concentration, slides with GFP-tagged proteins should be sealed on all four sides to minimize oxygenation of the sample. In contrast, in our experience many of the RFPs seem to exhibit optimal fluorescence when well oxygenated, and hence slides of RFP-tagged proteins should be only minimally sealed with VALAP at the four corners of the slide. Photoconversion of GFP to

RFP has been observed upon prolonged illumination with blue light in low-oxygen conditions (Elowitz *et al.*, 1997; Sawin and Nurse, 1997). This photoconversion is clearly undesirable when performing double-label imaging experiments with GFP and RFP. To minimize GFP photoconversion, a compromise on sample oxygenation must be reached: we have found imaging cells toward the edge of the sample, and thereby well oxygenated (to reduce photoconversion), but at the edge of dense areas patches of cells and thereby oxygen depleted (to optimize fluorescence) to be effective.

- To be sure that cells are imaged under physiological conditions, each sample should generally be discarded after 45 min to 1 h (unless longer term imaging is required).

B. Culture dishes allow easier manipulation of growth medium than agarose pads and thus can be useful in experiments requiring drug treatment followed by washout (A. Grallert, pers. comm.). However, one cannot achieve high cell densities with this mounting method, which may present problems if the event or cell cycle stage of interest is underrepresented in the sample.

- Prepare 0.2 mg/ml solution of soybean lectin in water (SIGMA L1395).
- Assemble appropriate imaging chamber with clean, untreated 25 mm round glass coverslip.
- Spin down about 200–300 μl of cell culture with approximately OD of 0.5 and resuspend cell pellet in 5–10 μl fresh media. If the cell pellet is bigger than 3 μl, increase resuspension volume accordingly.
- Mix 5 μl 0.2 mg/ml lectin with 5 μl of cell suspension and place all of cell/lectin mixture directly on coverslip in chamber.
- Incubate cells on coverslip for 3–4 min to allow cells to adhere to glass.
- Wash off nonadhered cells from coverslip with 3 × 1 ml media, aspirating media between washes.
- Immediately cover cells with 0.5–1 ml fresh media and image.

Acknowledgments

We thank D. Kelly for help with image processing, C. Bicho and E. Lynch for help with strain construction, and F. Chang and Y. Watanabe for strains. K.E.S. is a Wellcome Trust Senior Research Fellow in Basic Biomedical Sciences. This work was supported by a grant from the Wellcome Trust.

References

Adachi, Y., Toda, T., Niwa, O., and Yanagida, M. (1986). Differential expressions of essential and nonessential alpha-tubulin genes in *Schizosaccharomyces pombe*. Mol. Cell. Biol. **6**, 2168–2178.

Akhmanova, A., and Steinmetz, M. O. (2008). Tracking the ends: A dynamic protein network controls the fate of microtubule tips. Nat. Rev. Mol. Cell Biol. **9**, 309–322.

Alvarez-Tabares, I., Grallert, A., Ortiz, J. M., and Hagan, I. M. (2007). *Schizosaccharomyces pombe* protein phosphatase 1 in mitosis, endocytosis and a partnership with wsh3/tea4 to control polarised growth. J. Cell Sci. **120**, 3589–3601.

Anders, A., Lourenco, P. C., and Sawin, K. E. (2006). Noncore components of the fission yeast gamma-tubulin complex. Mol. Biol. Cell **17**, 5075–5093.

Bähler, J., Wu, J. Q., Longtine, M. S., Shah, N. G., McKenzie, A.3rd, Steever, A. B., Wach, A., Philippsen, P., and Pringle, J. R. (1998). Heterologous modules for efficient and versatile PCR-based gene targeting in *Schizosaccharomyces pombe*. Yeast **14**, 943–951.

Basi, G., Schmid, E., and Maundrell, K. (1993). TATA box mutations in the *Schizosaccharomyces pombe* nmt1 promoter affect transcription efficiency but not the transcription start point or thiamine repressibility. Gene **123**, 131–136.

Beinhauer, J. D., Hagan, I. M., Hegemann, J. H., and Fleig, U. (1997). Mal3, the fission yeast homologue of the human APC-interacting protein EB-1 is required for microtubule integrity and the maintenance of cell form. J. Cell Biol. **139**, 717–728.

Bratman, S. V., and Chang, F. (2007). Stabilization of overlapping microtubules by fission yeast CLASP. Dev. Cell **13**, 812–827.

Campbell, R. E., Tour, O., Palmer, A. E., Steinbach, P. A., Baird, G. S., Zacharias, D. A., and Tsien, R. Y. (2002). A monomeric red fluorescent protein. Proc. Natl. Acad. Sci. U.S.A. **99**, 7877–7882.

Chalfie, M., Tu, Y., Euskirchen, G., Ward, W. W., and Prasher, D. C. (1994). Green fluorescent protein as a marker for gene expression. Science **263**, 802–805.

Cody, C. W., Prasher, D. C., Westler, W. M., Prendergast, F. G., and Ward, W. W. (1993). Chemical structure of the hexapeptide chromophore of the Aequorea green-fluorescent protein. Biochemistry **32**, 1212–1218.

Ding, D. Q., Chikashige, Y., Haraguchi, T., and Hiraoka, Y. (1998). Oscillatory nuclear movement in fission yeast meiotic prophase is driven by astral microtubules, as revealed by continuous observation of chromosomes and microtubules in living cells. J. Cell Sci. **111**(Pt 6), 701–712.

Drummond, D. R., and Cross, R. A. (2000). Dynamics of interphase microtubules in *Schizosaccharomyces pombe*. Curr. Biol. **10**, 766–775.

Elowitz, M. B., Surette, M. G., Wolf, P. E., Stock, J., and Leibler, S. (1997). Photoactivation turns green fluorescent protein red. Curr. Biol. **7**, 809–812.

Garcia, M. A., Koonrugsa, N., and Toda, T. (2002). Two kinesin-like kin I family proteins in fission yeast regulate the establishment of metaphase and the onset of anaphase A. Curr. Biol. **12**, 610–621.

Garcia, M. A., Vardy, L., Koonrugsa, N., and Toda, T. (2001). Fission yeast ch-TOG/XMAP215 homologue alp14 connects mitotic spindles with the kinetochore and is a component of the mad2-dependent spindle checkpoint. EMBO J. **20**, 3389–3401.

Glynn, J. M., Lustig, R.J, Berlin, A., and Chang, F. (2001). Role of bud6p and tea1p in the interaction between actin and microtubules for the establishment of cell polarity in fission yeast. Curr. Biol. **11**, 836–845.

Grallert, A., Beuter, C., Craven, R. A., Bagley, S., Wilks, D., Fleig, U., and Hagan, I. M. (2006). *S. Pombe* CLASP needs dynein, not EB1 or CLIP170, to induce microtubule instability and slows polymerization rates at cell tips in a dynein-dependent manner. Genes Dev. **20**, 2421–2436.

Hagan, I. M. (1998). The fission yeast microtubule cytoskeleton. J. Cell Sci. **111**(Pt 12), 1603–1612.

Hauf, S., Biswas, A., Langegger, M., Kawashima, S. A., Tsukahara, T., and Watanabe, Y. (2007). Aurora controls sister kinetochore mono-orientation and homolog bi-orientation in meiosis-I. EMBO J. **26**, 4475–4486.

Heim, R., Cubitt, A. B., and Tsien, R. Y. (1995). Improved green fluorescence. Nature **373**, 663–664.

Hiraoka, Y., Toda, T., and Yanagida, M. (1984). The NDA3 gene of fission yeast encodes beta-tubulin: a cold-sensitive nda3 mutation reversibly blocks spindle formation and chromosome movement in mitosis. Cell **39**, 349–358.

Hoog, J. L., Schwartz, C., Noon, A. T., O'Toole, E. T., Mastronarde, D. N., McIntosh, J. R., and Antony, C. (2007). Organization of interphase microtubules in fission yeast analyzed by electron tomography. Dev. Cell **12**, 349–361.

Jones, R. H., Moreno, S., Nurse, P., and Jones, N. C. (1988). Expression of the SV40 promoter in fission yeast: Identification and characterization of an AP-1-like factor. Cell **53**, 659–667.

Kawashima, S. A., Yamagishi, Y., Honda, T., Ishiguro, K., and Watanabe, Y. (2010). Phosphorylation of H2A by bub1 prevents chromosomal instability through localizing shugoshin. Science **327**, 172–177.

Kerres, A., Jakopec, V., and Fleig, U. (2007). The conserved spc7 protein is required for spindle integrity and links kinetochore complexes in fission yeast. Mol. Biol. Cell **18**, 2441–2454.

Mallavarapu, A., Sawin, K., and Mitchison, T. (1999). A switch in microtubule dynamics at the onset of anaphase B in the mitotic spindle of *Schizosaccharomyces pombe*. Curr. Biol. **9**, 1423–1426.

Mandelkow, E. M., and Mandelkow, E. (1985). Unstained microtubules studied by cryo-electron microscopy. Substructure, supertwist and disassembly. J. Mol. Biol. **181**, 123–135.

Marks, J., Hagan, I. M., and Hyams, J. S. (1986). Growth polarity and cytokinesis in fission yeast: The role of the cytoskeleton. J. Cell Sci. Suppl. **5**, 229–241.

Mata, J., and Nurse, P. (1997). Tea1 and the microtubular cytoskeleton are important for generating global spatial order within the fission yeast cell. Cell **89**, 939–949.

Matsuyama, A., Shirai, A., and Yoshida, M. (2008). A series of promoters for constitutive expression of heterologous genes in fission yeast. Yeast **25**, 371–376.

Matsuyama, A., Shirai, A., Yashiroda, Y., Kamata, A., Horinouchi, S., and Yoshida, M. (2004). PDUAL, a multipurpose, multicopy vector capable of chromosomal integration in fission yeast. Yeast **21**, 1289–1305.

Maundrell, K. (1990). Nmt1 of fission yeast. A highly transcribed gene completely repressed by thiamine. J. Biol. Chem. **265**, 10857–10864.

Merzlyak, E. M., Goedhart, J., Shcherbo, D., Bulina, M. E., Shcheglov, A. S., Fradkov, A. F., Gaintzeva, A., Lukyanov, K. A., Lukyanov, S., Gadella, T. W., and Chudakov, D. M. (2007). Bright monomeric red fluorescent protein with an extended fluorescence lifetime. Nat. Methods **4**, 555–557.

Mitchison, T., and Kirschner, M. (1984). Dynamic instability of microtubule growth. Nature **312**, 237–242.

Moreno, S., Klar, A., and Nurse, P. (1991). Molecular genetic analysis of fission yeast schizosaccharomyces pombe. Methods Enzymol. **194**, 795–823.

Nabeshima, K., Kurooka, H., Takeuchi, M., Kinoshita, K., Nakaseko, Y., and Yanagida, M. (1995). P93dis1, which is required for sister chromatid separation, is a novel microtubule and spindle pole body-associating protein phosphorylated at the Cdc2 target sites. Genes Dev. **9**, 1572–1585.

Nabeshima, K., Nakagawa, T., Straight, A. F., Murray, A., Chikashige, Y., Yamashita, Y. M., Hiraoka, Y., and Yanagida, M. (1998). Dynamics of centromeres during metaphase-anaphase transition in fission yeast: Dis1 is implicated in force balance in metaphase bipolar spindle. Mol. Biol. Cell **9**, 3211–3225.

Nogales, E., Whittaker, M., Milligan, R. A., and Downing, K. H. (1999). High-resolution model of the microtubule. Cell **96**, 79–88.

Nogales, E., Wolf, S. G., and Downing, K. H. (1998). Structure of the alpha beta tubulin dimer by electron crystallography. Nature **391**, 199–203.

Örmo, M., Cubitt, A. B., Kallio, K., Gross, L. A., Tsien, R. Y., and Remington, S. J. (1996). Crystal structure of the Aequorea victoria green fluorescent protein. Science **273**, 1392–1395.

Pardo, M., and Nurse, P. (2005). The nuclear rim protein Amo1 is required for proper microtubule cytoskeleton organisation in fission yeast. J. Cell Sci. **118**, 1705–1714.

Patterson, G. H., Knobel, S. M., Sharif, W. D., Kain, S. R., and Piston, D. W. (1997). Use of the green fluorescent protein and its mutants in quantitative fluorescence microscopy. Biophys. J. **73**, 2782–2790.

Piel, M., and Tran, P. T. (2009). Cell shape and cell division in fission yeast. Curr. Biol. **19**, R823–R827.

Prasher, D. C., Eckenrode, V. K., Ware, W. W., Prendergast, F. G., and Cormier, M. J. (1992). Primary structure of the Aequorea victoria green-fluorescent protein. Gene **111**, 229–233.

Robertson, A. M., and Hagan, I. M. (2008). Stress-regulated kinase pathways in the recovery of tip growth and microtubule dynamics following osmotic stress in *S. Pombe*. J. Cell Sci. **121**, 4055–4068.

Rosenberg, J. A., Tomlin, G. C., McDonald, W. H., Snydsman, B. E., Muller, E. G., Yates, III, J. R., and Gould, K. L. (2006). Ppc89 links multiple proteins, including the septation initiation network, to the core of the fission yeast spindle-pole body. Mol. Biol. Cell **17**, 3793–3805.

Sagolla, M. J., Uzawa, S., and Cande, W. Z. (2003). Individual microtubule dynamics contribute to the function of mitotic and cytoplasmic arrays in fission yeast. J. Cell Sci. **116**, 4891–4903.

Samejima, I., Lourenco, P. C., Snaith, H. A., and Sawin, K. E. (2005). Fission yeast mto2p regulates microtubule nucleation by the centrosomin-related protein mto1p. Mol. Biol. Cell **16**, 3040–3051.

Samejima, I., Miller, V. J., Groocock, L. M., and Sawin, K. E. (2008). Two distinct regions of Mto1 are required for normal microtubule nucleation and efficient association with the gamma-tubulin complex in vivo. J. Cell Sci. **121**, 3971–3980.

Sandblad, L., Busch, K. E., Tittmann, P., Gross, H., Brunner, D., and Hoenger, A. (2006). The *Schizosaccharomyces pombe* EB1 homolog Mal3p binds and stabilizes the microtubule lattice seam. Cell **127**, 1415–1424.

Sato, M., Dhut, S., and Toda, T. (2005). New drug-resistant cassettes for gene disruption and epitope tagging in Schizosaccharomyces pombe. Yeast **22**, 583–591.

Sato, M., Toya, M., and Toda, T. (2009). Visualization of fluorescence-tagged proteins in fission yeast: The analysis of mitotic spindle dynamics using GFP-tubulin under the native promoter. Methods Mol. Biol. **545**, 185–203.

Sawin, K. E. (1999). GFP fusion proteins as probes for cytology in fission yeast. Methods Cell Biol. **58**, 123–138.

Sawin, K. E., Lourenco, P. C., and Snaith, H. A. (2004). Microtubule nucleation at non-spindle pole body microtubule-organizing centers requires fission yeast centrosomin-related protein mod20p. Curr. Biol. **14**, 763–775.

Sawin, K. E., and Nurse, P. (1996). Identification of fission yeast nuclear markers using random polypeptide fusions with green fluorescent protein. Proc. Natl. Acad. Sci. U.S.A. **93**, 15146–15151.

Sawin, K. E., and Nurse, P. (1997).Photoactivation of green fluorescent protein. Curr. Biol. **7**, R606–R607.

Sawin, K. E., and Tran, P. T. (2006). Cytoplasmic microtubule organization in fission yeast. Yeast **23**, 1001–1014.

Shaner, N. C., Campbell, R. E., Steinbach, P. A., Giepmans, B. N., Palmer, A. E., and Tsien, R. Y. (2004). Improved monomeric red, orange and yellow fluorescent proteins derived from *Discosoma* sp. red fluorescent protein. Nat. Biotechnol. **22**, 1567–1572.

Shaner, N. C., Lin, M. Z., McKeown, M. R., Steinbach, P. A., Hazelwood, K. L., Davidson, M. W., and Tsien, R. Y. (2008). Improving the photostability of bright monomeric orange and red fluorescent proteins. Nat. Methods **5**, 545–551.

Shaner, N. C., Patterson, G. H., and Davidson, M. W. (2007). Advances in fluorescent protein technology. J. Cell Sci. **120**, 4247–4260.

Shaner, N. C., Steinbach, P. A., and Tsien, R. Y. (2005). A guide to choosing fluorescent proteins. Nat. Methods **2**, 905–909.

Shcherbo, D., Merzlyak, E. M., Chepurnykh, T. V., Fradkov, A. F., Ermakova, G. V., Solovieva, E. A., Lukyanov, K. A., Bogdanova, E. A., Zaraisky, A. G., Lukyanov, S., and Chudakov, D. M. (2007). Bright far-red fluorescent protein for whole-body imaging. Nat. Methods **4**, 741–746.

Shcherbo, D., Murphy, C. S., Ermakova, G. V., Solovieva, E. A., Chepurnykh, T. V., Shcheglov, A. S., Verkhusha, V. V., Pletnev, V. Z., Hazelwood, K. L., Roche, P. M., Lukyanov, S., Zaraisky, A. G., *et al.* (2009). Far-red fluorescent tags for protein imaging in living tissues. Biochem. J. **418**, 567–574.

Shimomura, O., Johnson, F. H., and Saiga, Y. (1962). Extraction, purification and properties of aequorin, a bioluminescent protein from the luminous hydromedusan, Aequorea. J. Cell Comp. Physiol. **59**, 223–239.

Snaith, H. A., Samejima, I., and Sawin, K. E. (2005). Multistep and multimode cortical anchoring of tea1p at cell tips in fission yeast. EMBO J. **24**, 3690–3699.

Snaith, H. A., and Sawin, K. E. (2003). Fission yeast mod5p regulates polarized growth through anchoring of tea1p at cell tips. Nature **423**, 647–651.

Tatebe, H., Goshima, G., Takeda, K., Nakagawa, T., Kinoshita, K., and Yanagida, M. (2001). Fission yeast living mitosis visualized by GFP-tagged gene products. Micron **32**, 67–74.

Terenna, C. R., Makushok, T., Velve-Casquillas, G., Baigl, D., Chen, Y., Bornens, M., Paoletti, A., Piel, M., and Tran, P. T. (2008). Physical mechanisms redirecting cell polarity and cell shape in fission yeast. Curr. Biol. **18**, 1748–1753.

Toda, T., Adachi, Y., Hiraoka, Y., and Yanagida, M. (1984). Identification of the pleiotropic cell division cycle gene NDA2 as one of two different alpha-tubulin genes in Schizosaccharomyces pombe. Cell **37**, 233–242.

Tran, P. T., Marsh, L., Doye, V., Inoue, S., and Chang, F. (2001). A mechanism for nuclear positioning in fission yeast based on microtubule pushing. J. Cell Biol. **153**, 397–411.

Unsworth, A., Masuda, H., Dhut, S., and Toda, T. (2008). Fission yeast kinesin-8 Klp5 and Klp6 are interdependent for mitotic nuclear retention and required for proper microtubule dynamics. Mol. Biol. Cell **19**, 5104–5115.

Van Driessche, B., Tafforeau, L., Hentges, P., Carr, A. M., and Vandenhaute, J. (2005). Additional vectors for PCR-based gene tagging in *Saccharomyces cerevisiae* and *Schizosaccharomyces pombe* using nourseothricin resistance. Yeast **22**, 1061–1068.

Vardy, L., Fujita, A., and Toda, T. (2002). The gamma-tubulin complex protein Alp4 provides a link between the metaphase checkpoint and cytokinesis in fission yeast. Genes Cells **7**, 365–373.

Woods, A., Sherwin, T., Sasse, R., MacRae, T. H., Baines, A. J., and Gull, K. (1989). Definition of individual components within the cytoskeleton of Trypanosoma brucei by a library of monoclonal antibodies. J. Cell Sci. **93**(Pt 3), 491–500.

Yamashita, A., Sato, M., Fujita, A., Yamamoto, M., and Toda, T. (2005). The roles of fission yeast ase1 in mitotic cell division, meiotic nuclear oscillation, and cytokinesis checkpoint signaling. Mol. Biol. Cell **16**, 1378–1395.

Zimmerman, S., Daga, R. R., and Chang, F. (2004). Intra-nuclear microtubules and a mitotic spindle orientation checkpoint. Nat. Cell Biol. **6**, 1245–1246.

CHAPTER 10

Optical Trapping and Laser Ablation of Microtubules in Fission Yeast

Nicola Maghelli *and* Iva M. Tolić-Nørrelykke

Max Planck Institute of Molecular Cell Biology and Genetics (MPI-CBG), 01307 Dresden, Germany

Abstract

Manipulation has been used as a powerful investigation technique since the early history of biology. Every technical advance resulted in more refined instruments that led to the discovery of new phenomena and to the solution of old problems. The invention of laser in 1960 gave birth to what is now called optical manipulation: the use of light to interact with matter. Since then, the tremendous progress of laser technology made optical manipulation not only an affordable, reliable alternative to

978-0-12-381349-7
DOI: 10.1016/S0091-679X(10)97010-6

III. Optical Tweezing in Fission Yeast

A. Background

Optical forces can be exploited to displace the nucleus in fission yeast (Maghelli and Tolic-Norrelykke, 2008; Sacconi *et al.*, 2005b; Tolic-Norrelykke *et al.*, 2005). Since the refractive indexes of the cytoplasm and of the nucleoplasm do not differ significantly, it is not possible to directly apply optical forces to the nucleus. Therefore, it is necessary to trap a suitable particle and use it as a handle to apply forces on the nucleus. Due to the rigid cell wall of fungi, microinjecting particles that can be easily trapped using optical tweezers is cumbersome (Riveline and Nurse, 2009). However, it is possible to apply optical forces to the lipid granules naturally present in the cytoplasm (Maghelli and Tolic-Norrelykke, 2008; Sacconi *et al.*, 2005b; Tolic-Norrelykke *et al.*, 2005). The trapped granule can then be moved inside the cell and used to displace the nucleus.

B. Experiment

To demonstrate how optical tweezers can be used to perturb the intracellular arrangement, we set up an experiment using a *Schizosaccharomyces pombe* strain in which the nuclear envelope and the spindle pole body, a centrosome analogue in fission yeast, were labeled with GFP. The cells, fixed to the glass bottom of a Petri dish, were treated with a microtubule poison drug (MBC) to depolymerize the microtubules. By switching on the optical trap, and focusing inside the cell, it was possible to trap lipid granules in the cytoplasm (Fig. 1A and B). The trapped granule was successively moved against the nucleus (Fig. 1C). The manipulation was performed while simultaneously acquiring two-photon images of the cell. During the interaction, the nuclear envelope was deformed, clearly showing an indentation corresponding to the point where the trapped particle pressed against it (Fig. 1C, the trapped particle is marked by white arrows). After switching off the trap, the nuclear membrane relaxed into its original shape (Fig. 1C and D).

We next repeatedly applied forces on the nuclear envelope using a trapped lipid granule over a time interval of several minutes. As a result, the whole nucleus was displaced (Fig. 2A). After switching off the optical trap, and washing out the microtubule-depolymerizing drug, we imaged the manipulated cell, tracking the position of the nucleus (Fig. 2B).

C. Discussion

We used optical tweezers to deform and displace the nucleus in a living cell. Although displacing the nucleus in fission yeast has been achieved by other methods, such as centrifugation, working at a single cell level has several advantages as follows:

1. Allows for following of the displacement and relaxation processes
2. Permits a comparison of the same cell before and after the manipulation

3. Makes it possible to change the position of the nucleus only, without affecting other organelles or molecular gradients

When used *in vivo*, care must be taken to avoid inducing damage while trapping the lipid granules. We choose to perform trapping using a CW, near-infrared laser emitting at 970 nm. According to different studies (Ashkin and Dziedzic, 1987; Liang *et al.*,

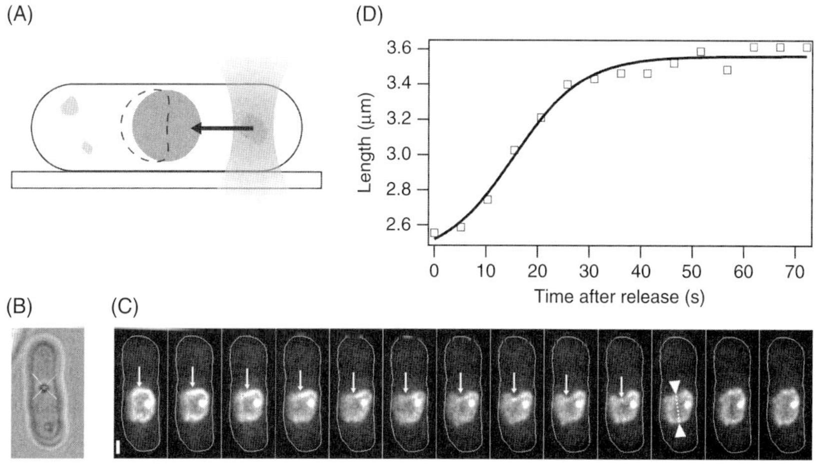

Fig. 1 (A) Schematics of the trapping experiment (side view). By focusing the optical trap inside the cell, it is possible to trap a lipid granule, naturally present in the cytoplasm. The trapped granule can then be used to apply forces on the nucleus. (B) Transmitted light image of a cell in which a lipid granule is trapped (white cross, the trapping laser is perpendicular to the image plane). In the same cell, the lipid granule was used to apply forces on the nuclear membrane: the time series (C) shows two-photon fluorescence images of the nucleus (labeled with Cut11-GFP) during the interaction with the trapped granule (white arrows). The optical trap is switched off (third last image) and the relaxation of the nuclear membrane, measured between the white triangles, is plotted as a function of time in (D). Scale bar is 1 μm, time between images in (C) is 25 s.

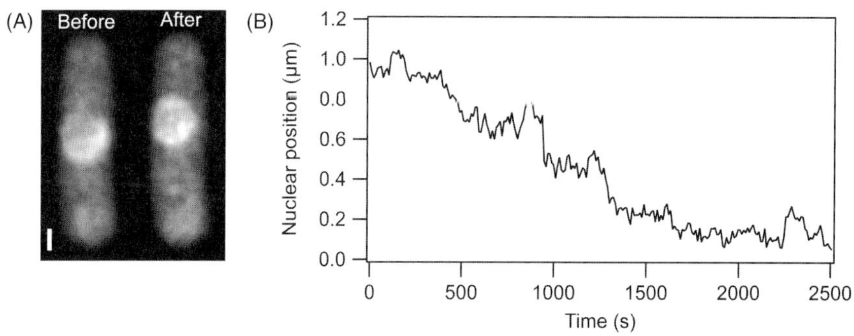

Fig. 2 (A) Displacement of the nucleus achieved by optical manipulation. Scale bar is 1 μm. After allowing for microtubule repolymerization by washing out MBC, the nucleus relaxes back to its original position in around 20 min (B).

1996; Liu *et al.*, 1995; Neuman *et al.*, 1999), wavelengths in the 950–1000 nm range appear to minimize the damage induced to the sample during the trapping process. In our experiments, we could not notice any changes in the two-photon signal from the nucleus during the manipulation. Moreover, after being deformed the nucleus always relaxed back to its original shape. Taken together, these results suggest that manipulating fission yeast cells with optical tweezers as described does not induce significant damage.

Analyzing how the nucleus of living cells reacts to such manipulation can provide information about its mechanical properties. In Fig. 1C it is possible to follow the length of the deformed nucleus, taken along the dotted line, after the optical trap has been switched off. The nucleus relaxes back to its original shape: fitting the data with a Sigmoid function $f(t) = y_0 + y_{\max}/(1 + \exp((t_0 - t)/\tau))$, solid line in Fig. 1D) gives a value of ~7 s for the characteristic rate τ. This characteristic time is determined by the mechanical properties of the nucleus and of the cytoplasm.

Optical forces can be used to displace the whole nucleus away from its natural position (Fig. 2A). We first depolymerized the microtubules using MBC and then employed the trapped granule to alter the geometrical arrangement of the cell. After washing out MBC, we were able to follow the position of the nucleus in time (Fig. 2B). We noticed that the nucleus came back to its original position, implying the existence of centering forces that actively keep the nucleus around the geometrical center of the cell. Previous studies (Sacconi *et al.*, 2005b; Tolic-Norrelykke *et al.*, 2005) have shown that the centering forces depend on microtubules, which microtubules exerting pushing forces on the nucleus (Daga *et al.*, 2006; Tolic-Norrelykke, 2008, 2010; Tolic-Norrelykke *et al.*, 2005; Tran *et al.*, 2001). Therefore, analyzing the time evolution of the nuclear position can yield information about the pushing forces generated by the microtubules.

IV. Laser Ablation of Microtubules

A. Background

Similarly to optical tweezers, laser ablation may be used to perturb the internal force balance of a cell. Laser ablation does not require any particle to be present inside a cell but can directly target cytoskeletal elements (Colombelli *et al.*, 2005; Khodjakov *et al.*, 2004; Maghelli and Tolic-Norrelykke, 2008; Raabe *et al.*, 2009; Sacconi *et al.*, 2005a; Tolic-Norrelykke *et al.*, 2004b; Vogel *et al.*, 2009) or organelles (Amy and Storb, 1965; Berns *et al.*, 1977; Sacconi *et al.*, 2007; Stiess *et al.*, 2010). It is possible to use laser ablation to investigate, e.g., the specific function of an organelle in a cell, or to study how a cell or an organism reacts to a modification of its structure. Laser ablation experiments can hence provide information about cellular processes that are complementary to the data that can be collected using genetic approaches. Technically, laser ablation can be implemented in any optical microscopy setup: by carefully designing the optical path, laser ablation can be used to manipulate the sample at a higher spatial and temporal resolution in comparison with optical tweezers. To perform laser ablation experiments *in vivo*, it is necessary to minimize any unspecific damage that could result from the ablation process. By controlling the laser power at the sample and the exposure time during ablation, it is possible to find the optimal parameter set that guarantees high ablation efficiency while minimizing the unspecific damage (Maghelli and Tolic-Norrelykke, 2008; Raabe *et al.*, 2009; Sacconi *et al.*, 2005a).

B. Experiment

To illustrate how laser ablation can contribute to the understanding of different biological processes, we ablated GFP-tagged microtubules in fission yeast during different phases of the cell cycle. The experiments were performed on a custom-built two-photon setup, using a pulsed femtosecond laser both for imaging and ablation (Maghelli and Tolic-Norrelykke, 2008). We started by searching for the ablation parameters maximizing the ablation efficiency while keeping the unspecific damage as low as possible. By keeping the power at the sample (100 mW) and the wavelength (895 nm) constant, we tried to ablate interphase microtubules using different exposure times [Fig. 3A, data taken from Maghelli and Tolic-Norrelykke (2008)]. We concluded that the optimal exposure time is between 20 and 30 ms. Using exposures in this range, the ablation efficiency was above 50% while no cell died as a consequence of the ablation.

We next used these settings to ablate interphase microtubules. During interphase, fission yeast microtubules are organized into antiparallel bundles. In each bundle, the plus ends point toward the cell periphery (scheme in Fig. 3B I). The ablation (Fig. 3B II, ablation marked by a white arrow) cuts the bundle creating a new plus and a new minus end. The newly created minus end is unstable; as a consequence, the severed fragment depolymerized mainly by shrinking from its minus end. As a control, we observed that the other, nonablated microtubules were not affected (Fig. 3B II–IV).

In our third experiment, we ablated the mitotic spindle. Fission yeast has a closed mitosis, i.e., the nuclear envelope does not break down. During mitosis, the two spindle

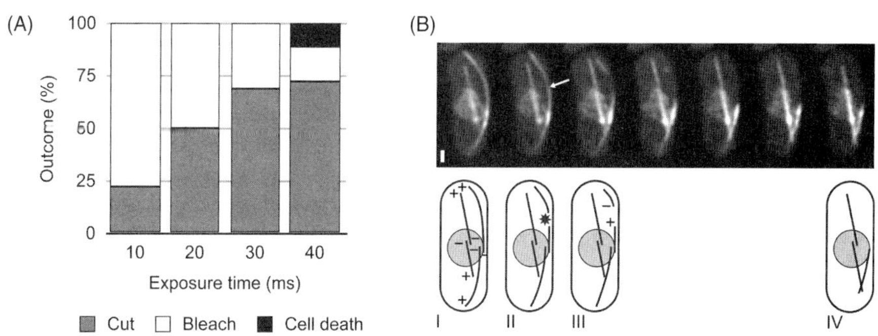

Fig. 3 Outcome of ablations performed on interphase microtubules using different exposure times (A) [data taken from Maghelli and Tolic-Norrelykke (2008)]. The optimal exposure time is between 20 and 30 ms. (B) By ablating interphase microtubules a free, unstable minus-end is created. As a result, the fragment depolymerizes mainly by shrinking from its minus end. Scale bar is 1 μm, time between frames is 5 s.

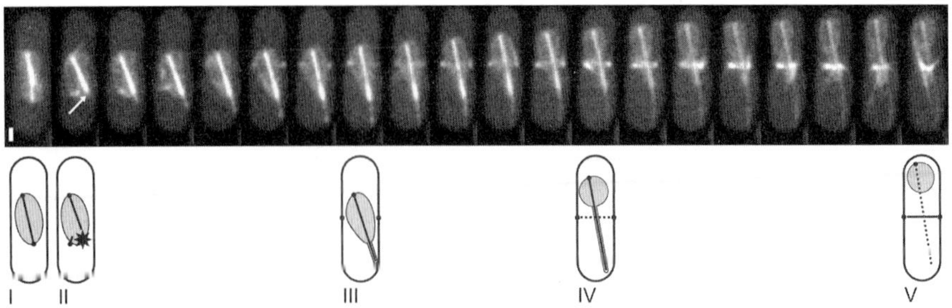

Fig. 4 Ablating the mitotic spindle near one spindle pole body induces asymmetric division. The ablated spindle continues to grow, deforming the nuclear envelope (ablation marked by the white arrow). The polymerization forces displace the nucleus, eventually leading to an abnormal mitosis, in which one daughter cell inherits all the nuclear material. Scale bar is 1 μm, time between frames is 100 s.

poles remain embedded in the nuclear membrane (Fig. 4 I). We performed the ablation near a spindle pole (Fig. 4 II, ablation spot marked by a white arrow). The ablation cut the spindle asymmetrically: one spindle pole remained embedded in the nuclear membrane (upper pole in Fig. 4 III), while near the other pole the microtubules forming the spindle deformed the nuclear membrane creating a protrusion (lower side in Fig. 4 III). As the spindle elongated, the outgrowing protrusion contacted the cell cortex: further polymerization of the spindle displaced the nucleus upward (Fig. 4 IV). As the cell started to divide, the nucleus had already been pushed across the septum. As a result, the cell segregated the nuclear material asymmetrically: one sibling retained the whole nucleus (the upper cell), while the other one was deprived of the nucleus (the lower cell in Fig. 4 V) (Raabe *et al.*, 2009).

We next used laser ablation to study the force balance during fission yeast meiosis. During meiosis, two cells fuse forming a zygote and their nuclei merge. In fission yeast, the meiotic prophase is accompanied by prominent oscillations of the fused nuclei, termed horsetail nuclear movement (HNM) (Chikashige *et al.*, 1994). By using laser ablation, it has been demonstrated that dynein generates the nuclear movement by pulling on cytoplasmic microtubules (Vogel *et al.*, 2009; Yamamoto *et al.*, 1999). To further investigate the mechanism underlying the nuclear oscillations, we cut a microtubule bundle during the HNM in front of the moving spindle pole (Fig. 5 II, ablation spot marked by a white arrow). The ablation disconnected a microtubule bundle from the moving nucleus, creating a free fragment. We observed that the severed microtubule bundle continued its movement along the cell cortex (Fig. 5 III–V).

C. Discussion

In our experiments, we used laser ablation to manipulate the cytoskeleton during different phases of the cell cycle. The ablations perturbed the mechanical equilibrium or the geometrical arrangement of the cytoskeleton. We then inferred information about the forces acting in the cell by observing how the cell reacted to the modifications. When performing ablation in living specimens, the major pitfall is to

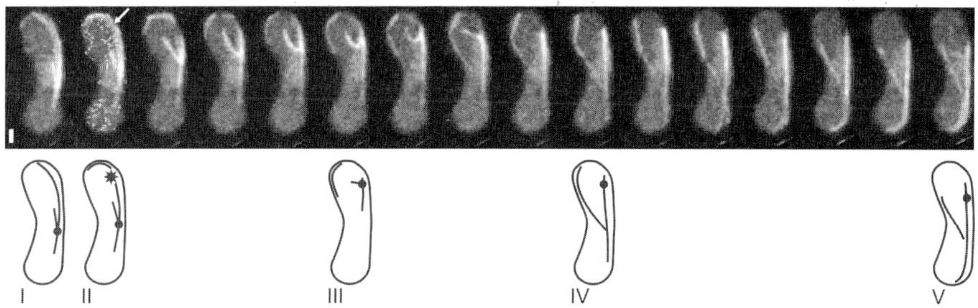

Fig. 5 Ablation of microtubules during meiosis. A microtubule bundle is cut (ablation marked by the white arrow) and a fragment detaches from the spindle pole body (black circle in the scheme). The fragment moves along the cell cortex, independently from the spindle pole body. Scale bar is 1 μm, time between frames is 12 s.

mistake an ablation-induced artifact for a real, physiological reaction of the cell. It is therefore necessary to perform controls to rule out any spurious effect. We therefore started our experiments by calibrating the ablation efficiency, trying to find the parameter set allowing us to perform ablations without inducing significant damage (Fig. 3). Using these parameters, we then performed ablations to perturb the force balance during mitosis (Fig. 4), exploiting the cell internal forces to asymmetrically segregate the nuclear material in a dividing cell. Compared with creating enucleated cells by centrifugation (Carazo-Salas and Nurse, 2006), the ablation-based method described here has several advantages: in the first place it permits one to follow the displacement process, allowing for insight into the intracellular forces leading to the asymmetric segregation. In the second place, since no external forces are used, the spatial arrangement of other organelles is not perturbed.

In our next experiment, we used laser ablation to study the force generators responsible for the nuclear movement during meiosis (Fig. 5). Previous work has shown that the nuclear movement is driven by dynein (Yamamoto *et al.*, 1999). A recent study employed laser ablation to dissect the force balance during the nuclear movement and put forward a model in which the force generators self-organize (Vogel *et al.*, 2009). Here, we observed independent movement of the microtubule bundle, disconnected from the moving spindle pole body (Fig. 5). The observed movement fits well with the model proposed in Vogel *et al.* (2009). The movement is most likely a consequence of the pulling forces exerted by dynein motors anchored at the cell cortex, which pull independently on the fragment and on the microtubules connected to the spindle pole body (Fig. 5 III–V).

V. Methods

A. Cell Culture and Sample Preparation

Fission yeast cells were grown in liquid yeast extract medium at 25° C. During imaging and manipulation, the cells were attached to the glass bottom of a Petri dish using ~2 μl of

2 mg/ml lectin. The Petri dish was filled with ∼ 3 ml of liquid minimal medium. On the microscope stage, the sample was kept at 25° C. When needed, carbendazim (MBC) at a concentration of 25 μ g/ml was used to depolymerize microtubules.

B. Microscopy

Laser ablation and optical trapping were performed using a custom-built two-photon setup (Maghelli and Tolic-Norrelykke, 2008). For ablation, the pulsed laser of the microscope was tuned to a wavelength of 895 nm. The power of the laser at the sample plane was ~5 mW during imaging and ~100 mW during ablation. Optical trapping was achieved using a near-infrared (970 nm) CW laser. To perform manipulation using the optical trap, the focused beam was either kept fixed with respect to the objective while moving the sample, or the sample was kept fixed with respect to the objective while steering the optical trap using a pair of computer-controlled galvanometer mirrors.

The objectives used were either a 63× 1.0 N.A. water dipping lens or a 100× 1.4 N.A. oil immersion objective.

The image acquisition and manipulations were performed using a custom-written software (LabView) controlling the setup. Image analysis was performed using ImageJ, data analysis using Igor Pro or Matlab.

References

Amy, R., and Storb, R. (1965). Selective mitochondrial damage by a ruby laser microbeam: An electron microscopic study. *Science* **150**, 756–758.

Ashkin, A. (1992). Forces of a single-beam gradient laser trap on a dielectric sphere in the ray optics regime. *Biophys. J.* **61**, 569–582.

Ashkin, A. (1998). Forces of a single-beam gradient laser trap on a dielectric sphere in the ray optics regime. *Methods Cell Biol.* **55**, 1–27.

Ashkin, A., and Dziedzic, J. M. (1987). Optical trapping and manipulation of viruses and bacteria. *Science* **235**, 1517–1520.

Ashkin, A., Dziedzic, J. M., and Yamane, T. (1987). Optical trapping and manipulation of single cells using infrared laser beams. *Nature* **330**, 769–771.

Berns, M. W., Rattner, J., Brenner, S., and Meredith, S. (1977). The role of the centriolar region in animal cell mitosis. A laser microbeam study. *J. Cell Biol.* **72**, 351–367.

Carazo-Salas, R. E., and Nurse, P. (2006). Self-organization of interphase microtubule arrays in fission yeast. *Nat. Cell Biol.* **8**, 1102–1107.

Chikashige, Y., Ding, D. Q., Funabiki, H., Haraguchi, T., Mashiko, S., Yanagida, M., and Hiraoka, Y. (1994). Telomere-led premeiotic chromosome movement in fission yeast. *Science* **264**, 270–273.

Colombelli, J., Reynaud, E. G., Rietdorf, J., Pepperkok, R., and Stelzer, E. H. (2005). In vivo selective cytoskeleton dynamics quantification in interphase cells induced by pulsed ultraviolet laser nanosurgery. *Traffic* **6**, 1093–1102.

Daga, R. R., Yonetani, A., and Chang, F. (2006). Asymmetric microtubule pushing forces in nuclear centering. *Curr. Biol.* **16**, 1544–1550.

Heisterkamp, A., Maxwell, I. Z., Mazur, E., Underwood, J. M., Nickerson, J. A., Kumar, S., and Ingber, D. E. (2005). Pulse energy dependence of subcellular dissection by femtosecond laser pulses. *Opt. Express* **13**, 3690–3696.

Khodjakov, A., La Terra, S., and Chang, F. (2004). Laser microsurgery in fission yeast; role of the mitotic spindle midzone in anaphase B. *Curr. Biol.* **14**, 1330–1340.

Liang, H., Vu, K. T., Krishnan, P., Trang, T. C., Shin, D., Kimel, S., and Berns, M. W. (1996). Wavelength dependence of cell cloning efficiency after optical trapping. *Biophys. J.* **70**, 1529–1533.

Liu, Y., Cheng, D. K., Sonek, G. J., Berns, M. W., Chapman, C. F., and Tromberg, B. J. (1995). Evidence for localized cell heating induced by infrared optical tweezers. *Biophys. J.* **68**, 2137–2144.

Maghelli, N., and Tolic-Norrelykke, I. M. (2008). Versatile laser-based cell manipulator. *J. Biophotonics* **1**, 299–309.

Neuman, K. C., Chadd, E. H., Liou, G. F., Bergman, K., and Block, S. M. (1999). Characterization of photodamage to *Escherichia coli* in optical traps. *Biophys. J.* **77**, 2856–2863.

Raabe, I., Vogel, S. K., Peychl, J., and Tolic-Norrelykke, I. M. (2009). Intracellular nanosurgery and cell enucleation using a picosecond laser. *J. Microsc.* **234**, 1–8.

Riveline, D., and Nurse, P. (2009). "Injecting" yeast. *Nat. Methods* **6**, 513–4.

Sacconi, L., O'Connor, R. P., Jasaitis, A., Masi, A., Buffelli, M., and Pavone, F. S. (2007). In vivo multiphoton nanosurgery on cortical neurons. *J. Biomed. Opt.* **12**, 050502.

Sacconi, L., Tolic-Norrelykke, I. M., Antolini, R., and Pavone, F. S. (2005a). Combined intracellular three-dimensional imaging and selective nanosurgery by a nonlinear microscope. *J. Biomed. Opt.* **10**, 14002.

Sacconi, L., Tolic-Norrelykke, I. M., Stringari, C., Antolini, R., and Pavone, F. S. (2005b). Optical micromanipulations inside yeast cells. *Appl. Opt.* **44**, 2001–2007.

Simmons, R. M., Finer, J. T., Chu, S., and Spudich, J. A. (1996). Quantitative measurements of force and displacement using an optical trap. *Biophys. J.* **70**, 1813–1822.

Stiess, M., Maghelli, N., Kapitein, L. C., Gomis-Ruth, S., Wilsch-Brauninger, M., Hoogenraad, C. C., Tolic-Norrelykke, I. M., and Bradke, F. (2010). Axon extension occurs independently of centrosomal microtubule nucleation. *Science* **327**, 704–707.

Tolic-Norrelykke, I. M. (2008). Push me pull you: How microtubules organize the cell interior. *Eur. Biophys. J.* **37**, 1271–1278.

Tolic-Norrelykke, I. M. (2010). Force and length regulation in the microtubule cytoskeleton: Lessons from fission yeast. *Curr. Opin. Cell Biol.* **22**, 21–28.

Tolic-Norrelykke, I. M., Munteanu, E. L., Thon, G., Oddershede, L., and Berg-Sorensen, K. (2004a). Anomalous diffusion in living yeast cells. *Phys. Rev. Lett.* **93**, 078102.

Tolic-Norrelykke, I. M., Sacconi, L., Stringari, C., Raabe, I., and Pavone, F. S. (2005). Nuclear and division-plane positioning revealed by optical micromanipulation. *Curr. Biol.* **15**, 1212–1216.

Tolic-Norrelykke, I. M., Sacconi, L., Thon, G., and Pavone, F. S. (2004b). Positioning and elongation of the fission yeast spindle by microtubule-based pushing. *Curr. Biol.* **14**, 1181–1186.

Tran, P. T., Marsh, L., Doye, V., Inoue, S., and Chang, F. (2001). A mechanism for nuclear positioning in fission yeast based on microtubule pushing. *J. Cell Biol.* **153**, 397–411.

Vogel, A., Noack, J., Hüttman, G., and Paltauf, G. (2005). Mechanisms of femtosecond laser nanosurgery of cells and tissues. *Appl. Phys. B.* **81**, 1015–1047.

Vogel, S. K., Pavin, N., Maghelli, N., Julicher, F., and Tolic-Norrelykke, I. M. (2009). Self-organization of dynein motors generates meiotic nuclear oscillations. *PLoS Biol.* **7**, e1000087.

Vogel, A., and Venugopalan, V. (2003). Mechanisms of pulsed laser ablation of biological tissues. *Chem. Rev.* **103**, 577–644.

Yamamoto, A., West, R. R., McIntosh, J. R., and Hiraoka, Y. (1999). A cytoplasmic dynein heavy chain is required for oscillatory nuclear movement of meiotic prophase and efficient meiotic recombination in fission yeast. *J. Cell Biol.* **145**, 1233–1249.

CHAPTER 11

A Fast Microfluidic Temperature Control Device for Studying Microtubule Dynamics in Fission Yeast

Guilhem Velve-Casquillas*, Judite Costa†, Frédérique Carlier-Grynkorn*, Adeline Mayeux*, *and* Phong T. Tran*,†

*Institut Curie, UMR 144 CNRS, Paris 75005, France

†Cell & Developmental Biology, University of Pennsylvania, Philadelphia, Pennsylvania 19104

METHODS IN CELL BIOLOGY, VOL. 97

978-0-12-381349-7
DOI: 10.1016/S0091-679X(10)97011-8

Abstract

Recent development in soft lithography and microfluidics enables biologists to create tools to control the cellular microenvironment. One such control is the ability to quickly change the temperature of the cells. Genetic model organism such as fission yeast has been useful for studies of the cell cytoskeleton. In particular, the dynamic microtubule cytoskeleton responds to changes in temperature. In addition, there are temperature-sensitive mutations of cytoskeletal proteins. We describe here the fabrication and use of a microfluidic device to quickly and reversibly change cellular temperature between 2°C and 30°C. We demonstrate the use of this device while imaging at high-resolution microtubule dynamics in fission yeast.

I. Introduction

The microtubule cytoskeleton is essential for cellular processes such as cell polarity or mitosis. Microtubules are dynamic biopolymers composed of α β-tubulin heterodimers. The fission yeast *Schizosaccharomyces pombe* has been effectively used to study the microtubule cytoskeleton. Historically, drugs have been used to modulate microtubule dynamics in fission yeast. For example, carbendazim (methyl benzimidazol-2-yl carbamate) is commonly used to depolymerize microtubules. Repolymerization is achieved upon drug washout. Microtubules also respond to temperature. Cells incubated in ice bath of below 6°C will completely depolymerize their microtubules. Repolymerization is achieved by heating up the cells. Microtubule dynamics in fission yeast are relatively fast—a typical microtubule has approximately 2 μm/min growth rate, 8 μm/min shrinkage rate, 0.02 min^{-1} catastrophe frequency, and little or no rescue. Thus, it is useful to be able to change drugs or temperature faster than the 1 min timescale to precisely observe microtubule dynamic responses.

The thermal time constant of a system decreases with decreasing size. Thus, miniaturized devices can achieve very fast temperature changes. Microfluidic systems, which enable fluid manipulation at the micron scale, are good candidates for fast temperature changes. Moreover, with recent development of technology based on the molding of PDMS (Polydimethylsiloxane), which is relatively inexpensive and easy to handle, microfluidics show a strong potential for fabrication of tools dedicated to cell biological experiments (Belanger *et al.*, 2001, Charati and Stern, 1998, Duffy *et al.*, 1998).

We present here a detailed protocol to fabricate and use a microfluidic temperature control device that enables temperature changes in the range of 2–50°C in less than 10 s. This device can be coupled to an oil immersion objective lens for high-resolution imaging. The device has been optimized for fission yeast studies, but can easily be adapted for other types of cells or organisms. Further, this device can be coupled with other microfluidic functionalities such as mechanical deformation or chemical perfusion. This device has been used to depolymerize microtubules at low temperature, and it has been also used to deactivate proteins at high temperature in the temperature-sensitive mutants widely available in fission yeast.

II. Device and Setup Presentation

The temperature control device presented here is a bilayer PDMS device bonded onto a 150 μm thick glass coverslip. The bottom channel containing cells are in contact with the coverslip glass surface for imaging and are topped by a larger channel dedicated to temperature control (Fig. 1A). A thin 15 μm PDMS membrane separates the top and bottom channels to avoid direct fluid contact and serves to conduct temperature. Flowing water at controlled temperature through the top channel will,

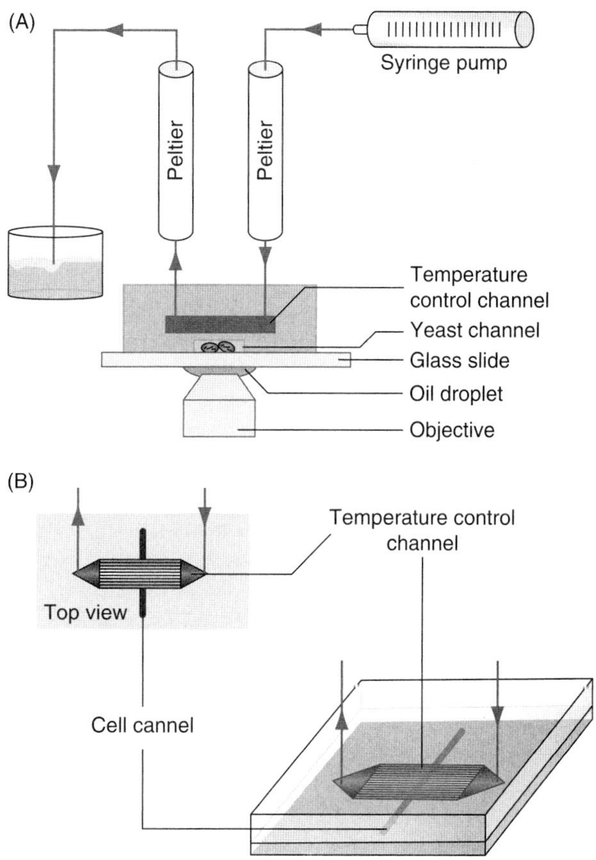

Fig. 1 Temperature control device schematic and setup. (A) Schematic of the complete setup. (B) Schematic of the bilayer PDMS device. The top layer contains microchannels for temperature changes. The bottom layer contains microchannels for holding cells.

by heat diffusion that occurs through the thin PDMS membrane, change the temperature of the bottom channel containing cells. The top temperature control channel is simply a parallel network of 200 × 100 μm cross-section channel (1 cm long) for water circulation. In contrast, channels containing cells can be fabricated with different shapes and thicknesses depending on the type of cells involved (yeast, bacteria, etc.) and the required applications (cell deformation, drug screening, etc.). The temperature control setup is composed of two Peltier modules, a syringe pump, and a microscope (Fig. 1B). We use the Peltier module to control water temperature before pumping it into the microfluidic device. A typical Peltier can switch from 0°C to 50°C in less than 1 min. For our purpose, one Peltier module is plugged upstream (inlet) of the device and the second downstream (outlet). Once Peltier modules are at the desired temperatures, the temperature in the cells channel can be quickly changed from one to the other. This is possible because changing the direction of the water flow changes the Peltier in which the water goes through before entering the microfluidic device. The temperature change of the device is then limited by the time required to reverse the flow and not by the Peltier time constant.

III. Mold and Device Fabrication

Devices are fabricated using soft lithography of PDMS. This method enables fabricating hundreds of microfluidic devices using a single mold. PDMS has several advantages for the fabrication of microfluidic devices dedicated to cell biology. First, this elastomer is transparent, biocompatible (McDonald and Whitesides 2002), and permeable to gas (Unger *et al.*, 2000). Those characteristics enable easy cell culture and microscope imaging. Second, from a technological point of view, PDMS material is cheap, easy to mold, has a low Young's modulus, and can be easily covalently bonded to itself or glass using plasma ionization treatment. The small Young's modulus of PDMS is particularly interesting for implementation of fluidic valves (Velve-Casquillas *et al.*, 2010). Moreover, two or more PDMS replicas can be bonded together using plasma treatment that enables the fabrication of multilayer microfluidic devices.

A. Preliminary Step: Microfluidic Mold Fabrication

The first step required for microfluidic device fabrication is the mold fabrication. Microfluidic molds are fabricated using photolithography of SU8 onto silicon substrate. Briefly, a mold is made from transferring the channel designs from a mask onto a photoresist-coated silicon wafer substrate via UV photolithography. This wafer serves as the future mold. Since this method is extensively documented and well described in the Microchem SU8 datasheet, we will not describe it in detail here. Since our device is composed of two layers of channels, then two different molds are required. Before spin coating the proper photoresist onto the silicon substrate, a layer of omnicoat is used to promote subsequent adhesion between the photoresist and the silicon. For the temperature control channel mold, we used a 100 μm thick SU8-2050 (or 3050) photoresist. Similarly, for cell channel mold, we used a 5 μm thick SU8-2005 photoresist. Different types of SU8 enable different channel thickness.

Once the molds are made, it is then necessary to coat them with an anti-adhesive treatment to avoid removing the photoresist patterns during subsequent PDMS replication. For this purpose, we place the mold inside a closed Petri dish and add a 10 μl droplet of chlorotrimethylsilane for 3 min. Natural evaporation of the silane in the Petri dish will coat a thin layer of silane onto the mold surface. Since silane is harmful and volatile, this operation should be performed under a fume hood. Once molds are treated with silane, they can be used for repeated PDMS device fabrication (10–100 replications) before requiring a renewal of anti-adhesive treatment or a new mold.

Photolithography of features as small as 2 μm does not necessarily require clean room facilities such as found in physics, engineering, or material sciences labs. We find that for cellular dimensions, photolithography equipment can be installed in a classical biology fume hood. To fabricate our molds we used the OAI UV lamp (Fig. 2B), Laurell spin coater (Fig. 2C), and two Barnstead hotplates. With the exception of isopropanol, all mold making reagents are from Microchem.

Mask design for the temperature control channels is an array of 12 parallel 1 cm long × 200 μm wide × 100 μm thick channels, spaced 100 μm apart. Mask design for the cell channels varies depending on the desired biological sample and experiment.

The temperature control device fabrication procedure is schematically described in Fig. 2A. The two layers of channels are fabricated independently and are then covalently bonded using plasma treatment. The full fabrication process is described below.

B. Step 1: PDMS Preparation

This step produces a homogenized mixture of PDMS and curing agent without air bubbles.

1. Pour into a cup liquid PDMS and then add curing agent to a ratio 9:1.
2. Stir vigorously for about 3 min (with a plastic fork or spoon) to homogenize the mixture. The stirring will generate small air bubbles, turning the clear mixture white.
3. Put the cup in a vacuum chamber for 30 min to degas to remove the air bubbles. The mixture should be transparent at the end.

1. Remarks!

The ratio of PDMS to curing agent affects the stiffness of the final PDMS product. More curing agent leads to stiffer PDMS products. We find that 9:1 ratio is optimal for our work, but small changes (7:1 to 12:1) will not be critical for our application.

C. Step 2a: Fabrication of the Temperature Control Channels

This step describes how to make PDMS replica of the microchannel pattern from the mold.

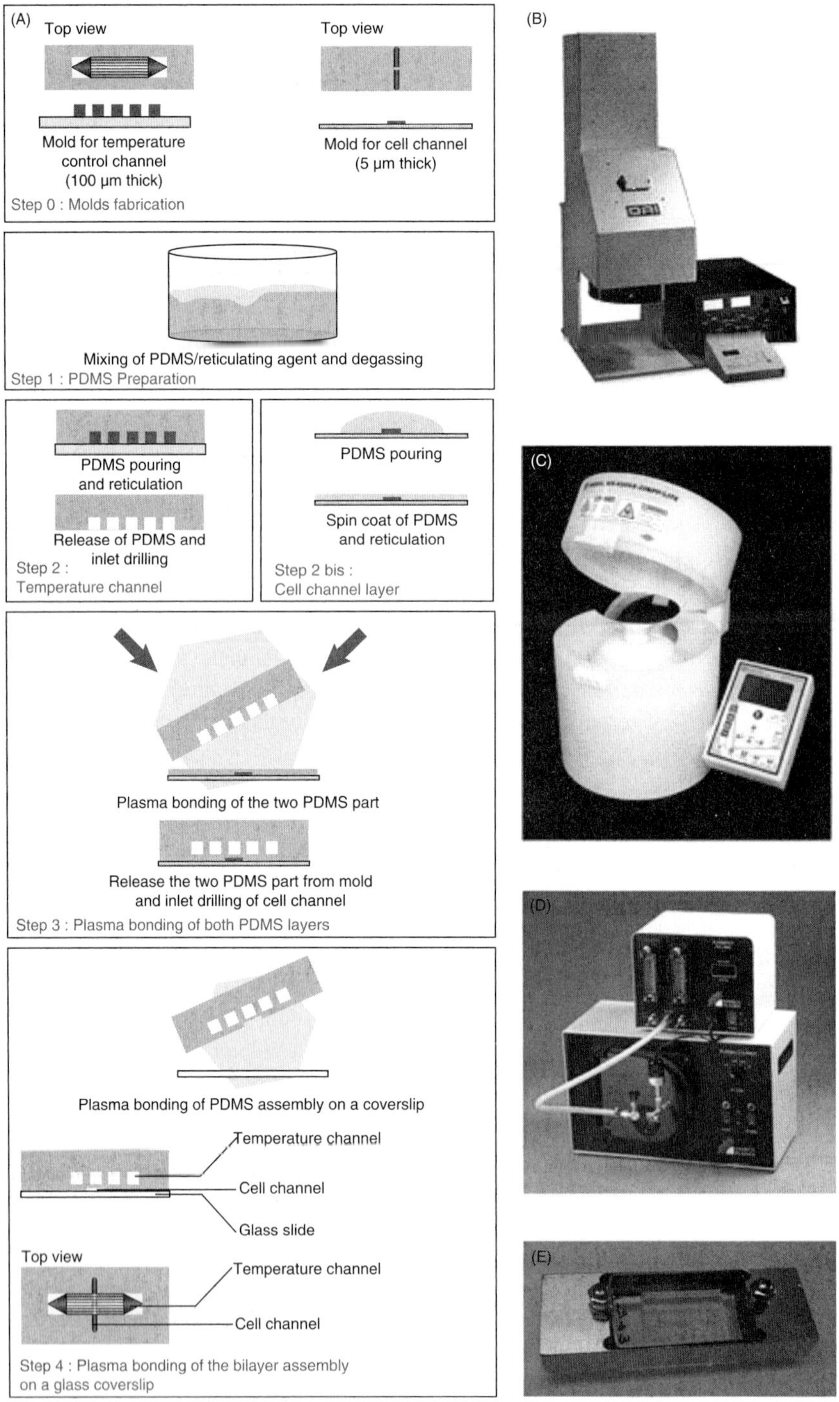

Fig. 2 The fabrication process. (A) Detailed fabrication procedure of the PDMS microfluidic temperature control device. (B) Photograph of the UV light source used to expose the patterns from the mask onto the wafer coated with photoresist (the future mold). (C) Photograph of the spin coater used to coat materials evenly onto surfaces. (D) Photograph of the plasma cleaner used to oxidize surfaces for subsequent bonding. (E) Photograph of a homemade coverslip holder.

1. Once the PDMS mixture is degassed and there are no bubbles, pour it gently onto the temperature control mold (which is glued to the bottom of a Petri dish) to a height of approximately 3 mm. Wait a few minutes to allow any new air bubbles to reach the surface, then gently blow them away. Once no air bubbles remain on the PDMS surface, put the dish into an oven at 65°C for 2–4 h to cure, and harden the PDMS mixture.
2. Once the PDMS is cured or hardened, cut using a surgical scalpel a region of interest around the microchannel pattern and lift it up from the mold surface. This piece should look like a block of clear and flexible material.
3. Drill inlet and outlet holes onto the PDMS block using a 20-gauge needle.
4. You can store the PDMS block in a closed Petri dish with the channel side up.

1. Remarks!

The recommended 2–4 h curing time is not critical for our application. However, too long a curing time could lead to PDMS aging and brittleness.

During the cut out step, one should take a minimum of 2–3 mm margin around the microchannels. This will yield a large surface for better bonding and water tightness for the subsequent device assembly.

Avoid touching the PDMS microchannels with your fingers, since the surface properties of the material are critical for plasma treatment and bonding.

During inlet and outlet drilling, when the needle goes through the PDMS block a little PDMS cylinder remains at the end of the needle and should be taken off before needle removal. Needles used for drilling should be smaller than the steel tube adaptor to allow a tight seal between the PDMS block and the steel tube adaptor used during the subsequent injection procedure. Moreover, to avoid PDMS cracking during drilling, the needle edge should be previously smoothed using sandpaper.

D. Step 2b: Fabrication of Cell Channel Layer

To fabricate a bilayer PDMS device, in which the top temperature control channels are separated from the bottom cell channels by just a few microns, the bottom layer has to be a very thin PDMS layer. This step describes how to fabricate a 15 μm thin PDMS membrane on the mold containing the cell microchannels.

1. Place the cell microchannel mold onto a spin coater (Fig. 2C), and then pour degassed PDMS mixture on top to cover about 30% of the surface.
2. Launch the spin coater at 500 rpm for 10 s (acceleration 100 rot^2/min), followed by 6000 rpm for 30 s (acceleration 500 rot^2/min). The spinning will spread the PDMS mixture onto the mold as an even layer approximately 15 μm thin.
3. Put the mold onto a hotplate at 95°C for 30 min to cure the PDMS mixture.
4. You can store the mold with the PDMS membrane on top in a closed Petri dish with the PDMS side up.

1. Remarks!

The relative thinness of the membrane is important since it plays a critical role in heat transfer between the temperature and the cell channels. The thinness of the membrane depends mainly of two parameters: the viscosity of the PDMS and the rotation speed during spin coating. The PDMS mixture should be freshly made and degassed (<1 h) before spin coating, because the PDMS mixture will slowly cure even at ambient room temperature and change its viscosity.

The PDMS surface containing the microchannels should not be touched when handling, since the surface properties of PDMS are crucial for proper plasma treatment and bonding.

E. Step 3: Plasma Treatment and Bonding of Both PDMS Layers (Temperature Control and Cell Microchannels)

To fabricate the PDMS bilayer it is necessary to covalently stick the top temperature control PDMS block (Step 2a) onto the bottom cell PDMS mold (Step 2b). For this purpose, the most common technique is plasma ionization treatment. We use a Plasma Cleaner (Fig. 2D), with a plasma flow module to control the pressure inside the plasma chamber.

1. Place both top PDMS block (microchannel side up) and bottom PDMS mold into the plasma chamber. Start the vacuum.
2. Once the air pressure inside the plasma chamber is stabilized between 500 mTorr and 1000 mTorr, turn the radio frequency RF power on high for 30 s. You should see a purple glow inside the chamber. This glow indicates that the surfaces of the PDMS are being ionized.
3. Release the vacuum and immediately take out the PDMS blocks (take care not to touch the surfaces). Then place the top temperature block (microchannel side down) directly on top of the bottom PDMS mold. The two PDMS surfaces should start to bond covalently.
4. Put the bilayer assembly on a hotplate at 95°C for 30 min. You now have a bilayer PDMS block on top of the cell microchannel mold.
5. Use a surgical scalpel to cut out bottom PDMS layer (which is now bonded to the top PDMS layer) and peel up the complete bilayer PDMS block.
6. Drill inlet and outlet holes onto the cell microchannels of the PDMS bilayer block using a 20-gauge needle.
7. You can store this bilayer PDMS device in a closed Petri dish with the microchannel side up.

1. Remarks!

To eliminate possible dust settling on the surface of PDMS block prior to plasma treatment and bonding, one can use an air gun to blow on the PDMS. An alternative to the air gun is to quickly stick on and then remove Scotch tape (3M) on the block surfaces just before plasma treatment.

Plasma treatment is critical for microfluidic fabrication. Mistake in the exposure time, the pressure, or the presence of impurities in the vacuum chamber can lead to inefficient plasma treatment. In good normal condition, the plasma should have a purple color. White or pink plasma indicates too high a pressure, and evanescent or clear plasma indicates too low a pressure.

The presence of refluxing oil from the vacuum pump into the plasma vacuum chamber also leads to inefficient plasma treatment and subsequent bonding. If the plasma treatment time is too long (> 1 min) subsequent bonding between PDMS surfaces will be inefficient.

Contact between the two PDMS blocks has to be done within 1 min after plasma treatment. Longer waiting time will lead to less efficient bonding. Moreover, once the two blocks are in contact, do not try to reposition or readjust them.

The bottom cell channels should be positioned under the center of the top temperature control channels to reach optimal temperature uniformity.

After the bilayer PDMS block is cut from the mold, one can clean the mold of residual PDMS coating by using a tweezer or rolling on the surface with a gloved finger.

F. Step 4: Plasma Bonding of the Bilayer PDMS Assembly onto a Glass Coverslip

This step will bond the bilayer PDMS block onto the glass coverslip to create the final enclosed device.

1. Place both the glass coverslip and the bilayer PDMS block (microchannels facing up) into the plasma chamber. Start the vacuum.
2. Once the air pressure inside the plasma chamber is stabilized between 500 mTorr and 1000 mTorr, turn the radio frequency RF power on high for 30 s. You should see a purple glow inside the chamber. This glow indicates that the surfaces of the PDMS are being ionized.
3. Release the vacuum and immediately take out the glass coverslip and PDMS block (take care not to touch the surfaces). Then place the PDMS block (microchannel side down) directly on top of the glass coverslip. The two surfaces should start to bond covalently immediately.
4. Put the device on a hotplate at 95°C for 30 min. You now have a bilayer PDMS block on top of the glass coverslip.

1. Remarks!

If the plasma treatment works well the contact area between PDMS block and glass coverslip should spread and bond within seconds. If this is not the case, and some noncontact regions remained (white area), one can push the PDMS block gently down onto the glass coverslip with a pair of tweezers. Do not apply too much force or you risk collapsing the microchannels.

To facilitate device handling and microscope imaging, we use a homemade glass coverslip holder as shown in Fig. 2E.

IV. Setup Installation

The experimental setup is composed of an inverted microscope (Nikon TE2000e, 100×/1.4NA oil immersion objective), two Peltier modules, a water tank, and a syringe pump. Beginning from this basic setup it is necessary to add two Peltier temperature controllers and two peristaltic pumps to control and maintain the Peltier temperature (Fig. 3A and B).

A. Step 1: Peltier Module Microfluidic Connection

In this step we will prepare the Peltier modules to fit with the PDMS microfluidic device. Each Peltier has an inlet–outlet metal tube. Water is pumped through the inlet, is heated up or cooled down as it travels through the Peltier, and then exit through the outlet into the PDMS device. The Peltier itself requires cooling.

1. Connect the Peltier outlet to the PDMS device. To ensure good fitting, we connect the Peltier metallic outlet to a 2 cm (1.14 mm ID) polyethylene (PE) tubing and then connect this to a 4 cm "Microline" tubing (0.5 mm ID). This is terminated by a stainless steel tube adaptor that will fit directly into the inlet hole of the PDMS device. This 6 cm long assembly is long enough to easily handle and plug into the device, but short enough to limit heat dissipation prior to reaching the cells.
2. Repeat for the second Peltier. With the two configurations, the first Peltier is used for heating and the second for cooling.
3. Connect the Peltier inlet to the water source. The tubing assembly is the same as above; however, the length of tubing is not important here.
4. Repeat for the second Peltier. Note that the first Peltier will be connected to a syringe pump, while the second will be submerged into a water bath. The syringe pump acts to push water through the first Peltier for heating and to pull water through the second Peltier for cooling.

1. Remarks!

To avoid accidental lifting of tubings (and therefore water leakage), one should tape the tubings at different strategic points on or around the microscope stage.

At low temperature settings, condensation can appear on the Peltier module. To avoid water leakage onto the microscope, one can tie a sponge or paper towel around the Peltier module.

To position the Peltier modules directly near the PDMS microfluidic device inlet, we used a common stand.

This microfluidic temperature control system can be used with all types of inverted microscope.

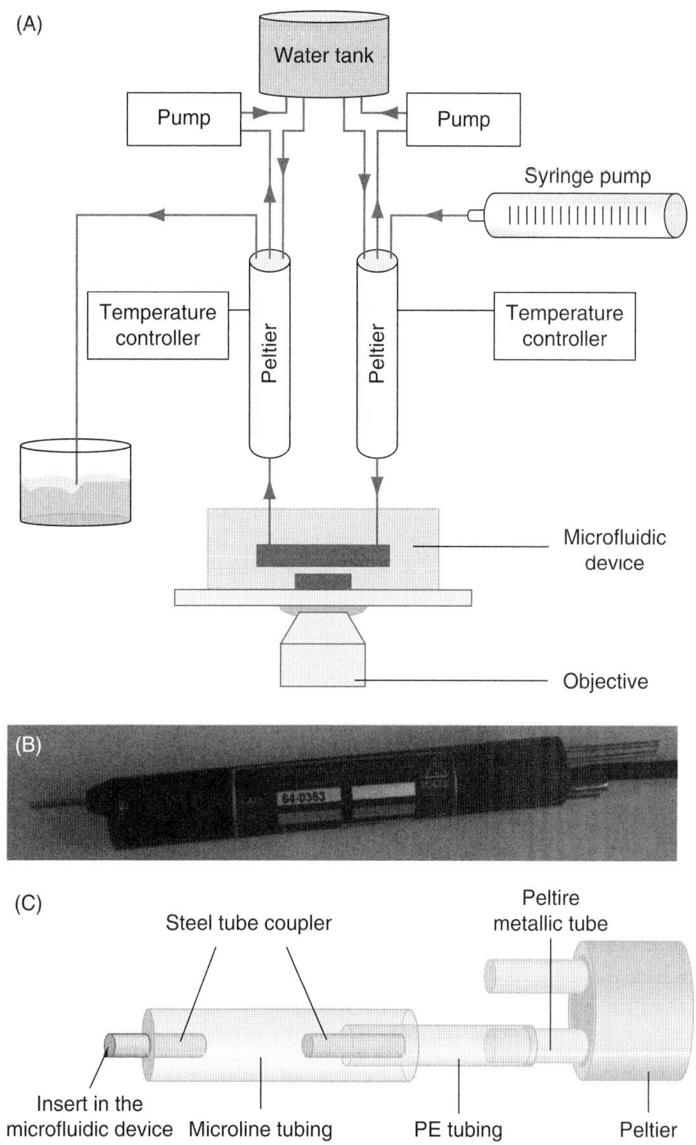

Fig. 3 Peltier setup. (A) Schematic of the complete setup for the Peltier modules. (B) Photograph of the Peltier module used for fast temperature changes. (C) Schematic of the tubing assembly between the Peltier outlet and the PDMS microfluidic device.

B. Step 2: Connection of Peristaltic Pump to Peltier Module

The Peltier modules need internal cooling to function properly. We directly cool the two Peltier modules using a closed-loop 2 l water reservoir bottle connected to Tygon R-1000 silicon tubing driven by peristaltic pumps.

1. Remarks!

When using Peltier as low as 1°C, the Peltier module may not be able to maintain this temperature. To solve this problem one can increase the peristaltic pump flow rate or decrease the reservoir water temperature by putting the water bottle on ice.

For low temperature experiments lasting several hours, the water reservoir should be at least 2 l, immersed in ice, and with a peristaltic pump rate of 6 l/h.

Other solutions for cooling the Peltier module are to use the thermal cooling module TCM1 from Warner Instrument or use a chiller.

Because of the high water flow rate involved for Peltier cooling, one should take care that tubings are connected tightly and securely to avoid leakage. Moreover, one should take care to examine frequently the ageing of the silicon tubing in contact with the rotary part of the peristaltic pump.

For more security against leakage, peristaltic pumps should be placed inside plastic bowls.

V. Biological Experiments

A. Step1: Device Preparation and Cell Injection

1. Use a small 2–3 ml syringe and a 24-gauge needle connected to a "Microline" tube with a stainless steel tubing adaptor for cell handling.
2. Fill the syringe with cells. One should avoid the presence of air bubbles in the syringe or tubing.
3. Plug the steel tubing end into the PDMS device cell inlet and inject gently the cells into the cell channels. Be sure that no air bubbles remain in the device at the end of the operation.

1. Remarks!

If the cell channels are not designed to allow media renewal and if the experiment will run for more than 2–3 h, then the cells may begin to miss fresh nutriment and the media in the channels may begin to dry up due to evaporation through PDMS. To limit this, one should plug a 1 cm long "Microline" tube at the outlet and inject media from the inlet to fill the outlet tubing without bubble. Once the outlet tube is full of media then inject cells from the inlet and cut the inlet tube at the same length as the outlet one.

This extra tubings and media will help renew the cell media and enable experiments up to 9 h depending on the cell concentration.

While injecting cells, one could put the PDMS device onto a black background piece of paper to facilitate visualization of air bubbles. In contrast to water (which has little refractive index difference compared to PDMS), air has a high refractive index difference compared to PDMS. Thus, air bubbles appear white inside the PDMS device against the black background. Water would be transparent, like PDMS.

B. Step 2: Installing the PDMS Device on the Peltier Setup

1. Install a large 200-ml syringe filled with water onto the syringe pump.
2. Connect both Peltier modules to the PDMS device, the syringe pump, and the peristaltic pump cooling system (refer to Section IV Setup Installation).
3. Start the syringe pump until all tubings connected to the temperature channels are filled with water. Be sure that the entire tubing capillary is filled, which is essential for fast temperature change. During this step, the Peltier modules should be at ambient temperature to avoid temperature changes in the device during capillary filling. Once the capillary is filled, the syringe pump can be stopped.
4. Start the peristaltic pumps at 100 ml/min rate.
5. Set the Peltier modules at the desired temperatures.

1. Remarks!

At this stage, if no leakage occurs, you are ready to put the PDMS device onto the microscope to search for cells.

C. Step 3: Performing Temperature Changes

To change the temperature in the capillary, it is necessary to push or pull water through the Peltier modules and the PDMS device by the syringe pump (Fig. 4A). The upstream Peltier (connected to the syringe pump) should be hot and the downstream Peltier should be cold. They are set within the range of temperature changes desired.

1. For heating, the syringe pump should push at flow rate of 2.5 ml/min to thermalize the PDMS microfluidic device with water passing through the upstream Peltier. There is a fluidic delay of 10–15 s between the action of the syringe pump and the beginning of the temperature change experienced by the cells. This delay should be taken into consideration for precisely timed experiments. Once the temperature begins to change, 10 more seconds is necessary to reach the desired temperature value (with a precision $<1°C$).
2. For cooling, the syringe pump should pull (reverse direction) at flow rate of 2.5 ml/min to thermalize the PDMS microfluidic device with water passing through the downstream Peltier.

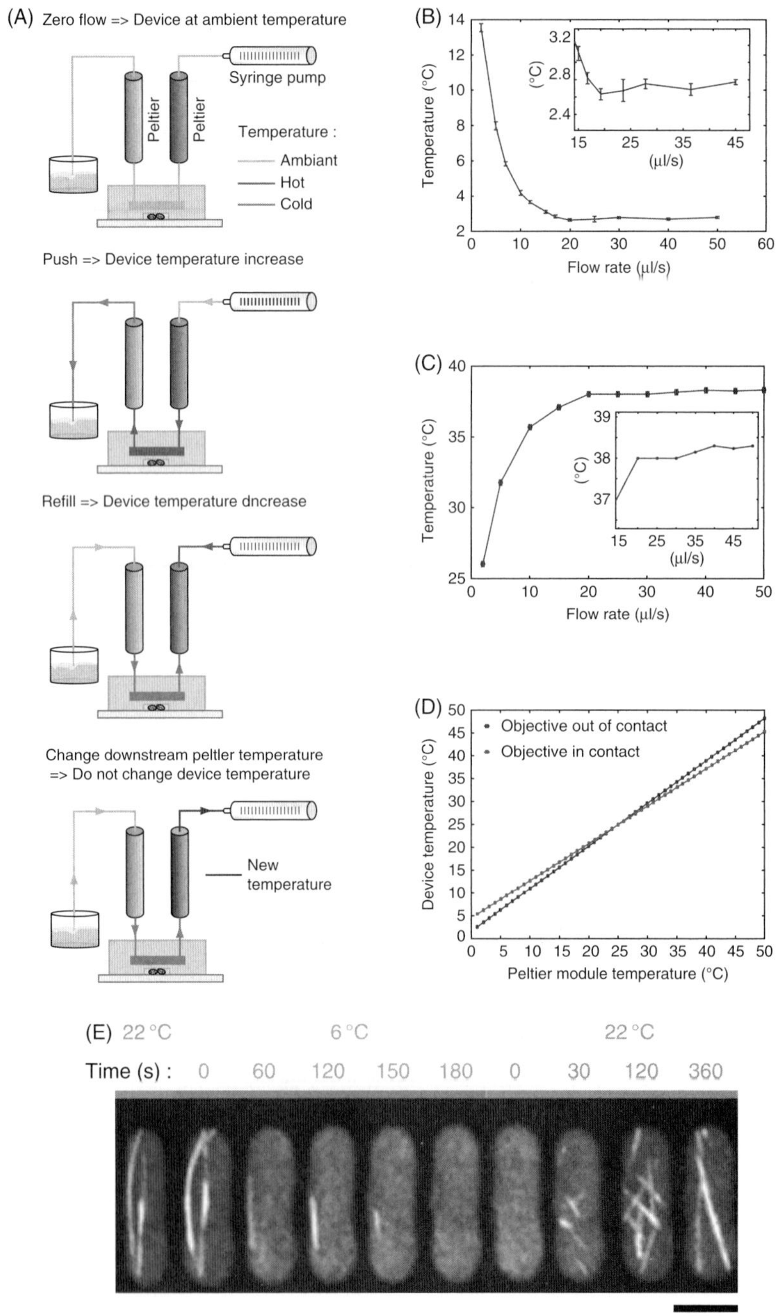

Fig. 4 (continued)

1. Remarks!

To prepare further temperature change, one can change the temperature of the downstream Peltier module without influencing the microfluidic device temperature. Peltier modules used here generally require 1–3 min to reach the desired temperature. Reaching Peltier temperature close to 0°C could take longer than 3 min depending on the peristaltic pump flow rate and temperature.

Water flow rate from the syringe pump is a critical parameter for temperature control. Figure 4B and C shows the dependence of temperature as a function of water flow rate with the Peltier modules set at 1°C and 39°C.

The temperature measured at the cells will be different from the Peltier module temperature setting. This is due to heat dissipation throughout the tubings and the PDMS device. For example, with the Peltier modules set at 1°C and ambient 24°C, the temperature experienced by the cells will be 2.7°C. Figure 4D gives the correlation between Peltier and device temperature.

The presence of the oil immersion objective acts as a strong heat sink and further increases the PDMS microfluidic device temperature. For example, with the cool Peltier module set at 1°C, the oil immersion objective will shift the cell temperature to 4.5°C. One can limit this problem by moving the position of the objective away from the cells being imaged during the interval between time points.

When using oil immersion objective lenses, temperature changes will generate materials dilation/contraction leading to focus drifting. The resulting drift is about 0.5 μm/°C. Although temperature changes inside the PDMS microfluidic device is very fast, a transient temperature gradient remains at the objective lens for 2–3 min before reaching steady state, leading to focus drift during this period. Nevertheless, because temperature changes obtained with this setup is reproducible, the extend of focus drift is thus predictable and therefore can be corrected during image acquisition.

During syringe pulling, bubbles may appear in the tubing and the syringe, leading to a higher fluidic time constant. Moreover, the presence of air in the tubing may

Fig. 4 Device characterization. (A) Schematic detailing the temperature control procedure. When the syringe pump is OFF, the device remains at ambient temperature, independent of Peltier temperature. When the syringe pump is pushing or pulling (refilling), the temperature change is dependent on the Peltier setting upstream of the water flow. Changing Peltier settings downstream of the water flow does not affect the device temperature. (B) Plot of device temperature versus syringe pump flow rate—cold. Peltier set at 1°C. The objective is not in contact with the coverslip. (C) Plot of device temperature versus syringe pump flow rate—hot. Peltier set at 39°C. The objective is not in contact with the coverslip. (D) Calculation of device temperature as a function of Peltier module temperature with 30 μl/s flow rate with ambient temperature of 25°C. Device temperature is given by the following equation: $T_{\text{device}} = T_{\text{peltier}} + A(T_{\text{ambient}} - T_{\text{peltier}})$, A, 0.071 when the objective is in contact with the coverslip; A, 0.187 when objective is not in contact with the coverslip. (E) A time-lapse montage of microtubule dynamics in fission yeast responding to temperature changes. The cell is expressing GFP-atb2p (tubulin). Imaging is done with a 100×/1.4NA oil immersion objective. At 6°C the microtubules depolymerized to completion. At ambient room temperature of 23°C the microtubules repolymerized. Bar, 5 μm.

lead to the microchannels filling up with air bubbles, leading to temperature nonuniformity.

Figure 4E shows an example of a cooling experiment. A fission yeast *S. pombe* cell expressing GFP-atb2p (tubulin) is cooled down to depolymerize the microtubules and then heated up to repolymerize the microtubules. The cooling Peltier was set at 1°C, and the effective temperature experienced by cells is 6°C. There is minor heat dissipation along the tubings, the PDMS device, and the oil contact between the glass coverslip and the microscopy objective.

VI. Conclusion

We described here a protocol which enables fabrication and use of a fast microfluidic temperature control device and setup. This kind of system allows fine control of microtubule polymerization dynamics. This system can also be used with temperature-sensitive mutant strains. The ability to couple this temperature control device with other microfluidic functionalities such as cell deformation and chemical perfusion will open new opportunities for cell biological experiments.

VII. Materials

A. Mold Fabrication

Spincoater, Laurell, CZ-650 series. (www.laurell.com)
Hotplate, Barnstead, model HP131720-33
UV lamp, OAI, model 30 with OAI intensity controller model 2105C2 (www.oainet.com)
Photoresist, Microchem, SU8 2005 (www.microchem.com)
Photoresist, Microchem, SU8 2050
Photoresist, Microchem, SU8 developer
Photoresist, Microchem, omnicoat
Isopropanol

B. Device Fabrication

Hotplate, Barnstead int, model HP131720-33
Oven, MEMMERT, 14L UNB100
Plasma cleaner, Harrick plasma, "Extended plasma" cleaner with "plasmaflow" pressure controller (www.harrickplasma.com)
Inlet and outlet drilling: 20-gauge needle smoothed with sandpaper
PDMS, Sylgard, 184
Coverglass, Dow Corning, 24 × 40 mm ref 2940-244

C. Temperature Control Setup

Peltier, Warner SC-20 Dual In-line Solution Heater/Cooler (www.warneronline.com)

Peltier controller, CL-100 Bipolar Temperature Controller (www.harvardapparatus.ciom)

Syringe Pump, Harvard Apparatus Remote Infuse/Withdraw PHD 4400 Hpsi Programmable

Peristaltic pump, Harvard Apparatus, Peristaltic Pump 66

Peristaltic pump tubing: Tygon, R-1000 1/8 in. ID * 1/4 in. OD

Syringe pump/Peltier tubing, Harvard Apparatus, "Microline" tubing 0.5 mm ID * 1.5 mm OD

Peltier metal tube/"Microline" tubing interface, Warner, Polyethylene tubing 1.57OD * 1.14ID (furnished with Peltier module)

Microfluidic device/"Microline" tubing interface, Harvard Apparatus, Stainless steel tubing coupler, 23-Gauge, 8 mm

Syringe for water injection, Monoject, 140cc Syringe with Luer Lock Tip

D. Cell Injection

2 ml syringe

24-gauge needle

"Microline" tubing 0.5 mm ID * 1.5 mm OD, Harvard Apparatus

Microfluidic device/"Microline" tubing interface, Harvard Apparatus, Stainless steel tubing coupler, 23-gauge, 8 mm

Acknowledgments

G.V-C. is supported by a postdoctoral fellowship from ARC; J.C. is supported by a predoctoral fellowship from FCT and ED Complexite du Vivant. This work is supported by grants from NIH, ACS, HFSP, FRM, ANR, LaLigue, and MarieParis.

References

Belanger, M. C., and Marois, Y. (2001). Hemocompatibility, biocompatibility, inflammatory and in vivo studies of primary reference materials low-density polyethylene and polydimethylsiloxane: a review. *J. Biomed. Mater. Res.* **58**, 467–477.

Charati, S. G., and Stern, S. A. (1998). Diffusion of Gases in Silicone Polymers: Molecular Dynamics Simulations. *Macromolecules* **31**, 5529–5535.

Duffy, D. C., McDonald, J. C., Schueller, O. J. A., and Whitesides, G.M. (1998). Rapid prototyping microfluidics systems in poly(dimetylsiloxane). *Anal. Chem.* **70**, 4974–4984.

McDonald, J. C., and Whitesides, G. M. (2002). Poly(dimethylsiloxane) as a material for fabricating microfluidic devices. *Acc. Chem. Res.* **35**, 491–499.

Unger, M. A., Chou, H. P., Thorsen, T., Scherer, A., and Quake, S. R. (2000). Monolithic microfabricated valves and pumps by multilayer soft lithography. *Science* **288**, 113–116.

Velve-Casquillas, G., Le Berre, M., Piel, M., Tran, P. T. (2010). Microfluidic tools for cell biological research. *Nano Today* **5**, 28–47.

CHAPTER 12

Microtubule-Dependent Spatial Organization of Mitochondria in Fission Yeast

Maitreyi Das[*], **Stéphane Chiron**[†], *and* **Fulvia Verde**[*]

[*]Department of Molecular and Cellular Pharmacology (R-189), University of Miami Miller School of Medicine, Miami, Florida 33101

[†]INSERM, U974, Université Pierre et Marie Curie-Paris, UMR-S974, CNRS, UMR-7215, Institut de Myologie, IFR14, Paris, F-75013, France

978-0-12-381349-7
DOI: 10.1016/S0091-679X(10)97012-X

Abstract

The microtubule cytoskeleton has an important role in the control of mitochondrial distribution in higher eukaryotes. In humans, defects in axonal mitochondrial transport are linked to neurodegenerative diseases. This chapter highlights fission yeast *Schizosaccharomyces pombe* as a powerful genetic model system for the study of microtubule-dependent mitochondrial movement, dynamics and inheritance.

I. Introduction

The cytoskeleton has a fundamental role in the control of mitochondrial distribution, dynamics, and inheritance in eukaryotic cells. Interaction with the cytoskeleton modulates mitochondrial respiration, fusion and fission, and localization to cellular sites of high energetic demand. In humans, defects in axonal mitochondrial transport are linked to neurodegenerative diseases, such as Charcot-Marie-Tooth (CMT 2A) disease (Palau *et al.*, 2009) and Huntington's disease (Trushina *et al.*, 2004). The molecular mechanisms mediating mitochondrial positioning and inheritance in different organisms, cell types, and tissues remain poorly understood (Boldogh and Pon, 2007).

Microtubules (MTs) play a critical role in providing positional information and in regulating cell shape in the fission yeast *Schizosaccharomyces pombe* (Martin, 2009; Piel and Tran, 2009). During interphase, MTs are organized in bundles of three to six MTs and are distributed along the main cell axis (Fig. 1A). The minus ends of MTs are localized around the area of the nucleus, where anti-parallel bundles overlap (Hagan, 1998; Sawin and Tran, 2006). The more dynamic plus ends extend toward the cell tips, where they are involved in the deposition of cell polarity marker proteins, such as Tea1, which define the site of polarized cell growth (La Carbona *et al.*, 2006; Sawin and Tran, 2006). During cell division astral MTs form, which mediate the correct positioning of the spindle, and the mitotic spindle assembles to promote chromosome segregation (Hagan, 2008).

In fission yeast, MTs are also involved in the correct spatial positioning of mitochondria. Only a few studies addressing the mechanisms of mitochondrial localization have been completed in fission yeast. These observations have shown that the majority of mitochondria align along the MT cytoskeleton during interphase, suggesting that mitochondria physically interact with MTs (Yaffe *et al.*, 1996) (see Fig. 1A and H). This association is clearly visualized by electron tomography (see Fig. 2), which shows mitochondria stretched along MTs and often localized between MTs of a splayed bundle (Höög *et al.*, 2007) (Fig. 2A and B). Consistent with a role for MTs in the spatial organization of mitochondria, the mitochondrial network becomes asymmetrically localized within the cell and fragmented when the MT cytoskeleton is disrupted (Yaffe *et al.*, 1996) (Section III. A.). Mitochondrial fragmentation in response to MT depolymerization is dependent on the function of dynamin-related protein Dnm1 (Jourdain *et al.*, 2009).

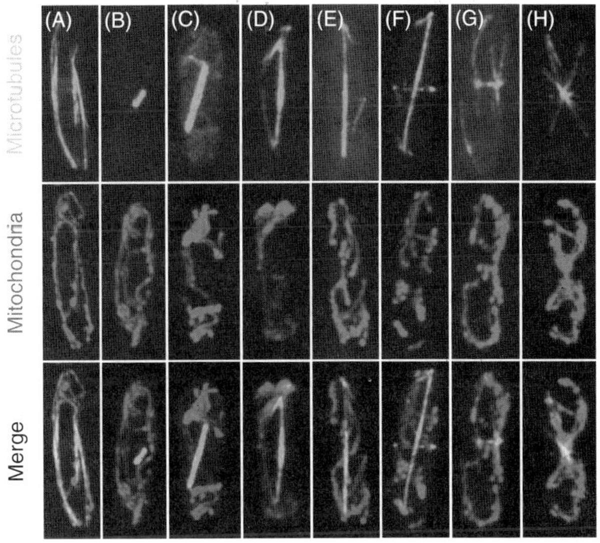

Fig. 1 Distribution of microtubules and mitochondria in fission yeast live cells during the cell cycle. Microtubules are visualized by expression of Atb1-GFP, while mitochondria are visualized by expression of COX4-RFP. For methods, see description in Section II. E. (See Plate no. 3 in the Color Plate Section.)

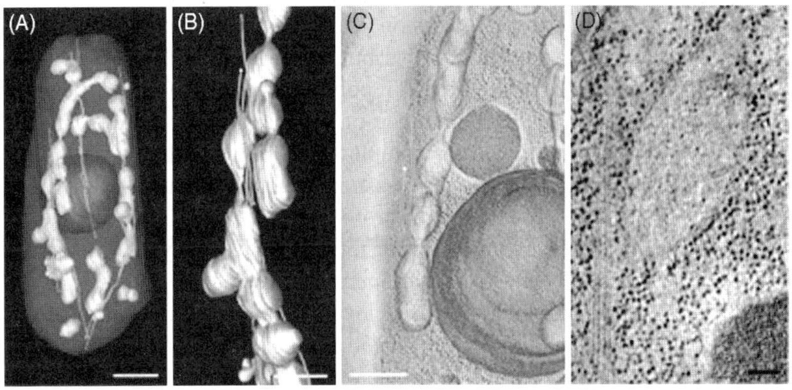

Fig. 2 (A) 3D model of a full cell reconstruction, showing the mitochondria in light blue, the nuclear envelope in pink, microtubules in green, and the plasma membrane in transparent green. Scale bar: 1 micrometer. (B) Enlargement of image shown in (A). Scale bar: 0.5 μm. (C) A tomographic slice (1 nm thick) with the 3D model superimposed. Scale bar: 0.5 μm. (D) A 23 nm thick tomographic slice showing a microtubule in close proximity to a mitochondrion. Scale bar: 50 nm. Images are a courtesy of Dr. Johanna Höög and Dr. Claude Antony (Höög *et al.*, 2007). (See Plate no. 4 in the Color Plate Section.)

During mitosis, mitochondria interaction with the MT cytoskeleton changes, when interphasic MTs depolymerize and the mitotic spindle assembles during metaphase (see Fig. 1B). In most cells, an interaction of spindle poles bodies and astral MTs with

A. Growth of Fission Yeast Cells

YE media	Yeast extract 5 g/l Glucose 30 g/l Supplements: 225 mg/l of adenine, histidine, leucine, uracil, and lysine hydrochloride (see below) pH should be adjusted to 5.6
Minimal Media (Edinburgh Minimal Media)	Potassium hydrogen phthalate 3.0 g/l Na_2HPO4 2.2 g/l NH_4Cl 5.0 g/l Glucose 20 g/l Salts Stock (×50) 20 ml/l Vitamin Stock (×1000) 1.0 ml/l Mineral Stock (×10K) 0.1 ml/l Supplements should be added when required (see below)
Salt Stock (×50)	$MgCl_2$ $6H_2O$ 53.5 g/l $CaCl_2$ $2H_2O$ 0.74 g/l KCl 50 g/l Na_2SO4 2.0 g/l
Vitamins (×1000)	Na pantothenate 1.0 g/l Nicotinic acid 10 g/l Inositol 10 g/l Biotin 10 mg/l Dissolve each component separately and autoclave
Minerals (×10,000)	H_3BO_3 5.0 g/l $MnSO_4$ 4.0 g/l $ZnSO_4$ $7H_2O$ 4.0 g/l $FeCl_3$ $6H_2O$ 2.0 g/l H_2MOO_4 H2O 0.4 g/l KI 1.0 g/l $CuSO_4$ 5H2O 0.4 g/l Citric acid 10 g/l Filter sterilize

Supplements (Stock)	Adenine 250 mg/l Histidine 250 mg/l Leucine 250 mg/l Uracil 250 mg/l Lysine 250 mg/l

1. Protocol

1. Grow fission yeast cells in 10 ml of YE or MIN media at 32°C (or suitable permissive temperature, for heat sensitive mutants, 25°C) overnight while shaking at 180 rpm, to prepare a preculture.
2. Inoculate freshly growing cells in 50 ml YE or MIN media and grow at 32°C, shaking at 180 rpm for 8 generations to optical density (OD) at 600 nm $\leq$ 0.5. (Wild-type cells have a generation time of 2 h at 32°C in YE media. The actual generation time of the cells depends on strain and growth conditions and will have to be determined accordingly).

Note: Minimal media is a defined media that is used for experimental reproducibility and to maintain plasmid selection. Supplements are added according to the auxotrophic markers present in the strain and the selection markers present in the plasmid. For more information on *S. pombe* growth conditions, media, and plasmids, see Moreno *et al.* (1991) and the Web site http://www-rcf.usc.edu/~forsburg/pombeweb.html.

B. Visualization of Mitochondria Using Vital Dyes

1. Staining with MitoTracker

MitoTracker Red CMXRos (Invitrogen) 5 μM in DMSO (Stock solution)

1. Grow cells in suitable media with sufficient aeration (180 rpm on shaker) at the optimum temperature (32°C for wild-type cells) for 8 generations.
2. Take 1 ml of cells (O.D. at 600 nm $\leq$ 0.5) in a microfuge tube and add 10 μl of 5 μM MitoTracker dye prepared in DMSO.
3. Cover the microfuge tube with foil and shake at 32°C for 20 min.
4. Centrifuge the cells for 3 min at 800×*g*.
5. Discard the supernatant but leave enough residual supernatant in order to resuspend the pelleted cells. Excess of dye can be removed by an optional wash with fresh medium.
6. Immediately mount 1–2 μl of the cell culture onto a clean glass slide and cover with coverslip. Avoid air bubbles between coverslip and slide.
7. Observe mitochondrial staining under a fluorescent microscope using a filter of suitable wavelength (depending on the type MitoTracker dye, see below).

8. Resuspend cells in 1 ml solution of 0.075 mg/ml Zymolyase 20T, 0.1 mg/ml Novozyme in PEM.
9. Incubate 5–10 min at 37°C to obtain 80% cell wall digestion. Take 9.5 μl of cells and add 0.5 μl 20% SDS. Look for percentage of ghost cells by light microscopy to determine extent of digestion. DO NOT OVER-DIGEST.
10. Subsequent washes by resuspension and pelleting:
    a. 1× PEM
    b. 2× PEMS
    c. 1× PEMS with 1% TritonX-100
    d. 3× PEM
11. Resuspend cells in 1 ml PEMBAL.
12. Incubate 1 h, room temperature, on wheel.
13. Pellet cells.
14. Resuspend in 100 μl PEMBAL + primary antibodies.
15. Incubate overnight, room temperature, on wheel.
16. Wash cells three times with PEMBAL.
17. Resuspend cells in 100 μl PEMBAL + secondary antibodies.
18. Incubate 1–2 h, room temperature, on a rotating wheel. Cover the microfuge tube with foil.
19. Wash three times with PEMBAL.
20. Cells can be kept at 4°C for days. Leaving cells in PEMBAL for 24–48 h will reduce background.

Note: Rabbit F1-ATPase antibody and mouse Tat1 antibody have been successfully used to visualize mitochondria and MT simultaneously in fission yeast (see Fig. 3) (Yaffe *et al.*, 1996).

E. Microscopic Analysis of MT and Mitochondrial Dynamics

As mentioned in Section II. C., fluorescently tagged tubulin (Atb1-GFP) can be visualized simultaneously with Cox4-RFP using a fluorescent microscope to understand MT-dependent mitochondrial dynamics (Yaffe *et al.*, 2003). Time-lapse imaging of these proteins can visualize the movement of these cellular structures in the cell.

For live cell microscopy of MTs and mitochondria, the strain expressing the transgenes *COX4-RFP* and *atb1-GFP* under the control of *nmt1* thiamine repressible promoter is used (MYP101: *h90 ade6-M? ura4-D18 leu1-32::nmt1::COX4-DsRFP: leu1$^+$nmt1::atb1-GFP:LEU2*, Yaffe *et al.*, 2003). To obtain expression and localization of the two fluorescent proteins, cells are grown in minimal media in the absence of thiamine for 48 h and then are shifted to minimal media with thiamine (15 μM) for 16–24 h. Growing cells for less than 48 h without thiamine will not permit mitochondrial labeling by Cox4-RFP, whereas growing cells for longer than 48 h in the absence of thiamine will induce excessive expression of tubulin-GFP, causing MT disruption.

Cells are then collected and seeded onto a pad (YE, with supplements, and gelatine 25%) on a slide. The coverslip is applied and then sealed with valap (1:1:1 vaseline:

lanolin:paraffin). In these conditions, cells are viable for 48 h without any mitochondrial morphology defects, allowing time-lapse imaging of mitochondria and MT dynamics. However, as with all live cell fluorescent imaging, care should be taken to limit cell exposure time, since high imaging frequency and long exposure time will be damaging to cellular structures.

To visualize the cells different microscope setups can be used. An Axiovert 200 M; CarlZeiss, Inc. equipped with a Plan-Apochromat 100 × NA 1.4 oil objective (Carl Zeiss, Inc.), a spinning disk confocal head (QLC-100; Yokogawa), and an argon/krypton laser (Melles Griot) coupled to an acousto-optical tunable filter (Neos Technologies) was previously used to describe regulation of mitochondrial distribution by CLASP (Chiron *et al.*, 2008). An extensive description of different imaging technologies used to visualize mitochondria has been previously published (Swayne *et al.*, 2007).

F. Electron Tomography Analysis of MT and Mitochondrial Organization

Recent developments in electron tomography and computing science allow large-scale 3D reconstruction at electron microscopy resolution. Fission yeast *S. pombe* is a biological model system that is particularly suitable for the study of MTs by electron tomography. First, MT dynamics have been very well described by fluorescence microscopy (Asakawa *et al.*, 2006; Brunner and Nurse, 2000; Drummond and Cross, 2000; Janson *et al.*, 2005; Sawin, 2004). Second, the small size of fission yeast cell allows the reconstruction of subcellular structures throughout the entirety of the cell volume.

The application of electron tomography and methods for acquiring, calculating tomograms, and reconstructing large cell volumes in the fission yeast has been published elsewhere (Giddings *et al.*, 2001; Höög and Antony, 2007; Höög *et al.*, 2007; McIntosh *et al.*, 2005; O'Toole *et al.*, 2002). In brief, following high-pressure freezing cryoimmobilization, freeze substitution, and serial sectioning, multiple two dimensional (2D) projection images are obtained from each section at different tilt increments. Each 2D image is then back-projected, with appropriated weighting, to form a 3D density distribution of the original section (Baumeister *et al.*, 1999). Serial tomograms are joined to create reconstructions of large volumes. An extensive description of methods, materials, and computer programs used for whole-cell tomographic investigations of fission yeast cytoskeleton architecture has been previously published in Methods in Cell Biology (Höög and Antony, 2007).

This approach has confirmed that MTs and mitochondria closely associate with each other (Höög *et al.*, 2007). Mitochondrial networks clustered with MTs were found to be more extensively branched and larger than non-MT-associated mitochondria. A preferred minimal distance of approximately 20 nm was detected between MTs and mitochondria, consistent with a close association. Further, mitochondria were often found where MT bundles splayed apart, suggesting that mitochondria influence MT bundle morphology. Thus, electron tomography can be employed to further understand the molecular basis of mitochondria and MT association.

III. Functional Analysis of MT–Mitochondria Interaction in Live Cells

A. Pharmacological Disruption of MT Organization

Pharmacological	Carbendazim (MBC) (Sigma); Thiabendazole (TBZ) (Sigma)
Genetic	Temperature sensitive *nda3-311* mutants (beta tubulin, cold sensitive) and *ban5-4* (alpha tubulin, *atb2*, heat sensitive) (Hiraoka *et al.*, 1984; Yaffe *et al.*, 1996)

1. Reagents

Thiabendazole (TBZ; Sigma Aldrich): 5 mg/ml in DMSO, freshly prepared.
Methyl benzimidazol-2-yl carbamate (carbendazim, MBC; Sigma Aldrich): 5 mg/ml in DMSO, freshly prepared.

2. Protocol

1. 30 μl of 5 mg/ml of MBC is mixed with 970 μl of media, and centrifuged at 13,000 rpm for 5 min to precipitate crystals. Use the supernatant for treating the cells.
2. Take 2 ml of freshly growing cells in a microfuge tube. Add either MBC to a final concentration of 50 μg/ml or TBZ to a final concentration of 20 μg/ml, respectively.
3. Incubate for 30 mins at 32°C (25°C for temperature sensitive cells).
4. The cells can be treated further as per the requirement of the experiment, e.g., immunostaining of MTs and mitochondria, or live cell imaging of mitochondria distribution.

Note: TBZ has to be used with caution since it causes transient delocalization of the actin cytoskeleton and arrest of cell elongation (Sawin and Snaith, 2004).

B. Identification of Mutants that Disrupt Mitochondrial Distribution

To identify mutants that disrupt fission yeast mitochondria distribution, classical mutagenesis techniques were employed using ethylmethanesulfonate (EMS; Sigma Aldrich, St. Louis, MO) as described previously (Moreno *et al.*, 1991; Weir and Yaffe, 2004). Cells were grown at 25°C on YE agar medium, and temperature sensitive mutants were identified by screening colonies for lack of growth at 36°C, following replica plating. Temperature sensitive strains were then analyzed by fluorescence microscopy for abnormal mitochondrial morphology and/or distribution, either by DASPMI staining or by Cox4-RFP expression (Yaffe *et al.*, 1996; Weir and Yaffe, 2004). This approach led to the identification of *atb2/ban5* (Yaffe *et al.*, 1996), *mmd1* (Weir and Yaffe, 2004) and *mmd4* (Chiron *et al.*, 2008) temperature-sensitive mutants.

Recent technical advances, including the sequencing of the whole *S. pombe* genome (Wood *et al.*, 2002) and the creation of an *S. pombe* haploid deletion library (Deshpande *et al.*, 2009; Kim *et al.*, 2010), will further facilitate the genetic dissection of MT-dependent mechanisms of mitochondrial spatial organization. For more details, see the corresponding Web sites http://www.sanger.ac.uk/Projects/S_pombe/genome_stats.shtml and http://pombe.bioneer.co.kr/. Genes of interest can be easily deleted in fission yeast by PCR-mediated deletion approaches (Bähler *et al.*, 1998). Using fluorescent microscopy each deletion mutant can be systematically analyzed to determine mitochondria distribution and morphology.

IV. Purification and Subfractionation of Fission Yeast Mitochondria

The techniques described above are useful for identifying novel cellular functions involved in the interaction between mitochondria and MTs. The techniques described in this section can be employed to assay the mitochondrial localization of proteins potentially involved in MT–mitochondrial interaction and to test if these proteins are peripherally associated with the mitochondria or if they are localized to the inner membrane or mitochondrial matrix. An extensive description of respiratory physiology and mitochondrial genome structure in fission yeast and of methods for mitochondrial purification and subfractionation has been previously published (Chiron *et al.*, 2007; Gouget *et al.*, 2008).

A. Growth of Fission Yeast Cells

1. Protocol

1. Grow fission yeast cells in 10 ml of YE (see Section II. A.) media at 32°C (or suitable temperature) overnight with shaking at 180 rpm to prepare a preculture.
2. Inoculate freshly growing cells in 50 ml YE media and grow at 32°C with shaking at 180 rpm for 8 generations to a final OD at 600 nm of 0.5. (Wild-type cells have a generation time of 2.5 h at 32°C in YE media. The actual generation time of the cells depends on strain and growth conditions and will have to determined accordingly).
3. Inoculate cells in 1 l YE media and grow at 32°C with shaking at 180 rpm for 8 generations to a final OD at 600 nm of 0.5.

Please Note: For fission yeast cells transformed with plasmids the cells have to be grown in minimal media. Care should be taken during cell growth, as the fission yeast cells are highly sensitive to changes in environmental and nutritional conditions, which can subsequently affect the isolation process. Fission yeast cells do not grow well under anaerobic conditions, therefore proper oxygenation and aeration of the cells during growth is essential. Also fission yeast cells should be harvested at an early stage of the exponential growth phase, as overgrown cells cannot be efficiently digested to generate protoplasts thereby decreasing the yield of mitochondria.

B. Isolation of Mitochondria

This protocol is a modification of Glick and Pon (1995); see also Chiron *et al.* (2007).

1. Reagents and Equipment

β-Mercaptoethanol 98%
BSA <0.02% fatty acid
Zymolyase 100T (Seikagaku Co.)
EDTA (ethylenediaminetetraacetic acid) 10 mM
PMSF (phenylmethylsulfonyl fluoride), freshly made 0.1 M stock solution in ethanol
Protease inhibitor tablets (Roche 11873580001)
Homogenizer with tight glass pestle (Wheaton).

2. Buffers

Digestion buffer: 1.2 M sorbitol, 10 mM sodium citrate, pH 5.8, 0.2 mM EDTA
Lysis buffer: 0.6 M sorbitol, 10 mM imidazole-HCl, pH 6.4, 2 mM EDTA

3. Protocol

1. Harvest cells by centrifugation at 2000×*g* for 10 min at room temperature. Discard the supernatant and wash the cell pellet with 200 ml distilled water.
2. Resuspend the pellet in 50 ml 10 mM EDTA, transfer to pre-weighed 50 ml falcon tube, and centrifuge for 10 min at 2000×*g* at room temperature.
3. Remove supernatant and determine the wet weight of the cell pellet by weighing the tube.
4. Resuspend the pellet in digestion buffer at 3 ml/gm of cells (add 0.3%(v/v) β-mercaptoethanol to digestion buffer immediately before resuspension).
5. Add 1 mg/ml Zymolyase 100T and incubate for 30 min (or until 80% cells are converted to spheroplasts) at 37°C under gentle shaking to generate protoplasts.
6. Place tubes in ice to stop the digestion reaction. Note: All steps here on have to be conducted in ice including prechilled buffers.
7. Pellet the protoplast by centrifugation for 15 min at 2000×*g* and 4°C. Remove the supernatant.

8. Resuspend the protoplast pellet in 15–20 ml lysis buffer to break the protoplasts. Pipette the mix up and down 10 times with a 10-ml pipet or with 10 strokes of a glass–glass homogenizer.
9. Incubate for 15 min in ice.
10. Remove cell debris by centrifugation for 15 min at 2500×*g* at 4°C.
11. Transfer the supernatant to a fresh tube; remove the pellet.
12. Centrifuge again for 5 min at 2500×*g* at 4°C. Transfer the supernatant to a fresh tube and remove the pellet. In case of any disturbance to the pellet during cell transfer of supernatant repeat centrifugation and removal of pellet.
13. Collect the mitochondria from the supernatant by centrifugation for 15 min at 12,000×*g* at 4°C.
14. Resuspend the mitochondria in 2 ml lysis buffer supplemented with 0.5% (w/v) BSA. Transfer to a tube and spin for 2 min at 800×*g* at 4°C.
15. To collect the mitochondrial pellet, transfer the supernatant to a fresh microfuge tube and spin for 15 min at 12,000×*g* at 4°C.
16. Discard the supernatant and remove the floating lipids with careful pipetting and cleaning the sides of the tube with a clean paper tissue.
17. To obtain a more purified fraction of mitochondria repeat steps 14–16 twice.
18. Resuspend the mitochondrial pellet in 100 μl lysis buffer with 0.5% (w/v) BSA. The pellet should appear brownish due to the presence of mitochondrial cytochromes. The darker the pellet, the better the yield.
19. To determine yield measure the protein concentration. Take 10 μl of mitochondrial prep and mix with 990 μl of 0.6% SDS. Measure OD at 280 nm. Absorbance of 0.21 corresponds to a protein concentration of 10 mg/ml. (A 1 l cell culture should ideally yield about 0.3 ml at 20 mg/ml).
20. Aliquot the mitochondrial prep and freeze in liquid nitrogen. The mitochondrial prep can be stored at –70°C until further use.

Note: The level of purification of the mitochondrial preparation can be determined by western blot analysis of using antibody probes against different subcellular components.

C. Mitochondrial Compartments and Protein Localization

Mitochondrial localization of proteins that influence mitochondrial organization and dynamics can be further analyzed by testing for the presence of the protein of interest in different subfractions of mitochondria. Subfractionation of fission yeast mitochondria has been described previously (Chiron *et al.*, 2007). The following protocol is used to establish the integrity of the purified mitochondria and to determine if the protein of interest associates with the outer membrane of the mitochondria or if it is situated inside the mitochondria. With this method, purified mitochondria are exposed to proteolysis by proteinase K. In the absence of sonication, only the outer membrane proteins are digested. Following sonication, the inner membrane proteins and the matrix proteins become also accessible to proteinase K. Accessibility of different

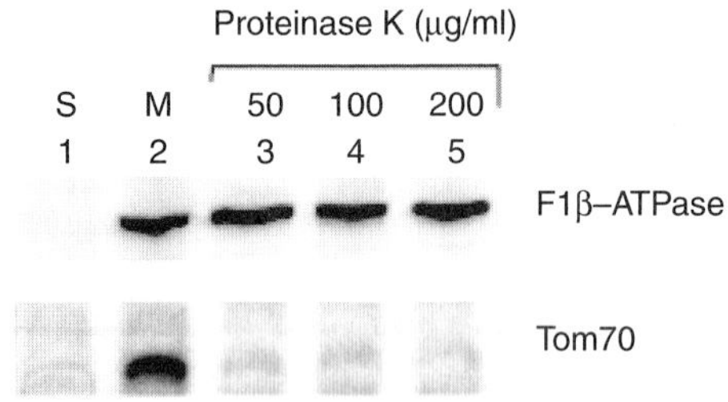

Fig. 4 Proteins localized to mitochondrial outer membrane and mitochondrial matrix. Mitochondria were purified by differential centrifugation. Lane 1 (S) supernatant; Lane 2 (M) mitochondria. Lanes 3, 4, 5, digestion of mitochondrial fraction (M) by proteinase K. Tom70, a mitochondrial outer membrane protein, is digested. F_1-β ATPase is protected, since it is localized to the internal mitochondrial matrix.

mitochondrial compartments by proteinase K can be followed by assaying the presence of appropriate marker proteins by western blot (see Fig. 4).

1. Buffers and Reagents

Sonication buffer: 0.6 M sorbitol-10 mM imidazole-HCl, pH 6.4, 2 mM EDTA
Proteinase K
PMSF 100 mM in ethanol
Laemmli buffer 1X: 1% sodium dodecyl sulfate, 50 mM Tris–HCl (pH 6.8), 4% glycerol, 0.4% β mercaptoethanol
12% SDS-polyacrylamide gel
Antibodies for western blot analysis: Rabbit polyclonal Anti-F1-ATPase 1:5000
Rabbit polyclonal Anti-Tom70 1:5000 (Jensen and Yaffe, 1988; P. Belenguer)

2. Protocol

1. Resuspend mitochondria at a protein concentration of 8 mg/ml in sonication buffer. Divide the mitochondrial preparation into 4 fractions.
2. Fraction 1 will not be treated further. Place on ice.
3. Disrupt mitochondria of fractions 3 and 4 by sonic irradiation using a VirSonic 100 sonicator at intensity 4 for 5 s. Place fraction 3 on ice and proceed further with fractions 2 and 4.
4. To fractions 2 and 4 add proteinase K to 250 μl mitochondrial particle suspensions at a final concentration of 50 μg/ml and incubate on ice for 60 min. Stop the reaction with PMSF at a final concentration of 2 mM.
5. Recover the mitochondria from all the four fractions by centrifugation at 100,000×*g*.
6. Resuspend the pellet in 1× Laemmli buffer.
7. Separate the different protein fractions (40 μg) by sodium dodecyl sulfate-polyacrylamide gel electrophoresis on a 12% polyacrylamide gel.

8. Western blot analysis of the mitochondrial proteins is then carried out with antibodies against subunit β of the membrane-associated F_1 portion of the F_1–F_0 ATPase (F_1-β) and Tom70 and the antibody against the protein, or the tagged fusion protein under study.
9. Proteins localized inside the mitochondria (for example, F_1-ATPase) are not digested by proteinase K in the absence of sonication since the mitochondria are intact. Such proteins are therefore detected by western blot upon proteinase K treatment alone but are lost following sonication and proteinase K treatment. Proteins localized on the outer membrane of the mitochondria (for example, Tom70) are digested by proteinase K even without sonication and are not detected by western blot in all proteinase K-treated fractions. If the mitochondrial membrane is disrupted during the procedure for mitochondria purification, the F_1-ATPase will also disappear in fraction 2 that is treated with proteinase K in the absence of sonication (Fig. 4). Note that matrix soluble proteins will remain in the supernatant after centrifugation (step 5) and will not be detected even in the absence of proteinase K.

Acknowledgments

We thank Dr. Johanna Höög and Dr. Claude Antony for providing unpublished tomographic images of microtubules and mitochondria in fission yeast. We thank Dr. Gennaro D'Urso, Dr. Flavia Fontanesi, and Dr. Antoni Barrientos (University of Miami) for critically reading the manuscript and the Yeast Club at the University of Miami for useful suggestions. M.D. and F.V. are supported by the National Science Foundation grant (NSF) 0745129 and by the Sylvester Comprehensive Cancer Center at the University of Miami Miller School of Medicine. S.C. was previously supported by the United Mitochondrial Disease Foundation.

References

Asakawa, K., Kume, K., Kanai, M., Goshima, T., Miyahara, K., Dhut, S., Tee, W. W., Hirata, D., and Toda, T. (2006). The V260I mutation in fission yeast alpha-tubulin Atb2 affects microtubule dynamics and EB1-Mal3 localization and activates the Bub1 branch of the spindle checkpoint. *Mol. Biol. Cell* **17**(3), 1421–1435.

Bähler, J., Wu, J. Q., Longtine, M. S., Shah, N. G., McKenzie, 3rd, A., Steever, A. B., Wach, A., Philippsen, P., and Pringle, J. R. (1998). Heterologous modules for efficient and versatile PCR-based gene targeting in *Schizosaccharomyces pombe*. *Yeast* **14**(10), 943–951.

Baumeister, W., Grimm, R., and Walz, J. (1999). Electron tomography of molecules and cells. *Trends Cell Biol.* **9**(2), 81–85.

Boldogh, I. R., and Pon, L. A. (2007). Mitochondria on the move. *Trends Cell Biol.* **17**(10), 502–510.

Brazer, S. C., Williams, H. P., Chappell, T. G., and Cande, W. Z. (2000). A fission yeast kinesin affects Golgi membrane recycling. *Yeast* **16**(2), 149–166.

Brunner, D., and Nurse, P. (2000). CLIP170-like tip1p spatially organizes microtubular dynamics in fission yeast. *Cell* **102**(5), 695–704.

Chiron, S., Bobkova, A., Zhou, H., and Yaffe, M. P. (2008). CLASP regulates mitochondrial distribution in *Schizosaccharomyces pombe*. *J. Cell Biol.* **182**(1), 41–49.

Chiron, S., Gaisne, M., Guillou, E., Belenguer, P., Clark-Walker, G. D., and Bonnefoy, N. (2007). Studying mitochondria in an attractive model: *Schizosaccharomyces pombe*. *Methods Mol. Biol.* **372**, 91–105.

Deshpande, G. P., Hayles, J., Hoe, K. L., Kim, D. U., Park, H. O., and Hartsuiker, E. (2009). Screening a genome-wide *S. Pombe* deletion library identifies novel genes and pathways involved in genome stability maintenance. *DNA Repair (Amst.)* **8**(5), 672–679.

Diot, A., Guillou, E., Daloyau, M., Arnaune-Pelloquin, L., Emorine, L. J., and Belenguer, P. (2009). Transmembrane segments of the dynamin Msp1p uncouple its functions in the control of mitochondrial morphology and genome maintenance. *J. Cell Sci.* **122**(Pt 15), 2632–2639.

Drummond, D. R., and Cross., R. A.(2000). Dynamics of interphase microtubules in *Schizosaccharomyces pombe. Curr. Biol.* **10**(13), 766–775.

Giddings, Jr., T. H., O'Toole, E. T., Morphew, M., Mastronarde, D. N., McIntosh, J. R., and Winey, M. (2001). Using rapid freeze and freeze-substitution for the preparation of yeast cells for electron microscopy and three-dimensional analysis. *Methods Cell Biol.* **67**, 27–42.

Glick, B. S., and Pon, L. A. (1995). Isolation of highly purified mitochondria from *Saccharomyces cerevisiae. Meth. Enzymol.* **260**, 213–223.

Gouget, K., Verde, F., and Barrientos, A. (2008). In vivo labeling and analysis of mitochondrial translation products in budding and in fission yeasts. *Methods Mol. Biol.* **457**, 113–124.

Hagan, I. M. (1998). The fission yeast microtubule cytoskeleton. *J. Cell Sci.* **111(Pt 12)**, 1603–1612.

Hagan, I. M. (2008). The spindle pole body plays a key role in controlling mitotic commitment in the fission yeast *Schizosaccharomyces pombe. Biochem. Soc. Trans.* **36**(Pt 5), 1097–1101.

Hiraoka, Y., Toda, T., and Yanagida, M. (1984). The NDA3 gene of fission yeast encodes beta-tubulin: A cold-sensitive nda3 mutation reversibly blocks spindle formation and chromosome movement in mitosis. *Cell* **39**(2 Pt 1), 349–358.

Höög, J. L., and Antony, C. (2007). Whole-cell investigation of microtubule cytoskeleton architecture by electron tomography. *Methods Cell Biol.* **79**, 145–167.

Höög, J. L., Schwartz, C., Noon, A. T., O'Toole, E. T., Mastronarde, D. N., McIntosh, J. R., and Antony, C. (2007). Organization of interphase microtubules in fission yeast analyzed by electron tomography. *Dev. Cell* **12**(3), 349–361.

Janson, M. E., Setty, T. G., Paoletti, A., and Tran, P. T. (2005). Efficient formation of bipolar microtubule bundles requires microtubule-bound gamma-tubulin complexes. *J. Cell Biol.* **169**(2), 297–308.

Jensen, R. E., and Yaffe, M. P. (1988). Import of proteins into yeast mitochondria: The nuclear MAS2 gene encodes a component of the processing protease that is homologous to the MAS1-encoded subunit. *EMBO J.* **7**(12), 3863–3871.

Jourdain, I., Gachet, Y., and Hyams, J. S. (2009). The dynamin related protein dnm1 fragments mitochondria in a microtubule-dependent manner during the fission yeast cell cycle. *Cell Motil. Cytoskeleton* **66**(8), 509–523.

Kim, D. U., Hayles, J., Kim, D., Wood, V., Park, H. O., Won, M., Yoo, H. S., Duhig, T., Nam, M., Palmer, G., Han, S., Jeffery, L., Baek, S. T., Lee, H., Shim, Y. S., Lee, M., Kim, L., Heo, K. S., Noh, E. J., Lee, A. R., Jang, Y. J., Chung, K. S, Choi, S. J., Park, J. Y., Park, Y., Kim, H. M., Park, S. K., Park, H. J., Kang, E. J., Kim, H. B., Kang, H. S., Park, H. M., Kim, K., Song, K., Song, K. B., Nurse, P., and Hoe, K. L. (2010). Analysis of a genome-wide set of gene deletions in the fission yeast Schizosaccharomyces pombe. *Nat. Biotechnol.* **28**(6), 617–623.

Krüger, N., and Tolic-Norrelykke, I. M. (2008). Association of mitochondria with spindle poles facilitates spindle alignment. *Curr. Biol.* **18**(15), R646–R647.

Kuznetsov, A. V., Hermann, M., Saks, V., Hengster, P., and Margreiter, R. (2009). The cell-type specificity of mitochondrial dynamics. *Int. J. Biochem. Cell Biol.* **41**(10), 1928–1939.

La Carbona, S., Le Goff, C., and Le Goff, X. (2006). Fission yeast cytoskeletons and cell polarity factors: Connecting at the cortex. *Biol. Cell* **98**(11), 619–631.

Martin, S. G. (2009). Microtubule-dependent cell morphogenesis in the fission yeast. *Trends Cell Biol.* **19**(9), 447–454.

Maundrell, K. (1993). Thiamine-repressible expression vectors pREP and pRIP for fission yeast. *Gene* **123**(1), 127–130.

McIntosh, R., Nicastro, D., and Mastronarde, D. (2005). New views of cells in 3d: An introduction to electron tomography. *Trends Cell Biol.* **15**(1), 43–51.

Mikawa, T., Kanoh, J., and Ishikawa, F. (2010). Fission yeast Vps1 and Atg8 contribute to oxidative stress resistance. *Genes Cells.* **15**(3), 229–243.

Moreno, S., Klar, A., and Nurse, P. (1991). Molecular genetic analysis of fission yeast *Schizosaccharomyces pombe*. *Meth. Enzymol.* **194**, 795–823.

O'Toole, E. T., Winey, M., McIntosh, J. R., and Mastronarde, D. N. (2002). Electron tomography of yeast cells. *Meth. Enzymol.* **351**, 81–95.

Palau, F., Estela, A., Pla-Martin, D., and Sanchez-Piris, M. (2009). The role of mitochondrial network dynamics in the pathogenesis of Charcot-Marie-Tooth disease. *Adv. Exp. Med. Biol.* **652**, 129–137.

Pelloquin, L., Belenguer, P., Menon, Y., and Ducommun, B. (1998). Identification of a fission yeast dynamin-related protein involved in mitochondrial DNA maintenance. *Biochem. Biophys. Res. Commun.* **251**(3), 720–726.

Piel, M., and Tran, P. T. (2009). Cell shape and cell division in fission yeast. *Curr. Biol.* **19**(17), R823–R827.

Roque, H., and Antony, C. (2010). Electron microscopy of model systems: the fission yeast Schizossacharomyces pombe. *Methods Cell Biol.* In press.

Sawin, K. E. (2004). Microtubule dynamics: Faint speckle, hidden dragon. *Curr. Biol.* **14**(17), R702–R704.

Sawin, K. E., and Snaith, H. A. (2004). Role of microtubules and tea1p in establishment and maintenance of fission yeast cell polarity. *J. Cell Sci.* **117**(Pt5), 689–700.

Sawin, K. E., and Tran, P. T. (2006). Cytoplasmic microtubule organization in fission yeast. *Yeast* **23**(13), 1001–1014.

Swayne, T. C., Gay, A. C., and Pon, L. A. (2007). Visualization of mitochondria in budding yeast. *Methods Cell Biol.* **80**, 591–626.

Takeda, K., Yoshida, T., Kikuchi, S., Nagao, K., Kokubu, A., Pluskal, T., Villar-Briones, A., Nakamura, T., and Yanagida, M. (2010). Synergistic roles of the proteasome and autophagy for mitochondrial maintenance and chronological lifespan in fission yeast., Proc. Natl. Acad. Sci. U.S.A. **107**(8), 3540–3545.

Trushina, E., Dyer, R. B., Badger, J. D., Ure, D., Eide, L., Tran, D. D., Vrieze, B. T., Legendre-Guillemin, V., McPherson, P. S., Mandavilli, B. S., Van Houten, B., Zeitlin, S., *et al.* (2004). Mutant huntingtin impairs axonal trafficking in mammalian neurons in vivo and in vitro. *Mol. Cell. Biol.* **24**(18), 8195–8209.

Weir, B. A., and Yaffe, M. P. (2004). Mmd1p, a novel, conserved protein essential for normal mitochondrial morphology and distribution in the fission yeast schizosaccharomyces pombe. *Mo.l Biol. Cell* **15**(4), 1656–1665.

Wiley, D. J., Catanuto, P., Fontanesi, F., Rios, C., Sanchez, N., Barrientos, A., and Verde, F. (2008). Bot1p is required for mitochondrial translation, respiratory function, and normal cell morphology in the fission yeast *Schizosaccharomyces pombe*. *Eukaryotic Cell* **7**(4), 619–629.

Wood, V., Gwilliam, R., Rajandream, M. A., Lyne, M., Lyne, R., Stewart, A., Sgouros, J., Peat, N., Hayles, J., Baker, S., Basham, D., Bowman, S., *et al.* (2002). The genome sequence of *Schizosaccharomyces pombe*. *Nature* **415**(6874), 871–880.

Woods, A., Sherwin, T., Sasse, R., MacRae, T. H., Baines, A. J., and Gull, K. (1989). Definition of individual components within the cytoskeleton of Trypanosoma brucei by a library of monoclonal antibodies. J. Cell Sci. **93(Pt 3)**, 491–500.

Yaffe, M. P., Harata, D., Verde, F., Eddison, M., Toda, T., and Nurse, P. (1996). Microtubules mediate mitochondrial distribution in fission yeast.Proc. Natl. Acad. Sci. U.S.A. **93**(21), 11664–11668.

Yaffe, M. P., Stuurman, N., and Vale, R. D. (2003). Mitochondrial positioning in fission yeast is driven by association with dynamic microtubules and mitotic spindle poles. Proc. Natl. Acad. Sci. U.S.A..**100**(20), 11424 11428.

CHAPTER 13

Microscopy Methods for the Study of Centriole Biogenesis and Function in *Drosophila*

Ana Rodrigues Martins[*], **Pedro Machado**[*], **Giuliano Callaini**[†], *and* **Monica Bettencourt-Dias**[*]

[*]Instituto Gulbenkian de Ciência, Rua da Quinta Grande, P-2780-156 Oeiras, Portugal

[†]Department of Evolutionary Biology, University of Siena, I-53100 Siena, Italy

Abstract

Centrosomes regulate cell motility, adhesion, and polarity in interphase and participate in spindle formation in mitosis. They are composed of two centrioles, which are microtubule-based structures, and a proteinaceous matrix recruited by those, called pericentriolar material. Centrioles are also necessary for the nucleation of the axoneme, the microtubule inner structure of cilia and flagella. The fruit fly, *Drosophila melanogaster*, has played an important role in the study of cell biology

METHODS IN CELL BIOLOGY, VOL. 97

978-0-12-381349-7
DOI: 10.1016/S0091-679X(10)97013-1

processes and their contextualization in a variety of developmental phenomena. In this chapter, we describe immunofluorescence and electron microscopy methods used to study *Drosophila* early embryogenesis and spermatogenesis. These methods have been widely used to study centriole assembly and its function as a centrosome organizer during mitotic and meiotic cell divisions and as an axoneme nucleator in the formation of flagella.

I. Introduction

Centrioles are essential for the formation of several microtubule-organizing structures including cilia and centrosomes (Fig. 1). Centrosomes regulate cell motility, adhesion, and polarity in interphase and participate in the formation of the spindle in mitosis. Centrosome abnormalities in mitosis, both in number and in structure, are present in many cancers and are linked to genomic instability and problems in stem cell homeostasis (Bettencourt-Dias and Glover, 2007; Zyss and Gergely, 2009). Centrosomes found in animals are most often composed of two components: a pair of centrioles and a surrounding cloud of electron-dense pericentriolar material (PCM). The older centriole in a centrosome is called mature or mother and the younger one, the daughter centriole. Mother and daughter centrioles generally display an orthogonal

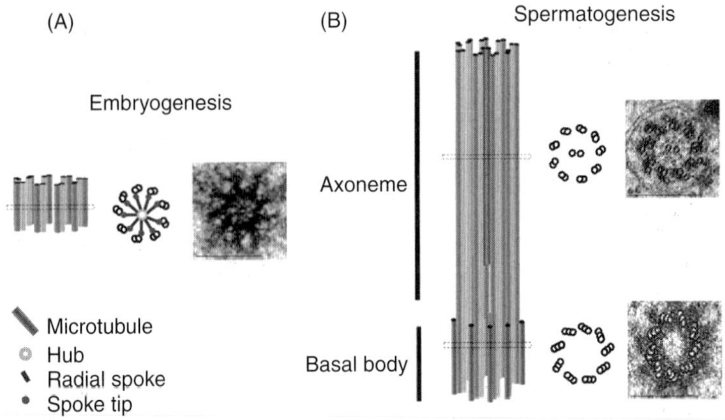

Fig. 1 Centriole structure in *Drosophila* embryos and testes. (A) Schematic representation and cross-section electron microscope image of centrioles in *Drosophila* embryos. Note the presence of the cartwheel structure composed of a central hub attached to the radial spokes through spoke tips. The embryonic centriole is approximately 200 nm in length and 200 nm in diameter and is composed of nine microtubule doublets displaying a radial symmetry. Each microtubule doublet is composed of one complete microtubule, the A-tubule, with a second partial microtubule grown on the side of the first, the B-tubule. (B) Schematic representation and cross-section electron microscope image of basal bodies and axonemes in *Drosophila* spermatogenesis. The basal body is approximately 2 μm in length and 200 nm in diameter and is composed of nine microtubule triplets displaying a radial symmetry. The axoneme can reach up to 1.8 mm and is composed of nine outer microtubule doublets and a central pair of microtubules. Scale bars represent 200 nm.

configuration to each other (Bettencourt-Dias and Glover, 2007; Bornens, 2002; Ou *et al.*, 2004). Centrioles have another distinct function as basal bodies that template the growth of axonemes, the microtubule-based skeleton of cilia and flagella (Dirksen, 1991). These are evolutionarily conserved eukaryotic organelles that extend from, and are continuous with, the cell membrane. Cilia and flagella are indispensable in a variety of cellular and developmental processes such as cell motility, propagation of morphogenetic signals, and sensory reception (Badano *et al.*, 2005; Plotnikova *et al.*, 2008). Moreover, one of the most surprising discoveries in cell biology over the last 5–10 years is the increasing number of human conditions resulting from defects in ciliary assembly and/or motility (Fliegauf *et al.*, 2007). Despite their importance, it is impressive how little is known about the control of centriole structure and number.

The number of centrioles in a cell is normally controlled through a "canonical duplication cycle" in coordination with the chromosome cycle. "One and only one" new centriole forms orthogonally to each preexisting centriole in a conservative fashion. In some cells in our body the regulation of the arithmetics of centriole biogenesis is different, with one of the most emblematic cases occurring during gametogenesis. Oocytes are devoid of centrioles. On the other hand, sperm cells contain a basal body that is responsible for nucleating the sperm tail. Upon fertilization, it is this basal body that becomes the centriole in the zygote. Centrioles can also be formed in the absence of preexisting centriolar structures. This *de novo* biogenesis is known to occur in insect species with parthenogenetic development, as well as in human cells upon ablation of their centrosomes and in *Drosophila* unfertilized embryos when master regulators of centriole biogenesis are overexpressed (La Terra *et al.*, 2005; Peel *et al.*, 2007; Riparbelli and Callaini, 2003; Rodrigues-Martins *et al.*, 2007b).

The fruit fly *Drosophila melanogaster* has been used as a model organism for more than 100 years. Thomas Hunt Morgan was the pioneer biologist studying *Drosophila* in the 1900s. Due to its small size, ease of culture, and short generation time, *Drosophila* is a widely used research model organism. Its potential for combining genetic and molecular approaches to questions of gene expression, developmental biology, and cell biology is very useful. A variety of collections of mutants have been generated (Greenspan, 2004; Ryder and Russell, 2003), and at least 40% of the *Drosophila* genes have available transposon insertions within 500 bp of the ATG which disrupts their expression (Bellen *et al.*, 2004; White-Cooper, 2009). RNAi collections are also available that allow specific depletion of most predicted *Drosophila* genes (Dietzl *et al.*, 2007). Information about flies and available mutant collections can be found in Flybase (www.flybase.org) and the *Drosophila* Stock collection at Bloomington (http://flystocks.bio.indiana.edu/).

The *D. melanogaster* life cycle is similar to that of many other insects: it includes an egg form, a larval form, and a pupal stage before it finally emerges as a flying adult. After the eggs hatch, approximately 1 day after being laid, small larvae start to be detected in the growing medium. Larvae go through three different instars, each 1 day long, and after the third instar they begin to migrate up the culture vial in order to pupate. More or less 3 days later, eclosion occurs and the adults emerge from the pupal case. This cycle takes on average 10 days if the temperature is kept at 25°C (Greenspan, 2004).

During the *Drosophila* life cycle, there are developmental stages particularly useful for studying centriole biogenesis and function within the context of a centrosome or of an axoneme nucleator in cilia/flagella formation. Canonical centriole biogenesis and centrosome function have been studied in the context of several tissues, including early embryos, neurogenesis, and spermatogenesis (Bettencourt-Dias and Glover, 2007; Januschke and Gonzalez, 2008; Yamashita, 2009). Early embryos provide the advantage of displaying very fast cell cycles (10 min) occurring in the context of a syncytium which is easily studied through a variety of techniques such as immunofluorescence, transmission electron microscopy, biochemical fractionation, injection of proteins and drugs, and live imaging (Foe *et al.*, 1993). Neurogenesis provides a thoroughly studied example of centrosome behavior in asymmetric stem cell divisions and in the formation of ciliated mechanosensory cells (Gogendeau and Basto, 2009; Januschke and Gonzalez, 2008). Spermatogenesis provides a great example of centrosome behavior in asymmetric stem cell division and in the differentiation of the centriole into a basal body to form the flagella of the sperm (Yamashita, 2009; Yamashita and Fuller, 2008). While most somatic cells in the fly can form a normal spindle without centrioles (Basto *et al.*, 2006; Bettencourt-Dias *et al.*, 2005), both mitosis in early embryogenesis and meiosis in spermatogenesis rely on centrosomes for accurate division (Rodrigues-Martins *et al.*, 2008). *De novo* centrosome formation has been studied in the context of unfertilized eggs that do not have centrioles (Peel *et al.*, 2007; Rodrigues-Martins *et al.*, 2007b).

This chapter focuses on microscopy methods, both immunofluorescence and transmission electron microscopy, used to study centrosomes during *Drosophila* early embryogenesis and spermatogenesis. For protocols to study centrosome function in *Drosophila* tissue culture cells please see Bettencourt-Dias and Goshima (2009). Other stages, such as neurogenesis and germ cell formation are very useful for studying the role of centrosomes in asymmetric cell divisions. For more detail on those stages please refer to Januschke and Gonzalez (2008) and Yamashita (2009). For other *Drosophila* protocols please check Sullivan *et al.* (2000).

For general *D. melanogaster* protocols including food recipes see http://fruitfly4.aecom.yu.edu/labmanual/contents.html and http://www.ceolas.org/VL/fly/protocols.html.

II. Centrioles in *Drosophila* Early Embryogenesis

Early *Drosophila* embryogenesis has been widely used for the study of centrosomes (Glover, 1991; Raff, 2004). After fertilization the female pronucleus migrates and joins the male pronucleus on the first mitotic spindle (Fig. 2) (Foe *et al.*, 1993; Raff, 2004). The embryo then proceeds through an extremely rapid series of synchronous nuclear divisions in a common cytoplasm (Foe *et al.*, 1993). The first 13 cell cycles in the *Drosophila* embryo are composed of only S- and M-phases, with no gap phases. An early nuclear division takes around 9 min and interphase gradually lengthens during late syncytial cycles (Foe *et al.*, 1993). Thus, the embryo forms several thousands of centrosomes in a short period of time. As this stage of development does not require

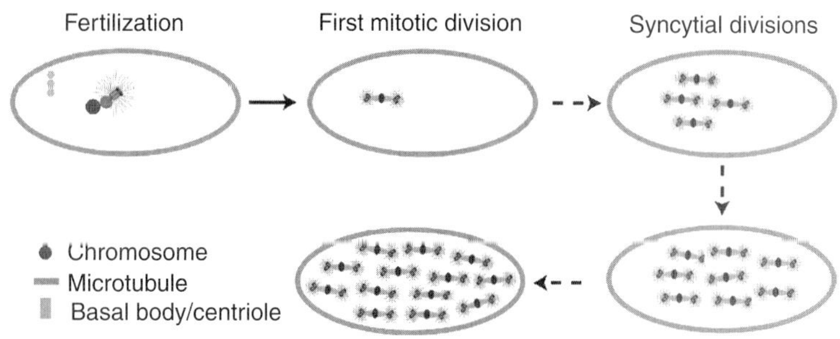

Fig. 2 *Drosophila* early embryogenesis. Embryogenesis starts at fertilization when the female pronucleus meets the male pronucleus. The other three haploid nuclei, products of meiosis II, become polar bodies and degenerate. The zygotic nucleus produced at fertilization is invariably positioned toward the anterior end of the embryo. After the first mitotic division syncytial divisions begin and the dividing nuclei spread evenly throughout the syncytial embryo. After that axial expansion, the nuclei migrate to a uniform monolayer at the cortex. Most migrating nuclei reach the cortex at nuclear cycle 10, where they proceed through four more rounds of mitosis until they cellularize and all cells become individually separated (Adapted from Foe *et al.*, 1993).

zygotic transcription, all the components needed to form both centrosomes and spindles are maternally provided.

During embryogenesis, centrioles are considerably shorter than their mammalian counterparts (Fig. 1A) (200 vs 500 nm approximately) (Gonzalez *et al.*, 1998) not showing the distal centriolar structures, such as distal appendages (Callaini *et al.*, 1997). Separation of centrosomes, which occurs in G2 in mammalian cells, occurs early in the cell cycle in early *Drosophila* embryos. Centrioles lose their orthogonal arrangement in the metaphase to anaphase transition, move apart during anaphase, and become widely separated at telophase. Concomitantly, the PCM expands and flattens, splitting into two units at late telophase (Callaini and Riparbelli, 1990). PCM components such as γTubulin and D-PLP (Drosophila Pericentrin Like Protein) are always present at the centrosome but other centrosomal components such as polo and CP190 are only recruited during centrosome maturation (Raff, 2004).

More recently *Drosophila* embryos and unfertilized eggs have been used to study *de novo* centrosome formation (Peel *et al.*, 2007; Rodrigues-Martins *et al.*, 2007b). *De novo* centrosome formation can be studied in *Drosophila* unfertilized eggs that are laid by virgin females. During *Drosophila* oogenesis centrioles are lost, hence the oocyte does not contain any centrioles which are only provided by the sperm upon fertilization. If fertilization does not occur, there is no centrosome formation and unfertilized eggs do not develop. In contrast, it was observed that upon overexpression of certain centriolar proteins, such as SAK/PLK4, centrioles are formed, helping to understand intermediate steps on centriole formation (Rodrigues-Martins *et al.*, 2007a).

A. Immunofluorescence of Embryos/Eggs

Embryos and eggs are normally collected from 4- to 5-day-old females kept at 25°C. Set up embryos/eggs collecting cages with at least 50 adult flies (30 females: 20 males

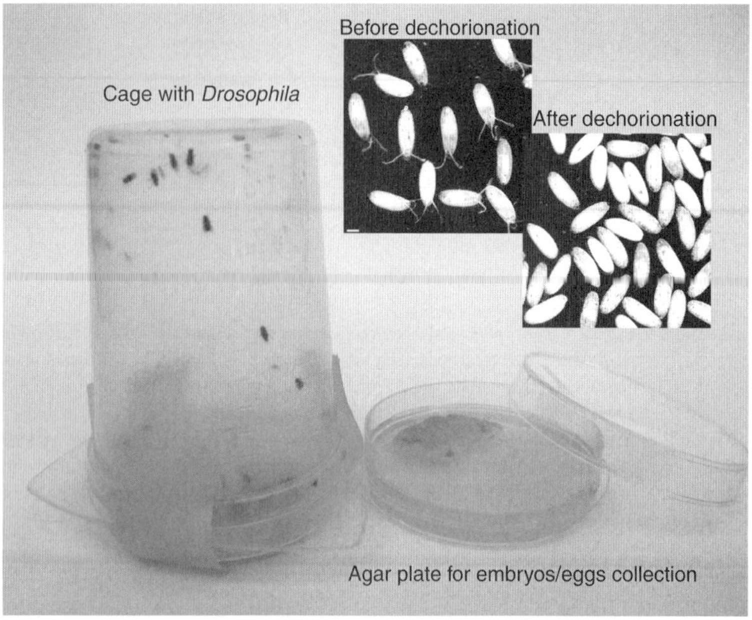

Fig. 3 Collecting and dechorionating *Drosophila* embryos. Example of an embryo/egg collecting cage and of one agar plate used for collection of *Drosophila* embryos/eggs. After dechorionation, embryos/eggs lose dorsal appendages and become glossy. Scale bar represents 100 μm.

for embryo collection and around 75 virgin females for egg collection) as the one shown in Fig. 3. The cages contain agar plates made off fruit juice, where the females lay their embryos/eggs. Agar plates are normally supplemented with freshly made yeast paste (mix baker's yeast with water to make a creamy but solid paste) so that females can be continuously fed and attracted to lay their embryos/eggs on the agar plate.

1. The Canonical Centriole Cycle

In order to observe centrosome behavior in early mitotic divisions, collect embryos at 10 min intervals. If collections are longer than 2 h embryos will be cellularized and in overnight collections most of the embryos will have gastrulated. As females retain embryos inside, a synchronization step is needed to ensure correct timing. For synchronization, replace collection plates, as the ones shown in Fig. 3, at least four times at 15 min intervals before starting the experiment. Embryos can then be collected at different time points according to the experimental needs.

2. *De Novo* Centrosome Formation

In order to test for *de novo* centrosome formation after misexpression of a gene of interest, collect unfertilized eggs from virgin females. Do not collect females coming from vials where there could be males older than 8 h at 25°C or 16 h at 18°C. Virgin females are less willing to lay eggs, so make sure to feed them with plenty of freshly made yeast paste before starting egg collection. Overnight collections may be useful to check whether *de novo* centrosome formation has occurred.

For immunofluorescence analysis start by collecting the embryos/eggs from the agar plate with a paintbrush wet in water and transfer those to a sieve (mesh diameter 0.125 mm). The embryos/eggs in the sieve are washed thoroughly with distilled water to remove any yeast paste. Dechorinate the embryos/eggs under a dissection scope by immersing the sieve in a 50% freshly made bleach solution. Gently shake the sieve to get homogenous dechorionation. Dechorionation finishes when the embryos/eggs dorsal appendages are no longer visible, which normally takes around 2 min (Fig. 3). At that point remove the sieve from the bleach solution and wash the embryos/eggs thoroughly with distilled water for at least 2 min. Poor washes can lead to DNA staining artifacts. After dechorionation, the vitelline membrane that surrounds the embryos/eggs has to be removed for proper antibody penetration. With the help of a clean paintbrush transfer the embryos/eggs from the sieve to a 1.5 ml microcentrifuge tube. From now on the embryos/eggs are kept in 1.5 ml microcentrifuge tubes. Remove vitelline membrane in 1 ml of a 1:1 solution of methanol and heptane for 3 min with vigorous hand shaking. Embryos/eggs should sink at this point. Remove the heptane–methanol solution without disturbing the embryos/eggs and add 1 ml of dry ice-cold methanol to fix them. Put the tubes on a wheel rotator for 10 min at room temperature. Fixed embryos can be kept in methanol at –20°C for several days. After fixation and before antibody staining embryos/eggs have to be rehydrated. For that, the methanol has to be removed. Wash embryos/eggs two times, 5 min each, with 1 ml of PBST (0.1% Tween-20 in PBS). Blocking is done in 1 ml of PBSTB (1% BSA, 0.1% Tween-20 in PBS) for 30 min at room temperature on a wheel rotator. Dilute primary antibodies in 500 μL of PBSTB and incubate embryos/eggs either for 2 h at room temperature or overnight at 4°C on a wheel rotator. Table I indicates commonly used centrosomal-related primary antibodies and their working dilutions. After primary antibody incubation wash the embryos/eggs three times, 20 min each, with 1 ml of PBSTB. Dilute secondary antibodies 1:200 in 500 μL of PBSTB and incubate for 2 h at room temperature. Incubation has to be performed in the dark on a wheel rotator. Wash embryos/eggs three times, 20 min each, with 1 ml of PBSTB followed by one more wash with Phosphate Buffer Saline (PBS) only. DNA staining can be done by incubating embryos/eggs with 500 μL of Toto-3-Iodide (Molecular Probes) in 500 μL of PBS for 10 min at room temperature on a wheel rotator. Finally wash the embryos/eggs in 1 ml of PBS for 5 min and transfer them to glass slides containing Vectashield mounting media for fluorescence (Vector Laboratories). To transfer the embryos/eggs remove all the PBS from the eppendorf, add around 15 μL of Vectashield to the eppendorf, and carefully pipette the embryos/eggs from the eppendorf to the glass slide. Gently cover with a coverslip and seal with nail polish.

This immunofluorescence protocol was derived from Riparbelli and Callaini (2005) and Warn and Warn (1986) and is particularly useful for observations of first mitotic divisions as shown in Fig. 4.

Table I
List of Centrosomal-Related Primary Antibodies Used for Immunofluorescence of *Drosophila* Embryos and Testes.

Primary antibodies	Labeled structure	Working dilution	Supplier/reference
Rat anti-α-Tubulin (YL1/2	Interphase mts, mitotic/meiotic spindle, flagella	1:50	Oxford Biosciences (MAB1864)
Mouse anti-γ-Tubulin (GTU88)	Centrioles, centrosomes	1:25	Sigma-Aldrich (T6557)
Rabbit anti-centrosomin (CNN)	Centrosomes	1:300	Bettencourt-Dias *et al.* (2004)
		1:400	Vaizel-Ohayon and Schejter, (1999)
Rabbit anti-CP190 (RB188)	Centrosomes	1:300	Whitfield *et al.* (1988)
Chicken anti-D-PLP	Centrosomes,	1:1000	Bettencourt-Dias *et al.* (2005)
Rabbit anti-D-PLP	centrosomes		Martinez-Campos *et al.* (2004)
Rabbit anti-DSAS-4	Centrioles	1:500	Basto *et al.* (2006)
Rabbit anti-DSPD-2	Centrioles	1:500	Rodrigues-Martins *et al.*, (2007a)
		1:500	Dix and Raff 2007
Mouse anti-acetylated tubulin (6-11B-1)	Interphase mts, mitotic/meiotic spindle, flagella	1:1000	Sigma-Aldrich (T7451)
Mouse anti-glutamylated tubulin (GT335)	Glutamylated sperm	1:500	Kavlie *et al.* (2010)

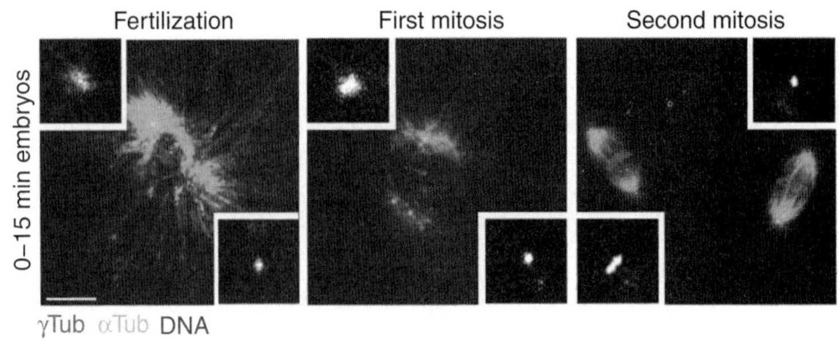

Fig. 4 Immunofluorescence of early *Drosophila* embryos. 0–15 min embryo collections allow the observation of fertilization, first mitosis, and second mitosis. Centrosomes can be analyzed using γTubulin antibody. Scale bar represents 10 μm. Insets are 3× magnification of γTub channel.

B. Transmission Electron Microscopy of Embryos/Eggs

Collection and dechorionation of embryos/eggs for transmission electron microscopy can be done in the same way as described above for the immunofluorescence protocol. When using electron microscopy, it is easier to find cortically located centrosomes and therefore it is better to collect embryos aged between 2 and 3 h. After dechorionation, transfer the embryos/eggs to a 5 ml glass vial. For electron microscopy, the vitelline membrane has to be removed in 3–5 ml of 25% glutaraldehyde in PBS with an equal volume (3–5 ml) of heptane for 3 min with vigorous hand shaking. Carefully pipette the embryos/eggs to a new glass vial containing 1 ml of 2.5% glutaraldehyde in PBS and incubate for 30 min at room temperature on a wheel rotator. Remember to always adjust the pH to between 7.2 and 7.4 to diminish fixation artifacts. The samples should always be incubated in volumes 15–20 times greater than the tissue volume. To remove the vitelline membrane, it is helpful to use double-sided scotch tape with some drops of PBS so that the vitelline membrane is easily removed while the embryos/eggs do not dry. Use tungsten needles to scratch the surface of the vitelline membrane and roll the embryos/eggs on the tape so that the vitelline membrane remains adhered to it. With a clean paintbrush transfer the embryos/eggs to a new glass vial. Fixation is performed overnight at 4°C with 1–2 ml of 2.5% glutaraldehyde in PBS (pH 7.2–7.4) (Rodrigues-Martins *et al.*, 2007b). Rinse three times in 1 ml of PBS (30 min each) to remove all traces of fixative, as glutaraldehyde and osmium tetroxide (to be used next) in solution form an intermediate compound which can break down to osmium black, a precipitate that can compromise the quality of the sample (Allen, 2008; Kuo, 2007; Stoward, 1973). Wash with 1 ml of distilled water three times and postfix in 1% osmium tetroxide for 2 h at 4°C. Dehydrate in 1 ml of a graded series of alcohol (70%, 90%, and absolute) incubating three times, 15 min each. Note that absolute ethanol incubation at 4°C can also be done overnight. After dehydration, an intermediate solvent is normally used as ethanol and embedding media are not readily miscible. Incubate embryos/eggs in 2 ml of propylene oxide solution [(EM grade, electron microscopy sciences (EMS)] three times for 10 min each. Be careful not to let your samples dry as this solution is very volatile. Incubate the samples in 2 ml of graded mixtures of propylene oxide and embedding resin (propylene oxide 2:1 resin; propylene oxide 1:1 resin, propylene oxide 1:2 resin) agitating the samples very slowly on a wheel rotator. Incubate the samples for a minimum of 1 h in 2 ml of pure embedding resin (EMS Embed 812) on a wheel rotator at slow speed. Transfer each one of the embryos/eggs to a flat embedding mold from EM (EMS Embedding Mold). Orient the embryos in order to maximize the cortical area exposure to easily screen for centrioles. Let the resin polymerize at 60°C for 48 h and trim the block. Cut thin sections (60–80 nm thick) of uniform thickness and as free of compression, scratches, vibrations, and wrinkles as possible (Callaini *et al.*, 1997). Collect the sections on grids (200–300 mesh). For section staining transfer the grids into a drop of 2% (w/v) uranyl acetate in a 70% methanol solution for 3–4 min (Hayat, 2000). Methanol has been shown to enhance the contrast given by uranyl acetate. Wash the grid in two drops of 70% methanol followed by five drops of distilled water. Transfer the grid to a drop of Reynold's lead citrate for 1–2 min. Staining duration should be optimized (normally uranyl acetate staining can vary between 3 and 5 min, lead citrate staining between 1 and 5 min) and staining reagents should always be fresh and filtered before use, to avoid precipitation on the sections (Hayat, 2000; Reynolds, 1963). Wash the grids in five drops of distilled water and let them dry at room temperature. Grids can be stored at room temperature in gelatin capsules or inside grid boxes.

C. Immunoelectron Microscopy of Embryos

Electron microscopy analysis is a useful tool to understand the ultrastructural aspects of the centriole. However, whereas a lot of detailed information on the architecture of this organelle is made available by this technique, no data on the spatial localization of centriole-related proteins are provided. Immunoelectron microscopy is a powerful tool very useful for antigen subcellular localization due to its high-resolution power (that cannot be obtained using conventional immunofluorescence).

When using immunoelectron microscopy, fixation is a critical step, as aldehyde fixatives can reduce significantly the tissue immunoreactivity (which is not the case for conventional immunofluorescence where methanol fixation does not significantly affect the antigens).

Another issue that must be taken into account is that the secondary antibodies used for immunoelectron microscopy are conjugated with gold particles varying from 5 to 10 nm diameter. This way, it is required to have the best possible membrane permeabilization and, at the same time, maintain cytological integrity.

There are two approaches that can be used for immunoelectron microscopy: labeling with antibodies either preembedding or postembedding. Briefly, the preembedding method consists of the recognition of the antigen followed by the inclusion of the sample. The postembedding method requires inclusion of the material and posterior localization of the antigen on thin sections. The postembedding method has the advantage of avoiding antibody penetration problems, since immunolocalization is done on thin sections. On the other hand, the preembedding method has the advantage of being more comparable to conventional immunofluorescence as the antigenic properties of the proteins are less affected, provided that the immunolabeling is performed before the tissue undergoes strong fixation and embedding.

Collection and dechorionation of embryos for immunoelectron microscopy can be done in the same way as described above for the immunofluorescence protocol.

1. Preembedding Immunoelectron Microscopy of Embryos

After dechorionation, transfer embryos to a glass vial containing 0.4 ml of 37% formaldehyde (EM grade), 0.6 ml of distilled water, and 5 ml of heptane and shake vigorously for 20 min on a wheel rotator. Transfer embryos to a small drop of PBS and remove the vitelline membrane with tungsten needles as described above for the electron microscopy protocol. Devitellinized embryos are then transferred to a glass vial containing 4% formaldehyde and 0.1% Triton-X in PBS and incubated for 20 min at room temperature. Carefully rinse the embryos in PBS and stain DNA for 3 min with 1 μg/ml of Hoechst 33258 (Sigma-Aldrich) in PBS to select the desired stages under fluorescence microscopy. Cut the embryos longitudinally in two halves. Cut again these two half-embryos longitudinally in order to obtain four thin strips to facilitate antibody penetration. Immerse the embryo strips for 2 h in a blocking solution of PBSB (1% BSA in PBS) in glass vials. Dilute primary antibodies in PBSB and incubate overnight or for 1 or 2 days on a wheel rotator at 4°C. Wash embryos stripes three times, 20 min each, with PBSB. Incubate the embryos stripes for 2 h with the secondary antibodies conjugated with 5 nm colloidal gold diluted in PBSB. The small gold diameter allows for better penetration of the secondary antibody within the sample. Carefully rinse embryo strips in PBS and fix overnight at 4°C in Karnovsky solution (2.5% glutaraldehyde and 1% paraformaldehyde in PBS) (Karnovsky, 1965). After washing in PBS, the embryo strips are postfixed for 1 h in 0.5% osmium tetroxide and then dehydrated and embedded as described above for electron microscopy analysis. Nuclei-associated centrioles localize just underneath the cell membrane. As such, grazing sections are adequate to detect the antibodies interacting with their antigens. Thin sections are counterstained as usual with uranyl acetate and lead citrate as described above.

2. Postembedding Immunoelectron Microscopy of Embryos

Devitellinization can be performed as described above for the preembedding technique. Embryos are then transferred to a glass vial and fixed in 4% formaldehyde for 1 h. Postfixation with osmium should be avoided or done in lower concentration (0.5% in PBS) since the antigenic properties of molecules are noticeably reduced by this fixative (Kellenberger, 1991). Embryos are then dehydrated and embedded in a hydrophilic resin. One of three resins is usually used: Lowicryl K4M (Electron Microscopy Sciences; Hatfield, USA), L.R. White (Ted Pella inc., USA), and L.R. Gold (Polysciences Inc., Warrington, USA).

L.R. White is a good choice as it is nontoxic and can be polymerized in a conventional oven. The dehydration step can be done as previously described for electron microscopy. After dehydration, transfer embryos into the embedding mold and allow the resin to polymerize at 55°C.

Cutting thin sections is not easy with hydrophilic resins, as water often penetrates the block making sectioning a challenging task. Sections must be collected on nickel or gold grids as these metals do not react with the reagents used in this technique. Transfer grids with thin sections into small drops of PBSB for 2 h, followed by incubation on 20 mM glycine in PBS for 20 min. Dilute the primary antibodies in 0.5% BSA in PBS and incubate overnight in a moist chamber at 4°C. Following primary antibody incubation wash the grids three times, 10 min each. The first wash is done in 0.5% Tween-20 in PBS and the following two in PBS (without Tween-20). Incubate the grids for 2 h at room temperature with the secondary antibodies conjugated to 10 nm colloidal gold. Wash the grids in PBS three times, 10 min each, and fix briefly in 0.5% glutaraldehyde for 5 min. Wash the grids two times in PBS, 10 min each, followed by three more washes in distilled water, 10 min each. The grids are finally counterstained with uranyl acetate and lead citrate as described above.

III. Centrioles in *Drosophila* Spermatogenesis

A variety of cell divisions characterize spermatogenesis (Fig. 5) (Fuller, 1993). During *Drosophila* spermatogenesis, centrioles become basal bodies and elongate from 0.5 to 2.3 μm (Fig. 1B) (Tates, 1971), making this an extremely useful system

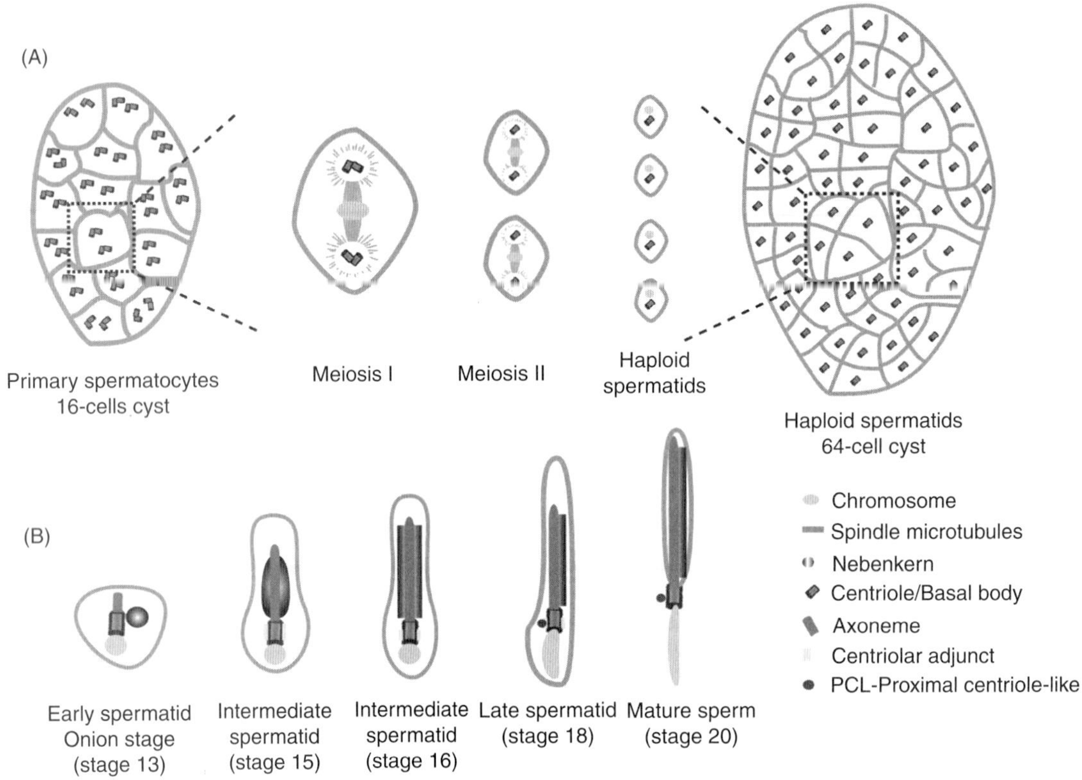

Fig. 5 *Drosophila* spermatogenesis. (A) *Drosophila* spermatogenesis begins with the asymmetric division of a germ line stem cell to produce a primary spermatogonial cell and another stem cell [(Tates, 1971; reviewed in Fuller (1993)]. Each primary spermatogonial cell proceeds through four rounds of mitosis to generate a cyst of 16 primary spermatocytes, which remain interconnected by intracellular bridges, due to an incomplete cytokinesis. These cysts stay in G_2-phase for about 90 h. During this extended G_2-phase period, both the cells and the centrioles undertake growth. Cells in G_2-phase grow 25 times in volume and transcribe most of the gene products known to be required during meiosis. After the long G_2-phase, primary spermatocytes enter meiosis I with two centrosomes, each one composed of two V-shaped centrioles that are responsible for the organization of the meiotic spindle. The period between the two meiotic divisions is short and as centrioles do not duplicate, meiosis II cells also contain two centrosomes but each one of them is composed of only one centriole. At the end of meiosis II a cyst containing 64 mature haploid spermatids is formed. As cytokinesis is incomplete the spermatids within a cyst remain interconnected by cytoplasmaic bridges and ring canals can be seen in the cytoplasm. (B) Haploid spermatids enter an extensive differentiation program with several morphological changes where long flagellar axonemes are formed, most cytoplasmic material is discarded, and the nuclei are transformed into needle-shaped sperm heads. Spermatid differentiation has been described and categorized into different stages based on the morphological changes of the mitochondria. In an early spermatid stage, called onion stage (stage 13), the mitochondria have fused and formed a spherical mass called Nebenkern or mitochondrial derivative that is located to one side of the nucleus and has the same size as the nucleus. It is at the onion stage that axoneme nucleation from the basal body begins in the cytoplasm. Note the presence of a recently identified proximal centriole-like structure close to the basal body at the end of spermatid differentiation (Blachon *et al.*, 2009) (Adapted from Fuller, 1993; Tates, 1971).

to study both centriole biogenesis and axoneme formation (Gonzalez *et al.*, 1998). For example, in the spermatogenesis of male mutants for the centriolar protein SAS-6 shorter centrioles were observed, suggesting that SAS-6 is required for centriole elongation (Peel *et al.*, 2007; Rodrigues-Martins *et al.*, 2007a). The centrosomal protein D-PLP was shown to be essential for the integrity of the centrioles/basal bodies, which often partially fragment during spermatogenesis in *D-PLP* mutants (Martinez-Campos *et al.*, 2004).

Centrioles and basal bodies in *Drosophila* spermatogenesis are composed of nine triplets of microtubules. Axonemes are composed of nine doublets of microtubules as the C-tubule present in the basal body stops growing. In addition, flagellar axonemes contain a central pair of microtubules required for sperm motility (Fig. 1B).

Spermatogenesis is also useful to study centrosome function as centrosomes are needed for meiotic divisions (Rodrigues-Martins *et al.*, 2008). Meiosis progression can be determined using live imaging or immunofluorescence. Alternatively, phase-contrast imaging of the meiotic products (Nebenkern stage) is also very informative (Fig. 5B). At this stage of spermatid differentiation, the volume of the nucleus is proportional to its DNA content (Gonzalez *et al.*, 1989). Hence, variations in DNA content resulting from abnormal DNA segregation, such as micronuclei, diploid, or tetraploid nuclei, are easily recognizable because of altered nuclear size or number. Also, if mitochondria assemble along the spindle and cytokinesis occurs correctly, each daughter spermatid cell receives the same amount of mitochondria and has the same size of mitochondrial derivative. Hence, different Nebenkern sizes are indicative of cytokinesis abnormalities.

A. Phase Contrast and Immunofluorescence of Testes

Testes are normally dissected from pharate adults in order to obtain a higher proportion of G_2-phase cells and primary spermatocytes. Pharate adult males can be easily distinguished because their legs have black sex combs. The easiest way of dissecting testes is exemplified in Fig. 6.

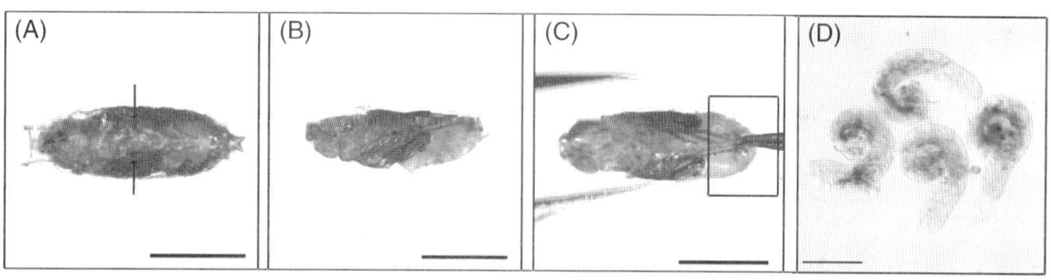

Fig. 6 Dissecting testes from *Drosophila* pharate adults. (A) Identify the male pharate adults by the sex combs in their legs (arrows). (B) Hold the pharate adult under a dissection scope and pull the fly out the pupal case very carefully. (C) Hold the fly with its legs facing up and pull out the testes that are localized in the lower part of the abdomen (defined by the square). From (A) to (C) scale bar represents 1 mm. (D) Example of two pairs of testes taken from two pharate adults. Scale bar represents 300 μm.

Drosophila testes are very long closed tubes (Fig. 6D) that contain at the apical part a cluster of somatic cells, the hub. Male stem cells are attached to the hub and whenever a germ cell divides the stem cell remains at the hub, whereas the differentiated cell moves along the testes tube leading to a spatial-temporal array of spermatogenic stages from an apical to a basal localization (Fuller, 1993; White-Cooper, 2009).

For phase-contrast analysis, testes are dissected in a drop of 0.7% NaCl under a dissection scope. Testes are then transferred to a 15 μL drop of 0.7% NaCl placed in the middle of glass slide and covered with a coverslip. Wipe off the excess of liquid under a dissection scope and stop wiping as soon as sperm tails become visible. Start imaging right away before the sample dries. Figure 7A shows some examples of cells at the Nebenkern stage.

For immunofluorescence perform testes dissection in a drop of TB buffer (183 mm KCl, 47 mm NaCl, 1 mm EDTA, and 10 mm Tris–HCl, pH 6.8), under a dissection

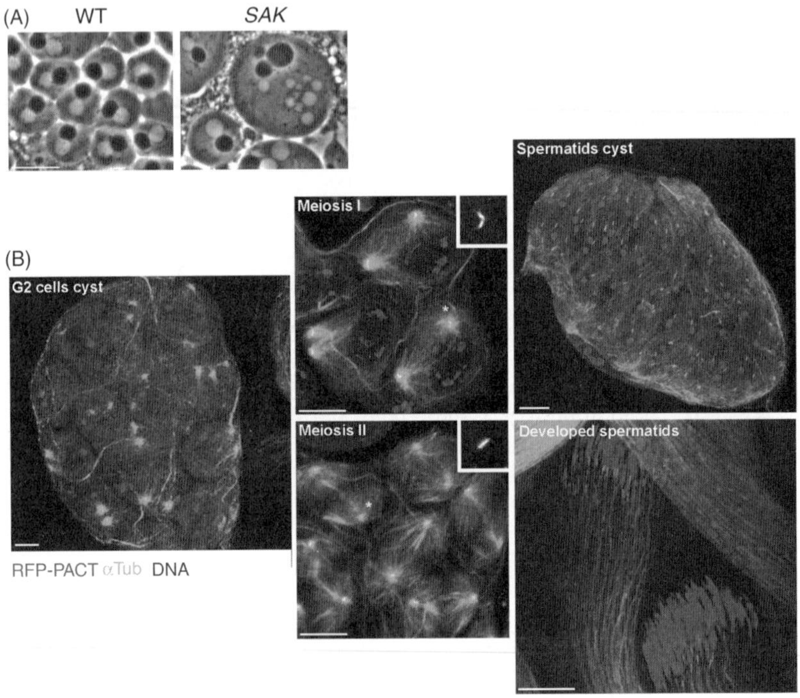

Fig. 7 Phase contrast and immunofluorescence of *Drosophila* spermatogenesis. (A) Phase contrast of *Drosophila* onion stage showing a 1:1 ratio of similar sized nuclei (white) and Nebenkern (black) in the wild-type image. In the *SAK* centriolar mutant these features are no longer kept. Scale bar represents 10 μm. (B) Immunofluorescence of testes from pharate adults allows the observation of G2-cell cysts, of both meiosis I and meiosis II cells and also of spermatids cysts and developed spermatids. Note the presence of two centrioles per centrosome in meiosis I cells but only a single centriole per centrosome in meiosis II [centrioles are labeled by RFP-PACT (Lucas and Raff, 2007)]. Scale bar represents 10 μm. Insets are 2× magnification of RFP-PACT channel.

scope. Use three pairs of testes per slide and three slides per antibody. After dissection, transfer the testes to a 4 μL TB buffer drop placed in the middle of a glass slide treated with poly-L-Lysine (Sigma). Poly-L-Lysine-coated slides and siliconized coverslips help to keep the dissected testes in place. Randomly open the testes with the forceps (twice in a testes pair) to allow antibody penetration. Cover the glass slide with an 18 × 18 siliconized coverslip (you can use Sigmacote from Sigma to siliconize normal coverslips) and freeze the slide immediately by immersing it in liquid nitrogen. Do not leave the slides in liquid nitrogen for more than 1 h as this may cause artifacts. Take the slide out of the liquid nitrogen and remove the coverslip by flipping it off with a scalpel. Fix the slides in dry ice-cold methanol in an upright slide box for 8 min. After methanol fixation, transfer the slides to another upright box containing dry ice-cold acetone and incubate it for 10 min. Keep both upright slide boxes in dry ice during the incubation times. Transfer the slides to a clean upright slide box and wash the slides three times, 5 min each, with PBS at room temperature. Blocking is done in the upright slide box in PBSB (1% BSA in PBS) for 1 h at room temperature. Transfer the slides to a moist chamber prewetted with PBS and incubate with primary antibodies diluted in PBSB. Use 50 μL of antibody solution per slide. To avoid drying of the antibody solution, carefully wipe off the blocking solution from the slide with tissue paper leaving just a small square where the dissected testes are. This cleaning technique can be used for all incubations. Incubate primary antibodies for 2 h at room temperature or overnight at 4°C. No agitation is needed. Table I indicates commonly used centrosomal-related primary antibodies and their working dilutions. After incubation, transfer the slides back again to an upright slide box and wash three times, 15 min each, with PBSB at room temperature. Dilute secondary antibodies 1:100 (or according to manufacturers instructions) in PBSB and incubate slides again in a moist chamber prewetted with PBS, for 2 h at room temperature and in the dark. Wash three times, 15 min each, with PBSB in an upright slide box. Perform a fourth wash with PBS. DNA staining can be achieved by incubating testes with 1 μL of Toto-3-Iodide (Molecular Probes) in 1 ml of PBS for 10 min at room temperature. Use the moist chamber prewetted with PBS for this incubation too. Wash testes in PBS for 5 min in an upright slide box. Let the slides dry and add an 8 μL drop of Vectashield mounting media for fluorescence (Vector Laboratories). Gently cover with a coverslip and seal with nail polish.

Note that for some antibodies a preextraction step is required to allow better visualization of centrioles. In this case, incubate testes, after dissection, in a drop of 0.2% Tween-20 and 0.04% sodium dodecyl sulfate (SDS) in PBS for 3 min at room temperature. Do not incubate for longer than 3 min as it can cause artifacts.

This immunofluorescence protocol was derived from Cenci *et al.* (1994) and allows the observation of testes as shown in Fig. 7B.

B. Transmission Electron Microscopy of Testes

Dissection of testes can be done in PBS as described in Fig. 6. Dissected testes are transferred to glass vials and fixed by immersion in 1 ml of 2.5% glutaraldehyde in

PBS for 2 h at 4°C on a wheel rotator. Glutaraldehyde is a dialdehyde and very efficient in protein cross-linking and ultrastructural preservation (Allen, 2008; Kuo, 2007; Stoward, 1973). On the other hand, its tissue penetration is not very fast and it is possible to start the protocol by prefixing testes for 15–20 min in 4% paraformaldehyde (EM grade) in PBS, which has a lower molecular weight and will penetrate faster (Kuo, 2007). After fixation rinse three times, 30 min each, in 1 ml of PBS in order to remove all traces of fixative. Wash three times with distilled water and postfix the samples in 1% aqueous solution of osmium tetroxide for 1 h at 4°C on a wheel rotator. Note that longer exposures to osmium tetroxide may result in extraction of both the nucleus and the cytoplasm (Stoward, 1973). Wash in distilled water three times for 10 min each. Incubate testes in 1 ml of 1% uranyl acetate aqueous solution for 1 h. This step, also called block staining, enhances the contrast of the samples thus improving the quality of cytological details. Wash in distilled water three times, 10 min each (Hayat, 2000; Kuo, 2007). Dehydrate in a graded series of alcohol (70%, 90%, and absolute) so that later the tissue can be properly infiltrated by the embedding media. This stepwise dehydration minimizes overall tissue shrinkage, as well as differential shrinkage of different cell components. The dehydration steps should take at least 10 min each and dehydration in absolute alcohol should be repeated at least three times in order to efficiently remove all water molecules. After dehydration, an intermediate solvent is normally used, as ethanol and embedding media are not readily miscible. Propylene oxide is the most used solvent for electron microscopy purposes. Incubate testes in 1 ml of propylene oxide three times for 10 min each. Be careful not to let the testes dry as this solution is very volatile. Incubate testes in a graded mixture of propylene oxide and embedding resin (propylene oxide 2:1 resin; propylene oxide 1:1 resin, propylene oxide 1:2 resin) agitating the samples very slowly on a wheel rotator. Incubate the samples for a minimum of 1 h in 2 ml of pure embedding resin (EMS Embed 812) at slow speed on a wheel rotator. Transfer each testis to a flat embedding mold for EM (EMS Embedding Mold). At this point, position the testes in the resin according to experimental needs. For example, it is possible to orient the testes perpendicularly to the cutting face in order to have good cross sections of cysts as the ones shown in Fig. 8.

Allow the resin to polymerize at 60°C for 48 h and trim the block following the standard technique. Cut thin sections (60–80 nm thick) of uniform thickness and as free of compression, scratches, vibrations, and wrinkles as possible. Collect the sections on grids (200–300 mesh). For section staining transfer the grids into a drop of 2% uranyl acetate in 70% methanol for 3–4 min. Methanol has been shown to enhance the contrast given by uranyl acetate. Wash the grid in two drops of 70% methanol followed by five drops of distilled water. Transfer to a drop of Reynold's lead citrate for 1–2 min using clean tweezers. Staining duration should be optimized (normally uranyl acetate staining can vary between 3 and 5 min, lead citrate staining between 1 and 5 min), and the staining reagents should always be freshly made and filtered to avoid precipitation on the sections. Wash in five drops of water and let it dry at room temperature. Grids can be stored at room temperature in gelatin capsules or inside grid boxes.

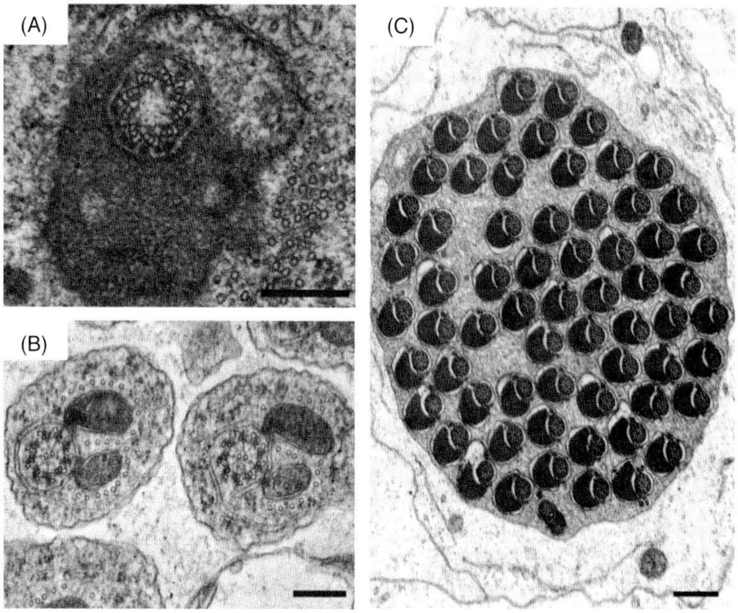

Fig. 8 Transmission electron micrographs of *Drosophila* testes. (A) Spermatid basal body showing accumulation of an electron-dense material, the centriole adjunct. Scale bar represents 200 nm. (B) Transversal section of axonemes at early stage 6 showing accessory microtubules and the presence of a paracrystalline body in the primary mitochondrial derivative (Fuller, 1993). Scale bar represents 200 nm. (C) Spermatids arranged in a cyst at stage 10 of maturation, where the axonemal sheath is absent and the packaging process has initiated. Scale bar represents 500 nm.

C. Immunoelectron Microscopy of Testes

The immunoelectron microscopy procedures for embryos and testes are virtually similar. However, whereas in a syncytial embryo all the centrioles are contained within the same cytoplasm, in larval, pupal, and adult testes, centrioles are surrounded by an external lamina and germ cells. Therefore, antibodies have to pass through all these barriers to reach the germ cell centrioles. While this may not be a problem if immunolocalization is done on thin sections, it may compromise the results if pre-embedding is done.

1. Preembedding Immuno-Electron Microscopy of Testes

The preembedding labeling protocol used in testes samples is similar to the one described above for embryos. Testes from pharate adults can be dissected in PBS as shown in Fig. 6. Testes from larvae, pupae, or adult males can also be used. Fixation is done in 4% formaldehyde in PBS for 20 min. For better penetration of the fixative cut testes into small pieces. To enable antibody penetration into germ cells, two alternative strategies can be used to permeabilize the plasma membrane. A mild permeabilization can be obtained by a 10 min prefixation in 4% formaldehyde and 0.1% Triton-X in PBS. On the other hand, the cell membranes may be cracked by freezing and thawing cycle. For this, testes fragments have to be infiltrated with sucrose for cryoprotection. Immerse testes in a glass vial, first in 10% sucrose solution in PBS for 1 h, then in 20% sucrose solution in PBS for 1 h, and finally in 30% sucrose solution in PBS overnight at 4°C. Transfer testes into a glass slide containing a drop of a 30% sucrose solution in PBS, put the slide on top of a copper bar precooled with liquid nitrogen, and freeze it. After freezing transfer the samples on the slides into a heating plate at 37°C for 5 min. Repeat this process two to three times. After the last heating, wash the testes samples in PBS. Testes samples are now ready for incubation with antibodies and postfixation procedures as described above for the preembedding immunoelectron microscopy of embryos.

2. Postembedding Immunoelectron Microscopy of Testes

The postembedding immunoelectron microscopy technique does not require tissue fragmentation or membrane cracking. Dissected testes can therefore be processed as early embryos and the same protocol described above for postembedding immunoelectron microscopy of embryos can be used.

Acknowledgments

We are thankful to Daniela Brito and Filipe Leal for critically reading this chapter. Work in MBD laboratory is funded by FCT (Portugal), EMBO, and IGC. A.R.M and P.M. have fellowships from FCT (Portugal). Work in GC laboratory is funded by PRIN and PAR (University of Siena).

References

Allen, T. (2008). "Introduction to Electron Microscopy for Biologists." Academic Press, San Diego.

Badano, J. L., *et al.* (2005). The centrosome in human genetic disease. *Nat. Rev. Genet.* **6**, 194–205.

Basto, R., *et al.* (2006). Flies without centrioles. *Cell* **125**, 1375–1386.

Bellen, H. J., *et al.* (2004). The BDGP gene disruption project: Single transposon insertions associated with 40% of *Drosophila* genes. *Genetics* **167**, 761–781.

Bettencourt-Dias, M., *et al.* (2004). Genome-wide survey of protein kinases required for cell cycle progression. *Nature* **432**, 980–987.

Bettencourt-Dias, M., *et al.* (2005). SAK/PLK4 is required for centriole duplication and flagella development. *Curr. Biol.* **15**, 2199–2207.

Bettencourt-Dias, M., and Glover, D. M. (2007). Centrosome biogenesis and function: Centrosomics brings new understanding. *Nat. Rev. Mol. Cell Biol.* **8**, 451–463.

Bettencourt-Dias, M., and Goshima, G. (2009). RNAi in Drosophila S2 cells as a tool for studying cell cycle progression. *Methods Mol. Biol.* **545**: 39–62.

Blachon, S., *et al.* (2009). A proximal centriole-like structure is present in Drosophila spermatids and can serve as a model to study centriole duplication. *Genetics* **182**(1): 133–44.

Bornens, M. (2002). Centrosome composition and microtubule anchoring mechanisms. *Curr. Opin. Cell Biol.* **14**, 25–34.

Callaini, G., *et al.* (1997). Centriole and centrosome dynamics during the embryonic cell cycles that follow the formation of the cellular blastoderm in *Drosophila*. *Exp. Cell Res.* **234**, 183–190.

Callaini, G., and Riparbelli, M. G. (1990). Centriole and centrosome cycle in the early *Drosophila* embryo. *J. Cell. Sci.* **97**(Pt 3), 539–543.

Cenci, G., *et al.* (1994). Chromatin and microtubule organization during premeiotic, meiotic and early postmeiotic stages of *Drosophila melanogaster* spermatogenesis. *J. Cell Sci.* **107**(Pt 12), 3521–3534.

Dietzl, G., *et al.* (2007). A genome-wide transgenic RNAi library for conditional gene inactivation in Drosophila. *Nature* **448**, 151–156.

Dirksen, E. R. (1991). Centriole and basal body formation during ciliogenesis revisited. *Biol. Cell* **72**, 31–38.

Dix, C. I., and Raff, J. W. (2007). *Drosophila* spd-2 recruits PCM to the sperm centriole, but is dispensable for centriole duplication. *Curr. Biol.* **17**, 1759–1764.

Fliegauf, M., *et al.* (2007). When cilia go bad: Cilia defects and ciliopathies. *Nat. Rev. Mol. Cell Biol.* **8**, 880–893.

Foe, V., Odell, G. M., and Edgar, B. (1993). Mitosis and morphegenesis in the *Drosophila* embryo. Point and counterpoint. In "The Development of *Drosophila melanogaster*" (M. Bate and Martinez Arias, eds.), pp. 149–287. Cold Spring Harbor Laboratory Press, Long Island, NY.

Fuller, M. T. (1993). Spermatogenesis. In "The Development of *Drosophila melanogaster*" (M. Bate and A. M. Arias, eds.), pp. 71–147. Cold Spring Harbor Laboratory Press, Long Island, NY.

Glover, D. M. (1991). Mitosis in the *Drosophila* embryo–in and out of control. *Trends Genet.* **7**, 125–132.

Gogendeau, D., and Basto, R. (2009). Centrioles in flies: The exception to the rule? *Semin. Cell Dev. Biol.* **21**(2): 163–73.

Gonzalez, C., *et al.* (1989). Relationship between chromosome content and nuclear diameter in early spermatids of *Drosophila melanogaster*. *Genet. Res.* **54**, 205–212.

Gonzalez, C., *et al.* (1998). Centrosomes and microtubule organisation during *Drosophila* development. *J. Cell. Sci.* **111**(Pt 18), 2697–2706.

Greenspan, R. J. (2004). "Fly Pushing: The Theory and Practice of *Drosophila Genetics*." Cold Spring Harbor Laboratory Press, Long Island, NY.

Hayat, M. (2000). "Principles and Techniques of Electron Microscopy: Biological Applications." Cambridge University Press, Cambridge.

Januschke, J., and Gonzalez, C. (2008). *Drosophila* asymmetric division, polarity and cancer. *Oncogene* **27**, 6994–7002.

Karnovsky, M. J. (1965). A formaldehyde-glutaraldehyde fixative of high osmolarity for use in electron microscopy. *J Cell Biol.* **27**.

Kavlie, R. G., *et al.* (2010). Hearing in Drosophila requires TilB, a conserved protein associated with ciliary motility. *Genetics* **185**(1): 177–88

Kellenberger, H. (1991). Some basic concepts for the choice of methods. *In* "Colloidal Gold: Principles, Methods and Applications" (M. A. Hayat, ed.), Vol. 3, pp. 1–30. Academic Press, San Diego, CA.

Kuo, J. (2007). "Electron Microscopy: Methods and Protocols." Humana Press, Totowa, NJ.

La Terra, S., *et al.* (2005). The de novo centriole assembly pathway in HeLa cells: Cell cycle progression and centriole assembly/maturation. *J. Cell Biol.* **168**, 713–722.

Lucas, E. P., and Raff, J. W. (2007). Maintaining the proper connection between the centrioles and the pericentriolar matrix requires Drosophila centrosomin. *J. Cell Biol.* **178**, 725–732.

Martinez-Campos, M., *et al.* (2004). The Drosophila pericentrin-like protein is essential for cilia/flagella function, but appears to be dispensable for mitosis. *J. Cell Biol.* **165**, 673–683.

Ou, Y., *et al.* (2004). The centrosome: The centriole-PCM coalition. *Cell Motil. Cytoskeleton* **57**, 1–7.

Peel, N., *et al.* (2007). Overexpressing centriole-replication proteins in vivo induces centriole overduplication and de novo formation. *Curr. Biol.* **17**, 834–843.

Plotnikova, O. V., *et al.* (2008). Cell cycle-dependent ciliogenesis and cancer. *Cancer Res.* **68**, 2058–2061.

Raff, J. W. (2004). Centrosomes in a developing organism: Lessons from Drosophila. *In* "Centrosomes in Development and Disease" (E. A. Nigg, ed.), pp. 251–278. Wiley-Vch, Weinheim.

Reynolds, E. (1963). The use of lead citrate at high pH as an electron-opaque stain for electron microscopy. *J. Cell Biol.* **17**, 208.

Riparbelli, M. G., and Callaini, G. (2003). Drosophila parthenogenesis: A model for de novo centrosome assembly. *Dev. Biol.* **260**, 298–313.

Riparbelli, M. G., and Callaini, G. (2005). The meiotic spindle of the *Drosophila* oocyte: The role of centrosomin and the central aster. *J. Cell. Sci.* **118**, 2827–2836.

Rodrigues-Martins, A., *et al.* (2007a). DSAS-6 organizes a tube-like centriole precursor, and its absence suggests modularity in centriole assembly. *Curr. Biol.* **17**, 1465–1472.

Rodrigues-Martins, A., *et al.* (2007b). Revisiting the role of the mother centriole in centriole biogenesis. *Science* **316**, 1046–1050.

Rodrigues-Martins, A., *et al.* (2008). From centriole biogenesis to cellular function: Centrioles are essential for cell division at critical developmental stages. *Cell Cycle* **7**, 11–16.

Ryder, E., and Russell, S. (2003). Transposable elements as tools for genomics and genetics in *Drosophila. Brief Funct. Genomic Proteomic* **2**, 57–71.

Stoward, P. (1973). "Fixation in Histochemistry." Chapman and Hall, London.

Sullivan, W., Ashburner, M., and Hawley, R. S. (eds.) 2000. *Drosophila Protocols. Cold Spring Harbor Laboratory Press*, Cold Spring Harbor, N.Y.

Tates, A. D. (1971). Cytodifferentiation during spermatogenesis in *Drosophila melanogaster.* PhD thesis, Department of Radiation Genetics. Transitorium voor Geneeskunde, Leiden, Netherlands.

Vaizel-Ohayon, D., and Schejter, E. D. (1999). Mutations in centrosomin reveal requirements for centrosomal function during early *Drosophila* embryogenesis. *Curr. Biol.* **9**, 889–898.

Warn, R. M., and Warn, A. (1986). Microtubule arrays present during the syncytial and cellular blastoderm stages of the early Drosophila embryo. *Exp. Cell Res.* **163**, 201–210.

White-Cooper, H. (2009). Studying how flies make sperm–investigating gene function in *Drosophila* testes. *Mol. Cell Endocrinol.* **306**, 66–74.

Whitfield, W. G., *et al.* (1988). Cloning of a gene encoding an antigen associated with the centrosome in *Drosophila*. J. Cel. Sci. **89**, 467–480.

Yamashita, Y. M. (2009). The centrosome and asymmetric cell division. Prion **3**, 84–88.

Yamashita, Y. M., and Fuller, M. T. (2008). Asymmetric centrosome behavior and the mechanisms of stem cell division. *J. Cell Biol.* **180**, 261–266.

Zyss, D., and Gergely, F. (2009). Centrosome function in cancer: Guilty or innocent? *Trends Cell Biol.* **19**, 334–346.

CHAPTER 14

Drosophila S2 Cells as a Model System to Investigate Mitotic Spindle Dynamics, Architecture, and Function

Sara Moutinho-Pereira[*], **Irina Matos**[*], *and* **Helder Maiato**[*,†]

[*]IBMC—Instituto de Biologia Molecular e Celular, Universidade do Porto, 4150-180 Porto, Portugal

[†]Laboratory of Cell and Molecular Biology, Faculdade de Medicina, Universidade do Porto, 4200-319 Porto, Portugal

Sara Moutinho-Pereira and Irina Matos have contributed equally to this work

Abstract

In order to perpetuate their genetic content, eukaryotic cells have developed a microtubule-based machine known as the mitotic spindle. Independently of the system studied, mitotic spindles share at least one common characteristic—the dynamic nature of microtubules. This property allows the constant plasticity needed to assemble a bipolar structure, make proper kinetochore–microtubule attachments, segregate chromosomes, and finally disassemble the spindle and reform an interphase microtubule array. Here, we describe a variety of experimental approaches currently used in

978-0-12-381349-7
DOI: 10.1016/S0091-679X(10)97014-3

our laboratory to study microtubule dynamics during mitosis using *Drosophila melanogaster* S2 cells as a model. By using quantitative live cell imaging microscopy in combination with an advantageous labeling background, we illustrate how several cooperative pathways are used to build functional mitotic spindles. We illustrate different ways of perturbing spindle microtubule dynamics, including pharmacological inhibition and RNA interference of proteins that directly or indirectly impair microtubule dynamics. Additionally, we demonstrate the advantage of using fluorescent speckle microscopy to investigate an intrinsic property of spindle microtubules known as poleward flux. Finally, we developed a set of laser microsurgery-based experiments that allow, with unique spatiotemporal resolution, the study of specific spindle structures (e.g., centrosomes, microtubules, and kinetochores) and their respective roles during mitosis.

I. Introduction

In order to maintain their inherited genetic background, cells have developed a specialized microtubule-based structure called the mitotic spindle, which mediates the segregation of chromosomes during cell division. Many cancer therapies currently in clinical practice employ the use of drugs that target the mitotic spindle (e.g., taxanes and vinca alkaloids) with the aim of preventing cell division by blocking (and killing) cells in mitosis or leading to unviable progeny. Given its essential role, mitotic spindle assembly in animal cells is a highly conserved and redundant process that is thought to involve multiple parallel pathways (O'Connell and Khodjakov, 2007). Among these, the most widespread has been proposed by Kirschner and Mitchison who hypothesized that centrosomes drive spindle morphogenesis by generating astral microtubules that continuously grow and shrink, randomly "searching" for chromosomes after nuclear envelope breakdown (Kirschner and Mitchison, 1986). The increase in astral microtubule dynamics at this stage results from the phosphorylation of several regulatory microtubule-associated proteins (MAPs) as a result of increased CDK1 activity (Maiato *et al.*, 2004b; Murray, 2004). When, by chance, astral microtubules encounter a kinetochore, they are captured and become stabilized. With time, more microtubules will be gradually incorporated forming a mature kinetochore fiber (K-fiber). Despite being very attractive in its essence, this "search-and-capture" model cannot explain mitosis in cells that lack centrosomes, such as higher plants and some oocytes (Gadde and Heald, 2004; Wadsworth and Khodjakov, 2004). However, it should be noted that some centrosome-independent "search-and-capture" events may be present in these systems (Lloyd and Chan, 2006). Moreover, a recent computational analysis has shown that, in a purely random fashion, the "search-and-capture" model is not sufficiently efficient to explain the capture of all kinetochores by approximately 20 microtubules in the rapid time frame of dividing animal cells (McEwen *et al.*, 1997; Wollman *et al.*, 2005).

Several other studies have shown that in the absence of centrosomes, animal somatic cells are able to assemble anastral spindles, either when they lack proteins that are necessary to build a functional centrosome, such as asterless or centrosomin (Cnn), or when this structure is physically destroyed or removed (Bonaccorsi *et al.*, 1998; Hinchcliffe *et al.*, 2001; Khodjakov *et al.*, 2000a; Mahoney *et al.*, 2006; Megraw *et al.*, 1999, 2001). In addition, a *Drosophila* cell line lacking centrioles could also be maintained in culture for several years (Debec *et al.*, 1982, 1995), whereas mutants for *cnn* (a protein required to recruit γ-tubulin to centrosomes) or *Dsas-4* (a protein required for centriole duplication) are able to develop into adult flies (Basto *et al.*, 2006; Megraw *et al.*, 2001). One possibility could be that in these cases the self-organization of microtubules in the vicinity of chromosomes is responsible for spindle assembly, as it happens in some female meiotic systems. In this context, kinetochores would be responsible for the formation of their own K-fibers and spindle bipolarity ensured by the action of microtubule motors and cross-linking proteins (Karsenti and Vernos, 2001; Rieder, 2005). In fact, in vertebrate cells, K-fibers were observed forming in association with chromosomes without a direct connection to the centrosome or chromatin (Khodjakov *et al.*, 2003; O'Connell *et al.*, 2009). In agreement, studies in live *Drosophila* S2 cells reported that unattached kinetochores that are not facing a centrosome are able to form microtubules *de novo* and that K-fiber growth likely occurs by microtubule addition at their kinetochore-associated end (Maiato *et al.*, 2004a). Kinetochore-driven microtubule formation appears to rely on microtubule nucleating/stabilizing factors such as γ-tubulin, TPX2, and chromosomal passenger proteins, as well as localized RanGTP at kinetochores (Mishra *et al.*, 2010; Torosantucci *et al.*, 2008; Tulu *et al.*, 2006). In conclusion, both centrosomes and kinetochores contribute to spindle assembly and acentrosomal mechanisms might always be present even in those systems normally relying on centrosomes.

More recently, it was proposed that microtubule propagation within the spindle itself could represent another mechanism to take into account in spindle morphogenesis (Mahoney *et al.*, 2006). This idea has been reinforced after the discovery of the augmin complex and its interaction with γ-tubulin and associated factors found in the spindle region (Goshima *et al.*, 2007, 2008; Uehara *et al.*, 2009; Zhu *et al.*, 2008). What remains unclear is whether this microtubule propagation pathway represents an independent molecular mechanism or whether it works together with more conventional centrosome- or kinetochore-mediated pathways (Bucciarelli *et al.*, 2009; Lawo *et al.*, 2009).

Upon its initial formation the mitotic spindle maintains a steady-state length and shape as cells reach metaphase. However, microtubules, including those attached at kinetochores, remain dynamic and can be recycled due to turnover (Gorbsky and Borisy, 1989; Zhai *et al.*, 1995) and poleward flux (Mitchison, 1989). This requires a labile interface that enables microtubules to slip and eventually detach from the kinetochore in response to a poleward force, whose origin remains controversial (Cameron *et al.*, 2006; Dumont and Mitchison, 2009; Ganem *et al.*, 2005; Matos *et al.*, 2009; Miyamoto *et al.*, 2004; Rogers *et al.*, 2004). The importance of a labile kinetochore–microtubule interface is reflected in the capacity to correct mistakes inherent to the stochastic nature of mitotic spindle assembly and the interaction between microtubules and chromosomes (Bakhoum *et al.*, 2009b; Ganem *et al.*, 2005; Matos *et al.*, 2009). In fact, increasing the stability of kinetochore–microtubule attachments is, by itself, sufficient to induce chromosomal instability in once stable diploid cells (Bakhoum *et al.*, 2009a).

A. Rationale

Drosophila culture cells offer a powerful set of experimental solutions to dissect the molecular basis of mitotic spindle dynamics, architecture, and function. Among these, we highlight the availability of a fully sequenced and well-annotated genome and the conservation of more than 60% of the genes with humans (Adams *et al.*, 2000). Of particular interest for high-throughput genome-wide screenings, specific gene silencing can be easily achieved by RNA interference (RNAi) (Goshima *et al.*, 2007) and commercial dsRNA libraries are available (see also Chapter 15 for a discussion of interpretations of siRNA-derived spindle phenotypes). Additionally, *Drosophila* culture cells have a low chromosome number (typically between 4 and 12) and details of mitotic spindle morphogenesis can be easily observed at the light microscopy level by the use of available stable cell lines expressing fluorescent components of the mitotic apparatus (Mahoney *et al.*, 2006; Maiato *et al.*, 2004a). Finally, *Drosophila* culture cells are amenable for high-resolution live cell microscopy and micromanipulation techniques, such as Fluorescent Speckle Microscopy (FSM) and laser microsurgery (Maiato *et al.*, 2004a, 2005; Matos *et al.*, 2009). In the following sections, we describe in detail some of the state-of-the-art methodologies currently in practice in our laboratory for the study and micromanipulation of microtubule organization, dynamics, and function during mitosis in live *Drosophila* culture cells.

II. Methods

A. Live Cell Microscopy Analysis of Mitotic Spindle Assembly

The availability of several *Drosophila* S2 cell lines expressing various fluorescently tagged components of the mitotic apparatus allows live imaging studies in a wide-range of labeling backgrounds. This, combined with the use of RNAi [for detailed protocols see (Maiato *et al.*, 2003; Pereira *et al.*, 2009)], provides a powerful tool to follow different processes with a high spatiotemporal resolution in a living dividing cell. To monitor microtubule organization during spindle assembly, we use a *Drosophila* S2 cell line stably expressing both GFP-α-tubulin and the inner-kinetochore protein *C*entromere *Id*entifier (CID) fused to mCherry. This cell line was created by transfecting S2 cells already expressing GFP-α-tubulin (Goshima and Vale, 2003) with a pMT-CID-mCherry-BLAST vector (Coelho *et al.*, 2008), which was derived from the original pMT-CID-GFP (Heun *et al.*, 2006), pCo-Blast (Invitrogen), and pRSET-B-mCherry (Invitrogen) vectors. Primers used to clone mCherry and blasticidin genes with appropriate restriction sites are detailed in Table I. Alternatively, virtually any cell line expressing fluorescently tagged components of interest can be easily generated using the Gateway Cloning System (Invitrogen) after full-length cDNA amplification by PCR, using cDNA clones as template (or genomic DNA, when cDNA clones are not available) and subsequently transfected with Cellfectin reagent (Invitrogen) according to an optimized protocol (Pereira *et al.*, 2009). For better results on stable cell line generation, it should be noted that transfection with one plasmid containing the resistance gene for selection

Table I
List of Oligonucleotides Used and Respective Sequences

Primer	Sequence (5′→3′)
mCherry-*Xho*I F	CCGCTCGAGCGGTATGGTGAGCAAGGGCGAGGAGG
mCherry-*Sac*II R	TCCCCGCGGGGATTACTTGTACAGCTCGTCCATGC
Blast-*Sal*I F	ACGCGTCGACGTCTGTTGGAATATACTATTCAACC
Blast-*Sal*I R	ACGCGTCGACGTCCCGATCCAGACATGATAAGATA

(usually blasticidin or hygromycin) is significantly better than co-transfection strategies (where pCoBlast or pCoHygro are used) to avoid the problem of having cells transfected only with pCoHygro/pCoBlast but without the gene of interest. The latter can be cloned into a plasmid under the control of an inducible or constitutive promoter, depending on the desired expression levels for each protein. In our particular case, the copper-inducible metallothionein promoter [pMT vector (Invitrogen)] was used, but several others are available (see Invitrogen catalogue of insect expression vectors).

To investigate how *Drosophila* cells assemble a mitotic spindle with or without centrosomes, control and Cnn-depleted S2 cells (Mahoney *et al.*, 2006) stably expressing GFP-α-tubulin and CID-mCherry are plated into 0.25 mg/ml concanavalin A-coated 22 × 22 mm coverslips and mounted in modified Rose chambers (Pereira *et al.*, 2009) for live cell analysis with a Spinning Disc confocal system, as described below. In control S2 cells most spindle microtubules are seen forming from the centrosomes (Fig. 1A, time 00:00). In cells with reduced Cnn after RNAi (therefore without functional centrosomes), microtubules are in close association with chromosomes and kinetochores since early prometaphase (Fig. 1B, time 00:00) and the spindle is built from chromosomes outwards. In both cases, fully functional mitotic apparatuses are assembled, which efficiently segregate chromosomes into two daughter cells (Fig 1A, time 20:00 and B, time 25:00). Importantly, the use of an agar overlay to flatten S2 cells (Pereira *et al.*, 2009) may be beneficial to reveal undisclosed aspects of spindle assembly, such as acentrosomal microtubule organization around chromatin in cells containing functional centrosomes (Fig. 1C–C′) or *de novo* microtubule formation from kinetochores (Fig. 2).

For image acquisition, four-dimensional data sets are collected with an Andor Revolution Spinning Disc confocal system (Andor) equipped with an Electron Multiplying CCD iXonEM+ camera and a Yokogawa CSU-22 unit based on an Olympus IX81 inverted microscope. Two laser lines (488 and 561 nm) are used for near simultaneous excitation of GFP and mCherry and the system is driven by Andor IQ software. Typically, laser intensity range is 8–12 (arbitrary units) for GFP and 35–40 for mCherry, depending on the protein expression level. Time-lapse image stacks are collected every 30 s with 0.5 μm z-steps and projected as maximum pixel intensities. For combination of GFP fluorescence with DIC we collect image series every 30 s using a Nikon Eclipse TE2000U DIC inverted wide-field microscope equipped with a CoolSnap HQ2 camera (Photometrics, Tucson, AZ). Time-lapse data sets are subsequently blind deconvolved with AutoDeblur X2 software (Media Cybernetics).

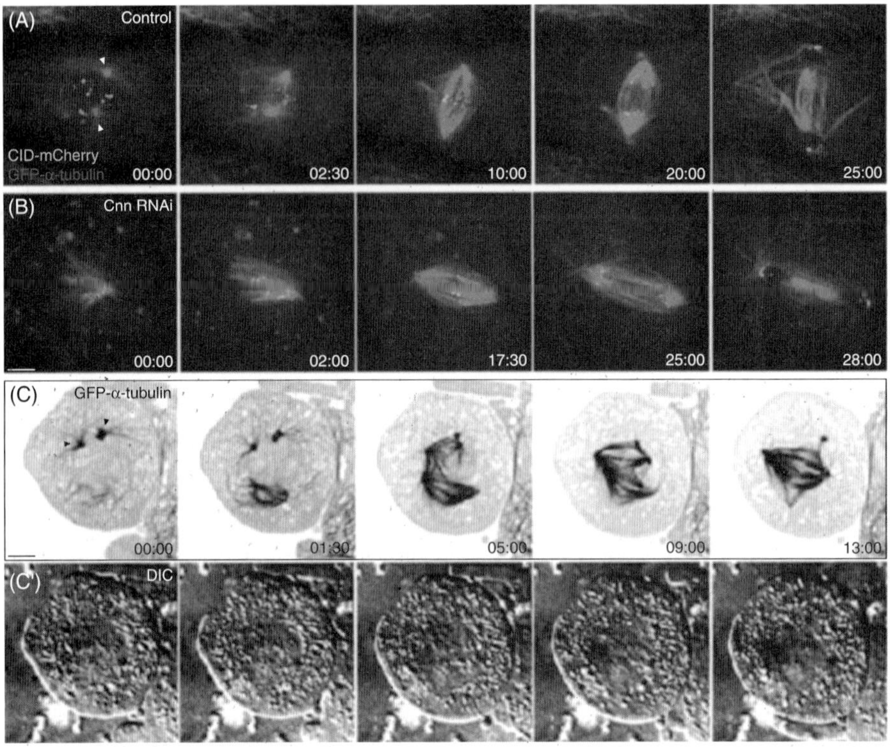

Fig. 1 Live cell imaging of spindle formation in control and Cnn-depleted S2 cells. (A–B) Time-lapse sequence showing mitotic spindle assembly in S2 cells stably expressing CID-mCherry (green) and GFP-α-tubulin (red). Note that GFP channel is here shown in red to allow an easier visualization of CID. (A) Control S2 cells assemble a mitotic spindle in a "dominant" centrosomal fashion, whereas Cnn-depleted cells (B) build a spindle through a purely chromatin/kinetochore–microtubule-mediated pathway, with a normal timing. (C–C′) Control S2 cell stably expressing GFP-α-tubulin under an agar overlay, which in this case leads to the physical exclusion of centrosomes (arrowheads) from a cellular region assembling a spindle in the vicinity of chromatin visualized with DIC. Note that over time both centrosome and acentrosomal spindles merge into a unified structure. Time is in min:s; scale bar, 5 μm. (See Plate no. 6 in the Color Plate Section.)

Additional image processing steps are performed using Image J 1.38× software (http://rsb.info.nih.gov/ij; NIH, Bethesda, MA) and Adobe Photoshop CS3 and Illustrator 8.0 (Adobe Systems, San Jose, CA).

B. Ways of Altering Mitotic Spindle Dynamics

Microtubules are macromolecular hollow cylinders formed by basic building blocks of α-tubulin and β-tubulin heterodimers. Several mitotic poisons are known to disrupt/alter microtubule properties and are commonly used as chemotherapeutic agents. Additionally, the use of these substances in mitotic cells can prove useful to study several aspects of microtubule dynamics during mitosis (Jordan and Wilson, 1999).

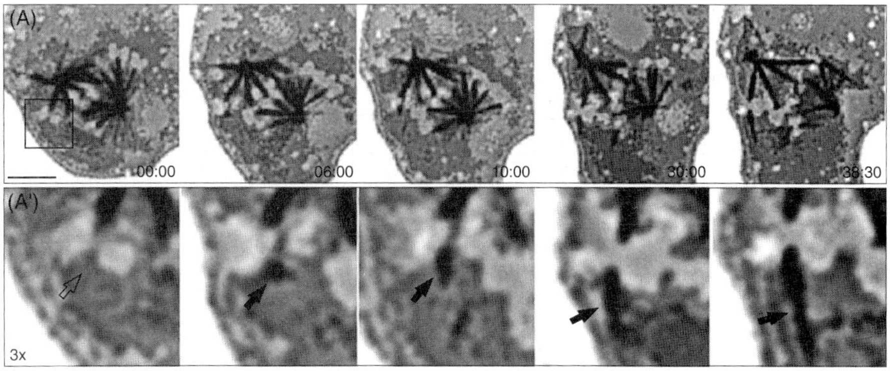

Fig. 2 Live cell imaging of *de novo* microtubule formation from kinetochores. (A) S2 cell stably expressing GFP-α-tubulin illustrating the kinetics of formation of K-fibers from naked kinetochores without the contribution of centrosomes. (A′) Higher magnification of the selected chromosome in A. Chromosomes can be inferred as fluorescent exclusion bodies. Time is in min:s; scale bar, 5 μm.

While some of these molecules block microtubule polymerization by binding to soluble tubulin (e.g., colchicine, colcemid, nocodazole, and vinca alkaloids), thus preventing new polymer formation, others act by blocking microtubule disassembly through stabilization of GDP-bound tubulin (e.g., taxanes). When late prometaphase/metaphase *Drosophila* S2 cells are exposed to 200 μM of colchicine (Sigma), microtubules that are initially present in the mitotic spindle can no longer assemble new polymer and start to depolymerize (Fig. 3A). Within approximately 20 min after drug addition, microtubules can no longer be detected, with K-fibers being the most resistant, and cells block in prometaphase.

Taxol binds directly to microtubules in cells (Manfredi *et al.*, 1982), more specifically to β-tubulin, causing hyperstabilization of microtubules. In *Drosophila* S2 cells exposed to 10 μM of taxol (Sigma), microtubules become hyperstable and two large asters form by recruiting microtubules from the spindle to the centrosome, while blocking cells in mitosis (Fig. 3B). However, in the presence of lower doses such as 10 nM, taxol's action creates more permissive conditions, allowing cells to enter anaphase (Fig. 3C) (Maresca and Salmon, 2009), despite a strong inhibition of microtubule dynamics (Fig 4B).

Entry and exit into mitosis is orchestrated by a series of biochemical "switches" that strongly impact on microtubule dynamics and consequently on mitotic spindle properties. One of the key regulators of early mitotic events during spindle assembly is CDK1. On the other hand, CDK1 inactivation leads to a series of changes responsible for mitotic spindle disassembly and late mitotic events that remain less well understood. When cells blocked in metaphase (by adding MG132, a proteasome inhibitor; Sigma) are treated with roscovitine (seliciclib; Sigma), a broad CDK inhibitor, we expect to see the specific effect of roscovitine action on CDK1 (rather than on other CDKs that are not operating during mitosis) (Skoufias *et al.*, 2007). In fact, immediately upon roscovitine addition to *Drosophila* S2 cells arrested in metaphase

45 years ago Inoue and Forer followed the movement of areas of reduced birefringence generated with an ultraviolet microbeam (locally ablated microtubules) (Forer, 1965; Inoue, 1964), poleward flux has been studied for several years following a fiduciary mark within a fluorescent-labeled spindle. This mark, originally generated by photobleaching of rhodamine-labeled tubulin, would move toward one pole and the velocity of the displacement measured relative to a fixed point (Buster *et al.*, 2007; Gorbsky and Borisy, 1989). However, the use of this approach poses some problems: as spindle microtubules turnover the fluorescence recovery may be too fast to allow the detection of movement by the bleached mark (Gorbsky and Borisy, 1989; Labbe *et al.*, 2004). In order to circumvent this limitation, photoactivation tools were developed, where fluorescent marks are generated on microtubules by activation of caged-fluorescent tubulin on a dark background (Mitchison, 1989). As nonkinetochore–microtubules more rapidly lose fluorescence, the remaining signal can be used as a fiduciary mark on kinetochore–microtubules (Ganem *et al.*, 2005; Maffini *et al.*, 2009; Mitchison, 1989; Waters *et al.*, 1996; Zhai *et al.*, 1995). With the development of FSM, today's strategies favor the tracking of small groups of fluorescent molecules and attempt to track single fluorophores along spindle microtubules (Yang *et al.*, 2007). Essentially FSM requires the existence of very low amounts of fluorescently labeled tubulin that will be mixed with the endogenous unlabeled monomers and randomly co-assemble the microtubule polymer. The stochasticity of fluorescence incorporation will define the speckled appearance of microtubules, generating numerous intrinsic fiduciary marks that can be tracked with reference to the entire spindle structure (Waterman-Storer *et al.*, 1998). Many important discoveries have been made due to the visualization of speckle dynamics, especially when assisted by computational tools for automatic speckle analysis (Danuser and Waterman-Storer, 2003; Yang *et al.*, 2007, 2008).

All the above-mentioned approaches are currently used in different laboratories to study the nature of spindle microtubule flux. In our laboratory we mostly employ FSM to investigate this process in *Drosophila* S2 cells. Accordingly, we use a *Drosophila* S2 cell line stably expressing GFP-α-tubulin and CID-mCherry, both under the control of an inducible metallothionein promoter, whose leakiness produces only very low levels of fluorescent tubulin without induction (Fig. 4). The CID-mCherry construct is used to guarantee the analysis of speckles specifically within K-fibers and has been described in a previous section. Cells are grown in concanavalin A-coated coverslips (0.25 mg/ml) and mounted in modified Rose chambers with Schneider's medium (Sigma-Aldrich) containing 10% of FBS (Pereira *et al.*, 2009). It is advisable to use conditioned media so that cells do not activate stress responses, which increases the probability of finding mitotic cells. Moreover, due to variable levels of expression, it is important to look for mitotic cells that have the lowest expression of GFP-α-tubulin detectable by the human eye.

To successfully achieve high-resolution FSM, it is important to be able to image a small number of fluorophores (2–10) within diffraction limited regions (~0.25 μm), while preventing photobleaching. This requires a sensitive imaging system, with efficient light collection and a low noise/high quantum efficiency camera. In our setup, time-lapse images are collected in a single plane every 5 s with a Nikon

TE2000U inverted wide-field microscope equipped with a Coolsnap HQ2 CCD camera (Photometrics, Tucson, AZ) and Brightline fluorescence filters (Semrock, Rochester, NY) using a 100× 1.4 NA plan-Apochromatic DIC objective and driven by NIS-elements software (Nikon, Japan). Because microtubule dynamics are highly dependent on temperature we perform all our recordings at 25°C, the physiological temperature for *Drosophila*, by using a temperature controlled Perspex case. In order to improve the signal-to-noise ratio and eliminate the out-of-focus blur, images are subsequently blind deconvolved with AutoDeblur X2 software (Media Cybernetics). Another possibility would be to use a similar setup coupled to a spinning-disk confocal microscope and appropriate laser lines as described before (see also (Reis *et al.*, 2009)).

Flux velocities are measured by kymograph analysis, in which a thin rectangular region along the axis of speckle movement within a K-fiber is extracted from each image in the time-lapse series and aligned sequentially to make a montage of the region over time (Fig. 4A and B). In these 2D kymographs, oblique lines corresponding to the movement of bright microtubule speckles are drawn over time (Fig. 4A′ and B′). The slopes of those lines reveal the velocity of speckle movement (Waterman-Storer *et al.*, 1998). It should be noted that before kymograph analysis spindles are aligned using a guided kymography tool written in Matlab (TheMathWorks, Inc.) to minimize the effects of cellular and spindle translation (Pereira and Maiato, 2010).

D. Laser Microsurgery

In recent years, laser microsurgery has been established as one of the most important tools to investigate mitosis. The mitotic apparatus is an appealing context for the use of such spatial domain techniques because it contains discrete structures that can be manipulated to address their respective roles in chromosome movement and signaling during mitosis (Fig. 5). Indeed, laser microsurgery has been seminal to elucidate the mechanistic basis of the spindle assembly checkpoint (Rieder *et al.*, 1995), the role of the centrosome in spindle assembly (Khodjakov *et al.*, 2000b), and the role of kinetochore–microtubules in chromosome movement (Khodjakov and Rieder, 1996; Khodjakov *et al.*, 1996; McNeill and Berns, 1981). *Drosophila* culture cells offer the additional possibility of combining powerful live cell imaging and laser microsurgery with molecular tools such as RNAi. This has been particularly successful in the study of the molecular mechanism regulating kinetochore–microtubule dynamics (Maiato *et al.*, 2005), where laser microsurgery of K-fibers in S2 cells stably expressing GFP-α-tubulin generates a reproducible assay characterized by K-fiber growth from their kinetochore-associated end at near flux rates (Fig. 5A) (Maiato *et al.*, 2004a; Matos *et al.*, 2009). Another useful application is the laser-mediated ablation of centrosomes, which allows one to investigate the molecular basis of acentrosomal spindle formation in animal somatic cells, as well as to dissect how acentrosomal spindles are maintained by ablating centrosomes after spindle assembly (Fig. 5B). For this purpose we use an S2 cell line stably expressing γ-tubulin fused with dsRed or GFP at its C-terminus. The latter can be combined with stable expression of mCherry-α-tubulin, which allows the

CHAPTER 15

Assessment of Mitotic Spindle Phenotypes in *Drosophila* S2 Cells

Gohta Goshima

Division of Biological Science, Graduate School of Science, Nagoya University, Nagoya 464-8602, Japan

978-0-12-381349-7
DOI: 10.1016/S0091-679X(10)97015-5

Abstract

The *Drosophila* S2 cell line is popularly used to study mitosis. In this cell line, multiple genes can be easily and efficiently knocked down by RNA interference (RNAi), and the associated mitotic phenotypes can be assessed with high-resolution microscopy after immunofluorescence or in a living cell. However, compared to untransformed cells in wild-type organisms such as yeasts or worms, mitosis in the S2 cell line is more variable and often looks abnormal even in RNAi-untreated cells. Therefore, in order to judge whether a phenotype is derived from RNAi of the target gene or is simply a variation of control cells, it is critical to prepare proper control samples and perform objective imaging and image analysis. Here, we discuss how *bona fide* mitotic phenotypes associated with RNAi can be identified, avoiding selecting false positives, in S2 cells.

I. Introduction

Cell division is a multistep process that requires a number of genes for segregating sister chromatids equally into two daughter cells. The central macromolecular structure that executes this task is the mitotic spindle, consisting of microtubules and associated proteins, and its assembly mechanism has been one of the most fascinating research topics in cell biology (Gadde and Heald, 2004; Goshima and Kimura, 2010; Inoue, 2008; Karsenti and Vernos, 2001; Kwon and Scholey, 2004; McIntosh *et al.*, 2002; Mitchison and Salmon, 2001; Scholey *et al.*, 2003; Walczak and Heald, 2008). An important research goal for decades has been the identification of components responsible for building the spindle, segregating chromosomes, and dividing cells. The mitotic function of a gene has been inferred through observation of cells after perturbation of the gene or by identifying biochemical activity of the gene product using *in vitro* assays.

The *Drosophila* S2 cell line, which was derived from embryos that were approximately 1 day old (Schneider, 1972), is the most popular cell line to study mitosis in insect species for several reasons. The diamond-shaped spindle seen at metaphase is similar in structure to its mammalian counterpart, and genes involved in spindle formation/function are mostly conserved (Goshima and Vale, 2003; Goshima *et al.*, 2007). It is very easy to culture and manipulate S2 cells, and RNA interference (RNAi) almost always works very efficiently with any constructs not only for one gene but also for two to three genes when dsRNAs targeting two to three genes are simultaneously added to the culture medium (Clemens *et al.*,

2000; Goshima and Vale, 2003; Goshima *et al.*, 2007; Laycock *et al.*, 2006; Rogers *et al.*, 2003; Zhang *et al.*, 2007). Unlike mammalian cells, which usually require carefully designed siRNAs for RNAi, for S2 RNAi, dsRNAs 300–1000 bp in length can be prepared through polymerase chain reaction followed by *in vitro* transcription and are directly added to the culture medium (Bettencourt-Dias and Goshima, 2009; Clemens *et al.*, 2000; Rogers and Rogers, 2008). Another advantage of this cell line is that it allows high-resolution spatiotemporal observations through fluorescence microscopy (Goshima and Vale, 2003; Mahoney *et al.*, 2006; Maiato *et al.*, 2005; Rogers *et al.*, 2002; 2003). The recent accumulation of cell lines in which the mitotic components are tagged by green fluorescent protein (GFP) or other fluorophores has rendered this cell line suitable for high-resolution live cell imaging as well [e.g., (Goshima and Vale, 2005; Goshima *et al.*, 2008; Rogers *et al.*, 2002)].

One of the difficulties in using S2 cells is that not all the cells exhibit "textbook" mitosis, in which the mitotic spindle is assembled with microtubules nucleated at two centrosomes and anaphase occurs after all the chromosomes are aligned at the center of the bipolar spindle. Instead, mitosis in this cell line often starts with >2 centrosomes, and centrosome clustering/fusion takes place during spindle assembly (see Fig. 3) (Goshima and Vale, 2003). The clustering process is occasionally incomplete; cells can trigger anaphase and succeed in cell division without metaphase chromosome congression in the multipolar spindle. The presence of this non-textbook mitosis is also the case in other immortal cell lines, such as human HeLa, but is a very rare event in, for example, fission yeast or *Caenorhabditis elegans* in which a genuine wild-type strain is utilized as the master strain.

II. Rationale

In this chapter, I describe the current procedure I use to identify mitotic phenotypes after RNAi in *Drosophila* S2 cells. Any deviation from the control should be defined as a "phenotype" in an RNAi study. However, since S2 mitosis is variable even in the control population, drawing conclusions is not as easy as it is in other model organisms such as yeasts or *C. elegans* where variability is much lower. In particular, there is a greater possibility of selecting a false positive after RNAi, in which abnormal mitosis generally seen in S2 cells at a certain frequency is misinterpreted as an RNAi-derived phenotype. The false positive in RNAi studies is a serious issue because it is extremely difficult to correct the previously obtained positive conclusion: the result of not obtaining a phenotype later on in the study might be due to poorer RNAi efficiency and more abundant residual proteins that are sufficient for mitosis. The aim of this chapter is to discuss how a mitotic phenotype can be correctly identified in S2 cells, based on my 8-year experience with observing >4 million spindles in this cell line (Goshima and Vale, 2003; Goshima *et al.*, 2007).

III. Material Check

A. Grasping *Your* S2 Cell Line

The S2 cell line was originally isolated nearly 40 years ago (Schneider, 1972). However, perhaps due to accumulation of mutations and/or epigenetic changes in the transcription pattern, current S2 cells in each lab are divergent in several aspects, such as cell morphology, cell spreading efficiency, RNAi efficiency, or cell aging. For example, S2 cells available at Invitrogen (termed S2-I here) generally adhere more to the culture dish and spread more uniformly on concanavalin-A (Con-A)-coated dish than S2-U, obtained ~10 years ago from the University of California San Francisco (UCSF) cell culture facility. The clear advantage of using S2-I was when cell morphology was assessed after spreading on Con-A (Rogers *et al.*, 2003) but the disadvantage is that this cell line is not tolerant of large number of passages (~30). The S2-U line is more tolerant to repeated cell passaging, the reason for which is unclear. On the other hand, S2-U does not adhere to the plate as well and spreading on Con-A-coated glass is less efficient and nonuniform.

Since 2002 I have mostly used S2-U cells, and a number of S2-U-derived lines useful for studying mitosis, such as those expressing GFP-tubulin (Goshima and Vale, 2003), mCherry-tubulin/HistoneH2B-GFP (Goshima *et al.*, 2007), and GFP-tubulin/Mis12-mCherry (Goshima *et al.*, 2008), have been distributed to other laboratories. The description below is generally for S2-U cells, and it is possible that S2 cells with different origins behave differently.

B. Checking *Your* Culture Medium

The methods of S2 cell culturing have been described elsewhere [e.g., (Bettencourt-Dias and Goshima, 2009; Rogers and Rogers, 2008)]. However, occasionally, there are reports on the difficulties associated with maintaining healthy S2 cells during passaging. I believe that the major reason is the use of inappropriate batches of medium and/or serum. Here, our method to test the batch of a medium and serum is described.

When a new medium (or serum) bottle is purchased, it is important to check cell growth and RNAi efficiency by using the medium since batch-to-batch variability is significant (at least for S2-I and S2-U cells). For this reason, S2 cells growing in the old medium are cultured for three passages in the new medium (the bad medium would cause cell death at this stage). Subsequently, RNAi is performed against the Pavarotti (Pav) gene (essential for cytokinesis) (Goshima and Vale, 2003) and control dsRNA (e.g., sequences against the pBluescript vector) (Bettencourt-Dias and Goshima, 2009). After 4–5 days, the cell number in the control well and cell size in the Pav RNAi cells are checked under a conventional microscope with a 10× objective lens. If the cells are not healthy in a given batch of medium, many cells die and the resultant cell debris is scattered in the culture plate (see next section and Fig. 1B). RNAi efficiency is verified by the predominant presence of gigantic cells, which is derived from cytokinesis failure, in Pav RNAi samples. It is also important to set control samples in which the

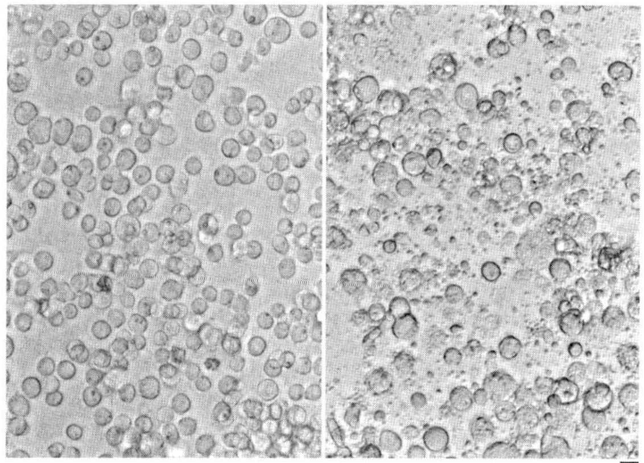

Fig. 1 Cell death after RNAi treatment, likely caused by the use of bad medium/serum batch. (A) Cells are in a steady growing state. (B) Cell debris is visible in the entire image due to cell death during culturing. Both pictures were taken after an RNAi treatment (day 3), using a 20× objective lens and transmission light. See text for tips to avoid cell death during RNAi.

old medium is utilized for the entire procedure. If one of the above two categories is unsatisfactory, the medium batch should not be used and another batch should be tested. It is often desirable to order several to dozens of verified batches of medium (serum) at the same time. They can be stored for long times; Schneider's medium from Invitrogen can be stored at 4°C for ~1 year.

C. RNAi Toxicity

Occasionally, RNAi treatment itself leads to complete cell death after 2–3 days, regardless of the types of target genes, and cell debris is predominantly observed prior to the fixation of cells (Fig. 1). We were unable to determine a single cause for this toxicity but found that the combination of dsRNA, medium, and cell lines was critical (Nico Stuurman (UCSF) and G.G. unpublished). In one case, we discovered that two dsRNAs, synthesized on different days, had different effects on cell growth even when the same cells and medium were used (we overcame this problem by using different batches of the medium). Another possible approach is to increase the cell density at the time of RNAi treatment, although it is unclear why the increase in cell number reduces cell death. We recommended 2×10^6 cells/ml during RNAi treatment [see Section 3.6 in (Bettencourt-Dias and Goshima, 2009)] but the cell density could be increased up to 5×10^6/ml. However, increasing the initial cell density might lead to a lower mitotic index after several days of culturing (assuming that some cells enter the stationary phase).

Note that the cell debris seen in Fig. 1B is scarcely observed even after the essential genes are knocked down (e.g., ribosome or mitotic kinesins); after RNAi of these genes, the cell number is lower than that of the controls, but the cell shape looks normal.

IV. RNAi and Cell Imaging

A. RNAi with Appropriate Controls

RNAi procedures are described in multiple reports [e.g., (Bettencourt-Dias and Goshima, 2009; Rogers and Rogers, 2008)] and have not been repeated here. In brief, we treated S2 cells with serum-free medium supplemented with dsRNA for ~1 h, followed by adding serum.

RNAi can be performed at the desired scale. Usually, 96-well plates are used in our laboratory since sufficient numbers of mitotic cells are obtained and dsRNA is significantly saved (it requires only 1 μg per well). Moreover, 12- or 8-channel Pipetman can be used for the 96-well plate. However, 24-well (5 μg dsRNA) or 6-well (20 μg dsRNA) plates are also used when larger scales are necessary. In addition, the RNAi toxicity effect described above is generally less frequent when performed on a larger scale.

Regardless of the type of plates, it is critical to seal the lid to the plate tightly with parafilm after RNAi in order to avoid concentration of the medium due to evaporation during RNAi treatment (3–7 days). If not, the medium gradually evaporates during incubation. A serious consequence is that wells at the edge of the plate have more evaporation than those at the center, and subsequent subtle differences in concentration of the medium lead to not only cell density changes but also alteration of the mitotic index, the mechanism of which is uncertain. In fact, when a large-scale RNAi screen was performed without sealing 96-well plates, we noted that the mitotic index is generally the highest in the row A/H and column 1/12 (Nico Stuurman and G.G., unpublished). The use of edge wells should be avoided as much as possible (Fig. 2A; note that the cells are not plated in row A or column 1/12).

Every RNAi experiment needs negative controls. In *Drosophila* S2 cells, sequences that do not exist in *Drosophila* have been chosen for control RNAi, such as pBluescript or GFP. Perhaps in many cell types, one control sample is sufficient in RNAi experiments (or mutant analysis) of multiple genes. However, in S2 cells with > 3 samples, it is recommended that the control is prepared in multiple wells in the same plate. For example, when seven samples are to be stained, it is preferable to distribute two to three controls in a 96-well plate, as shown in Fig. 2A. Comparison between controls would be a good quality check for the RNAi experiment. If plate sealing is imperfect, for example, the control at the edge (columns 2 and 11 in the case of Fig. 2A) and others might be quite different in terms of cell density or mitotic index. Having a control at one well of the plate alone is risky.

B. Immunostaining of Microtubules and Mitotic Proteins

The protocol of immunostaining followed in our lab is described in Section 3.6 in Bettencourt-Dias and Goshima (2009). For immunofluorescence after RNAi, I recommend using the multiwell plate since all the wells can be treated identically during the whole procedure. We typically use an 8-well glass-bottom plate for <8 samples (IWAKI 5232-008), and a 96-well glass-bottom plate for 9–96 samples (IWAKI 5866-096).

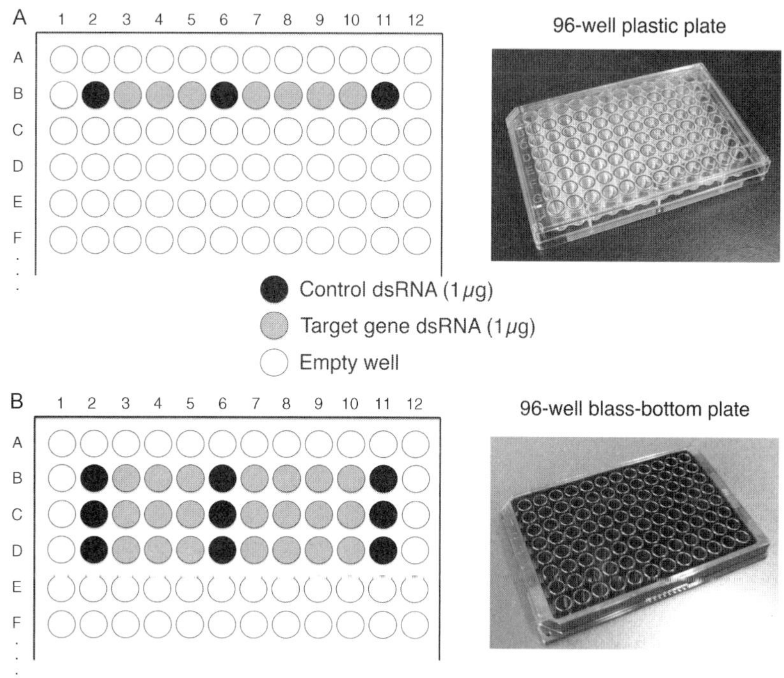

Fig. 2 Setting up RNAi with multiple samples. (A) An example of an RNAi experiment for multiple genes. When RNAi phenotypes of seven genes are assessed, I typically use one row of a 96-well plate and have two to three control RNAi distributed in the row. The use of 96-well plates enables the use of a 12-channel Pipetman, which would reduce the probability of well-to-well variation during RNAi and immunostaining. By comparing multiple controls, it is checked that RNAi is consistent in every well independent of its position. If plate sealing is imperfect during RNAi culture, for example, the cell density or mitotic index might significantly differ between wells in the middle and those at the edge because the medium would be more concentrated at the edge due to evaporation. (B) Transfer of 10 samples to a 96-well glass-bottom plate for immunofluorescence microscopy. Using a 12-channel Pipetman, cells can be distributed easily to multiple rows (e.g., 30 μl of a 100-μl cell culture to each well). Use of multiple rows is essential when a very transient phase of mitotic process (e.g., anaphase A) is screened for because not many of those cells can be found in a well.

Using a 22 × 22 mm coverslip for each treatment does work as well (Goshima and Vale, 2003; Rogers *et al.*, 2003) but the use of a multiwell plate and multichannel Pipetman ensures the same condition for each sample. Furthermore, automated microscopy available from some vendors (e.g., IXmicro of Molecular Devices) can be used for plate imaging (Goshima *et al.*, 2007; Guo *et al.*, 2008).

Several days after RNAi, the cells are transferred to a multiwell glass-bottom plate for imaging. As shown in Fig. 2B, it is possible to use multiple wells per treatment to increase the number of cells. Each well is pre-coated with either poly-lysine or Con-A, so that the cells are not peeled off during immunostaining (Bettencourt-Dias and Goshima, 2009). Con-A (2 μg/well for the 96-well plate and 4 μg/well for the 8-well plate) is useful because it enables cell spreading, and therefore, two poles of the spindle can often be detected in the same focal plane (Goshima *et al.*, 2007; Rogers *et al.*, 2002). On the other hand, Con-A inhibits completion of cytokinesis, and the long-term culturing on Con-A increases multinucleated cells (Goshima and Vale, 2003). Note that cell density affects spreading efficiency; spreading is not robust when too many or too few cells are plated. After 2–3 h on Con-A plates, the cells are fixed with 6.4 % paraformaldehyde.

To identify mitotic spindle phenotypes, it is helpful to stain α-tubulin and γ-tubulin together with DAPI (or Hoechst dye) because these antibodies are commercially available. If an additional channel is available, staining of phospho-HistoneH3 (rabbit polyclonal) helps to identify condensed chromosomes in mitosis (Bettencourt-Dias and Goshima, 2009; Goshima *et al.*, 2007). α-Tubulin staining can be done with DM1A (mouse monoclonal) or YOL1/34 (rat monoclonal) (Bettencourt-Dias and Goshima, 2009). DM1A is particularly good for astral microtubule staining, whereas YOL1/34 is suitable for kinetochore microtubule staining. Since the γ-tubulin antibody (GTU-88) is mouse monoclonal, costaining of GTU-88 (mouse) and YOL1/34 (rat) after paraformaldehyde fixation is recommended at the first screening of the spindle morphology, unless the focus is astral microtubules. Note that γ-tubulin staining by GTU-88 is dramatically improved after SDS treatment (Bettencourt-Dias and Goshima, 2009).

For some proteins, such as microtubule end-tracking EB1 or some kinetochore proteins, methanol fixation is more suitable than paraformaldehyde fixation. Even in the case of methanol fixation, however, addition of paraformaldehyde (final 90% methanol, 3.4% paraformaldehyde, 5 mM sodium bicarbonate [pH 9.0]) often improves the staining (Goshima *et al.*, 2008; Rogers *et al.*, 2002).

C. Phenotype Observation and Imaging

How many cells should we observe to ascertain a spindle phenotype? In the previous large RNAi screen using *C. elegans* embryos, Sonnichsen *et al.* (2005) successfully obtained "hit" genes by assessing five time-lapse movies of mitosis. However, for most genes, except the highly robust ones (e.g., Kinesin-5/Klp61F), this number is too small for immunostained S2 cells, in which generally ~30% of the cells have nontextbook type of spindles (Goshima and Vale, 2003; Goshima *et al.*, 2007) (see Fig. 3 for randomly picked up mitotic cells).

In our previous RNAi screening for spindle morphology, we acquired, on an average, ~200 metaphase spindles for evaluation per RNAi treatment (Goshima *et al.*, 2007). This was achieved through the use of automated microscopy (IXmicro, Molecular Devices) and computer-based identification of mitotic cells (characterized by strong phospho-histone staining). The 200 spindle images contained slightly out-of-focus ones and those with an abnormal centrosome number, which were not derived from the RNAi of the particular gene. Perhaps, 200 is more than necessary when the phenotypes are manually imaged and inspected. For example, manual inspection of cells under microscopy could easily clear the focusing problem and could therefore significantly reduce the number of the cells needed to evaluate phenotypes. Manual inspection also allows us to focus only on the spindles with two centrosomes, ignoring multipolar spindles that would also be found during observation. As discussed below, some

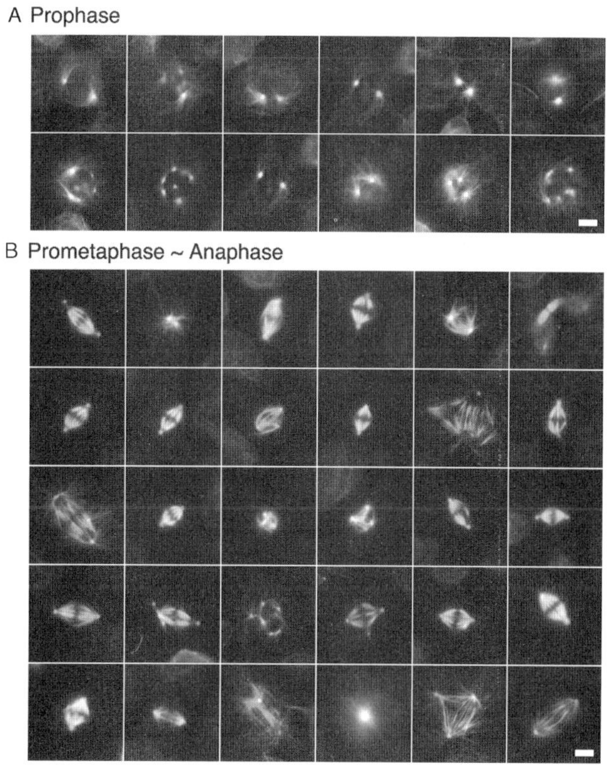

Fig. 3 Variability of S2 cell spindles. (A) 12 randomly selected prophase cells imaged under a wide-field microscope (40× 1.30 NA lens). Growing cells were plated on the Con-A-coated glass. Note that many S2 cells have more than two centrosomes during prophase. (B) 36 mitotic cell images were randomly acquired. Not all the spindles have the textbook look, and abnormal spindles such as multipolar or monopolar spindles are frequently observed. However, by comparing dozens of spindles side-by-side, it is reasonably easy to determine if an RNAi treatment truly induced a spindle abnormality (Goshima *et al.*, 2007). Red, γ-tubulin (GTU-88 antibody); green, microtubule (YOL1/34 antibody); blue, chromosome (DAPI). Bars, 5 μm. (See Plate no. 8 in the Color Plate Section.)

parameters are highly variable between cells and others are not. Therefore, there are no common criteria on how many spindles should be inspected to conclude a phenotype. However, roughly speaking, ~200 spindles acquired randomly or ~50 manually selected spindles would suffice to assess most of the spindle phenotypes.

V. Typical Phenotypes

Through a genome-wide RNAi screening, 12 were classified as major phenotypes for the metaphase spindle (including chromosomal phenotypes) (Fig. 4A) (Goshima *et al.*, 2007). Importantly, in many cases, more than one phenotype was found for a spindle; for example, chromosome misalignment was usually accompanied with abnormally long spindles.

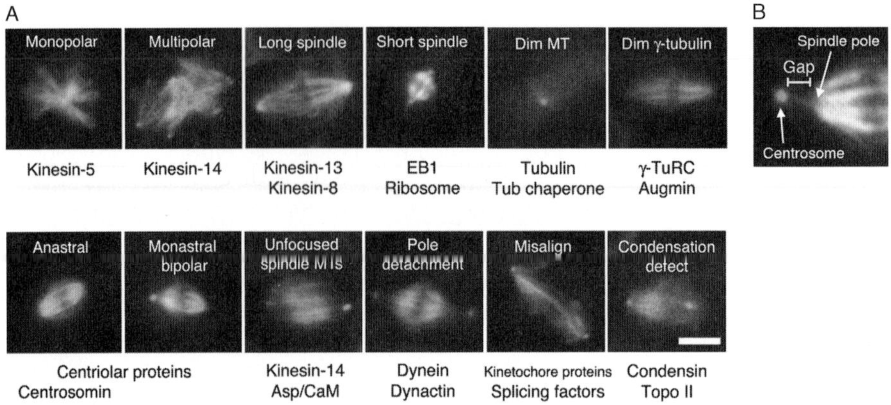

Fig. 4 12 major phenotypes of the metaphase spindle in S2 cells. (A) A representative spindle image and some genes that generate the phenotype upon RNAi are shown for each phenotype. Note that more than one phenotype is associated with a spindle in many cases. For example, the "dim microtubule" phenotype, which is seen after tubulin chaperone RNAi, is accompanied by monopolar or short spindle formation. Red, γ-tubulin (GTU-88 antibody); green, microtubule (YOL1/34 antibody); blue, chromosome (phospho-HistoneH3 antibody). Bar, 5 μm. (B) The centrosome and the spindle pole are distinct, and the gap between them is often visible in S2 spindles. (See Plate no. 9 in the Color Plate Section.)

A. Monopolar Spindle

Centrosomes are clustered into one site and chromosomes are scattered around (fan shape) or clustered as in a metaphase configuration. This is clearly seen after Kinesin-5 (Klp61F), CLASP (Mast/Orbit), or augmin RNAi (Goshima and Vale, 2003; Goshima *et al.*, 2007, 2008; Lemos *et al.*, 2000). It should be noted that monopolar spindles are also observed in the control cells (~5%) and these are often converted to monastral bipolar types (see below for this phenotype) (Goshima and Vale, 2003). Therefore, live cell imaging is essential to determine if the monopolar phenotype is the terminal or intermediate phenotype.

B. Multipolar Spindle

More than two centrosomes are present. The size of each centrosome (as judged by γ-tubulin staining) may vary. Each centrosome is present at a random location relative to the center of the spindle. This is mostly caused by failure in centrosome clustering/fusion, and Ncd and actin-related factors are reported to be involved in the clustering process (Goshima and Vale, 2003; Goshima *et al.*, 2005a; Kwon *et al.*, 2008).

C. Anastral Spindle

There are no centrosomes in the spindle. This is a unique phenotype for the genes required for centriole duplication (e.g., Sak, Sas-4) or centrosome maturation (e.g., Cnn) and is very rarely seen in the control spindles (Goshima *et al.*, 2007).

D. Monastral Bipolar Spindle

Bipolar spindles are formed with only one centrosome due to excessive centrosome clustering or failure in centriole duplication. This interesting phenotype is quite commonly seen in S2 cells and, although much less frequently, even in the wild-type fly (Goshima and Vale, 2003; Goshima *et al.*, 2007; Wilson *et al.*, 1997). Monopolar spindles are often converted to this type of bipolar spindle, and cells can execute chromosome segregation properly (Goshima and Vale, 2003). The RNAi of centriole duplication genes exhibit this phenotype prior to acentrosomal spindle formation in the next round of the cell cycle.

E. Pole Detachment

Centrosomes are detached from the main body of the spindle. Note that the gap between the centrosome and the spindle pole (focused point of the minus ends of the microtubules) is generally visible in S2 cells, especially when the cells are spread on Con-A [~1 μm; (Goshima *et al.*, 2005a)]. Therefore, the terms centrosome and spindle pole should correspond to different locations in the spindle (Fig. 4B). The centrosome detachment is most obvious after the RNAi of the dynein–dynactin or Asp–CaM complex (Goshima *et al.*, 2005a, 2007). However, since the gap is also detectable in untreated cells (Fig. 4B), quantification of the distance between centrosome and spindle pole is necessary to conclude that a mutant phenotype is present (Goshima *et al.*, 2005a).

F. Pole Unfocusing

The minus ends of microtubules are unfocused, as frequently seen after Ncd or Asp–CaM complex depletion (Goshima *et al.*, 2005a, 2007). This is also a quantitative difference, and therefore, the width of the spindle pole should be measured after microtubule staining, and the mean value should be compared to control spindles (Goshima *et al.*, 2005a).

G. Longer Spindle

In most cases, this phenotype accompanies chromosome misalignment, likely due to disruption of the force balance between microtubule sliding and opposing chromatid tension (e.g., due to kinetochore defect) (Goshima *et al.*, 2005b, 2007). Note that the centrosome–centrosome distance is not always a good marker of spindle length since centrosomes can be detached after certain treatments (see Section V.E.). Therefore we also recommend measuring the pole-to-pole distance (between focusing points of the microtubule minus ends).

H. Shorter Spindle

A short spindle phenotype is observed after the RNAi of microtubule plus-end-tracking protein EB1, microtubule polymerase TOG (Msps), ribosomes, etc. (Cullen *et al.*, 1999; Goshima *et al.*, 2005b, 2007; Rogers *et al.*, 2002). As in the case of longer spindles, measuring the pole-to-pole rather than centrosome-to-centrosome distance is valuable (e.g., EB1 RNAi detaches centrosomes, and the shorter centrosome-to-centrosome phenotype is not as clear as the pole-to-pole one).

I. Dim Microtubules

Microtubule signals become dim, as is seen after the RNAi of α/β-tubulin or tubulin cofactors (Goshima *et al.*, 2007). Although rare, the staining is somehow poorer in one well than in another in an experiment. However, these false positives based on experimental errors can be easily recognized by repeating the experiment.

J. Dim γ-Tubulin

This refers to a dim γ-tubulin on the centrosomes, spindles, or both. γ-tubulin has served as a centrosome marker, but it has become clear that γ-tubulin is also present on the spindle and that it is very important for spindle function (Goshima *et al.*, 2007, 2008; Uehara *et al.*, 2009). Loss of γ-tubulin from the centrosomes is observed after the knockdown of pericentriolar protein Cnn, whereas spindle γ-tubulin is specifically decreased after the knockdown of the eight-subunit complex augmin or outer subunits of the γ-tubulin ring complex (Dgrip75, 128, 163, 71WD). γ-Tubulin is decreased from both the centrosomes and the spindles upon RNAi depletion of the core γ-tubulin complex subunits Dgrip84 and Dgri91 (or γ-tubulin itself). The dim spindle γ-tubulin phenotype is not a very easy phenotype to identify but the longer spindle phenotype is associated with augmin and γ-tubulin RNAi, perhaps due to decreased nucleation sites of the microtubules within the spindle (Goshima and Kimura, 2010).

K. Chromosome Misalignment

Chromosomes are not congressed to the metaphase plate. Dramatic chromosome misalignment usually accompanies metaphase spindle expansion due to force imbalance (Goshima *et al.*, 2005b, 2007). Misalignment is one of the most commonly observed phenotypes after mitotic gene RNAi, but it is also difficult to assign in S2 cells, because prometaphase cells naturally have unaligned chromosomes, and many control cells in metaphase also seem to have misaligned chromosomes, largely due to the presence of multiple poles. Observation of metaphase-arrested cells (see Section VI.B.) could overcome the first problem, but this treatment itself also increases chromosome misalignment, as discussed in the following section. Therefore, the misalignment phenotype based on immunostained images should be assessed through increased quantification; for example, we determined the frequency of the presence of misaligned chromosomes for ~100 bipolar spindles (Goshima and Vale, 2003). Multipolar spindles should not be included in quantification of this type. However, the best approach to assess the chromosome alignment defect would be time-lapse imaging of chromosomes or kinetochores in living cells (histone or kinetochore proteins serves as the marker) since it reveals not only the frequency of misaligned chromosome appearance but also the dynamics of those chromosomes (Goshima *et al.*, 2008).

L. Chromosome Condensation

Chromosome structure defect that can be seen after knockdown of condensin and topoisomerase II (Goshima *et al.*, 2007).

VI. How to Avoid Recording False Positives

A. Basis of False Positives

The major cause of "abnormal" control spindles is the presence of multiple centrosomes in this cell line (Fig. 3) (Goshima and Vale, 2003). S2 cells, like many other cell types in *Drosophila*, have mature centrosomes only during mitosis (Rogers *et al.*, 2008). However, the centrosome number is highly variable, and nearly 50% of the cells have more than two centrosomes at the time of nuclear envelope breakdown (Fig. 3A) (Goshima and Vale, 2003). As a result, half of the cells initially form multipolar spindles during prometaphase. The centrosomes then cluster and fuse during prometaphase and metaphase, but they are sometimes incomplete, resulting in the presence of metaphase or anaphase cells with > 2 centrosomes (Fig. 3B). The presence of > 2 centrosomes not only changes the polarity of the spindle but also increases the probability of having misaligned chromosomes. In some cases, centrosome clustering occurs exceedingly well, and monopolar spindles as well as monastral bipolar spindles are formed.

Cell size is a critical determinant of the metaphase spindle length and, in general, larger cells have longer spindles (Wuhr *et al.*, 2008). Therefore, when S2 cells are observed on Con-A-coated glass, the efficiency of cell spreading would alter the spindle length. If spindle length is the major parameter to be determined, the well-spread cells should be selected for length measurement. Alternatively, cells may be fixed and stained on poly-lysine-coated glass on which cells are not spread out and remain uniformly round in shape.

Some phenotypes are rarely seen in a control population. For example, if anastral spindle is seen in $> 1\%$ of the spindle in an RNAi-treated sample, it is likely that the phenotype is derived from RNAi. Dim microtubule staining or dim γ-tubulin staining, if not caused by a staining error, is also a specific phenotype of RNAi.

B. Metaphase Arrest to Reduce the Effect of Over-duplicated Centrosomes

Centrosomes are clustered and fused during prometaphase and metaphase in S2 cells (Goshima and Vale, 2003). Therefore, most of the cells arrested in metaphase have two centrosomes. This is achieved through RNAi knockdown of the subunits of APC/C (E3 ubiquitin ligase) that is required for the destruction of anaphase inhibitors (Goshima *et al.*, 2007) or inhibition of the proteasome by MG132 treatment (Kwon *et al.*, 2008). APC/C knockdown [e.g., Cdc27 or Cdc16 (Goshima *et al.*, 2007)] also increases the number of metaphase cells 5–10-fold, and therefore, it becomes dramatically easier to detect mitotic cells under the microscope. However, there are two precautions to bear in mind when using this method. First, later mitotic events, such as anaphase or cytokinesis, are very rarely seen due to strong metaphase arrest. Second, chromosome alignment is often impaired through prolonged arrest (Goshima *et al.*, 2007). The metaphase image galleries displayed at http://rnai.ucsf.edu/mitospindle-screen/index.html (Goshima *et al.*, 2007) suggest that only ~20% of the cells have perfectly aligned chromosomes in the Cdc27-arrested condition. Nevertheless, metaphase arrest method is a powerful approach for assessing chromosome alignment defects; when essential kinetochore components are knocked down, for example, the extent of misalignment becomes very drastic and can easily be identified.

C. Rescue Experiment

The rescue experiment is a powerful approach to ensure that the observed phenotype is derived from the targeted RNAi; here, the endogenous protein is depleted by RNAi while the exogenous, RNAi-insensitive protein is expressed (Fig. 5A) (Goshima and Vale, 2005). This is usually done by using a dsRNA construct that targets the UTR region of the endogenous genes, but in some cases, RNAi-insensitive genes can be expressed by changing their codon usage (Dean and Spudich, 2006). In the rescue assay, the exogenous gene should be tagged with GFP or some other epitope for specific detection, and a stable line should preferably be selected (but do not isolate a clonal line; see below) [methods described in Bettencourt-Dias and Goshima (2009)]. When an endogenous gene is depleted by UTR RNAi and the exogenous gene is expressed, some cells are expressed but others are not (Fig. 5B); this is because the "stable line" actually is a mixture of cells with various expression levels of the exogenous gene (from no expression to overexpression). Therefore, we can have an internal control in this experiment; GFP-expressing and nonexpressing cells coexist in the same sample after RNAi. If a phenotype is rescued only in GFP-expressing cells, it is highly probable that the phenotype is indeed derived from the targeted gene depletion (Fig. 5B).

D. Live Cell Imaging

Tracing the mitotic process by time-lapse microscopy is always informative (Goshima and Vale, 2003; Goshima *et al.*, 2008; Maiato *et al.*, 2005). Our method of sample preparation is described in Bettencourt-Dias and Goshima (2009). In the case of S2 cells, however, it still requires more sample numbers than a yeast or *C. elegans* embryo, which is much less variable in the mitotic process. Live imaging typically begins with a manual search of a cell of interest under the microscope. For example, if the spindle assembly process is to be traced, the GFP-tubulin cell line can be used and prophase cells (cells with mature centrosomes) are selected. As described above, the centrosome number is highly variable during prophase in S2 cells. However, in a live cell analysis, it is possible to target only cells that have two centrosomes. Therefore, it is not necessary to acquire dozens of cell images for evaluating the spindle phenotype. Alternatively, automated

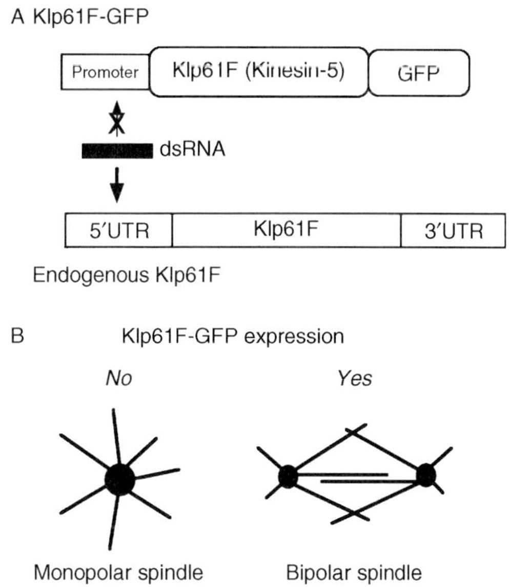

Fig. 5 Rescue experiment to verify spindle & phenotypes. (A) Schematic representation of the rescue assay in S2 cells. dsRNA targeting 5′UTR (or 3′UTR) is used for RNAi knockdown of an endogenous gene. An exogenous gene lacking UTR sequences, such as the GFP-fusion gene shown here, can be expressed, and it is resistant to the dsRNA. Alternatively, it is possible to use a dsRNA-targeting an ORF and an RNAi-resistant exogenous gene whose codons are synonymously altered. (B) Examples of a monopolar spindle after 5′UTR-based RNAi of the Klp61F gene and no Klp61F-GFP expression (left) and a bipolar spindle after being rescued by Klp61F-GFP expression (right). This result demonstrates that the monopolar phenotype is indeed derived from Klp61F knockdown and not by the off-target-effect of the dsRNAs used. In this assay, monopolar spindle appearance in "no GFP-expressing" cells in the same sample serves as a reliable internal control. In order to have this internal control, using a cell line in which all the cells have GFP expression [which is obtained through clonal isolation of the cell line (Bettencourt-Dias and Goshima, 2009)] should be avoided. One application of this assay is to quantify the total amount of Klp61F in a cell by measuring GFP intensity after endogenous Klp61F depletion (Goshima *et al.*, 2005b).

microscopy can also be applied for live cell imaging (Goshima *et al.*, 2007, 2008). In this case, cells to be analyzed can be selected after manual inspection of numerous image sequences that were automatically acquired.

VII. Summary

The *Drosophila* S2 cell line is a widely used cell line in mitosis research mainly because of ease of use and its robust RNAi. However, it is also true that its highly variable nature of mitosis makes it more difficult to identify a phenotype than in untransformed cell systems. However, years of experience in mitosis observation in S2 cells has established the phenotype that is robust and that which requires numerous imaging and repeats in this cell line. Therefore, if there is a gene of interest whose mitotic role needs to be assessed, I believe that the S2 cell line is still the primary choice to perform a loss-of-function analysis of the gene.

Acknowledgments

I thank Wenjing Li for providing the microscopic images. Our laboratory work is supported by grants from MEXT, Japan, and the Human Frontier Science Programe (HFSP) organization.

References

Bettencourt-Dias, M., and Goshima, G. (2009). RNAi in drosophila S2 cells as a tool for studying cell cycle progression. *Methods Mol. Biol.* **545**, 39–62.

Clemens, J. C., Worby, C.A., Simonson-Leff, N., Muda, M., Maehama, T., Hemmings, B. A., and Dixon, J. E. (2000). Use of double-stranded RNA interference in drosophila cell lines to dissect signal transduction pathways. *Proc. Natl. Acad. Sci. U.S.A.* **97**, 6499–6503.

Cullen, C. F., Deak, P., Glover, D. M., and Ohkura, H. (1999). Mini spindles: A gene encoding a conserved microtubule-associated protein required for the integrity of the mitotic spindle in drosophila. *J. Cell Biol.* **146**, 1005–1018.

Dean, S. O., and Spudich, J. A. (2006). Rho kinase's role in myosin recruitment to the equatorial cortex of mitotic drosophila S2 cells is for myosin regulatory light chain phosphorylation. *PLoS ONE* **1**, e131.

Gadde, S., and Heald, R. (2004). Mechanisms and molecules of the mitotic spindle. *Curr. Biol.* **14**, R797–R805.

Goshima, G., and Kimura, A. (2010). New look inside the spindle: Microtubule-dependent microtubule generation within the spindle. *Curr. Opin. Cell Biol.* **22**, 44–49.

Goshima, G., Mayer, M., Zhang, N., Stuurman, N., and Vale, R. D. (2008). Augmin: A protein complex required for centrosome-independent microtubule generation within the spindle. *J. Cell Biol.* **181**, 421–429.

Goshima, G., Nedelec, F., and Vale, R. D. (2005a). Mechanisms for focusing mitotic spindle poles by minus end-directed motor proteins. *J. Cell Biol.* **171**, 229–240.

Goshima, G., and Vale, R. D. (2003). The roles of microtubule-based motor proteins in mitosis: Comprehensive RNAi analysis in the drosophila S2 cell line. *J. Cell Biol.* **162**, 1003–1016.

Goshima, G., and Vale, R. D. (2005). Cell cycle-dependent dynamics and regulation of mitotic kinesins in drosophila S2 cells. *Mol. Biol. Cell.* **16**, 3896–3907.

Goshima, G., Wollman, R., Goodwin, N., Zhang, J. M., Scholey, J. M., Vale, R. D., and Stuurman, N. (2007). Genes required for mitotic spindle assembly in drosophila S2 cells. *Science* **316**, 417–421.

Goshima, G., Wollman, R., Stuurman, N., Scholey, J. M., and Vale, R. D. (2005b). Length control of the metaphase spindle. *Curr. Biol.* **15**, 1979–1988.

Guo, Y., Walther, T. C., Rao, M., Stuurman, N., Goshima, G., Terayama, K., Wong, J. S., Vale, R. D., Walter, P., and Farese, R. V. (2008). Functional genomic screen reveals genes involved in lipid-droplet formation and utilization. *Nature* **453**, 657–661.

Inoue, S. (2008). Microtubule dynamics in cell division: Exploring living cells with polarized light microscopy. *Annu. Rev. Cell Dev. Biol.* **24**, 1–28.

Karsenti, E., and Vernos, I. (2001). The mitotic spindle: A self-made machine. *Science* **294**, 543–547.

Kwon, M., Godinho, S. A., Chandhok, N.S., Ganem, N. J., Azioune, A., Thery, M., and Pellman, D. (2008). Mechanisms to suppress multipolar divisions in cancer cells with extra centrosomes. *Genes Dev.* **22**, 2189–2203.

Kwon, M., and Scholey, J. M. (2004). Spindle mechanics and dynamics during mitosis in drosophila. *Trends Cell Biol.* **14**, 194–205.

Laycock, J. E., Savoian, M. S., and Glover, D. M. (2006). Antagonistic activities of klp10a and orbit regulate spindle length, bipolarity and function in vivo. *J. Cell Sci.* **119**, 2354–2361.

Lemos, C. L., Sampaio, P., Maiato, H., Costa, M., Omel'yanchuk, L. V., Liberal, V., and Sunkel, C. E. (2000). Mast, a conserved microtubule-associated protein required for bipolar mitotic spindle organization. *EMBO J.* **19**, 3668–3682.

Mahoney, N. M., Goshima, G., Douglass, A. D., and Vale, R. D. (2006). Making microtubules and mitotic spindles in cells without functional centrosomes. *Curr. Biol.* **16**, 564–569.

Maiato, H., Khodjakov, A., and Rieder, C.L. (2005). Drosophila CLASP is required for the incorporation of microtubule subunits into fluxing kinetochore fibres. *Nat. Cell Biol.* **7**, 42–47.

McIntosh, J. R., Grishchuk, E. L., and West, R. R. (2002). Chromosome-microtubule interactions during mitosis. *Annu. Rev. Cell Dev. Biol.* **18**, 193–219.

Mitchison, T. J., and Salmon, E. D. (2001). Mitosis: A history of division. *Nat. Cell Biol.* **3**, E17–E21.

Rogers, S. L., and Rogers, G. C. (2008). Culture of drosophila S2 cells and their use for RNAi-mediated loss-of-function studies and immunofluorescence microscopy. *Nat. Protoc.* **3**, 606–611.

Rogers, S. L., Rogers, G. C., Sharp, D. J., and Vale, R. D. (2002). Drosophila EB1 is important for proper assembly, dynamics, and positioning of the mitotic spindle. *J. Cell Biol.* **158**, 873–884.

Rogers, G. C., Rusan, N. M., Peifer, M., and Rogers, S.L. (2008). A multicomponent assembly pathway contributes to the formation of acentrosomal microtubule arrays in interphase drosophila cells. *Mol. Biol. Cell.* **19**, 3163–3178.

Rogers, S. L., Wiedemann, U., Stuurman, N., and Vale, R. D. (2003). Molecular requirements for actin-based lamella formation in drosophila S2 cells. *J. Cell Biol.* **162**, 1079–1088.

Schneider, I. (1972). Cell lines derived from late embryonic stages of drosophila melanogaster. *J. Embryol. Exp. Morphol.* **27**, 353–365.

Scholey, J. M., Brust-Mascher, I., and Mogilner, A. (2003). Cell division. *Nature* **422**, 746–752.

Sonnichsen, B., Koski, L. B., Walsh, A., Marschall, P., Neumann, B., Brehm, M., Alleaume, A. M., Artelt, J., Bettencourt, P., Cassin, E., Hewitson, M., Holz, C., *et al.* (2005). Full-genome RNAi profiling of early embryogenesis in *Caenorhabditis elegans*. *Nature* **434**, 462–469.

Uehara, R., Nozawa, R. S., Tomioka, A., Petry, S., Vale, R. D., Obuse, C., and Goshima, G. (2009). The augmin complex plays a critical role in spindle microtubule generation for mitotic progression and cytokinesis in human cells. *Proc. Natl. Acad. Sci. U.S.A.* **106**, 6998–7003.

Walczak, C. E., and Heald, R. (2008). Mechanisms of mitotic spindle assembly and function. *Int. Rev. Cytol.* **265**, 111–158.

Wilson, P. G., Fuller, M. T., and Borisy, G. G. (1997). Monastral bipolar spindles: Implications for dynamic centrosome organization. *J. Cell. Sci.* **110**(Pt 4), 451–464.

Wuhr, M., Chen, Y., Dumont, S., Groen, A.C., Needleman, D. J., Salic, A., and Mitchison, T. J. (2008). Evidence for an upper limit to mitotic spindle length. *Curr. Biol.* **18**, 1256–1261.

Zhang, D., Rogers, G. C., Buster, D. W., and Sharp, D. J. (2007). Three microtubule severing enzymes contribute to the "pacman-flux" machinery that moves chromosomes. *J. Cell Biol.* **177**, 231–242.

CHAPTER 16

Analysis of Microtubules in Budding Yeast

Alexander Rauch[*,1], **Elena Nazarova**[†,1], *and* **Jackie Vogel**[†,‡,1]

[*]Institute of Biochemistry, ETH-Zurich, 8093 Zurich, Switzerland

[†]Department of Biology, McGill University, Montreal, Quebec, Canada H3G 0B1

[‡]School of Computer Science, McGill University, Montreal, Quebec, Canada H3A 2A7

I. Introduction

A. Tubulin and Microtubules

Microtubules are a major component of the cytoskeleton in all eukaryotic cells. Microtubules are built of repeating α,β-tubulin heterodimers, which assemble head to tail to form polarized linear protofilaments, 13 of which form the microtubule lattice

[1]Alexander Rauch, Elena Nazarova and Jackie Vogel have contributed equally to this study

METHODS IN CELL BIOLOGY, VOL. 97

978-0-12-381349-7
DOI: 10.1016/S0091-679X(10)97016-7

(Lowe *et al.*, 2001; Nogales, 2001). Lateral contacts between subunits in these protofilaments can form two types of lattices, the A-lattice form (α-β contact) and the B-type lattice (α-α or β-β contacts) (Amos and Klug, 1974).

In the classic *allosteric* model heterodimers can exist in two different conformations: GTP-bound α,β-tubulin, which exhibits a more straight conformation and GDP-bound tubulin, which is slightly more bent (Gigant *et al.*, 2000; Lowe *et al.*, 2001; Ravelli *et al.*, 2004). Hydrolysis of the GTP molecule bound to the β-subunit occurs within the lattice, and heterodimers at the growing end tend to be in the GTP-bound state (GTP cap). Dynamics of microtubules at the resolution of single tubulin dimers are highly variable as recently shown *in vitro* (Kerssemakers *et al.*, 2006; Schek *et al.*, 2007). The cause of these fluctuations in the growth rates is not known. Two alternative explanations have been proposed: the fluctuations could originate either from random addition of tubulin oligomers from solution or they could be based on variations in the length of the GTP-tubulin cap (Howard and Hyman, 2009).

B. Tubulin in Budding Yeast

Saccharomyces cerevisiae has three essential tubulin genes: one α (*TUB1*) (Schatz *et al.*, 1986a), one β (*TUB2*) (Neff *et al.*, 1983), and one γ (*TUB4*) (Sobel and Snyder, 1995). *TUB3* encodes a nonessential α-tubulin (Schatz *et al.*, 1986a) and deletion of *TUB3* results in increased sensitivity to benomyl, a microtubule-destabilizing drug (Neff *et al.*, 1983; Schatz *et al.*, 1986b). On a functional level, *TUB1* and *TUB3* are identical (Schatz *et al.*, 1986b). The ratio between α- and β-tubulin is tightly regulated at both the transcriptional and the translational level *in vivo* and overexpression of either subunit is highly toxic to cells (Katz *et al.*, 1990).

Tub4p is associated with the spindle pole bodies (SPBs) and does not assemble into microtubules (Marschall *et al.*, 1996; Sobel and Snyder, 1995; Spang *et al.*, 1996). Tub4p is part of the budding yeast γ-tubulin complex (Geissler *et al.*, 1996; Knop *et al.*, 1997; Rout and Kilmartin, 1990; Soues and Adams, 1998; Wigge *et al.*, 1998). This complex consists of Spc97p and Spc98p, two evolutionarily conserved proteins that are core components in γ-tubulin complexes in budding and fission yeast and metazoans. This complex anchors the minus ends of nuclear and cytoplasmic microtubules at both the inner and the outer plaque of the SPB (Knop *et al.*, 1999; Vogel and Snyder, 2000). Tub4p has a postnucleation role in organizing both spindle and cytoplasmic microtubules and is regulated by phosphorylation *in vivo* (Vogel *et al.*, 2001). Due to its key role in microtubule nucleation, depletion or mutation of *TUB4* leads to aberrant nuclear and cytoplasmic microtubule organization (Marschall *et al.*, 1996; Sobel and Snyder, 1995; Spang *et al.*, 1996). Tub4p function at the SPB is required for proper assembly of +TIP complexes on cytoplasmic microtubules (Cuschieri *et al.*, 2006).

C. Microtubule Organization by the Numbers

Numerous studies using combinations of fluorescent fusion proteins that label microtubules, SPBs, centromeres, or microtubule ends have provided quantitative

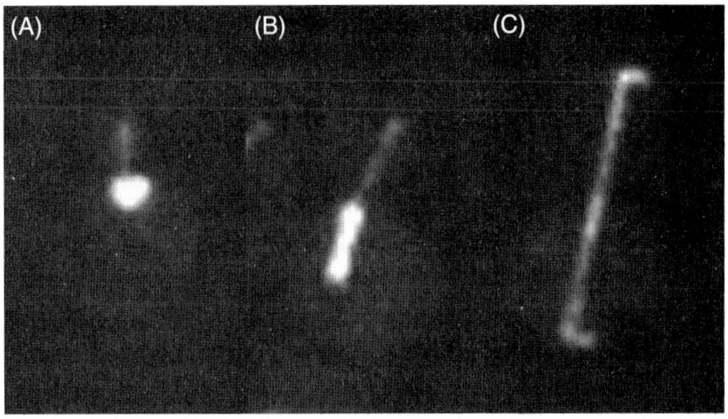

Fig. 1 Cytoplasmic and nuclear microtubule organization in budding yeast. Microtubules are labeled with EGFP-Tub1. Cytoplasmic microtubules project from the old (bud-bound) spindle pole into the bud (A, B) during preanaphase spindle alignment. Cytoplasmic microtubules extend from both poles of the anaphase spindle (C); note the increase in intensity at the spindle mid-zone, where antiparallel microtubules overlap. Images were collected using a spinning disk confocal microscope, 63X 1.4 NA objective, 493 solid-state laser (Coherent), and EM-CCD camera (Hamamatsu ImageEM).

information regarding microtubule number, organization, and dynamics during the budding yeast cell cycle (Fig. 1).

In G1, the astral microtubules are organized as a radial array of two to three astral microtubules from the single SPB (Kilmartin and Adams, 1984). Kinetochores remain attached to the SPB and cluster adjacent to the SPB. The astral microtubules from the SPB push against the cell cortex and propel the nucleus in the opposite direction (Maddox *et al.*, 2000; Shaw *et al.*, 1997, 1998). This movement has been proposed to be either directed (Adames and Cooper, 2000) or random (Shaw *et al.*, 1997), although a quantitative assessment of spindle trajectories in unbudded cells is still lacking. It has been proposed that back-pushing of the nucleus in G1 eventually leads the astral microtubules to the site of the incipient bud, followed by a capture-shrinkage event that results in pulling of the nucleus toward the site where the bud is formed (Huisman and Segal, 2005; Huisman *et al.*, 2004; Segal and Bloom, 2001).

SPB duplication and insertion into the nuclear envelope is completed during S-phase. The bipolar spindle is composed of ~40 spindle (32 pole–kinetochore, 8 pole–pole) microtubules. The number of astral microtubules is much lower and prior to anaphase is biased with respect to the old bud-bound SPB (~2) and new mother-bound (~1) SPB. In anaphase B the length of the antiparallel pole–pole microtubules increases ~5× and the poles separate at two distinct velocities: initially at ~0.50 μm/min and later at ~0.2 μm/min (Straight *et al.*, 1997, 1998). Anaphase B continues until the spindle attains an average curvilinear length of ~9 μm (Vogel *et al.*, 2001) and chromosomes have been cleared from the future plane of cytokinesis at the bud neck (Mendoza *et al.*, 2009). Throughout the vegetative cell cycle and during mating, individual astral microtubules can be detected and their dynamical properties and +end trajectories measured. Examples of spindle and astral microtubules labeled with GFP-Tub1p are shown in Fig. 1. Astral microtubules projecting from an unaligned (Fig. 1A) or an aligned (Fig. 1B) preanaphase spindle terminate in the bud. Astral microtubules project from both poles of an anaphase spindle (Fig. 1C).

The majority of studies published to date have been performed using manual or semi-manual pixel limited methods for measuring the curvilinear length of microtubules or spindles and the position of SPBs or centromeres with respect to the spindle axis or future plane of cytokinesis (bud neck). Manual measurement places a practical limit to the number of objects (microtubule ends, microtubule, and spindle length) measured, thus sample sizes in most studies are small ($n < 10$). Important advances in analysis of cellular objects have arisen from experimental condensed matter physics, in particular from two-dimensional and three-dimensional tracking of particles and rods in successive time steps. The advantages of these methods are that they are not limited by pixel size and are suited for data acquisition using sensitive cameras. The application of high-resolution, semi-automated, and automated methods to the analysis of microtubule dynamics and function in yeast cells will be discussed in Section IV.

When using GFP–tubulin fusions to study microtubule dynamics the following points have to be considered:

1. Endogenous tagging of both *TUB1* and *TUB2* leads to a synthetic sick phenotype. Therefore a second copy of a GFP–Tub1 protein fusion is introduced into the genome with expression driven by a weak promoter (Straight *et al.*, 1997).
2. Overexpression of *TUB2* is highly toxic for the cell. *TUB1* and *TUB3* overexpression is tolerated by the cell as a consequence of translational control, although this does not rule out possible changes in microtubule dynamics due to alterations in the stoichiometry of α and β tubulin and therefore microtubule composition. For this reason, it is recommended that the second copy of the labeled tubulin be maintained in the genome rather than on a plasmid.
3. When microtubule bending can be excluded, it is possible to analyze microtubule dynamics by staining only the ends of the microtubule. The length can be calculated as a three-dimensional vector between the minus and the plus-end of the microtubule. We will discuss possibilities further below.

II. The Cellular Toolbox for Analysis of Microtubules in Budding Yeast

A. The Uses (and Abuses) of Fluorescent Fusion Proteins and Mutations

Budding yeast has long been viewed as an excellent model organism due to its facile genetics, short cell cycle (90–120 min depending on growth conditions), and highly stereotyped correspondence between specific cell morphologies and cell cycle states. Extensive functional annotation of genes and proteins has occurred since the complete genome sequence was obtained in 1995. Currently, ~80% of nonessential genes are organized into functional subnetworks (Costanzo *et al.*, 2010) and thousands of proteins have been placed into interaction networks (Tarassov *et al.*, 2008). Budding yeast offers numerous tools and methods suitable for quantitative analysis of microtubule dynamics and function in living cells, including collections of strains bearing loss of function open reading frame (ORF deletion) mutations (Winzeler *et al.*, 1999) or expressing carboxyl terminal enhanced GFP (EGFP) fusion proteins (Huh *et al.*, 2003). These resources offer tremendous potential for discovery when used with care.

The following are key concepts for quantitative analysis of microtubules in living cells:

1. The fusion of a genetically encoded tag (e.g., GFP) to the protein of interest should not impact the protein's function. Two tests for the function of a fusion protein should be performed: (1) assessment of its ability to support normal growth and (2) its synthetic genetic interaction profile. An example of quality control using the Stu2–EGFP fusion is shown in Fig. 2. The progeny of a *STU2-EGFP:HIS3* strain crossed to a wild-type reference strain (BY4742) and to a *mad2*Δ strain (BY4742 background) are shown with genotypes for each spore. This test revealed that expression of the Stu2–GFP fusion did not perturb growth on YPAD. The Stu2–EGFP fusion did not exhibit synthetic sickness or lethality in combination with *mad2*Δ, indicating that the presence of the carboxyl terminal EGFP does not perturb Stu2 function with respect to spindle assembly.

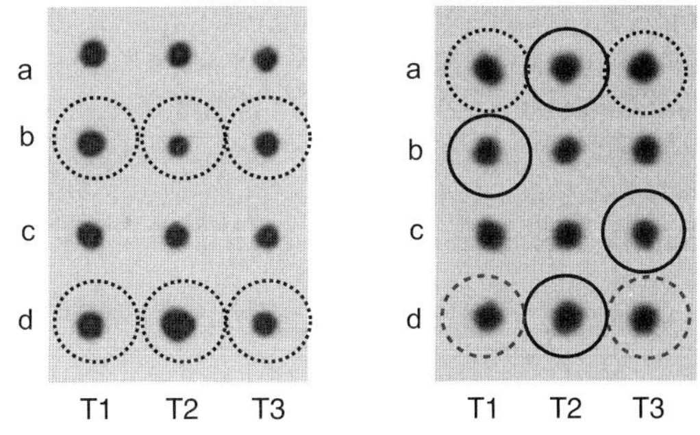

Fig. 2 Genetic analysis used to test the function of a fusion protein. A strain expressing Stu2-EGFP from the Huh *et al.* (2003) collection was tested for growth defects and synthetic lethality/sickness in combination with a *mad2*Δ mutation. The growth of cells expressing the fusion, and the fusion in combination with *mad2*Δ, does not differ from wild type. This analysis indicates Stu2-EGFP is functional with respect to the spindle microtubules.

2. The genetic background must not confound the cellular framework in which the protein will be analyzed. A common mutation in yeast strains that increases nonspecific fluorescence is *ade2*. Medium can be supplemented with adenine can be done, but cell-to-cell variation in nonspecific fluorescence remain an issue. Thus for single-cell analysis it is advised to use an *ADE+* strain background [e.g., YBR4741-2-3 (Brachmann *et al.*, 1998)].
3. The natural variation in intensities of the reporter(s) of interest cannot be exceeded by experimentally introduced noise. Noise can arise from medium or dirt in the imaging system, including glass slides used to prepare the sample. Commonly used media and reagents may interfere with fluorescent signals/measurements. Increasing the signal of a fusion protein through overexpression is not a solution, as this introduces artificial variation across cells in even a homogeneous population. When expressing fusion proteins from centromeric plasmids, variation in copy number is sufficient to confound many types of single-cell analyses. Thoughtful selection of fluorophores, the introduction of genetically encoded fluorescent tags at the chromosomal locus that are expressed at the endogenous level, and optimization of imaging parameters will provide the best data for quantitative analysis.
4. The process of imaging should not perturb either the function of microtubules or the cell viability. Optimization of imaging parameters will be discussed in greater detail later in this chapter.

B. Proteins Contributing to the Regulation of Microtubule Dynamics

Regulation of microtubule dynamics requires microtubule-associating proteins (MAPs) including microtubule plus-end tracking proteins (+TIPs) and plus- and minus-end-directed motor proteins (Akhmanova and Hoogenraad, 2005; Akhmanova and Steinmetz, 2008; Carvalho *et al.*, 2003; Howard and Hyman, 2003, 2007; Schuyler and Pellman, 2001; Wu *et al.*, 2006). Potential mechanisms for +TIP tracking of microtubule plus-ends include treadmilling, transport, and hitchhiking (Carvalho *et al.*, 2003). It is important to note that these mechanisms are not mutually exclusive and that many +TIPs have the capability to track plus-ends by different mechanisms (Caudron *et al.*, 2008). The following section introduces a few of the different +TIPs in budding yeast. We provide a short overview of each protein and discuss its applicability to study microtubule dynamics.

Depending on what aspect of microtubule dynamics is studied the following features of these proteins need to be considered:

1. Localization of MAPs may include nuclear and cytoplasmic microtubules and spindle poles. In addition, they may move along microtubules and preferentially localize to specific subsets of astral microtubules.
2. Expression levels may be constant or undergo cell cycle-dependent fluctuations.
3. Fluorophore fusions should ideally be made using the endogenous gene locus and may require various constructs, such as multifluorophore fusions.

Most of the proteins are available as part of a collection of functional EGFP carboxyl terminal fusions (Huh *et al.*, 2003). It is strongly suggested that strains from the GFP collection (provided as *MATa* haploids) be backcrossed to the reference wild-type *MAT*α strain BY4742 and reisolated by dissection of tetrads prior to their use. If additional fusions are needed, it is recommend that these be introduced into the diploid strain resulting from the backcross, as dissection of this diploid will verify the functionality of the new fusion protein and reveal synthetic genetic interactions that may exist between two or more fusion proteins. This "cleaning" approach is recommended prior to using any strain from a large collection (e.g., GFP, knockout collections).

1. Microtubule Assembly Promoters

a. BIM1 *Overview.* Bim1p belongs to the EB1 protein family, a ubiquitous group of evolutionarily conserved microtubule binding proteins (Tirnauer and Bierer, 2000). EB1 proteins are highly interactive and are key for various regulatory processes occurring at plus-ends of both nuclear and cytoplasmic microtubules (Lansbergen and Akhmanova, 2006; Morrison, 2007). EB1 proteins preferentially bind to the plus-ends of microtubules (Tirnauer *et al.*, 2002). *In vitro*, EB1 has been shown to bind both plus and minus ends of growing microtubules (Bieling *et al.*, 2007). Bim1p is transcriptionally regulated with its expression peaking during G1-S and decreasing during mitosis (Tirnauer *et al.*, 1999). Consequently, the effect of Bim1p on dynamic instability has been found to be most prominent during G1, where it has been shown to increase growth/shrinkage rates and transition frequencies, resulting in a net increase in microtubule dynamicity total tubulin turnover (Adames and Cooper, 2000; Tirnauer *et al.*, 1999; Wolyniak *et al.*, 2006).

Application to Analysis of Microtubules. Bim1p can track both growing and shrinking microtubules (Wolyniak *et al.*, 2006) and localizes to both sets of microtubules (nuclear and cytoplasmic) as well as to spindle poles. Bim1p can be tagged endogenously and a *bim1*Δ null mutant is viable. Single and multifluorophore constructs have been used extensively and are therefore available (Badin-Larcon *et al.*, 2004; Cuschieri *et al.*, 2006; Fridman *et al.*, 2009; Gardner *et al.*, 2008b; Khmelinskii *et al.*, 2007; Liakopoulos *et al.*, 2003; Schwartz *et al.*, 1997; Tirnauer *et al.*, 1999; Wolyniak *et al.*, 2006). Given Bim1 functions on both nuclear and cytoplasmic microtubules, Bim1 fusions must be tested for functionality with respect to both spindle and cytoplasmic contexts.

b. STU2 *Overview.* Stu2p was originally identified in a screen for *tub2-423* suppressors, a mutation in the β-tubulin gene that causes spindle defects at low temperature (Wang and Huffaker, 1997). Stu2p is a member of the Dis1/TOG protein or MAP215/TOG family (Gard *et al.*, 2004; Ohkura *et al.*, 2001) which is defined by the presence of two or more N-terminal TOG domains, each of which contains several HEAT repeats (Gard *et al.*, 2004; Ohkura *et al.*, 2001). Dimerization of Stu2p is essential for its proper function *in vivo* (Al-Bassam *et al.*, 2006; De Wulf *et al.*, 2003; Van Breugel *et al.*, 2003). Depletion of Stu2p decreases the number but not the length of cytoplasmic microtubules in G1 and metaphase (Kosco *et al.*, 2001) and reduces the dynamicity of microtubules (Kosco *et al.*, 2001; Wolyniak *et al.*, 2006).

The first TOG domain of Stu2p is required to stabilize cytoplasmic and nuclear microtubules *in vivo*. Contrary to its role in microtubule stabilization *in vivo*, recombinant Stu2p destabilizes microtubules and increases catastrophe frequency *in vitro*, clearly reducing microtubule length (Van Breugel *et al.*, 2003). The apparent complexity of Stu2p function in microtubule dynamics regulation has been discussed in Al-Bassam *et al.* (2006).

Application to Analysis of Microtubules. Stu2p localizes to spindle poles and to both nuclear and cytoplasmic microtubules (Kosco *et al.*, 2001; Wang and Huffaker, 1997). Stu2p was shown to track the plus ends of both growing and shrinking cytoplasmic microtubules throughout the cell cycle (Wolyniak *et al.*, 2006); therefore it makes a good tool for the analysis of microtubule dynamics. *STU2* is an essential gene and the endogenously tagged (single as well as multifluorophore fusions) Stu2p is viable (Al-Bassam *et al.*, 2006; Chen *et al.*, 1998; Kosco *et al.*, 2001; Ma *et al.*, 2007; Usui *et al.*, 2003; Wang and Huffaker, 1997; Wolyniak *et al.*, 2006).

c. BIK1 *Overview*. Bilateral karyogamy defect 1 (*BIK1*) was originally discovered as a gene required for nuclear fusion during mating (Berlin *et al.*, 1990; Kurihara *et al.*, 1994; Trueheart *et al.*, 1987). Structural analysis predicts that Bik1p contains three different structural domains (Berlin *et al.*, 1990). Both yeast CLIP-170 homologues ScBik1p and SpTip1 are proposed to hitchhike on a plus-end-directed motor (Busch and Brunner, 2004; Busch *et al.*, 2004; Carvalho *et al.*, 2004). Whether Bik1p binds tubulin dimers and thus can bind to plus-ends of microtubules by an end loading/treadmilling mechanism is not fully understood (Miller *et al.*, 2006). A recent study suggests that transport along microtubules and end tracking of Bik1p are two distinct processes (Caudron *et al.*, 2008). Deletion of *BIK1* results in short or undetectable cytoplasmic microtubules, defects in spindle elongation, and mispositioned nuclei (Berlin *et al.*, 1990). Bik1p is required for maintaining cytoplasmic microtubule attachment to the shmoo tip once the nucleus is properly oriented and may function in anchoring microtubule plus-ends to the shmoo tip similarly to Kar3p (Molk *et al.*, 2006). Consequently, in *bik1*Δ mutant cells, cytoplasmic microtubules are shorter and less stable, depolymerizing rapidly back to the spindle poles (Molk *et al.*, 2006). Extensive reviews on Bik1p structure and *in vivo* function have been recently published (Miller *et al.*, 2006).

Application to Analysis of Microtubules. Bik1p localizes to spindle poles, kinetochores, and both nuclear and cytoplasmic microtubules (Carvalho *et al.*, 2004; Lin *et al.*, 2001). Bik1p tracks both growing and shrinking astral microtubules throughout the cell cycle, although its backtracking ability is less robust in G1 (Carvalho *et al.*, 2004). Many different multifluorophore fusions are available for Bik1p providing a very bright signal for the microtubule plus-ends (Badin-Larcon *et al.*, 2004; Carvalho *et al.*, 2004; Caudron *et al.*, 2008; Grava *et al.*, 2006; Li *et al.*, 2005; Lin *et al.*, 2001; Molk *et al.*, 2006). The fact that Bik1p is transported along microtubules via the motor protein Kip2p (Carvalho *et al.*, 2004) must to be considered when using Bik1p as a reporter for plus-ends as its distribution along the length of a microtubule can give rise to difficulties for tracking Bik1p with automated tracking algorithms.

2. Microtubule Disassembly Promoters

a. KIP3 *Overview.* The *KIP3* gene was identified in a BLAST search for kinesin-related genes and encodes an 805 amino acid motor protein (91 kD) (DeZwaan *et al.*, 1997). Deletion of *KIP3* results in long cytoplasmic microtubules (Miller *et al.*, 1998; Yeh *et al.*, 2000). *In vitro*, Kip3p has been shown to be a highly processive plus-end-directed microtubule motor that depolymerizes microtubules specifically from their plus-ends (Gupta *et al.*, 2006; Varga *et al.*, 2006). The destabilizing activity of Kip3p is length dependent, leading to faster depolymerization of long microtubules (Varga *et al.*, 2006, 2009).

Application to Analysis of Microtubules. Endogenously expressed Kip3p localizes along nuclear and cytoplasmic microtubules with enrichment at plus-ends throughout the cell cycle (Gupta *et al.*, 2006; Varga *et al.*, 2006). However, Kip3p accumulates only on growing but not on shrinking microtubules (Gupta *et al.*, 2006; Varga *et al.*, 2006) limiting its use for microtubule dynamics analysis. Endogenous tagging of Kip3p is possible and multifluorophore fusions are available (Gardner *et al.*, 2008a; Gupta *et al.*, 2006; Tytell and Sorger, 2006; Varga *et al.*, 2006, 2009).

b. KAR3 *Overview.* Kar3p defines a subclass of kinesin-related minus-end-directed motor proteins that have their motor domains located at the C-terminus of the protein (Endow *et al.*, 1994; Meluh and Rose, 1990; Middleton and Carbon, 1994). Kar3p forms functionally different complexes with two proteins, Cik1p and Vik1p (Manning *et al.*, 1999; Page *et al.*, 1994). *In vitro*, Kar3p shortens microtubules in a plus-to-minus-end fashion, suggesting that Kar3p induces microtubule depolymerization by a sequential release of tubulin heterodimers while remaining attached to the microtubule (Sproul *et al.*, 2005). These observations confirm earlier *in vivo* observations in mating cells, where Kar3p depolymerizes microtubules specifically at the mating projection (Maddox *et al.*, 2003). Kar3p also depolymerizes microtubules in preanaphase cells (Saunders *et al.*, 1997). However, during anaphase, Kar3p is required for spindle assembly and stability both at the spindle poles, where it tethers and cross-links microtubule minus ends, and along the spindle where it bundles microtubules (Allingham *et al.*, 2007; Gardner *et al.*, 2008b).

Application to Analysis of Microtubules. Kar3p localizes to cytoplasmic microtubules and to spindle poles in mating cells (Meluh and Rose, 1990). In these cells Kar3 accumulates at the ends of shmoo tip microtubules (Maddox *et al.*, 2003; Zaichick *et al.*, 2009). During vegetative growth, Kar3p localizes to spindle poles and nuclear microtubule throughout the cell cycle but not to cytoplasmic microtubules (Gardner *et al.*, 2008b; Manning *et al.*, 1999). Thus the use of Kar3p as an end marker is therefore restricted to either mating cells or to the analysis of phenotypes in the spindle. Single fluorophore fusions have been reported (Gardner *et al.*, 2008b; Maddox *et al.*, 2003; Zaichick *et al.*, 2009) but no multifluorophore-Kar3p protein is currently available.

3. Microtubule Stabilization Through Cortical Interactions

a. KAR9 *Overview.* In *S. cerevisiae*, Kar9p is considered to be the functional homologue of APC because it shares limited similarity with a part with the C-terminal

EB1-binding region of APC (Bienz, 2001). Kar9p encodes a 644 AA, 74 kDa basic protein with no significant homology to any known protein (Miller and Rose, 1998). Deletion of *KAR9* results in spindle misorientation (Miller and Rose, 1998; Yeh *et al.*, 2000) but does not affect microtubule dynamics (Beach *et al.*, 2000). Kar9p binds to microtubules via Bim1p (Korinek *et al.*, 2000; Lee *et al.*, 2000). The type V myosin Myo2p directly interacts with Kar9p (Yin *et al.*, 2000) and is important for spatial organization of Kar9p, since mutating Myo2p results in Kar9p redistribution between mother and bud (Beach *et al.*, 2000). Kar9p is therefore important to link cytoplasmic microtubules to the actin cytoskeleton.

Application to Analysis of Microtubules. Endogenously expressed Kar9p-GFP localizes along microtubules and to the plus-ends of growing and shrinking microtubules (Kusch *et al.*, 2002; Lee *et al.*, 2000; Liakopoulos *et al.*, 2003; Maekawa *et al.*, 2003; Miller *et al.*, 2000). Localization of Kar9p within the cell is regulated temporally and spatially. During metaphase, Kar9p localizes asymmetrically to the bud-proximal spindle pole in a microtubule-independent manner (Liakopoulos *et al.*, 2003; Maekawa *et al.*, 2003). Endogenous fluorophore tagging of Kar9p has been widely used and multifluorophore fusions are now available (Beach *et al.*, 2000; Grava *et al.*, 2006; Huisman *et al.*, 2004; Korinek *et al.*, 2000; Kusch *et al.*, 2002; Lee *et al.*, 2000; Leisner *et al.*, 2008; Liakopoulos *et al.*, 2003; Maekawa and Schiebel, 2004; Maekawa *et al.*, 2003; Miller and Rose, 1998; Miller *et al.*, 1998, 1999; Miller *et al.*, 2000; Moore and Miller, 2007; Moore *et al.*, 2006; Sagot *et al.*, 2002; Segal *et al.*, 2000). One limitation of using Kar9p as a microtubule end marker comes from the observation that during metaphase Kar9p predominantly localizes to microtubules emanating from the bud-proximal SPB (Liakopoulos *et al.*, 2003; Maekawa *et al.*, 2003).

b. *DYNEIN* (DYN1, DHC1) *Overview. DHC1* was identified in a polymerase chain reaction-based search for dynein homologues in *S. cerevisiae* and it is the only dynein heavy chain isoform in budding yeast (Eshel *et al.*, 1993; Li *et al.*, 1993). Extensive reviews about dynein function in *S. cerevisiae* (Moore *et al.*, 2009) and different dynein isoforms in fungi (Yamamoto and Hiraoka, 2003) are available.

Dynein is critical for spindle elongation during anaphase. Different models have been provided to describe the process of dynein recruitment to plus-ends and off-loading at the cell cortex. In summary, dynein is recruited to plus-ends via Pac1p, Ndl1p, Kip2p, and Bik1p. Upon attachment, dynein and Pac1p are off-loaded from the microtubule tip and anchored at the cortex, where dynein is transferred to the membrane. Activation of dynein results in pulling on microtubules and the associated spindle pole (Lee, 2003; Li *et al.*, 2005; Sheeman *et al.*, 2003). In dynein mutant cells, astral microtubules spend more time at the cortex (Carminati and Stearns, 1997) and the average length of microtubules is increased (Carminati and Stearns, 1997; Knaus *et al.*, 2005). In these cells, astral microtubules are less dynamic, with decreased speeds and transition frequencies and an increase in pausing times (Carminati and Stearns, 1997).

Application to Analysis of Microtubules. Localization of dynein to SPBs and cytoplasmic microtubules is cell cycle dependent. While it localizes asymmetrically to the proximal SPB and astral microtubules during G1 and metaphase, its localization to SPBs and astral microtubules become symmetric during anaphase (Grava *et al.*, 2006; Lee, 2003; Sheeman *et al.*, 2003; Yeh *et al.*, 1995). Dyn1-3GFP is found along growing and shrinking astral microtubules (Sheeman *et al.*, 2003). Dynein can be tagged endogenously and different multifluorophore constructs are available (Carvalho *et al.*, 2004; Caudron *et al.*, 2008; Lee, 2003; Lee *et al.*, 2005; Li *et al.*, 2005; Sheeman *et al.*, 2003; Woodruff *et al.*, 2009).

III. Microscopy and Data Collection

A. Overview

In the past decade the field of microscopy and in particular fluorescence microscopy has evolved rapidly, providing a vast number of new microscope setups. Each system has different traits expanding spatial and temporal resolution and the effective amount of signal collected. Three parameters define which restrictions apply when choosing among these possible setups; the type of microtubules to be imaged (nuclear or cytoplasmic), whether a +end reporter (e.g., a +TIP fusion) or a reporter for curvilinear length (e.g., GFP-Tub1) is used, and the temporal/spatial constraints imposed by the biological process itself. Imaging single microtubules requires a different setup (confocal, sensitive camera) than observing a population of microtubules (speckle microscopy, Fluorescence Recovery after Photobleaching (FRAP), Fluorescence Loss in Photobleaching (FLIP)). Depending on what the biological question is, the required resolution on space and time may vary and thus the microscope setup as well.

Numerous reviews organize the vast amount of information available to biologists new to the field of microcopy (Combs, 2010; Lichtman and Conchello, 2005; Salmon *et al.*, 2005; Schulz and Semmler, 2008; Taylor and Salmon, 1989; Waters, 2007; Wolf, 2007). Acquisition optimization involves balancing signal-to-noise ratio and spatial and temporal resolution (Combs, 2010; Dorn *et al.*, 2008). Fluorophore bleaching and phototoxic effects ultimately restrict parameter optimization (Combs, 2010; Dorn *et al.*, 2008).

Microtubules can be imaged using different microscope setups including differential interference contrast (DIC), wide-field, or confocal microscopes. Single-beam confocal microscopes are widely used and provide an increase in spatial (but not temporal) resolution compared to wide-field microscopy. Multibeam confocal microscopes (e.g., spinning disk or swept-field confocal microscopes) are the preferred choice when acquiring images at high spatial and temporal resolution. A detailed comparison between the two different setups is available (Gräf *et al.*, 2005; Marcus, 2007) as well as a very good comparison between different microscope setups and their advantages/disadvantages (Combs, 2010).

B. Essential Microscope Parameters

In Section II of this chapter we introduced the reader to different plus-end tracking proteins. Careful selection of such a protein in combination with a multifluorophore tag will significantly influence the signal quality in recorded movies. Several other parameters need to be carefully considered when establishing a microscope system to ensure maximal signal readout at a high acquisition rate.

Spatial Resolution. The resolution of a light microscope defines the minimal distance by which two adjacent objects can be separated. This minimal distance d is defined by the diffraction limit of the optical system (criterions formulated by Raleigh) and depends primarily on the wavelength of the illumination light (λ) and the numerical aperture (NA) of the lens the light is passing through ($d = 0.61\lambda/\mathrm{NA}$) (Heintzmann and Ficz, 2007; Hell, 2003). The resolution of the microscope system will determine the sampling in space that is required to resolve the spindle poles and microtubules [see Dorn *et al.* (2008) and references therein].

Camera Type. In order to image live yeast cells charge-coupled device (CCD) or electron multiplying CCD (EM-CCD) cameras can be used. Guidelines for which type of camera should be used and a detailed discussion about their differences are provided by Hinsch (2007), Moomaw (2007), Rasnik *et al.* (2007), and Spring (2007). Significant amplification of low-light-level signals above the CCD read noise is achieved by EM-CCD technology. In addition to the total amplification of a signal and a background as in CCD camera, EM-CCD performs amplification of only detected photons before converting them to a digital signal. Two types of EM-CCD cameras are used for high-resolution microscopy: front-illuminated (FI) and back-illuminated (BI) EM-CCD cameras (Heemskerk *et al.*, 2007). FI EM-CCD cameras bear some limitations in signal detection resulting in a low stability of spatial and energy resolution upon decreasing the temperature. BI EM-CCD cameras (cooled to –65°C or lower) demonstrate relatively stable results and significantly improve the imaging capabilities for high-resolution microscopy.

Light Source. The choice of a light is very important for quantitative microscopy. The main requirements to a light source are stability and uniformity. Mercury and Xenon Arc lamps, representing slightly different spectral radiation profiles, were commonly used source of light for wide-field fluorescence microscopy as they are much cheaper than lasers. Gas or solid-state lasers produce very narrow light spectra, corresponding to almost one wavelength ± manufacture standard deviation Gas lasers are less expensive than solid-state lasers and usually produce multiple emission lines: 488 and 514.5 nm, which represent approximately 75% of the total laser power. However, gas lasers are less suitable when used for emissions in the red spectrum and require active cooling (air or water). Solid-state lasers are the most stable and accurate light source. Lasers do not require emission filters and therefore can be used to perform an ultra fast switch between different wavelengths.

C. Protocol: High-Resolution Imaging Using a Spinning Disk Confocal

Temperature influences the dynamical properties of microtubules *in vivo*. Using automated, subpixel tracking methods to identify the spindle poles (Fig. 3A) within a volume (Fig. 3B, C) and optimized parameters described in the protocol below, we measured the temperature dependence of the velocity of pole separation during the fast phase of anaphase. The mean velocity ($< v >$; µm/min) of pole separation is significantly slower at 25°C (0.74 ± 0.272, $n = 24$ cells) with respect to $< v >$

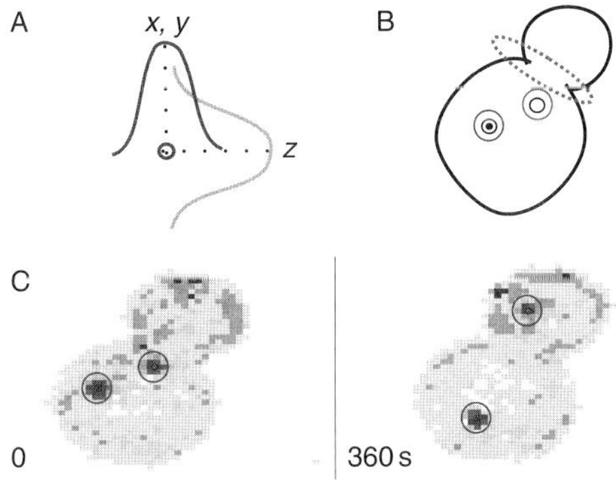

Fig. 3 Tracking spindle poles using automated feature finding. Spindle poles labeled with Spc42-CFP are a point-like feature. Pole intensities (in *xy* and *z*) are fit to a Gaussian as described in Section IV (A) to give the position of the poles in the mother and bud cavities (B). Images were collected using the protocol a spinning disk confocal microscope, 63× 1.4 NA objective, 493 solid-state laser (Coherent), and EM-CCD camera (Hamamatsu ImageEM).

measured at 30°C (1.08 ± 0.153, $n = 10$ cells). Similar temperature dependency has been found for kinetochore microtubule dynamics (Dorn *et al.*, 2005). Therefore temperature must be considered when planning the experimental regime and controlled during data acquisition.

Cells are stressed by imaging, and while it is impossible to avoid stress, phototoxicity (generally resulting from the formation of reactive oxygen species) must be minimized. We have observed that > 90% of preanaphase spindles ($n > 300$) fail to execute anaphase when high laser power is combined with rapid pulses (30 frames × 50 ms per stack, acquired at 5 s intervals). In cells that are minimally stressed, the trajectories of pole displacement are stereotyped and fast anaphase pole separation is concomitant with spindle movement into the bud (Fig. 4). Spindles <3 μm in length (defined as population *N*) are considered to be preanaphase spindles. The proportion of spindles that undergoes anaphase with respect to those expected to undergo anaphase (defined here as F_{ma}) is high when cells are pretreated for 60–90 min with 1 mM ascorbate in synthetic minimal (SD) medium, and image stacks are collected at 10 or 20 s intervals (Table I). Deletion of the checkpoint protein Mad2 (*mad2Δ*) did not reduce F_{ma} indicating that, under these conditions, anaphase onset is not delayed by activation of the spindle checkpoint. Sampling at 60 s intervals reduced F_{ma}. The distribution of values for F_{ma} obtained (0.70–0.45) is in part due to the intrinsic variability of the time between spindle assembly and the completion of biorientation of all 16 chromosomes.

The quality of images largely depends on the method of cell culture preparation and slide cleaning. In order to collect data from cells in all possible cell cycle stages, the samples must be prepared from exponentially growing culture (OD 0.4–0.6 at 600 nm). Cells are generally grown in low-fluorescence medium, for example, SD medium. Clean slides improve the signal-to-noise ratio, which is highly important for single-cell analysis. Uniform and appropriate thickness (0.17 mm) of coverslips improves results.

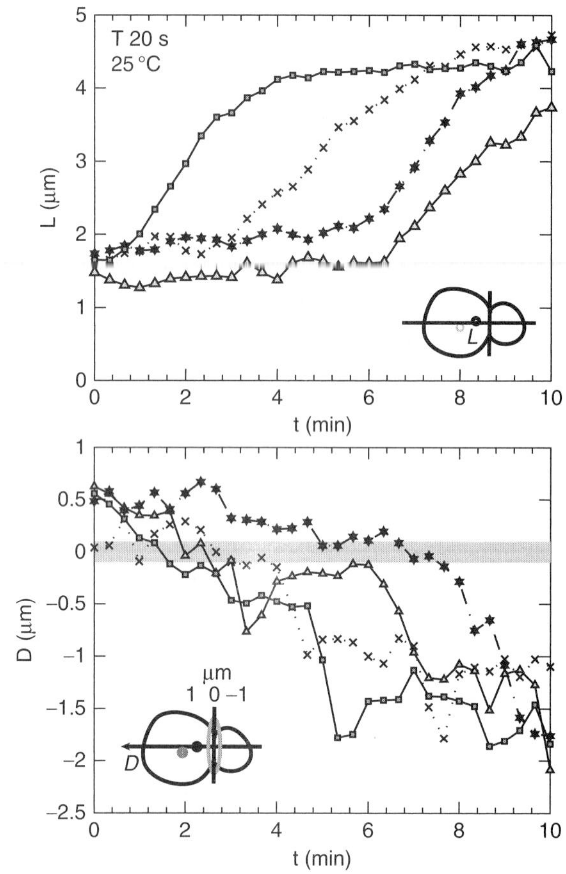

Fig. 4 Spindle dynamics during the metaphase–anaphase transition. High-resolution tracking of spindle poles (upper panel). Spindle entry into the bud (lower panel); negative values indicate pole is located in the bud. Spindle poles are labeled with Spc42-CFP (cerulean). Image stacks ($z = 6$ μm; 300 nm step size) were collected at 20 s intervals (described in Section III) using a spinning disk confocal microscope and Metamorph software, 63× 1.4 NA objective, 440 solid-state laser (Coherent), and EM-CCD camera (Hamamatsu ImageEM). Automated tracking was performed with particle tracking software implemented in MatLab.

1. Cleaning Microscope Slides and Coverslips

Slides and coverslips should be cleaned shortly before the beginning of the imaging process. Commercial precleaned slides are dirty enough to produce stochastic results from single-cell-based fluorescence measurements. Slides and coverslips should be rinsed thoroughly on both sides with a large amount of ultrapure water, then ethanol, acetone, and finally a second ethanol rinse. The washed slides and coverslips are dried with compressed air, using a cotton baffle to trap any particulates or oil that may exist in the compressed air line.

ConA is a lectin that will cause yeast cells to adhere tightly to glass. It is important to consider that ConA may alter the cytoskeleton and to perform appropriate control

Table I
Optimization of Imaging Conditions

	Laser pulse interval[a]					
	10 s		20 s			60 s
Condition	—	+ ascorbate	—	+ ascorbate	+ ascorbate *mad2Δ*	+ ascorbate
N	47	102	76	91	75	94
N_E	17	23	25	29	24	29
F_{ma}	0.29	0.70	0.65	0.58	0.58	0.45

[a] Laser power, integration time, stack depth (6 μm), and z-spacing (300 nm) were uniform. N is the population of cells with a preanaphase spindle <3 μm at t = 0. NE is defined as the population of cells expected to undergo the metaphase–anaphase transition in the 10 min window of acquisition, at 25°C. Fma, anaphase transitions observed for NE.

experiments. To prepare ConA coverslips, the glass must be first cleaned with 1 N NaOH for 2 h, washed with ultrapure water, and dried. ConA solution - (10 μl) is then applied to one side of the coverslip for 10–20 min (do not let it dry) and removed with two water washes. The coverslip is then air dried, marked to indicate the side coated with ConA, and used immediately. The coverslip *must be marked* such that the ConA side is indicated *clearly*, as ConA will damage objective lenses.

2. Preparing the Cells

1. Preincubate cells for 1 h in synthetic complete (SC) medium with 1 mM ascorbate.
2. Pellet cells in 1 ml of yeast cell culture (OD 0.4–0.6) by centrifugation (500 rpm, 30 s).
3. Vortex the cell suspension for 30–40 s.

   This sample can be used for *in vivo* microscopy for 20 min (max. 30 min). Keeping yeast cells concentrated in a small volume can change their metabolism and cell cycle progression.
4. Pipet 5 μl of the concentrated cell culture on to a clean slide and apply a clean coverslip (Ted Pella 0.16–0.19 mm). Avoid air bubbles.
5. Press gently on two of the coverslip edges with a Kimtech wipe to wick/remove excessive liquid. Avoid compressing the cells. The coverslip can be sealed with clear nail polish or VALAP (1:1:1 vasoline:lanolin:paraffin). Use minimal sealant and ensure the seal is dry before imaging to prevent damage to objective lenses. For experiments where data is collected for less than 10 min, sealing is not required.

3. Optimization of the Imaging regime

To optimize conditions for imaging the following parameters should be considered:

- The objective magnification and NA should satisfy experimental goals. Imaging +TIPs proteins requires 63× or 100× Plan-Apochromat objective with NA 1.4 or higher.

- Electron multiplication gain of the EM-CCD camera should be adjusted to optimize signal-to-noise ratio of the outcome image. The gain of EM-CCD and CCD cameras can also be adjusted to improve signal.
- Photon counting mode is available for some EM-CCD cameras. Low-intensity signals can be amplified with photon counting mode (set to "1"), without significant perturbation of background readout.
- The illumination source (laser) together with exposure time should be adjusted according to the temporal resolution required and to minimize photobleaching. Fast imaging (10+ fps) will require short exposure time and thus higher laser power. Applying higher laser power will produce more photon noise. High laser power or long exposures, e.g., the total number of photons the sample is exposed to during the experiment, contributes to fluorophore bleaching and phototoxicity and therefore both parameters should be optimized.
- It is essential to avoid signal saturation for quantification of a fluorescent signal. All parameters should be adjusted in a range satisfying characteristics of the camera used, typically 16 bit range.

IV. Methods of Analysis

In this section we provide an overview of methods suited for analysis of microtubule dynamics in yeast. Factors such as sample size, dimensionality, and accuracy determine the information content of the data set, therefore influence what analysis strategy is employed.

A. Ensemble (Large-Scale, Averaged) Versus Single-Cell Analysis

Microtubule growth and shrinkage are dynamic subcellular processes. Therefore any analysis of the dynamical properties of microtubules has to be performed at the single-cell level. Moreover, the process of dynamic instability is highly variable not only at the level of light microscopy but even at single-molecule resolution (Gardner *et al.*, 2008c; Kerssemakers *et al.*, 2006; Schek *et al.*, 2007). This has two important consequences: (1) microtubules may exhibit a variety of different behaviors, some of which can be rare and (2) individual comparisons, or direct averaging, of microtubule trajectories between cells (or microtubules) is meaningless.

1. Capturing the Full Dynamic Phenotype

The natural variation of microtubule behaviors implies the acquisition and analysis of large data sets. The size of the data set is largely determined by the tools available to obtain and analyze the data. In the case of microtubule tracking, use of (semi-) automated tracking software is required for the acquisition of a homogeneous large-scale data set (Dorn *et al.*, 2008). If a dynamic process exhibits only a limited number of stereotypical behaviors, manual analysis of a small (<100) number of cells can be used to obtain information on a variety of perturbations of the microtubule cytoskeleton—see, e.g., (Adames and Cooper, 2000; Carminati and Stearns, 1997; Cuschieri *et al.*, 2006; Shaw *et al.*, 1997; Tirnauer *et al.*, 1999; Wolyniak *et al.*, 2006) for manual tracking of astral microtubules. The minimal sample size increases for subtle phenotypes and the number of behaviors under study; more cells have to be analyzed in order to obtain an accurate view of a complex phenotype. In addition, if there is phenotypic heterogeneity between cells, large amounts of data are needed to be able to statistically discriminate between the different populations. It is therefore essential to minimize variation in experimental regime and in strain construction. Manual analysis of such numbers of cells is unrealistic, and thus, automated computational image and data analysis are required (Dorn *et al.*, 2005; Jaqaman *et al.*, 2006, 2007).

It should be noted that great care has to be taken when developing the analysis schemes, since rigid assumptions during design may make the software discard rare behaviors. Therefore, although the analysis of microtubule dynamics requires the development of automated analysis pipelines, a high level of interaction between software developers and experimenters is required to ensure that the full dynamic phenotype is captured (Jaqaman and Danuser, 2006).

2. Characterizing Microtubule Dynamics

The stochastic nature of microtubule dynamics makes it impossible to average individual microtubule length trajectories. To obtain information about populations of microtubules, it is therefore key to define salient descriptors of the dynamics and to average those as long as there is no population heterogeneity (Dorn *et al.*, 2008).

Microtubule analysis based on the characterization and comparison of the classic parameters such as transition frequencies and speeds is difficult because small variations are likely to be lost in the noise part of the data set. At least for kinetochore microtubules, these parameters are not unique; it is possible to obtain similar parameter sets for real and randomized data (Jaqaman *et al.*, 2006). Furthermore, the parameters are highly sensitive to undersampling and to missing data points (Dorn *et al.*, 2005; Jaqaman *et al.*, 2006).

Different approaches have been proposed to circumvent this problem. As an alternative to comparing speed distributions and transition frequencies, microtubule length life histories can be characterized by analyzing their oscillating behavior (Odde and Buettner, 1995; Odde *et al.*, 1996). Jaqaman *et al.* (2006) further expanded the application of regression analysis to study microtubule dynamics by developing an approach based on autoregressive moving average models (ARMA). These statistical models can be used to extract the coupling of the microtubule length state in time series. To access biologically meaningful information, an ARMA model can be used to help generate *in silico* microtubule length series. Since *in silico* microtubule length series do not suffer from data undersampling, they can be used to study microtubule speeds and transition frequencies (Jaqaman and Danuser, 2006; Jaqaman *et al.*, 2006).

When image data sets do not provide the required resolution in both space and time to address specific biological questions, *in silico* modeling of microtubule behavior can be used to gain insights into the mechanisms investigated. These models can be used to simulate microtubule distributions in cells in order to estimate physical parameters from the model distributions (Shariff *et al.*, 2010) or gain understanding on how specific microtubule organization patterns are achieved in cells (Gregoretti *et al.*, 2006; Janson *et al.*, 2007; Nédélec, 2002; Pinot *et al.*, 2009). Model development needs to be constantly tested against experimental data in order to validate the output of the simulation (Gardner *et al.*, 2007; Jaqaman and Danuser, 2006). Global analysis of microtubule behavior does not require detailed modeling of events at individual microtubule ends (Karsenti *et al.*, 2006), greatly simplifying the task of model building and evaluation. In budding yeast, where the dimension of the cell restricts analysis of events such as resolving individual spindle microtubules, modeling of spindle microtubule dynamics can be important to understand how these microtubules are organized (Pearson *et al.*, 2006; Sprague *et al.*, 2003). A detailed summary on how microtubule dynamics can be simulated has been provided recently (Gardner and Odde, 2010).

B. Considerations with Respect to Temporal and Spatial Resolution

Microtubule movements occur in three dimensions. Rates of growth and shrinkage and the frequency of switching between states are relatively fast. Capturing the dynamics at an accurate spatial and temporal resolution is key for accurate tracking. Likewise, accurate tracking requires a sufficient high signal-to-noise ratio throughout the entire experiment. Finally, the length of the experiment must be sufficient to capture all possible states of the microtubule. These prerequisites are tightly linked and optimizing one parameter will ultimately downgrade one or several of the other parameters. A detailed discussion on parameter optimization is given in Dorn *et al.* (2008).

1. Dimensional Data Analysis: Achieving High Spatial Resolution

In light microscopy resolution is diffraction limited to about 200 nm in the *x*/*y* plane and 600 nm in the *z* plane (Gustafsson, 1999). To overcome these limits and determine object positions below the diffraction limit different approaches to enhance spatial resolution have been developed. Methods available can be divided into approaches that either manipulate the point-spread function or the optical transfer function or techniques that apply prior knowledge such as shape and dynamical behavior of the analyzed objects in an image (Danuser, 2001; Gustafsson, 1999). Model-based resolution enhancement techniques are especially powerful when the tracked object has a simple geometrical shape such as a sphere (Thomann *et al.*, 2002, 2003). Although microscopy-based resolution-enhancing approaches [reviewed in Gustafsson (1999)] are not fast enough to fully capture dynamics of budding yeast microtubules, application of prior knowledge to achieve subpixel localization and enhanced resolution has been applied to study behavior of both astral and nuclear microtubules (Cardinale *et al.*, 2009; Dorn *et al.*, 2005). It is important to note that these studies provide positional uncertainties for each measurement, which is essential to statistically analyze the microtubule length series.

2. Dimensional Data Analysis: Temporal Sampling

The appropriate choice of sampling frequency requires *a priori* knowledge of the process that is studied (ICRU, 2006). Ideally, a noisy signal (spatial or temporal) should be sampled by at least three times its characteristic frequency (Stelzer, 1999). Limited information is available for rates of microtubule growth and shrinkage in cells. Reported rates of growth and shrinkage are ~1 μm/min and the frequency of switches between states (growth, shrinkage) are <1/min (Cuschieri *et al.*, 2006; Vogel *et al.*, 2001). For kinetochore microtubules these parameters are in the range of 5 μm/min for speeds and below 1/s for the transition frequencies (Dorn *et al.*, 2005). The sampling frequency and the timescale over which data must be collected must accommodate the spectrum of dynamical behavior of microtubules in the experimental regime (e.g., wild-type versus mutant or temperature). Simple tests can be used to determine whether sampling frequency is sufficient for accurate tracking and are described in Dorn *et al.* (2008)]. It is important to note that temporal undersampling results in an apparent reduction of both speeds and transition frequencies (Dorn *et al.*, 2005).

C. Methods of Analysis

1. Manual Tracking of Microtubules in Four Dimensions

Manual analysis of microtubule dynamics requires a careful experimental setup to ensure completeness and consistency of the data set. First, each movie must be analyzed several times and tracks must be averaged to minimize tracking bias introduced by the investigator (Tirnauer *et al.*, 1999). Then, the necessary amount of data points to estimate each descriptor of microtubule dynamics needs to be determined (Dorn *et al.*, 2008; Jaqaman *et al.*, 2006). Finally, movies need to be analyzed in an unbiased manner to avoid artificially enriching for a specific phenotype. This can be achieved by having different investigators preselect cells for analysis.

Manual tracking of microtubule lengths is generally done in three dimensions. To track microtubules in a single plane rather than in three-dimensional image stacks requires both ends to be in the same focal plane (Adames and Cooper, 2000; Sprague *et al.*, 2003). However, a single focal plane is rarely sufficient to accurately measure the length of a microtubule over its entire lifetime (Dorn *et al.*, 2005, 2008). The coordinates of the microtubule ends can be found and linked in an image stack or measured in a two-dimensional projection of the stack, followed by counting the planes with in-focus signal of the microtubule (Kosco *et al.*, 2001; Tirnauer *et al.*, 1999; Wolyniak *et al.*, 2006). The curvilinear length can then be reconstructed by calculating the hypotenuse. A method to calculate microtubule dynamics from kymograph images has been described in Mennella *et al.* (2005).

While individual astral microtubules can be resolved in the light microscope, this is not possible for individual spindle microtubules, which restricts manual analysis of microtubule dynamics to cytoplasmic microtubules. In anaphase, the ensemble dynamics are reported by increase in curvilinear length and velocity of pole displacement (shown in Fig. 4). In certain mutants and to a lesser extent in wild-type cells

[(Vogel *et al.*, 2001) and our observations], curvilinear length and pole separation become uncorrelated as spindle length increases $> 6\ \mu m$.

2. Semi-automated/Automated Tracking of Microtubules in Four Dimensions

Although automated tracking of microtubules requires a substantial initial investment to establishment of the necessary software tools, it provides several advantages over manual microtubule tracking. First, microtubules can be tracked at subpixel localization, increasing tracking precision significantly. Second intra- and interuser variability of tracking results is eliminated when using automated tracking methods. Third the amount of data that can be analyzed when using computer-aided tracking is unmatched. To increase tracking precision, resolution in space and time must be as high as possible. Depending on the experimental setup and the biological question behind the experiment different tracking approaches can be applied. In the following section we briefly discuss different methods available for microtubule tracking.

3. Tracking Methods—Microtubule Filament Tracking

Line tracking of microtubules relies on the successful detection of the curvilinear structure in images. Different approaches have been developed for line feature detection and subsequent microtubule segmentation in microscopic images (Brangwynne *et al.*, 2007; Danuser *et al.*, 2000; Hadjidemetriou *et al.*, 2005, 2008; Jiang *et al.*, 2004, 2005; Lichtenstein *et al.*, 2003; Shelden and Knecht, 1998). Many methods to detect and track curvilinear features rely on steps involving image enhancing using filter and threshold-based algorithms followed by morphological image processing to detect the tubules. A variety of other methods exist [for an overview see Hadjidemetriou *et al.* (2008) and references therein]. These methods work well when microtubule intersections are scarce such as in budding and fission yeasts (Tischer *et al.*, 2008). In addition to quantifying microtubule dynamics, shape-based tracking methods allow quantitative analysis of microtubule bending (Bicek *et al.*, 2007).

The greatest obstacle in filament tracking lies in the successful identification of the microtubule ends in successive time steps. In budding yeast, direct labeling of tubulin with a fluorophore is restricted to α-tubulin, in which an N-terminally tagged *TUB1* is introduced into the genome as a second copy (Straight *et al.*, 1997). Substoichiometric labeling of the microtubule lattice results in localized intensities ("speckling") distributed over the length of the microtube and lower intensity at the dynamic plus-end. This makes accurate detection of the microtubule end, and thus precise measurement of length, difficult. As an alternative, a +TIP protein that tracks growing and shrinking microtubules can be tagged with the same fluorophore to enhance signal-to-noise ratio at the plus-end (discussed below).

Tracking Methods—Feature Point Tracking

Feature point tracking (FPT) algorithms have been developed to address single and multiple particle tracking problems. These particles represent fluorescent speckles of individual or aggregated proteins with an object size smaller than the diffraction limit which results in their appearance in a point-spread function-based shape. Tracking algorithms are based on model fitting (Danuser *et al.*, 2000; Dorn *et al.*, 2005; Gao and Kilfoil, 2009; Kerssemakers *et al.*, 2009; Thomann *et al.*, 2002, 2003) and filtering methods (Cardinale *et al.*, 2009; Sage *et al.*, 2005; Sbalzarini and Koumoutsakos, 2005; Smal *et al.*, 2007, 2008a, b). Performance and accuracy comparisons can be found in Gao and Kilfoil (2009), Sbalzarini and Koumoutsakos (2005), and Thomann *et al.* (2002).

High particle density leads to several problems such as overlapping objects and/or object trajectories or temporary particle disappearance due to either out-of-focus shift of the imaged object or fluorophore blinking. Several attempts have been made to overcome these difficulties (Gao and Kilfoil, 2009; Jaqaman *et al.*, 2008; Smal *et al.*, 2007, 2008a, b). These algorithms are optimized for specific experimental conditions, and their suitability depends on the assays used to acquire the image. A recent quantitative comparison of these tracking algorithms can be found in Gao and Kilfoil (2009).

When particle density is low, FPT basically reduces to successful detection and localization of particles. Low particle density eliminates or simplifies trajectory linking and greatly reduces the complexity of the tracking algorithm and need for optimization strategies (Cardinale *et al.*, 2009).

5. Finding and Tracking Microtubule Ends Using +TIPs

In order to track yeast microtubules with FPT-based methods, both ends need to be visualized as spots. Microtubule minus ends are visualized by tagging components of the SPB (Cardinale *et al.*, 2009; Dorn *et al.*, 2005). Plus-end visualization differs depending on whether nuclear or cytoplasmic microtubules need to be stained.

In the yeast spindle, each of the 16 chromosomes is linked to 1 spindle microtubule. So far no treadmilling has been detected (Maddox *et al.*, 2000), and so movement of kinetochores relative to the spindle pole appears to directly reflect microtubule dynamics. In order to resolve dynamics of a single kinetochore microtubule, the centromeric region of one specific chromosome can be tagged with fluorophores (Dorn *et al.*, 2005). Note that measured dynamics may differ depending on the chromosome size (Jaqaman *et al.*, 2007). Staining of individual plus-ends is possible; however, since the yeast spindle is small, resolution of the individual ends for tracking has not yet been achieved and dynamics analysis is restricted to the entire kinetochore microtubule population using FRAP analysis or model convolution (Gardner *et al.*, 2005; Pearson *et al.*, 2006; Sprague *et al.*, 2003).

Tracking of cytoplasmic microtubules requires the direct labeling of the plus-end. With the emphasis on FPT-based tracking of microtubules in this chapter we do not recommend the labeling of *TUB1* because, in this case, *TUB1* is overexpressed and thus microtubule dynamics are altered. Instead, a suitable tip tracking protein can be selected for plus-end labeling. This method is not suitable to track the length of long microtubules, such as those observed during anaphase or specific mutants that cause microtubule bending. Proteins that are suitable to track growing and shrinking microtubules have been described above and include Bim1p, Bik1p, Kip2p, or Stu2p (Carvalho *et al.*, 2004; Caudron *et al.*, 2008; Molk and Bloom, 2006; Wolyniak *et al.*, 2006). For these proteins, triple-fluorophore-tagged fusion proteins exist that provide high signal-to-noise ratios. In addition, these proteins localize symmetrically on growing and shrinking microtubules emanating from both sides of the spindle. In principle, all of these proteins are present on growing and shrinking astral microtubules throughout the cell cycle, although small variations may occur (Carvalho *et al.*, 2004) which must be considered when deciding which end tracking protein to use. In addition, these proteins are transported along the microtubule lattice toward the plus end. Upon arrival of the cargo at the microtubule end the "fluorescent" shape of the tip will temporarily elongate, a process, which has to be taken into account when tracking microtubules using plus-end proteins.

Acknowledgments

The authors thank members of the Barral and Vogel lab for discussions; Chris Weirich, Khouloud Jaqaman, and Jonas Dorn for critical reading of the manuscript; and Vincent Pelletier, Kemp Plumb, and Maria Kilfoil (Department of Physics, McGill University) for their expertise and contributions to tracking methods used in this study. The authors acknowledge support from operating and equipment grants from CIHR and the Canadian Foundation for Innovation (CFI) to J.V. and to the Developmental Biology Research Initiative (Biology, McGill University) and from the ETH-Zurich to Yves Barral.

References

Adames, N. R., and Cooper, J. A. (2000). Microtubule interactions with the cell cortex causing nuclear movements in *Saccharomyces cerevisiae*. *J. Cell Biol.* **149**(4), 863–874.

Akhmanova, A., and Hoogenraad, C. C. (2005). Microtubule plus-end-tracking proteins: Mechanisms and functions. *Curr. Opin. Cell Biol.* **17**, 47–54.

Akhmanova, A., and Steinmetz, M. (2008). Tracking the ends: A dynamic protein network controls the fate of microtubule tips. *Nat. Rev. Mol. Cell Biol.* **9**, 309–322.

Al-Bassam, J., van Breugel, M., Harrison, S. C., and Hyman, A. (2006). Stu2p binds tubulin and undergoes an open-to-closed conformational change. *J. Cell Biol.* **172**, 1009–1022.

Allingham, J. S., Sproul, L. R., Rayment, I., and Gilbert, S. P. (2007). Vik1 modulates microtubule-Kar3 interactions through a motor domain that lacks an active site. *Cell* **128**, 1161–1172.

Amos, L., and Klug, A. (1974). Arrangement of subunits in flagellar microtubules. *J. Cell Sci.* **14**, 523–549.

Badin-Larcon, A. C., Boscheron, C., Soleilhac, J. M., Piel, M., Mann, C., Denarier, E., Fourest Lieuvin, A., Lafanechère, L., Bornens, M., and Job, D. (2004). Suppression of nuclear oscillations in *Saccharomyces cerevisiae* expressing Glu tubulin. *Proc. Natl. Acad. Sci. U.S.A.* **101**(15), 5577–5582.

Beach, D. L., Thibodeaux, J., Maddox, P., Yeh, E., and Bloom, K. (2000). The role of the proteins Kar9 and Myo2 in orienting the mitotic spindle of budding yeast. *Curr. Biol.* **10**, 1497–1506.

Berlin, V., Styles, C. A., and Fink, G. R. (1990). BIK1, a protein required for microtubule function during mating and mitosis in *Saccharomyces cerevisiae*, colocalizes with tubulin. *J. Cell Biol.* **111**, 2573–2586.

Bicek, A. D., Tüzel, E., Kroll, D. M., and Odde, D. J. (2007). Analysis of microtubule curvature. *Methods Cell Biol.* **83**, 237–268.

Bieling, P., Laan, L., Schek, H., Munteanu, E. L., Sandblad, L., Dogterom, M., Brunner, D., and Surrey, T. (2007). Reconstitution of a microtubule plus-end tracking system in vitro. *Nature* **450**, 1100–1105.

Bienz, M. (2001). Spindles cotton on to junctions, APC and EB1. *Nat. Cell Biol.* **3**, E.

Brachmann, C. B., Davies, A., Cost, G. J., Caputo, E., Li, J., Hieter, P., and Boeke, J. D. (1998). Designer deletion strains derived from *Saccharomyces cerevisiae* S288C: A useful set of strains and plasmids for PCR-mediated gene disruption and other applications. *Yeast* **14**, 115–132.

Brangwynne, C. P., Koenderink, G. H., Barry, E., Dogic, Z., MacKintosh, F. C., and Weitz, D. A. (2007). Bending dynamics of fluctuating biopolymers probed by automated high-resolution filament tracking. *Biophys. J.* **93**(1), 346–359.

Busch, K. E., and Brunner, D. (2004). The microtubule plus end-tracking proteins mal3p and tip1p cooperate for cell-end targeting of interphase microtubules. *Curr. Biol.* **14**, 548–559.

Busch, K. E., Hayles, J., Nurse, P., and Brunner, D. (2004). Tea2p kinesin is involved in spatial microtubule organization by transporting tip1p on microtubules. *Dev. Cell* **6**, 831–843.

Cardinale, J., Rauch, A., Barral, Y., Szekely, G., and Sbalzarini, I. (2009). Bayesian image analysis with on-line confidence estimates and its application to microtubule tracking. In Biomedical Imaging: From Nano to Macro, 2009 ISBI '09 IEEE International Symposium on, pp. 1091–1094.

Carminati, J. L., and Stearns, T. (1997). Microtubules orient the mitotic spindle in yeast through dynein-dependent interactions with the cell cortex. *J. Cell Biol.* **138**(3), 629–641.

Carvalho, P., Gupta, M. L., Hoyt, M. A., and Pellman, D. (2004). Cell cycle control of kinesin-mediated transport of bik1 (CLIP-170) regulates microtubule stability and dynein activation. *Dev. Cell* **6**, 815–829.

Carvalho, P., Tirnauer, J. S., and Pellman, D. (2003). Surfing on microtubule ends. *Trends Cell Biol.* **13**(5), 229–237.

Caudron, F., Andrieux, A., Job, D., and Boscheron, C. (2008). A new role for kinesin-directed transport of bik1p (CLIP-170) in *Saccharomyces cerevisiae*. *J. Cell. Sci.* **121**, 1506–1513.

Chen, X. P., Yin, H., and Huffaker, T. C. (1998). The yeast spindle pole body component spc72p interacts with stu2p and is required for proper microtubule assembly. *J. Cell Biol.* **141**, 1169–1179.

Combs, C. A. (2010). Fluorescence microscopy: A concise guide to current imaging methods. *Curr. Protoc. Neurosci.* Chapter 2, 2.1.1-2.1.14

Costanzo, M., Baryshnikova, A., Bellay, J., Kim, Y., Spear, E. D., Sevier, C. S., Ding, H., Koh, J. L., Toufighi, K., Mostafavi, S., et al., (2010). The genetic landscape of a cell. *Science* **327**, 425–431.

Cuschieri, L., Miller, R., and Vogel, J. (2006). Gamma-tubulin is required for proper recruitment and assembly of kar9-bim1 complexes in budding yeast. *Mol. Biol. Cell* **17**, 4420–4434.

Danuser, G. (2001). Super-resolution microscopy using normal flow decoding and geometric constraints. *J. Microsc.* **204**, 136–149.

Danuser, G., Tran, P. T., and Salmon, E. D. (2000). Tracking differential interference contrast diffraction line images with nanometre sensitivity. *J. Microsc.* **198**(Pt 1), 34–53.

De Wulf, P., McAinsh, A. D., and Sorger, P. K. (2003). Hierarchical assembly of the budding yeast kinetochore from multiple subcomplexes. *Genes Dev.* **17**, 2902–2921.

DeZwaan, T. M., Ellingson, E., Pellman, D., and Roof, D. M. (1997). Kinesin-related KIP3 of *Saccharomyces cerevisiae* is required for a distinct step in nuclear migration. *J. Cell Biol.* **138**, 1023–1040.

Dorn, J. F., Danuser, G., and Yang, G. (2008). Computational processing and analysis of dynamic fluorescence image data. *Methods Cell Biol.* **85**, 497–538.

Dorn, J. F., Jaqaman, K., Rines, D. R., Jelson, G. S., Sorger, P. K., and Danuser, G. (2005). Yeast kinetochore microtubule dynamics analyzed by high-resolution three-dimensional microscopy. *Biophys. J.* **89**, 2835–2854.

Endow, S. A., Kang, S. J., Satterwhite, L. L., Rose, M. D., Skeen, V. P., and Salmon, E. D. (1994). Yeast kar3 is a minus-end microtubule motor protein that destabilizes microtubules preferentially at the minus ends. *EMBO J.* **13**, 2708–2713.

Eshel, D., Urrestarazu, L. A., Vissers, S., Jauniaux, J. C., van Vliet-Reedijk, J. C., Planta, R. J., and Gibbons, I. R. (1993). Cytoplasmic dynein is required for normal nuclear segregation in yeast. *Proc. Natl. Acad. Sci. U.S.A.* **90**, 11172–11176.

Fridman, V., Gerson-Gurwitz, A., Movshovich, N., Kupiec, M., and Gheber, L. (2009). Midzone organization restricts interpolar microtubule plus-end dynamics during spindle elongation. *EMBO Rep.* **10**, 387–393.

Gao, Y., and Kilfoil, M. L. (2009). Accurate detection and complete tracking of large populations of features in three dimensions. *Opt. Express* **17**, 4685–4704.

Gard, D. L., Becker, B. E., and Josh Romney, S. (2004). MAPping the eukaryotic tree of life: Structure, function, and evolution of the MAP215/dis1 family of microtubule-associated proteins. *Int. Rev. Cytol.* **239**, 179–272.

Gardner, M. K., Bouck, D. C., Paliulis, L. V., Meehl, J. B., O'Toole, E. T., Haase, J., Soubry, A., Joglekar, A. P., Winey, M., Salmon, E. D., et al., (2008a). Chromosome congression by kinesin-5 motor-mediated disassembly of longer kinetochore microtubules. *Cell* **135**, 894–906.

Gardner, M. K., Haase, J., Mythreye, K., Molk, J. N., Anderson, M., Joglekar, A. P., O'Toole, E. T., Winey, M., Salmon, E. D., Odde, D. J., et al., (2008b). The microtubule-based motor kar3 and plus end-binding protein bim1 provide structural support for the anaphase spindle. *J. Cell Biol.* **180**, 91–100.

Gardner, M. K., Hunt, A. J., Goodson, H. V., and Odde, D. J. (2008c). Microtubule assembly dynamics: New insights at the nanoscale. *Curr. Opin. Cell Biol.* **20**, 64–70.

Gardner, M. K., and Odde, D. J. (2010). Stochastic simulation and graphic visualization of mitotic processes. *Methods* **51**(2), 251–256.

Gardner, M. K., Odde, D. J., and Bloom, K. (2007). Hypothesis testing via integrated computer modeling and digital fluorescence microscopy. *Methods* **41**, 232–237.

Gardner, M. K., Pearson, C. G., Sprague, B. L., Zarzar, T. R., Bloom, K., Salmon, E. D., and Odde, D. J. (2005). Tension-dependent regulation of microtubule dynamics at kinetochores can explain metaphase congression in yeast. *Mol. Biol. Cell* **16**, 3764–3775.

Geissler, S., Pereira, G., Spang, A., Knop, M., Soues, S., Kilmartin, J., and Schiebel, E. (1996). The spindle pole body component Spc98p interacts with the gamma-tubulin-like Tub4p of *Saccharomyces cerevisiae* at the sites of microtubule attachment. *EMBO J.* **15**, 3899–3911.

Gigant, B., Curmi, P. A., Martin-Barbey, C., Charbaut, E., Lachkar, S., Lebeau, L., Siavoshian, S., Sobel, A., and Knossow, M. (2000). The 4 A X-ray structure of a tubulin: Stathmin-like domain complex. *Cell* **102**, 809–816.

Gräf, R., Rietdorf, J., and Zimmermann, T. (2005). Live cell spinning disk microscopy. *Adv. Biochem. Eng. Biotechnol.* **95**, 57–75.

Grava, S., Schaerer, F., Faty, M., Philippsen, P., and Barral, Y. (2006). Asymmetric recruitment of dynein to spindle poles and microtubules promotes proper spindle orientation in yeast. *Dev. Cell* **10**, 425–439.

Gregoretti, I. V., Margolin, G., Alber, M. S., and Goodson, H. V. (2006). Insights into cytoskeletal behavior from computational modeling of dynamic microtubules in a cell-like environment. *J. Cell Sci.* **119**(Pt 22), 4781–4788.

Gupta, M. L., Jr., Carvalho, P., Roof, D. M., and Pellman, D. (2006). Plus end-specific depolymerase activity of kip3, a kinesin-8 protein, explains its role in positioning the yeast mitotic spindle. *Nat. Cell Biol.* **8**, 913–923.

Gustafsson, M. G. (1999). Extended resolution fluorescence microscopy. *Curr. Opin. Struct. Biol.* **9**, 627–634.

Hadjidemetriou, S., Toomre, D., and Duncan, J. S. (2005). Segmentation and 3D reconstruction of microtubules in total internal reflection fluorescence microscopy (TIRFM). *Med. Image Comput. Comput. Assist. Interv.* **8**(Pt 1), 761–769.

Hadjidemetriou, S., Toomre, D., and Duncan, J. (2008). Motion tracking of the outer tips of microtubules. *Med. Image Anal.* **12**(6), 689–702.

Heemskerk, J.W.T., Westra, A. H., Linotte, P. M., Ligtvoet, K. M., Zbijewski, W., and Beekman, F. J. (2007). Front-illuminated versus back-illuminated photon-counting CCD-based gamma camera: Important consequences for spatial resolution and energy resolution. *Phys. Med. Biol.* **52**, 8.

Heintzmann, R., and Ficz, G. (2007). Breaking the resolution limit in light microscopy. *Methods Cell Biol.* **81**, 561–580.

Hell, S. W. (2003). Toward fluorescence nanoscopy. *Nat. Biotechnol.* **21**, 1347–1355.

Hinsch, J. (2007). Mating cameras to microscopes. *Methods Cell Biol.* **81**, 55–61.

Howard, J., and Hyman, A. A. (2003). Dynamics and mechanics of the microtubule plus end. *Nature* **422**, 753–758.

Howard, J., and Hyman, A. A. (2007). Microtubule polymerases and depolymerases. *Curr. Opin. Cell Biol.* **19**, 31–35.

Howard, J., and Hyman, A. A. (2009). Growth, fluctuation and switching at microtubule plus ends. *Nat. Rev. Mol. Cell Biol.* **10**, 569–574.

Huh, W. K., Falvo, J. V., Gerke, L. C., Carroll, A. S., Howson, R. W., Weissman, J. S., and O'Shea, E. K. (2003). Global analysis of protein localization in budding yeast. *Nature* **425**, 686–691.

Huisman, S. M., Bales, O. A., Bertrand, M., Smeets, M. F., Reed, S. I., and Segal, M. (2004). Differential contribution of Bud6p and Kar9p to microtubule capture and spindle orientation in *S. cerevisiae*. *J. Cell Biol.* **167**, 231–244.

Huisman, S. M., and Segal, M. (2005). Cortical capture of microtubules and spindle polarity in budding yeast —where's the catch? *J. Cell Sci.* **118**, 463–471.

(2006). 6 Temporal and spatio-temporal sampling problems. *J. ICRU* **6**, 65–76.

Janson, M. E., Loughlin, R., Loi?odice, I.., Fu, C., Brunner, D., Nédélec, F. J., and Tran, P. T. (2007). Crosslinkers and motors organize dynamic microtubules to form stable bipolar arrays in fission yeast. *Cell* **128**(2), 357–368.

Jaqaman, K., and Danuser, G. (2006). Linking data to models: Data regression. *Nat. Rev. Mol. Cell Biol.* **7**, 813–819.

Jaqaman, K., Dorn, J. F., Jelson, G. S., Tytell, J. D., Sorger, P. K., and Danuser, G. (2006). Comparative autoregressive moving average analysis of kinetochore microtubule dynamics in yeast. *Biophys. J.* **91**(6), 2312–2325.

Jaqaman, K., Dorn, J. F., Marco, E., Sorger, P. K., and Danuser, G. (2007). Phenotypic clustering of yeast mutants based on kinetochore microtubule dynamics. *Bioinformatics* **23**, 1666–1673.

Jaqaman, K., Loerke, D., Mettlen, M., Kuwata, H., Grinstein, S., Schmid, S. L., and Danuser, G. (2008). Robust single-particle tracking in live-cell time-lapse sequences. *Nat. Methods* **5**, 695–702.

Jiang, M., Ji, Q., and McEwen, B. (2004). Enhancement of microtubules in EM tomography. In Biomedical Imaging: Nano to Macro, 2004 IEEE International Symposium on, Vol. **1122,** pp. 1123–1126.

Jiang, M., Ji, Q., and McEwen, B. (2005). Automated extraction of microtubules and their plus-ends. In Application of Computer Vision, 2005 WACV/MOTIONS '05 Volume 1 Seventh IEEE Workshops on, pp. 336–341.

Karsenti, E., Nédélec, F., and Surrey, T. (2006). Modelling microtubule patterns. *Nat. Cell Biol.* **8**(11), 1204–1211.

Katz, W., Weinstein, B., and Solomon, F. (1990). Regulation of tubulin levels and microtubule assembly in *Saccharomyces cerevisiae*: Consequences of altered tubulin gene copy number. *Mol. Cell. Biol.* **10**, 5286–5294.

Kerssemakers, J., Ionov, L., Queitsch, U., Luna, S., Hess, H., and Diez, S. (2009). 3D nanometer tracking of motile microtubules on reflective surfaces. *Small* **5**(15), 1732–1737.

Kerssemakers, J. W., Munteanu, E. L., Laan, L., Noetzel, T. L., Janson, M. E., and Dogterom, M. (2006). Assembly dynamics of microtubules at molecular resolution. *Nature* **442**, 709–712.

Khmelinskii, A., Lawrence, C., Roostalu, J., and Schiebel, E. (2007). Cdc14-regulated midzone assembly controls anaphase B. *J. Cell Biol.* **177**(6), 981–993.

Kilmartin, J. V., and Adams, A. E. (1984). Structural rearrangements of tubulin and actin during the cell cycle of the yeast *Saccharomyces*. *J. Cell Biol.* **98**(3), 922–933.

Knaus, M., Cameroni, E., Pedruzzi, I., Tatchell, K., De Virgilio, C., and Peter, M. (2005). The Bud14p-Glc7p complex functions as a cortical regulator of dynein in budding yeast. *EMBO J.* **24**, 3000–3011.

Knop, M., Pereira, G., Geissler, S., Grein, K., and Schiebel, E. (1997). The spindle pole body component Spc97p interacts with the gamma-tubulin of *Saccharomyces cerevisiae* and functions in microtubule organization and spindle pole body duplication. *EMBO J.* **16**, 1550–1564.

Knop, M., Pereira, G., and Schiebel, E. (1999). Microtubule organization by the budding yeast spindle pole body. *Biol. Cell.* **91**, 291–304.

Korinek, W. S., Copeland, M. J., Chaudhuri, A., and Chant, J. (2000). Molecular linkage underlying microtubule orientation toward cortical sites in yeast. *Science* **287**(5461), 2257–2259.

Kosco, K. A., Pearson, C. G., Maddox, P. S., Wang, P. J., Adams, I. R., Salmon, E. D., Bloom, K., and Huffaker, T. C. (2001). Control of microtubule dynamics by Stu2p is essential for spindle orientation and metaphase chromosome alignment in yeast. *Mol. Biol. Cell* **12**, 2870–2880.

Kurihara, L. J., Beh, C. T., Latterich, M., Schekman, R., and Rose, M. D. (1994). Nuclear congression and membrane fusion: Two distinct events in the yeast karyogamy pathway. *J. Cell Biol.* **126**, 911–923.

Kusch, J., Meyer, A., Snyder, M. P., and Barral, Y. (2002). Microtubule capture by the cleavage apparatus is required for proper spindle positioning in yeast. *Genes Dev.* **16**, 1627–1639.

Lansbergen, G., and Akhmanova, A. (2006). Microtubule plus end: A hub of cellular activities. *Traffic* **7**, 499–507.

Lee, W. (2003). The role of the lissencephaly protein Pac1 during nuclear migration in budding yeast. *J. Cell Biol.* **160**, 355–364.

Lee, W. L., Kaiser, M. A., and Cooper, J. (2005). The offloading model for dynein function: Differential function of motor subunits. *J. Cell Biol.* **168**, 201–207.

Lee, L., Tirnauer, J. S., Li, J., Schuyler, S. C., and Liu, J. Y. (2000). Positioning of the mitotic spindle by a cortical-microtubule capture mechanism. *Science* **287**(5461), 2260–2262.

Leisner, C., Kammerer, D., Denoth, A., Britschi, M., Barral, Y., and Liakopoulos, D. (2008). Regulation of mitotic spindle asymmetry by SUMO and the spindle-assembly checkpoint in yeast. *Curr. Biol.* **18**, 1249–1255.

Li, J., Lee, W. L., and Cooper, J. (2005). NudEL targets dynein to microtubule ends through LIS1. *Nat. Cell Biol.* **7**, 686–690.

Li, Y. Y., Yeh, E., Hays, T., and Bloom, K. (1993). Disruption of mitotic spindle orientation in a yeast dynein mutant. *Proc. Natl. Acad. Sci. U.S.A.* **90**, 10096–10100.

Liakopoulos, D., Kusch, J., Grava, S., Vogel, J., and Barral, Y. (2003). Asymmetric loading of Kar9 onto spindle poles and microtubules ensures proper spindle alignment. *Cell* **112**, 561–574.

Lichtenstein, N., Geiger, B., and Kam, Z. (2003). Quantitative analysis of cytoskeletal organization by digital fluorescent microscopy. *Cytometry A* **54**(1), 8–18.

Lichtman, J. W., and Conchello, J.-A. (2005). Fluorescence microscopy. *Nat. Methods* **2**(12), 910–919.

Lin, H., de Carvalho, P., Kho, D., Tai, C. Y., Pierre, P., Fink, G. R., and Pellman, D. (2001). Polyploids require Bik1 for kinetochore-microtubule attachment. *J. Cell Biol.* **155**, 1173–1184.

Lowe, J., Li, H., Downing, K. H., and Nogales, E. (2001). Refined structure of alpha beta-tubulin at 3.5 A resolution. *J. Mol. Biol.* **313**, 1045–1057.

Ma, L., McQueen, J., Cuschieri, L., Vogel, J., and Measday, V. (2007). Spc24 and Stu2 promote spindle integrity when DNA replication is stalled. *Mol. Biol. Cell* **18**, 2805–2816.

Maddox, P. S., Bloom, K. S., and Salmon, E. D. (2000). The polarity and dynamics of microtubule assembly in the budding yeast *Saccharomyces cerevisiae*. *Nat. Cell Biol.* **2**, 36–41.

Maddox, P. S., Stemple, J. K., Satterwhite, L., Salmon, E. D., and Bloom, K. (2003). The minus end-directed motor Kar3 is required for coupling dynamic microtubule plus ends to the cortical shmoo tip in budding yeast. *Curr. Biol.* **13**, 1423–1428.

Maekawa, H., and Schiebel, E. (2004). Cdk1-Clb4 controls the interaction of astral microtubule plus ends with subdomains of the daughter cell cortex. *Genes Dev.* **18**, 1709–1724.

Maekawa, H., Usui, T., Knop, M., and Schiebel, E. (2003). Yeast Cdk1 translocates to the plus end of cytoplasmic microtubules to regulate bud cortex interactions. *EMBO J.* **22**, 438–449.

Manning, B. D., Barrett, J. G., Wallace, J. A., Granok, H., and Snyder, M. (1999). Differential regulation of the Kar3p kinesin-related protein by two associated proteins, Cik1p and Vik1p. *J. Cell Biol.* **144**, 1219–1233.

Marcus, A. I. (2007). Visualization of spindle behavior using confocal microscopy. *Methods Mol. Med.* **137**, 125–137.

Marschall, L. G., Jeng, R. L., Mulholland, J., and Stearns, T. (1996). Analysis of Tub4p, a yeast gamma-tubulin-like protein: Implications for microtubule-organizing center function. *J. Cell Biol.* **134**, 443–454.

Meluh, P. B., and Rose, M. D. (1990). KAR3, a kinesin-related gene required for yeast nuclear fusion. *Cell* **60**, 1029–1041.

Mendoza, M., Norden, C., Durrer, K., Rauter, H., Uhlmann, F., and Barral, Y. (2009). A mechanism for chromosome segregation sensing by the NoCut checkpoint. *Nat. Cell Biol.* **11**(4), 477–483.

Mennella, V., Rogers, G. C., Rogers, S. L., Buster, D. W., Vale, R. D., and Sharp, D. J. (2005). Functionally distinct kinesin-13 family members cooperate to regulate microtubule dynamics during interphase. *Nat. Cell Biol.* **7**, 235–245.

Middleton, K., and Carbon, J. (1994). KAR3-encoded kinesin is a minus-end-directed motor that functions with centromere binding proteins (CBF3) on an in vitro yeast kinetochore. *Proc. Natl. Acad. Sci. U.S.A.* **91**, 7212–7216.

Miller, R. K., Cheng, S. C., and Rose, M. D. (2000). Bim1p/Yeb1p mediates the Kar9p-dependent cortical attachment of cytoplasmic microtubules. *Mol. Biol. Cell* **11**, 2949–2959.

Miller, R. K., D'Silva, S., Moore, J., and Goodson, H. V. (2006). The CLIP-170 orthologue Bik1p and positioning the mitotic spindle in yeast. *Curr. Top. Dev. Biol.* **76**, 49–87.

Miller, R. K., Heller, K. K., Frisen, L., Wallack, D. L., Loayza, D., Gammie, A. E., and Rose, M. D. (1998). The kinesin-related proteins, Kip2p and Kip3p, function differently in nuclear migration in yeast. *Mol. Biol. Cell* **9**, 2051–2068.

Miller, R., Matheos, D., and Rose, M. D. (1999). The Cortical Localization of the Microtubule Orientation Protein, Kar9p, Is Dependent upon Actin and proteins required for polarization. *J. Cell Biol.* **144**(5), 963–975.

Miller, R., and Rose, M. D. (1998). Kar9p Is a Novel Cortical Protein Required for Cytoplasmic Microtubule Orientation in Yeast. *J. Cell Biol.* **140**(2), 377–390.

Molk, J. N., and Bloom, K. (2006). Microtubule dynamics in the budding yeast mating pathway. *J. Cell. Sci.* **119**, 3485–3490.

Molk, J. N., Salmon, E. D., and Bloom, K. (2006). Nuclear congression is driven by cytoplasmic microtubule plus end interactions in *S. cerevisiae*. *J. Cell Biol.* **172**, 27–39.

Moomaw, B. (2007). Camera technologies for low light imaging: Overview and relative advantages. *Methods Cell Biol.* **81**, 251–283.

Moore, J. K., D'Silva, S., and Miller, R. K. (2006). The CLIP-170 homologue Bik1p promotes the phosphorylation and asymmetric localization of Kar9p. *Mol. Biol. Cell* **17**, 178–191.

Moore, J. K., and Miller, R. K. (2007). The cyclin-dependent kinase Cdc28p regulates multiple aspects of Kar9p function in yeast. *Mol. Biol. Cell* **18**, 1187–1202.

Moore, J. K., Stuchell-Brereton, M. D., and Cooper, J. A. (2009). Function of dynein in budding yeast: Mitotic spindle positioning in a polarized cell. *Cell Motil. Cytoskeleton* **66**, 546–555.

Morrison, E. E. (2007). Action and interactions at microtubule ends. *Cell Mol. Life Sci.* **64**, 307–317.

Nédélec, F. (2002). Computer simulations reveal motor properties generating stable antiparallel microtubule interactions. J. Cell Biol. **158**(6), 1005–1015.

Neff, N. F., Thomas, J. H., Grisafi, P., and Botstein, D. (1983). Isolation of the beta-tubulin gene from yeast and demonstration of its essential function in vivo. Cell **33**, 211–219.

Nogales, E. (2001). Structural insight into microtubule function. Annu. Rev. Biophys. Biomol. Struct. **30**, 397–420.

Odde, D. J., and Buettner, H. M. (1995). Time series characterization of simulated microtubule dynamics in the nerve growth cone. Ann. Biomed. Eng. **23**, 268–286.

Odde, D. J., Buettner, H. M., and Cassimeris, L. (1996). Spectral analysis of microtubule assembly dynamics. AIChE J. **42**, 1434–1442.

Ohkura, H., Garcia, M. A., and Toda, T. (2001). Dis1/TOG universal microtubule adaptors—one MAP for all? J. Cell. Sci. **114**(Pt21), 3805–3812.

Page, B. D., Satterwhite, L. L., Rose, M. D., and Snyder, M. (1994). Localization of the Kar3 kinesin heavy chain-related protein requires the Cik1 interacting protein. J. Cell Biol. **124**, 507–519.

Pearson, C. G., Gardner, M. K., Paliulis, L. V., Salmon, E. D., Odde, D. J., and Bloom, K. (2006). Measuring nanometer scale gradients in spindle microtubule dynamics using model convolution microscopy. Mol. Biol. Cell **17**(9), 4069–4079.

Pinot, M., Chesnel, F., Kubiak, J., Arnal, I., Nedelec, F., and Gueroui, Z. (2009). Effects of confinement on the self-organization of microtubules and motors. Curr. Biol. **19**(11), 954–960.

Rasnik, I., French, T., Jacobson, K., and Berland, K. (2007). Electronic cameras for low-light microscopy. Methods Cell Biol. **81**, 219–249.

Ravelli, R. B., Gigant, B., Curmi, P. A., Jourdain, I., Lachkar, S., Sobel, A., and Knossow, M. (2004). Insight into tubulin regulation from a complex with colchicine and a stathmin-like domain. Nature **428**, 198–202.

Rout, M. P., and Kilmartin, J. V. (1990). Components of the yeast spindle and spindle pole body. J. Cell Biol. **111**, 1913–1927.

Sage, D., Neumann, F., Hediger, F., Gasser, S., and Unser, M. (2005). Automatic tracking of individual fluorescence particles: Application to the study of chromosome dynamics. IEEE Trans. Image Process **14** (9), 1372–1383.

Sagot, I., Klee, S. K., and Pellman, D. (2002). Yeast formins regulate cell polarity by controlling the assembly of actin cables. Nat. Cell Biol. **4**, 42–50.

Salmon, E. D., von Lackum, K., and Canman, J. C. (2005). Proper alignment and adjustment of the light microscope. Curr. Protoc. Microbiol. Chapter 2, Unit 2, A 1.

Saunders, W., Hornack, D., Lengyel, V., and Deng, C. (1997). The Saccharomyces cerevisiae kinesin-related motor Kar3p acts at preanaphase spindle poles to limit the number and length of cytoplasmic microtubules. J. Cell Biol. **137**, 417–431.

Sbalzarini, I. F., and Koumoutsakos, P. (2005). Feature point tracking and trajectory analysis for video imaging in cell biology. J. Struct. Biol. **151**(2), 182–195.

Schatz, P. J., Pillus, L., Grisafi, P., Solomon, F., and Botstein, D. (1986a). Two functional alpha-tubulin genes of the yeast Saccharomyces cerevisiae encode divergent proteins. Mol. Cell. Biol. **6**, 3711–3721.

Schatz, P. J., Solomon, F., and Botstein, D. (1986b). Genetically essential and nonessential alpha-tubulin genes specify functionally interchangeable proteins. Mol. Cell. Biol. **6**, 3722–3733.

Schek, H. T., 3rd, Gardner, M. K., Cheng, J., Odde, D. J., and Hunt, A. J. (2007). Microtubule assembly dynamics at the nanoscale. Curr. Biol. **17**, 1445–1455.

Schulz, R. B., and Semmler, W. (2008). Fundamentals of optical imaging. Handb. Exp. Pharmacol. (**185** Pt 1), 3–22.

Schuyler, S. C., and Pellman, D. (2001). Microtubule "plus-end-tracking proteins": The end is just the beginning. Cell **105**, 421–424.

Schwartz, K., Richards, K., and Botstein, D. (1997). BIM1 encodes a microtubule-binding protein in yeast. Mol. Biol. Cell **8**, 2677–2691.

Segal, M., and Bloom, K. (2001). Control of spindle polarity and orientation in Saccharomyces cerevisiae. Trends Cell Biol. **11**, 160–166.

Segal, M., Bloom, K., and Reed, S. I. (2000). Bud6 directs sequential microtubule interactions with the bud tip and bud neck during spindle morphogenesis in Saccharomyces cerevisiae. Mol. Biol. Cell **11**, 3689–3702.

Shariff, A., Murphy, R. F., and Rohde, G. K. (2010). A generative model of microtubule distributions, and indirect estimation of its parameters from fluorescence microscopy images. Cytometry A **77**, 457–466.

Shaw, S. L., Maddox, P., Skibbens, R. V., Yeh, E., Salmon, E. D., and Bloom, K. (1998). Nuclear and spindle dynamics in budding yeast. Mol. Biol. Cell **9**(7), 1627–1631.

Shaw, S. L., Yeh, E., Maddox, P., Salmon, E. D., and Bloom, K. (1997). Astral microtubule dynamics in yeast: A microtubule-based searching mechanism for spindle orientation and nuclear migration into the bud. J. Cell Biol. **139**(4), 985–994.

Sheeman, B., Carvalho, P., Sagot, I., Geiser, J., Kho, D., Hoyt, M. A., and Pellman, D. (2003). Determinants of S. cerevisiae dynein localization and activation: Implications for the mechanism of spindle positioning. Curr. Biol. **13**, 364–372.

Shelden, E., and Knecht, D. A. (1998). Reconstruction and display of curvilinear objects from optical section data using 3-D curve fitting algorithms. J. Microsc. **191**(Pt 1), 97–107.

Smal, I., Draegestein, K., Galjart, N., Niessen, W., and Meijering, E. (2007). Rao-Blackwellized marginal particle filtering for multiple object tracking in molecular bioimaging. Inf. Process Med. Imaging **20**, 110–121.

Smal, I., Draegestein, K., Galjart, N., Niessen, W., and Meijering, E. (2008a). Particle filtering for multiple object tracking in dynamic fluorescence microscopy images: Application to microtubule growth analysis. IEEE Trans. Med. Imaging **27**(6), 789–804.

Smal, I., Meijering, E., Draegestein, K., Galjart, N., Grigoriev, I., Akhmanova, A., van Royen, M. E., Houtsmuller, A. B., and Niessen, W. (2008b). Multiple object tracking in molecular bioimaging by Rao-Blackwellized marginal particle filtering. Med. Image Anal. **12**(6), 764–777.

Sobel, S. G., and Snyder, M. (1995). A highly divergent gamma-tubulin gene is essential for cell growth and proper microtubule organization in Saccharomyces cerevisiae. J. Cell Biol. **131**, 1775–1788.

Soues, S., and Adams, I. R. (1998). SPC72: A spindle pole component required for spindle orientation in the yeast Saccharomyces cerevisiae. J. Cell. Sci. **111**(Pt 18), 2809–2818.

Spang, A., Geissler, S., Grein, K., and Schiebel, E. (1996). gamma-Tubulin-like Tub4p of Saccharomyces cerevisiae is associated with the spindle pole body substructures that organize microtubules and is required for mitotic spindle formation.J. Cell Biol. **134**, 429–441.

Sprague, B. L., Pearson, C. G., Maddox, P. S., Bloom, K. S., Salmon, E. D., and Odde, D. J. (2003). Mechanisms of microtubule-based kinetochore positioning in the yeast metaphase spindle. Biophys. J. **84** (6), 3529–3546.

Spring, K. R. (2007). Cameras for digital microscopy. Methods Cell Biol. **81**, 171–186.

Sproul, L. R., Anderson, D. J., Mackey, A. T., Saunders, W. S., and Gilbert, S. P. (2005). Cik1 targets the minus-end kinesin depolymerase kar3 to microtubule plus ends. Curr. Biol. **15**, 1420–1427.

Stelzer, E.H.K. (1999). "Practical Limits to Resolution in Fluorescence Light Microscopy."Cold Spring Harbor Laboratory PressNew York.

Straight, A. F., Marshall, W. F., Sedat, J. W., and Murray, A. W. (1997). Mitosis in living budding yeast: Anaphase A but no metaphase plate. Science **277**, 574–578.

Straight, A. F., Sedat, J. W., and Murray, A. W. (1998). Time-lapse microscopy reveals unique roles for kinesins during anaphase in budding yeast. J. Cell Biol. **143**, 687–694.

Tarassov, K., Messier, V., Landry, C. R., Radinovic, S., Serna Molina, M. M., Shames, I., Malitskaya, Y., Vogel, J., Bussey, H., and Michnick, S. W. (2008). An in vivo map of the yeast protein interactome. Science **320**, 1465–1470.

Taylor, D. L., and Salmon, E. D. (1989). Basic fluorescence microscopy. Methods Cell Biol. **29**, 207–237.

Thomann, D., Dorn, J., Sorger, P. K., and Danuser, G. (2003). Automatic fluorescent tag localization II: Improvement in super-resolution by relative tracking. J. Microsc. **211**, 230–248.

Thomann, D., Rines, D. R., Sorger, P. K., and Danuser, G. (2002). Automatic fluorescent tag detection in 3D with super-resolution: application to the analysis of chromosome movement. J. Microsc. **208**, 49–64.

Tirnauer, J. S., and Bierer, B. E. (2000). EB1 proteins regulate microtubule dynamics, cell polarity, and chromosome stability. J. Cell Biol. **149**, 761–766.

Tirnauer, J. S., Grego, S., Salmon, E. D., and Mitchison, T. J. (2002). EB1-microtubule interactions in Xenopus egg extracts: Role of EB1 in microtubule stabilization and mechanisms of targeting to microtubules. Mol. Biol. Cell **13**, 3614–3626.

Tirnauer, J. S., O'Toole, E., Berrueta, L., Bierer, B. E., and Pellman, D. (1999). Yeast Bim1p promotes the G1-specific dynamics of microtubules. J. Cell Biol. **145**, 993–1007.

Tischer, C., Brunner, D., and Dogterom, M. (2008). Chapter 20: Automated spatial mapping of microtubule catastrophe rates in fission yeast. Methods Cell Biol. **89**, 521–538.

Trueheart, J., Boeke, J. D., and Fink, G. R. (1987). Two genes required for cell fusion during yeast conjugation: Evidence for a pheromone-induced surface protein. Mol. Cell. Biol. **7**, 2316–2328.

Tytell, J. D., and Sorger, P. K. (2006). Analysis of kinesin motor function at budding yeast kinetochores. J. Cell Biol. **172**, 861–874.

Usui, T., Maekawa, H., Pereira, G., and Schiebel, E. (2003). The XMAP215 homologue Stu2 at yeast spindle pole bodies regulates microtubule dynamics and anchorage. EMBO J. **22**, 4779–4793.

Van Breugel, M., Drechsel, D., and Hyman, A. (2003). Stu2p, the budding yeast member of the conserved Dis1/XMAP215 family of microtubule-associated proteins is a plus end-binding microtubule destabilizer. J. Cell Biol. **161**(2), 359–369.

Varga, V., Helenius, J., Tanaka, K., Hyman, A., Tanaka, T. U., and Howard, J. (2006). Yeast kinesin-8 depolymerizes microtubules in a length-dependent manner. Nat. Cell Biol. **8**, 957–962.

Varga, V., Leduc, C., Bormuth, V., Diez, S., and Howard, J. (2009). Kinesin-8 motors act cooperatively to mediate length-dependent microtubule depolymerization. Cell **138**, 1174–1183.

Vogel, J., Drapkin, B., Oomen, J., Beach, D., Bloom, K., and Snyder, M. (2001). Phosphorylation of gamma-tubulin regulates microtubule organization in budding yeast. Dev. Cell **1**(5), 621–631.

Vogel, J., and Snyder, M. (2000). gamma-Tubulin of budding yeast. Curr. Top. Dev. Biol. **49**, 75–104.

Wang, P. J., and Huffaker, T. C. (1997). Stu2p: A microtubule-binding protein that is an essential component of the yeast spindle pole body. J. Cell Biol. **139**, 1271–1280.

Waters, J. C. (2007). Live-cell fluorescence imaging. Methods Cell Biol. **81**, 115–140.

Wigge, P. A., Jensen, O. N., Holmes, S., Soues, S., Mann, M., and Kilmartin, J. V. (1998). Analysis of the Saccharomyces spindle pole by matrix-assisted laser desorption/ionization (MALDI) mass spectrometry. J. Cell Biol. **141**, 967–977.

Winzeler, E. A., Shoemaker, D. D., Astromoff, A., Liang, H., Anderson, K., Andre, B., Bangham, R., Benito, R., Boeke, J. D., Bussey, H., et al., (1999). Functional characterization of the S. cerevisiae genome by gene deletion and parallel analysis. Science **285**(5429), 901–906.

Wolf, D. E. (2007). Fundamentals of fluorescence and fluorescence microscopy. Methods Cell Biol. **81**, 63–91.

Wolyniak, M. J., Blake-Hodek, K., Kosco, K., Hwang, E., You, L., and Huffaker, T. C. (2006). The regulation of microtubule dynamics in Saccharomyces cerevisiae by three interacting plus-end tracking proteins. Mol. Biol. Cell **17**, 2789–2798.

Woodruff, J. B., Drubin, D. G., and Barnes, G. (2009). Dynein-driven mitotic spindle positioning restricted to anaphase by She1p inhibition of dynactin recruitment. Mol. Biol. Cell **20**, 3003–3011.

Wu, X., Xiang, X., and Hammer, J. A. (2006). Motor proteins at the microtubule plus-end. Trends Cell Biol. **16**(3), 135–143.

Yamamoto, A., and Hiraoka, Y. (2003). Cytoplasmic dynein in fungi: Insights from nuclear migration. J. Cell Sci. **116**, 4501–4512.

Yeh, E., Skibbens, R. V., Cheng, J. W., Salmon, E. D., and Bloom, K. (1995). Spindle dynamics and cell cycle regulation of dynein in the budding yeast, Saccharomyces cerevisiae. J. Cell Biol. **130**, 687–700.

Yeh, E., Yang, C., Chin, E., Maddox, P., Salmon, E. D., Lew, D. J., and Bloom, K. (2000). Dynamic positioning of mitotic spindles in yeast: Role of microtubule motors and cortical determinants. Mol. Biol. Cell **11**, 3949–3961.

Yin, H., Pruyne, D., Huffaker, T. C., and Bretscher, A. (2000). Myosin V orientates the mitotic spindle in yeast. Nature **406**, 1013–1015.

Zaichick, S. V., Metodiev, M. V., Nelson, S. A., Durbrovskyi, O., Draper, E., Cooper, J. R., and Stone, D. E. (2009). The Mating-specific G{alpha} Interacts with a Kinesin-14 and Regulates Pheromone-induced Nuclear Migration in Budding Yeast. Mol. Biol. Cell **20**(12), 2820–2830.

CHAPTER 17

Imaging and Analysis of the Microtubule Cytoskeleton in *Giardia*

Scott C. Dawson *and* Susan A. House

Department of Microbiology, One Shields Avenue, UC Davis, Davis, CA 95616

Abstract

Giardia intestinalis, a common parasitic protist, possesses a complex microtubule cytoskeleton critical for cellular function and transitioning between the cyst and

978-0-12-381349-7
DOI: 10.1016/S0091-679X(10)97017-9

trophozoite life cycle stages. The giardial microtubule cytoskeleton is comprised of highly dynamic and stable structures. Novel microtubule structures include the ventral disc that is essential for the parasite's attachment to the intestinal villi to avoid peristalsis. The completed *Giardia* genome combined with new molecular genetic tools and live imaging will aid in the characterization and analysis of cytoskeletal dynamics in *Giardia*. Fundamental areas of giardial cytoskeletal biology remain to be explored and knowledge of the molecular mechanisms of cytoskeletal functioning is needed to better understand *Giardia's* unique biology and pathogenesis.

I. Introduction

The zoonotic parasite *Giardia intestinalis* is the causative agent of giardiasis, the most prevalent protozoan intestinal infection in the US and worldwide. Outbreaks of acute giardiasis frequently occur in areas with inadequate water treatment, resulting in several hundred million cases of malabsorptive diarrhea each year (Savioli *et al.*, 2006). In developed countries, acute giardiasis is most often found in travelers and among children at day care centers. Chronic giardiasis is a serious problem for children in developing countries, where it contributes to malnutrition, anemia, poor cognitive function, and failure to thrive.

Giardia has a complex microtubule (MT) cytoskeleton that is critical for growth, movement, and development during its two life cycle stages (Adam, 2001; Elmendorf *et al.*, 2003). In the environment, *Giardia* persists as a dormant, infectious cyst (Adam, 2001; Gillin *et al.*, 1996). Cysts are ingested by animals (humans or other mammals) from contaminated water or food and excyst in the small intestine of the animal host to become the trophozoite or flagellated form. The trophozoite has a flattened teardrop shape (~15 μm long by 5 μm wide and 5 μm thick) with a complex three-dimensional (3D) cytoskeletal ultrastructure. Using a specialized structure, the ventral disc, the trophozoite attaches to the intestinal villi to avoid peristalsis. It is in the small intestine that the trophozoite thrives and multiplies. Trophozoites that reach the colon encyst based on environmental cues and are released to infect new hosts (Roxstrom-Lindquist *et al.*, 2006).

Giardia is well understood in terms of disease (Adam, 2001; Savioli *et al.*, 2006), yet little is known about the function, assembly, and division of the many complex cytoskeletal structures essential to the parasite's life cycle. *Giardia* is bilaterally symmetrical and the cytoskeleton establishes anterior–posterior, ventral–dorsal, and left–right asymmetry. The elaborate MT cytoskeleton is also essential for motility, attachment, intracellular transport, cell division, and encystation/excystation, all of which are required for maintenance of infection. The study of the giardial cytoskeleton is thus both clinically relevant and informs basic cell biology, molecular biology, and cellular evolution (Elmendorf *et al.*, 2003).

II. Structural Elements of the Giardial MT Cytoskeleton

The MT cytoskeleton is composed of elements common to flagellated protists (eight flagella and two mitotic spindles), as well as stable and dynamic structures (the median body, ventral disc, funis, and axoneme-associated elements) that are unique to *Giardia*. Some homologs of MT-associated proteins have been identified in the *Giardia* genome (Morrison *et al.*, 2007) or otherwise characterized (see Table I); however, many specific proteins comprising MT-associated structures remain to be defined (Feely *et al.*, 1990). Ultimately, molecular genetic analysis of both structural and regulatory cytoskeletal proteins, combined with visualization of live cytoskeletal dynamics, will be pivotal in assessing dynamic processes in *Giardia* including attachment, cell division, intracellular protein trafficking, and encystation/excystation. While the majority of efforts to study the *Giardia* cytoskeleton have been cytological, live imaging is a critical step in understanding the molecular mechanisms of giardial cytoskeletal processes. Genomic and/or proteomic-based approaches, combined with new reverse genetic tools to generate dominant-negative mutants (Dawson *et al.*, 2007; Gaechter *et al.*, 2008), or antisense (Touz *et al.*, 2005), and morpholino-based knockdowns (Carpenter and Cande, 2009) will allow the identification of novel structural components and uncover the mechanisms of *Giardia*'s cytoskeletal dynamics (Ortega-Pierres *et al.*, 2009). A brief overview of the primary MT structures follows (see Fig. 1).

Giardia attaches to substrates using the **ventral disc**, which consists of three major structural elements: (1) the right-handed MT spiral array; (2) trilaminar microribbons, structures that attach perpendicularly to the MT spiral and extend into the cytoplasm; and (3) crossbridge structures that horizontally link the microribbons (Crossley and Holberton, 1983; 1985; Feely *et al.*, 1982; Holberton, 1973, 1981; Holberton and Ward, 1981). Using the ventral disc, trophozoites orient ventral side "down" to substrates via an undefined mechanism (Holberton, 1973, 1974) that might involve suction (Hansen *et al.*, 2006). The ventral disc is surrounded by the lateral crest, a fibrillar structure of unknown composition with purported contractile functions (Kulda and Nohynkova, 1995).

The median body is an enigmatic, semi-organized, and nonmembrane-bound MT array of unknown function (Piva and Benchimol, 2004) present on the dorsal side of trophozoites (Elmendorf *et al.*, 2003). Median body MTs are dynamic in interphase (Dawson *et al.*, 2007 a, b; Sagolla *et al.*, 2006), disappear completely after mitosis, and reappear after cytokinesis. Median body MTs may play a critical role in the giardial life cycle (Kabnick and Peattie, 1990), possibly serving as a reservoir of tubulin subunits for duplicating MT structures such as daughter ventral discs (Brugerolle, 1975; Feely *et al.*, 1990).

The **funis** is composed of sheets of MTs surrounding the caudal axonemes. Bands of linked MTs fan out laterally at the point of emergence of the ventral axonemes (Benchimol *et al.*, 2004). The funis is suggested either to have a structural role in maintaining giardial cell shape or to generate flexion of the posterior "tail" region (Benchimol *et al.*, 2004).

The eight **flagella** are organized into four bilaterally symmetrical pairs termed the anterior, the caudal, the posterolateral, and the ventral flagella. Each axoneme has long, cytoplasmic regions as well as an external membrane-bound portion; the length ratio of the cytoplasmic region to the membrane-bound portion varies between each flagellar

Table I
Giardial homologs of MT-associated proteins

	GiardiaDB	Gene name	Family	PFAM	Molecular function	References
Microtubules (MTs) (5)	GL50803_103676	alpha-tubulin 1	Tubulin	PF00091	MT metabolism	Morrison *et al.* (2007)
	GL50803_112079	alpha-tubulin 2	Tubulin	PF00091	MT metabolism	Morrison *et al.* (2007)
	GL50803_101291	beta-tubulin 1	Tubulin	PF00091	MT metabolism	Morrison *et al.* (2007)
	GL50803_136021	beta-tubulin 2	Tubulin	PF00091	MT metabolism	Morrison *et al.* (2007)
	GL50803_136020	beta-tubulin 3	Tubulin	PF00091	MT metabolism	Morrison *et al.* (2007)
Tubulin modification (7)	GL50803_95661	tubulin tyrosine ligase	Tubulin tyrosine ligase	PF03133		Morrison *et al.* (2007)
	GL50803_14498	tubulin tyrosine ligase	Tubulin tyrosine ligase	PF03133		Morrison *et al.* (2007)
	GL50803_8592	tubulin tyrosine ligase	Tubulin tyrosine ligase	PF03133		Morrison *et al.* (2007)
	GL50803_10382	tubulin tyrosine ligase	Tubulin tyrosine ligase	PF03133		Morrison *et al.* (2007)
	GL50803_9272	tubulin tyrosine ligase	Tubulin tyrosine ligase	PF03133		Morrison *et al.* (2007)
	GL50803_8456	tubulin tyrosine ligase	Tubulin tyrosine ligase	PF03133		Morrison *et al.* (2007)
	GL50803_10801	tubulin tyrosine ligase	Tubulin tyrosine ligase	PF03133		Morrison *et al.* (2007)
MT dynamics/ regulators (10)	GL50803_14373	dynamin	Dynamin	PF00350		Morrison *et al.* (2007)
	GL50803_14048	EBI	EBI	PF03271		Dawson *et al.* (2007)
	GL50803_96399	xmap215	XMAP215 family	none		Morrison *et al.* (2007)
	GL50803_91480	stu2	Stu2 family	none		Morrison *et al.* (2007)
	GL50803_11953	katanin (p80)	WD domain, G-beta repeat	PF00400		Morrison *et al.* (2007)
	GL50803_15368	katanin (p60)	AAA ATPase family	PF00004		Morrison *et al.* (2007)
	GL50803_16535	tubulin-specific chaperone E	cap_gly MT-binding domain	PF01302		Morrison *et al.* (2007)
	GL50803_5374	tubulin-specific chaperone B	cap_gly MT-binding domain	PF01302		Morrison *et al.* (2007)
	GL50803_16893	tip elongation aberrant protein 1	Kelch2 motif	PF0646		Morrison *et al.* (2007)
	GL50803_15054	kelch-repeat-containing protein	Kelch2 motif	PF0646		Morrison *et al.* (2007)
Gamma Turc/ Tusc complex (3)	GL50803_17429	gcp-2	Spc97_Spc98	PF04130	MT metabolism	Morrison *et al.* (2007)
	GL50803_12057	gcp-3	Spc97_Spc98	PF04130	MT metabolism	Morrison *et al.* (2007)
	GL50803_114218	gamma-tubulin	Tubulin	PF00091	MT metabolism	Morrison *et al.* (2007)
Basal-body associated (10)	GL50803_104685	caltractin	Caltractin	PF00036	Signalling	Meng *et al.* (1996)
	GL50803_6744	centrin	Centrin	PF00036	Signalling	Belhadri *et al.* (1995)
	GL50803_5462	delta-tubulin	Tubulin	PF00091	MT metabolism	Morrison *et al.* (2007)
	GL50803_6336	epsilon-tubulin	Tubulin	PF00091	MT metabolism	Morrison *et al.* (2007)
	GL50803_5167	VFL3	Centriole positioning	none		Merchant *et al.* (2007)
	GL50803_15956	FAP52	Basal body proteome BUG14	SSF50978	Protein interaction	Keller *et al.* (2005)
	GL50803_15455	PACRGI	Basal body proteome as BUG21 ; PACRG parkin co-regulated gene. E04F6.2 like protein	PF10274		Keller *et al.* (2005)

	GL50803_13372	FAP45	BUG28	SSF50978		Keller *et al.* (2005)
	GL50803_32375	POC18	Basal body proteome	none	Protein interaction	Keller *et al.* (2005)
	GL50803_33762	POC1	Basal body proteome WD repeat protein	SSF50978	Protein interaction	Keller *et al.* (2005)
BBSome (5)	GL50803_8738	BBS1		none		Morrison *et al.* (2007)
	GL50803_23934	BBS2		none		Morrison *et al.* (2007)
	GL50803_10529	BBS4	TPR_1 Tetratricopeptide repeat	PF01515	Protein interaction	Morrison *et al.* (2007)
	GL50803_8146	BBS5	DUF1448 domain of unknown function	PF07289	Protein interaction	Morrison *et al.* (2007)
	GL50803_8508	BBS8	TPR_1 Tetratricopeptide repeat	PF01515	Protein interaction	Morrison *et al.* (2007)
Axoneme structure (30)						
Axonemai dyneins (IDAs) (5)	GL50803_100906	IAD-1 alpha	IAD-1 alpha dynein heavy chain (DHC) family	PF03028	Flagellar structure	Wickstead *et al.* (2007)
	GL50803_94440	IAD-1 beta	IAD-1beta dynein heavy chain (DHC) family	PF03028	Flagellar structure	Wickstead *et al.* (2007)
	GL50803_40496	IAD-4	IAD-4 dynein heavy chain (DHC) family	PF03028	Flagellar structure	Wickstead *et al.* (2007)
	GL50803_37985	IAD-4	IAD-4 dynein heavy chain (DHC) family, partial	PF03028	Flagellar structure	Wickstead *et al.* (2007)
	GL50803_111950	IAD-5	IAD-5 dynein heavy chain (DHC) family	PF03028	Flagellar structure	Wickstead *et al.* (2007)
Axonemal dyneins (ODAs) (2)	GL50803_17265	OAD-alpha	OAD-alpha dynein heavy chain (DHC) family	PF03028	Flagellar structure	Wickstead *et al.* (2007)
	GL50803_17243	OAD-beta	OAD-beta dynein heavy chain (DHC) family	PF03028	Flagellar structure	Wickstead *et al.* (2007)
Radial spokes (3)	GL50803_16450	rsp3	Radial spoke protein 3	PF06098	Flagellar structure	Morrison *et al.* (2007)
	GL50803_17278	RSP9	Radial spoke protein 9; A subunit in the radial spoke head; (pf17)	none	Flagellar structure	Morrison *et al.* (2007)
	GL50803_114462	axonemal p66 (RSP6)	Outer dynein arm-docking complex subunit 2 (ODA-DC 2)	none	Flagellar structure	Morrison *et al.* (2007)
Dynein regulatory complex (1)	GL50803_16540	PF2	Component of dynein regulatory complex (DRC) of flagellar axoneme; trypanin	PD936484	Flagellar structure	Morrison *et al.* (2007)

Table I
(*Continued*)

	GiardiaDB	Gene name	Family	PFAM	Molecular function	References
Central pair (4)	GL50803_16500	PF20	Central Pair WD-repeat protein of the central pair; associates with the intermicrotubule bridge.	PF00400	Flagellar structure	Morrison *et al.* (2007)
	GL50803_16202	PF16	Central pair-associated protein	PF00514	Flagellar structure	Morrison *et al.* (2007)
	GL50803_137712	HY3 (FAP74)	(HYD3) Similar to mouse hydrocephaly protein hydin HY3	SSF52540	Flagellar structure	Morrison *et al.* (2007)
	GL50803_14568	PPI	Ser/Thr protein phosphatase PP 1-alpha 2 catalytic subunit	SSF56300	Signalling	Morrison *et al.* (2007)
Axoneme-associated (15)	GL50803_11867	RIB43a	Associated with protofilament ribbons of flagellar microtubules	PF05914	Flagellar structure	Morrison *et al.* (2007)
	GL50803_41512	RIB72	Novel component of the ribbon compartment of flagellar MTs	PS51336	Flagellar structure	Morrison *et al.* (2007)
	GL50803_16263	DIP13	Deflagellation inducible protein; Sjogren's syndrome nuclear autoantigen 1.	PD968187	MT metabolism	Morrison *et al.* (2007)
	GL50803_102248	MBO2	Move backward only mutant defective in ciliary waveform.	PD936484	Protein interaction	Morrison *et al.* (2007)
	GL50803_14048	EBI	EBI	PF03271	MT metabolism	Dawson *et al.* (2007)
	GL50803_5333	calmodulin	EF-hand domain	SSF47473	Signalling	Morrison *et al.* (2007)
	GL50803_7439	PP2A	Ser/Thr phosphatase 2A, 65kDa reg sub A; ARM repeat domain	SSF48371	Signalling	Morrison *et al.* (2007)
	GL50803_14004	long-flagella protein LF4	Protein kinase domain (MAP kinase)	PF00069	Flagellar regulation	Morrison *et al.* (2007)
	GL50803_137716	GASP-180	Ankyrin repeat family	PF00023	Protein interaction	Elmendorf *et al.* (2005)
	GL50803_5649	alpha10-giardin	Annexin	PF00191	Protein interaction	Weiland *et al.* (2005)
	GL50803_15097	alpha14-giardin	Annexin	PF00191	Protein interaction	Weiland *et al.* (2005)
	GL50803_15101	alpha17-giardin	Annexin	PF00191	Protein interaction	Weiland *et al.* (2005)
	GL50803_7796	alpha2-giardin	Annexin	PF00191	Protein interaction	Weiland *et al.* (2005)
	GL50803_7797	alpha5-giardin	Annexin	PF00191	Protein interaction	Weiland *et al.* (2005)
	GL50803_103437	alpha9-giardin	Annexin	PF00191	Protein interaction	Weiland *et al.* (2005)
Intraflagellar transport, IFT (18)						
IFT complex A (2)	GL50803_17251	IFT140	Intraflagellar transport protein IFT140	none	Flagellar transport	Hoeng *et al.* (2008)
	GL50803_16547	IFT122	Intraflagellar transport protein IFT122	none	Flagellar transport	Briggs *et al.* (2004)

IFT complex B (8)	GL50803_17105	IFT172	Intraflagellar transport protein IFT172	none	Flagellar transport	Briggs *et al.* (2004)
	GL50803_7664	IFT46	Intraflagellar transport protein IFT46; FAP32	none	Flagellar transport	Briggs *et al.* (2004)
	GL50803_112963	IFT52	Intraflagellar transport protein IFT52; osm-6.	none	Flagellar transport	Briggs *et al.* (2004)
	GL50803_14713	IFT57	Intraflagellar transport protein IFT57	none	Flagellar transport	Briggs *et al.* (2004)
	GL50803_9750	IFT74/72	Intraflagellar transport protein IFT74/72	none	Flagellar transport	Briggs *et al.* (2004)
	GL50803_17223	IFT80	Intraflagellar transport protein IFT80; WD domain, G-beta repeat	PF00400	Flagellar transport	Briggs *et al.* (2004)
	GL50803_15428	IFT81	Intraflagellar transport protein IFT81	none	Flagellar transport	Hoeng *et al.* (2008)
	GL50803_16660	IFT88	Intraflagellar transport protein IFT88	none	Flagellar transport	Briggs *et al.* (2004)
IFT complex B-associated (4)	GL50803_87202	DYF-1	PR protein; TPRS (FAP259)	SSF48452	Protein interaction	Merchant et al. (2007)
	GL50803_16707	DYF-3	D. rerio cystic kidney disease gene qilin; FAP22; Clusterin-associated protein	PF10234	Protein interaction	Merchant et al. (2007)
	GL50803_9098	DYF-11	MT-Associated TRAF3-Interacting Protein (FAP116)	PF10243	Trafficking	Merchant et al. (2007)
	GL50803_16375	DYF-13	(FBB2) required for ciliogenesis in C. *elegans.*	SSF48452	Protein interaction	Merchant et al. (2007)
IFT motors (4)	GL50803_114885	KAP	Kinesin-associated protein; non-motor subunit of kinesin-II complex	PF05804	Flagellar transport	Morrison *et al.* (2007)
	GL50803_16456	GiKIN2a	Kinesin-2	PF00225	Flagellar transport	Hoeng *et al.* (2008)
	GL50803_17333	GiKIN2b	Kinesin-2	PF00225	Flagellar transport	Hoeng *et al.* (2008)
	GL50803_93736	cytoDHC-1b	cytoDHC 1b, putative IFT cytoplasmic dynein 1b family	PF03028	Flagellar transport	Wickstead *et al.* (2007)
Ventral disc-associated (8)	GL50803_4812	beta-giardin	SF-assemblin	PF06705		Baker *et al.* (1988)
	GL50803_17230	gamma-giardin		none		Nohria *et al.* (1992)
	GL50803_86676	delta-giardin	SF-assemblin	PF06705		Elmendorf *et al.* (2001)
	GL50803_4410	SALP-1	SF-assemblin	PF06705		Palm *et al.* (2003)
	GL50803_7796	alpha2-giardin	Annexin	PF00191		Weiland *et al.* (2005)
	GL50803_11683	alpha3-giardin	Annexin	PF00191		Weiland *et al.* (2005)
	GL50803_7797	alpha5-giardin	Annexin	PF00191		Weiland *et al.* (2005)
	GL50803_15101	alpha17-giardin	Annexin	PF00191		Weiland *et al.* (2005)
Median body-associated (2)	GL50803_16343	median body protein	MBP	none		Holberton *et al.* (1995)
	GL50803_14048	EBI	EBI	PF03271		Dawson *et al.* (2007)

Table I
(Continued)

	GiardiaDB	Gene name	Family	PFAM	Molecular function	References
Spindle, kinetochore-associated (3)	GL50803_15248	Bub2	TBC domain	PF00566		Morrison *et al.* (2007)
	GL50803_100955	mad2	Mitotic checkpoint protein	PF00557		Morrison *et al.* (2007)
	GL50803_14048	EBI	EBI	PF03271		Dawson *et al.* (2007)
MT motor proteins						
Other Flagellar Kinesins (3)	GL50803_16945	GiKIN13	Kinesin-13	PF00225	MT metabolism	Dawson *et al.* (2007)
	GL50803_10137	GiKIN9a	Kinesin-9	PF00225	MT metabolism	Wickstead *et al.* (2006)
	GL50803_6404	GiKIN9b	Kinesin-9	PF00225	MT metabolism	Wickstead *et al.* (2006)
	GL50803_13825	GiKIN1	Kinesin-1	PF00225		
Other kinesins (21)	GL50803_6262	GiKIN3a	Kinesin-3	PF00225		
	GL50803_102101	GiKIN3b	Kinesin-3	PF00225		
	GL50803_112846	GiKIN3c	Kinesin-3	PF00225		
	GL50803_16650	GiKIN4	Kinesin-4	PF00225		
	GL50803_16425	GiKIN5	Kinesin-5	PF00225		
	GL50803_102455	GiKIN6a	Kinesin-6	PF00225		
	GL50803_15134	GiKIN6b	Kinesin-6	PF00225		
	GL50803_15962	GiKIN7	Kinesin-7	PF00225		
	GL50803_4371	GiKIN8	Kinesin-8	PF00225		
	GL50803_10137	GiKIN9a	Kinesin-9	PF00225		
	GL50803_6404	GiKIN9b	Kinesin-9	PF00225		
	GL50803_16945	GiKIN13	Kinesin-13	PF00225		
	GL50803_8886	GiKIN14a	Kinesin-14	PF00225		
	GL50803_13797	GiKIN14b	Kinesin-14	PF00225		
	GL50803_7874	GiKIN16a	Kinesin-16	PF00225		
	GL50803_16161	GiKIN16b	Kinesin-16	PF00225		
	GL50803_16224	GiKIN20	Orphan	PF00225		
	GL50803_17264	GiKIN21	Orphan	PF00225		
	GL50803_14070	GiKIN22	Orphan	PF00225		
	GL50803_112729	GiKIN23	Orphan	PF00225		
	GL50803_11442	GiKIN24	Orphan	PF00225		
	GL50803_11177	KLC	Kinesin light chain			
Other Dynein Heavy Chains (7)	GL50803_17478	cytoDHC	CytoDHC cytoplasmic dynein heavy chain family	PF03028	MT metabolism	Wickstead *et al.* (2007)
	GL50803_103059		Dynein heavy chain (DHC) family	PF03028	MT metabolism	Wickstead *et al.* (2007)
	GL50803_8172		Dynein heavy chain (DHC) family, partial	PF03028	MT metabolism	Wickstead *et al.* (2007)
	GL50803_16804		Dynein heavy chain (DHC) family	PF03028	MT metabolism	Wickstead *et al.* (2007)

	GL50803_101138		Dynein heavy chain (DHC) family	PF03028	MT metabolism	Wickstead *et al.* (2007)
	GL50803_10538		Dynein heavy chain (DHC) family	PF03028	MT metabolism	Wickstead *et al.* (2007)
	GL50803_29256		Axonemal dynein heavy chain, partial	PF03028	MT metabolism	Wickstead *et al.* (2007)
Dynein light chains (10)	GL50803_4236	DYNLTI (Tctex 1/LC9)	Tctex-1 family	PF03645	MT metabolism	Wickstead *et al.* (2007)
	GL50803_4463	LC1	Dynein light chain (DLC) family	PF01221	MT metabolism	Wickstead *et al.* (2007)
	GL50803_7578	LC5	Dynein light chain (DLC) family	PF01221	MT metabolism	Wickstead *et al.* (2007)
	GL50803_9848	LC8	Dynein light chain (DLC) family	PF03645	MT metabolism	Wickstead *et al.* (2007)
	GL50803_13575	DYNLT2 (Tctex2/LC19)	Tctex-1 family	PF01221	MT metabolism	Wickstead *et al.* (2007)
	GL50803_14270	roadblock/LC7	Roadblock-related dynein light chain	PD03259	MT metabolism	Wickstead *et al.* (2007)
	GL50803_27308	LC4	Dynein light chain (DLC) family	PF01221	MT metabolism	Wickstead *et al.* (2007)
	GL50803_15606	Tctex-1	Tctex-1 family	PF03645	MT metabolism	Wickstead *et al.* (2007)
	GL50803_17371	DYNLT1 (Tctex1/LC9)	Tctex-1 family	PF03645	MT metabolism	Wickstead *et al.* (2007)
	GL50803_15124	roadblock/LC7	Roadblock/LC7 domain family	PD03259	MT metabolism	Wickstead *et al.* (2007)
Dynein light intermediate chain (1)	GL50803_13273	axonemal DLIC	Axonemal dynein light chain family, p28	PF10211	MT metabolism	Wickstead *et al.* (2007)
Dynein intermediate chain (3)	GL50803_6939	IC70	Dynein intermediate chain (DIC) family	PF05783	MT metabolism	Wickstead *et al.* (2007)
	GL50803_10254	IC138	Dynein intermediate chain (DIC) family	PF05783	MT metabolism	Wickstead *et al.* (2007)
	GL50803_33218	IC78	Dynein intermediate chain (DIC) family, Flagellar outer dynein arm intermediate chain, ODA-ICI	PF05783	MT metabolism	Wickstead *et al.* (2007)

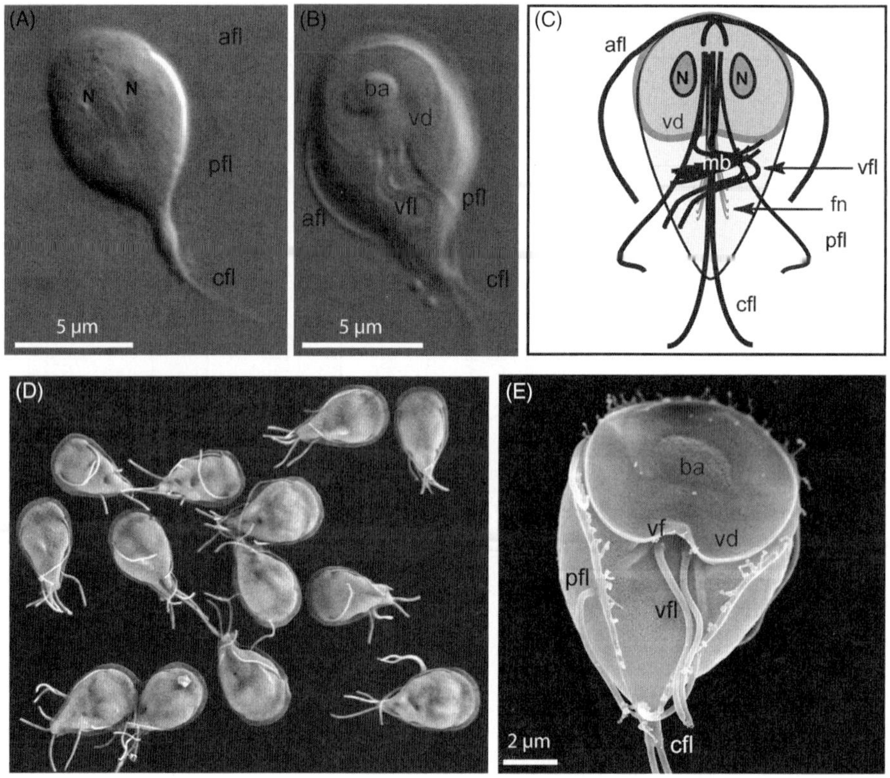

Fig. 1 The microtubule cytoskeleton and ventral disc of *Giardia*. Trophozoites are approximately 15 μm long by 5 μm wide and 5 μm thick and have a flattened teardrop shape (A = dorsal view using DIC microscopy, B = ventral view; N = nucleus, vd = ventral disc, ba = bare area). The microtubule cytoskeleton (C) is comprised of four primary elements: eight flagella organized in four pair (afl = anterior flagella, pfl = posterolateraral flagella, cfl = caudal flagella, and vfl = ventral flagella), the ventral disc (vd), the median body (mb), and the funis (fn). Trophozoites attach to surfaces via the ventral disc, thus orient with the ventral disc "down" (D). The ventral disc occupies the anterior and ventral portion of the cell (E) and is composed of a spiral array of microtubules surrounding a circular area of vesicles, the bare area (ba). Ventral flagella (vfl) exit at the posterior region of the ventral disc (E) under a region termed the ventral flange (vf). The median body (mb) and funis (not visible) are two other MT arrays of unknown function. SEM images are courtesy of Joel Mancuso (UC Berkeley).

pair. Specific structures are associated with, and distinguish, each flagellar pair—such as the "caudal complex," a jacket of MTs around the cytoplasmic region of the caudal axonemes. These **axoneme-associated structures** remain virtually uncharacterized (Elmendorf *et al.*, 2003) and confer a unique structural compositional identity and likely, unique functional roles, on each flagellar pair (Campanati *et al.*, 2002).

Giardia has a semi-open mitosis (Raikov, 1994) with two extranuclear **mitotic spindles** (Sagolla *et al.*, 2006). In anaphase B, each nucleus elongates as the spindle elongates during pole–pole separation. The two nuclei reposition to the center of the cell in prophase, followed by lateral chromosome segregation in anaphase (Sagolla *et al.*, 2006). Two extranuclear spindles access chromatin through polar openings in the nuclear membranes. Each spindle pole is associated with at least one axoneme. The giardial spindle MTs radiate from one of the flagellar basal bodies near each spindle pole, forming a sheath around the nuclear envelope. The nuclear envelope remains, forming a barrier between cytoplasmic MT arrays and chromatin; there is no evidence of mixing of the chromatin between nuclei (Sagolla *et al.*, 2006). Presumptive kinetochore MTs penetrate at the spindle poles through large polar openings in the nuclear membrane (Sagolla *et al.*, 2006). Following mitosis, the median body, the ventral disc, and the eight flagella are duplicated and partitioned to two daughter cells (Nohynkova *et al.*, 2006; Sagolla *et al.*, 2006).

III. Culture and Molecular Genetic Techniques

A. Axenic Culture of *Giardia* Trophozoites

Giardia (strain WBC6, ATCC 50803) is cultured axenically in TYI-S-33 medium, a complex (nondefined) medium supplemented with bile (Keister, 1983). TYI-S-33 medium is most easily prepared using stock solutions that are made in advance and can be frozen (–20 or –80°C) or stored at 4°C. The required stock solutions are 5X Basic Medium, 50X phosphate buffer (pH 7.2), 6.5% bile, and ferric ammonium citrate (FAC, 2.28 mg/ml). All are prepared using double-distilled water. Two other required solutions should be purchased: Donor Adult Bovine Serum (Thermo Scientific HyClone SH3007503) and 100X antibiotic–antimycotic (GIBCO 15420). 5X Basic Medium (500 ml) is prepared by adding 25 g D-glucose, 5 g NaCl, 25 g yeast extract (BD Bacto 212750), and 50 g casein peptone digest (Sigma P6838) to ~350 ml water. The medium is stirred on a stirplate for approximately 30 min until it is well dissolved (clear). The volume is adjusted to 500 ml, and the solution is filtered through a 0.22 μm PVDF filter (Millipore SCGVU05RE). Single-use (100 ml) aliquots of 5X Basic Medium can be stored for several weeks/months at 4°C or several years at –20°C. 50X phosphate buffer is prepared by dissolving 15 g KH_2PO_4 and 32.5 g $K_2HPO_4 \times 3H_2O$ in 500 ml water and adjusting the pH to ~7.2. The 6.5% bile and FAC solutions are made by dissolving 6.5 g bovine/ovine bile (Sigma B8381) and 228 mg ferric ammonium (III) citrate (Sigma F5879), respectively, in 100 ml water. Each solution is sterilized by passage through a 0.22 μm filter. These stocks may be stored frozen (FAC) or at 4°C (6.5% bile, 50X phosphate buffer) for several weeks to months. For Complete TYI-S-33 Medium supplemented with bile (~500 ml), 1 g L-cysteine hydrochloride (Sigma C7880) and 0.05 g ascorbic acid (Sigma A7506) are dissolved in 350 ml double-distilled water and the pH is adjusted to 7.0–7.2 with sodium hydroxide. In a laminar flow hood, the following are added to the cysteine/ascorbic acid solution: 100 ml 5X Basic Medium, 10 ml 50X phosphate buffer solution, 4 ml 6.5% bile, 1.5 ml FAC, 50 ml Adult Bovine Serum (ABS), and 5 ml antibiotic–antimycotic. The medium is mixed by swirling and filter sterilized with a 0.22 μm filter (Millipore PVDF works best). TYI-S-33 medium is clear and amber-colored. Fresh medium lasts about 7–10 days as the cysteine eventually crystallizes and precipitates; it is best to prepare medium weekly and store it at 4°C. Small aliquots of complete TYI-S-33 medium may also be frozen at –80°C for future use.

Giardia trophozoites are grown at 37°C in sterile 13 ml screw-capped disposable tubes (BD Falcon 352025) filled to 12 ml with warmed (37°C) TYI-S-33 medium and incubated upright without shaking. *Giardia* also grows well in filled, vertically incubated tissue culture flasks (50 ml, 100 ml, etc.). *Giardia* cultures are commonly maintained in culture tubes, but can also be grown and/or archived in 48-well or 96-well plates. 48-well plates are somewhat preferable to 96-well plates because they have a higher surface: volume ratio. *Giardia* grown in 48-well plates are placed in a sealed chamber (PlasLabs 850-LCS) and gassed with 100% N_2 to maintain a low oxygen atmosphere. The chamber, which can hold 10–15 plates, is then placed in a 37°C incubator. Plates can also be incubated anoxically in BioBags (Beckton Dickinson 261214).

When maintaining cultures in tubes or flasks, cells should be "split" or diluted when they become close to 100% confluent (typically every 3–4 days). To maintain sterility, cultures are transferred within a laminar flow hood. Trophozoites are detached from the inner walls of the culture tube (or flask) by placing the tube on ice for 15–30 min. They are then diluted approximately 1:10 (or more) in fresh, warmed TYI-S-33 medium (e.g., 1 ml cells in 11 ml fresh medium). Trophozoites can be pelleted by centrifugation in a clinical centrifuge at a maximum speed of 900×*g*. *Giardia* is considered a biosafety level 2 pathogen, and appropriate biosafety protocols for maintaining and disposing of cultures and contaminated materials should be employed (Canada, 2010).

Giardia cultures are best stored long term in liquid nitrogen in medium supplemented with dimethyl sulfoxide (DMSO). Cells from confluent cultures are detached by chilling tubes on ice for 15–30 min, then are pelleted for 5 min at 900×*g*. The medium is decanted, and the cells are resuspended in fresh TYI-S-33 medium containing 9% (final concentration) sterile DMSO. The cells are aliquoted into cryovials and are first frozen at –80°C. The cryovials should be transferred to liquid nitrogen the following day. *Giardia* strains can be revived by thawing frozen cryovials rapidly at 37°C, and immediately transferring the thawed cells to 12 ml of fresh, warmed TYI-S-33 medium using a sterile transfer pipette. After cells attach (about 1 h at 37°C), the medium is decanted and replaced with an equal volume of fresh warmed TYI-S-33 medium to remove residual DMSO. Multiwell plates containing strains in 9% DMSO can be repeatedly frozen and thawed and can be stored at –80°C for several years, allowing strain collections to be barcoded and archived.

B. Genetic Tools for Tagging and Expression of Proteins

Several novel MT-based structures, such as the ventral disc, the median body, and the funis, are critical to giardial biology and pathogenesis; however, many of the components comprising these structures have not yet been identified. To verify the localization of known (see Table I) or novel cytoskeletal proteins, expression of Green Fluorescent Protein (GFP) or epitope-tagged proteins in *Giardia* is becoming common. The pace of research in

Giardia has been slowed, however, by a lack of diverse or reliable molecular genetic tools. Specifically, there are few episomal vectors for expressing epitope-, GFP- or affinity purification-tagged fusion proteins or for tetracycline-inducible overexpression of dominant-negative or wild-type proteins (Sun and Tai, 2000; Sun *et al.*, 2005). Existing episomal vectors were designed for traditional restriction enzyme-based cloning with ligation; moreover, they have few multiple cloning sites and limited antibiotic resistance markers (Davis-Hayman and Nash, 2002).

1. Multisite GATEWAY Cloning Vectors and Modules for Protein Tagging in *Giardia*

To address the lack of reliable research tools in *Giardia* and to increase the speed and efficiency of protein tagging, overexpression, and localization, our laboratory has modified the GATEWAY ligation-independent cloning system (Invitrogen) for use in *Giardia*. GATEWAY cloning is based on the site-specific recombination system of the bacteriophage lambda, which facilitates the integration of lambda into the *Escherichia coli* chromosome (Landy, 1989). Recombination reactions occur in a well-defined and conservative fashion between specific recombination sequences (*att* sites) and occur only when catalyzed by specific mixtures of recombination proteins (Clonase II enzymes).

GATEWAY cloning requires the construction of module plasmids ("entry clones") and backbone plasmids ("destination vectors"), which are subsequently stitched together via the aforementioned recombination reactions. GATEWAY cloning can initially be time-consuming and moderately expensive to set up, but its main advantage lies in its modularity. Once a library of entry clones and destination vectors has been constructed, swapping out modules at given positions is simple, quick, and efficient. The GATEWAY strategy allows rapid swapping of fluorescent or protein purification tags, promoters, and resistance genes, and the archiving of different versions of constructs. This is advantageous when tagging numerous proteins, fusing a single protein to multiple tags, or when regulating protein expression with different promoters. We have developed GATEWAY module plasmids and compatible backbone plasmids for use in both episomal and integration vectors in *Giardia* (Table II), permitting the rapid, systematic localization of putative disc-associated proteins. The process of generating a desired episomal or integration vector suitable for *Giardia* transformation is as follows: GATEWAY polymerase chain reaction (PCR) primers for the giardial genes of interest are purchased in single use or 96-well plate format. After successful PCR amplification of the desired gene segment, module plasmids ("entry clones") are created in a ligation-independent manner by performing a BP reaction, in which the PCR product of the gene of interest (GOI) is mixed with the corresponding GATEWAY pDONR vector and BP Clonase II (both from Invitrogen). Following successful module plasmid generation, the creation of a suitable *Giardia* vector is straightforward; the final vector containing the GOI fused to the other modules via LR reaction is available within 1–3 days. Employing this GATEWAY-based system, a new *Giardia* gene fusion vector can be reliably produced in a week. To date our laboratory has developed 1-fragment, 2-fragment, and 4-fragment GATEWAY cloning vectors

Table II
GATEWAY modules available for the development of episomal or vectors in *Giardia*

Strategy	Destination Vector	X1	X2	X3	X4
1-Modular fragment	P1PRX1TGRP	1F1G-GOI			
2-Modular fragment	P2PRX1X2RP	2F1G-GOI	2F1G-GOI		
		2F1T-GFP1	2F2T-GFP		
		2F1T-GFP2	2F2T-MCHERRY		
		2F1T-GFP3			
4-Modular fragment	P4X1X2X3X4	4F1G-GOI	4F2G-GOI	4F3G-GOI	4F4G-GOI
	P4X1X2X3X4RP	4F1P-pRAN	4F2T- MCHERRY	4F3T- MCHERRY	4F4T- MCHERRY
		4F1P-pACTIN	4F2G-GFP	4F3G-GFP	4F4T-GFP
		4F1P-pARP			4F4R-NEO
					4F4R-PAC

Currently available GATEWAY entry clone modules (X1–X4) for each of the constructed destination vectors (1-fragment, 2-fragment, and 4-fragment) from our laboratory. Entry clones are named for their position in a particular destination vector and the type of gene or tag (GOI = gene of interest; GFP(1–3) = C-terminal GFP tag in each of three frames; MCHERRY = red fluorescent tag; NEO = neomycin resistance gene; PAC = puromycin resistance gene; pRAN = ran GTPase promoter; pACTIN = actin promoter; and pARP = actin-related protein (ARP) gene promoter.

that have been verified for use in *Giardia*. Four categories of modules have been constructed and validated (see Fig. 2 for the schematic and Table II for the list of currently constructed 1-, 2-, and 4-fragment module plasmids with their compatible destination vectors).

As proof of the efficacy of GATEWAY cloning in *Giardia*, a construct was created via a 4-fragment GATEWAY strategy, yielding a stably transformed, puromycin-resistant α7.3-giardin::GFP strain (see Fig. 3). For puromycin-resistant strains, 120 μl of 5 mg/ml puromycin per 12 ml culture are added to maintain selection in stably transformed strains, such as the α7.3-giardin::GFP strain. Over 90% of the cells have sufficiently high GFP expression for live microscopy. The viability of these strains indicates that the GFP tag does not adversely affect protein structure or localization nor does it affect functioning of the giardial cytoskeleton.

2. Stable Transformation of *Giardia* by Electroporation

In *Giardia*, episomal vectors or constructs to be integrated into the genome are introduced via electroporation. One confluent 13 ml culture tube is required for approximately four transformations. Cultures in log phase (nearly 100% confluent, or about 3–5 million cells/ml) appear to transform most effectively. Trophozoites are chilled for 15–30 min to cause them to detach from the tube and are pelleted at 900 × *g*. The cell pellet is washed once in cold fresh TYI-S-33 medium, the trophozoites are recentrifuged, and the medium is decanted. The cells are then resuspended in about 1.5 ml of fresh medium to a density of $\sim 50 \times 10^6$/ml and are kept chilled on ice until

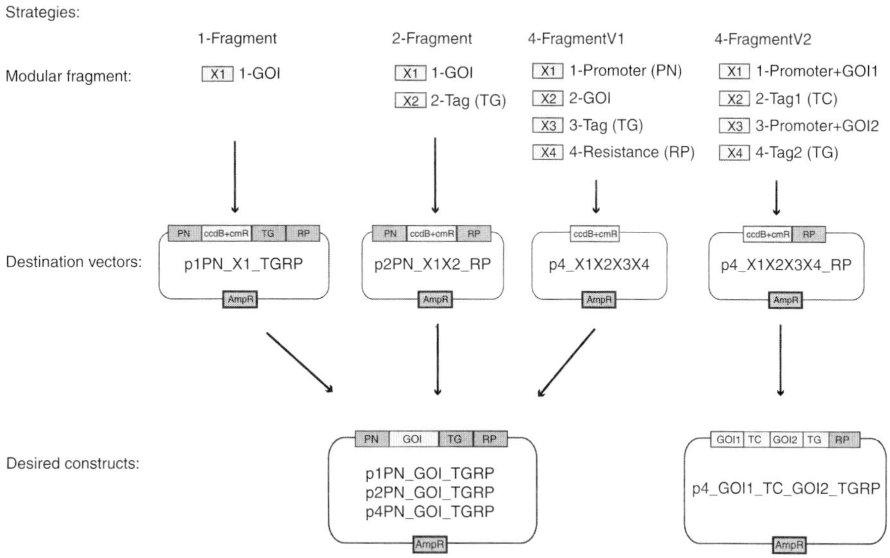

Fig. 2 Gateway cloning strategies for *Giardia*. Depending on the level of modularity desired, various entry clones (modular fragments) and destination vectors can be used to construct episomal or integrative vectors for transformation in *Giardia*. If only one variable module/fragment is needed, then 1-fragment destination vectors and 1-fragment entry clones can be used. A 1-fragment strategy is useful when tagging numerous proteins with the same tag. In this instance, the variable fragment would encode a gene of interest [1-GOI (X1)] that is recombined into the ccdB+cmR site in a vector that contains a promoter [e.g., *ran* (PN)] and an antibiotic resistance gene [e.g., puromycin resistance (RP)]. A 2-fragment GATEWAY strategy is useful when tagging numerous proteins with different tags and requires two variable modules. In this instance, fragment 1 (X1) would encode a gene of interest (1-GOI) and fragment 2 (X2) would encode a tag of interest (2-Tag), such as GFP (TG). Finally, a 4-fragment GATEWAY strategy is useful when flexibility with respect to the promoter, the GOI, the tag, and the antibiotic resistance marker is required. In the first example, fragment 1 encodes a promoter [i.e., ran (PN)], fragment 2 encodes a GOI, fragment 3 encodes a tag [e.g., GFP (TG)], and fragment 4 encodes an antibiotic resistance protein [e.g., puromycin resistance (RP)]. Alternatively, a 4-fragment strategy can be used for colocalization experiments. In the second example, fragment 1 (X1) encodes a GOI and its native promoter (GOI1), fragment 2 (X2) encodes a tag such as eGFP (TG), fragment 3 (X3) encodes a second GOI and its native promoter (GOI2), and fragment 4 encodes an alternate tag such as mCherry (TC). All constructs also contain an ampicillin resistance gene (AmpR) for selection in *E. coli*. The names of the destination vectors and desired constructs are descriptive; for example, the construct p1PN_GOI_TGRP was constructed using a 1-fragment strategy and contains the promotor for the giardial *ran* gene, a GOI tagged with GFP and a gene conferring resistance to puromycin.

use. For transformations of episomal plasmids, plasmid DNA is prepared using a Qiagen EndoFree Plasmid Maxi Kit to remove bacterial endotoxins. The DNA is then resuspended in double-distilled water, sterilized by filtration through a 0.22 μm Spin-X centrifuge tube filter (Costar 8160), and added to a 0.4 cm electroporation cuvette (BioRad 165-2088). To prevent arcing during electroporation, a maximum of 75 μl of plasmid DNA are added to 0.3 ml ($\sim 1 \times 10^7$) cells. Cuvettes are gently flicked to resuspend trophozoites and plasmid DNA and are chilled on ice for 15 min.

two giardial kinesins—kinesin-2 and kinesin-13. Overexpression of the mutant proteins resulted in cytoskeletal phenotypes that elucidated the conserved and novel functions of these motor proteins in *Giardia* (Dawson *et al.*, 2007; Hoeng *et al.*, 2008). *Giardia* kinesins are amenable to this strategy because studies in other organisms (Gelfand *et al.*, 2001; Lin-Jones *et al.*, 2003) have shown that point mutations in the kinesin ATPase domain permit the binding of ATP but prevent its hydrolysis. The *Giardia* kinesin ATP rigor mutations were designed based on a multiple sequence alignment of kinesin homologs from diverse eukaryotes. This strategy has been used successfully to disrupt kinesin function in many other experimental systems (Boleti *et al.*, 2001; Brown *et al.*, 2005; Gelfand *et al.*, 2001; Kline-Smith *et al.*, 2004).

Genes containing dominant-negative mutations are placed under the control of a tetracycline-inducible promoter (Sun and Tai, 2000); GFP tagging allows the identification and characterization of trophozoites with significant levels of overexpression. Induction of expression of constructs in inducible strains is achieved by using 250 μl 1 mg/ml tetracycline or doxycycline per 12 ml culture for 24–48 h. Maximal induction of transgenes occurs 6–8 h after induction and continues for over 48 h after the addition of doxycycline (Dawson *et al.*, 2007). The induction of genes following addition of doxycycline can be confirmed using RT-PCR. Phenotypic changes in cytoskeletal structures e.g., alterations in flagellar length (Hoeng *et al.*, 2008), median body volume (Dawson *et al.*, 2007), or disc size and morphology) can be assessed using epifluorescent imaging of individual high-expressing trophozoites and the Imaris software package (see Section III).

5. Flow Sorting of GFP-expressing Trophozoites

Chilled GFP-expressing strains can be sorted using a Cytomation MoFlo Cell Sorter (or similar). Strains are sorted into fresh sterile tubes at three GFP expression levels: high GFP expression, low GFP expression, and no expression. Negative controls are wild-type, nontransformed cells. Cells are flow sorted in 1X HEPES buffered saline (HBS) at 4°C at a rate of 1 ml/h. The sorted strains are grown overnight and dispensed into 48-well plates using serial dilution to a density of approximately 5–10 cells/well. These plates are grown to confluency and archived at –80°C. Stocks of flow-sorted cells are also archived and frozen at –80°C before dilution into plates.

6. Flow Cytometry of Ethanol-Fixed Trophozoites for Cell Cycle Analyses

Trophozoites are detached by incubation on ice and pelleted by centrifugation at 900×*g* for 5 min. The cell pellets are then washed twice in 2 ml of 1X HBS. The cells are resuspended in 300 μl 1X HBS, and 700 μl of ice-cold 100% ethanol is added drop by drop while vortexing gently to prevent clumping. Ethanol-fixed trophozoites can be stored at 4°C indefinitely. For flow cytometry, the ethanol-fixed trophozoites are centrifuged, and the pellet is rinsed with 50 mM sodium citrate. The cells are resuspended in 0.5 ml 50 mM sodium citrate containing fresh 0.1 mg/ml RNase A and incubated overnight at 4°C to digest RNA. RNAse-treated cells are again pelleted by centrifugation for 5 min at 900×*g*. The liquid is decanted and the pellet is resuspended in 0.5 ml of 50 mM sodium citrate containing a final concentration of 5 μM Sytox Green (Sigma) immediately before flow cytometry. Flow cytometry has been performed using a Beckman-Coulter EPICS XL flow cytometer, and results analyzed with FlowJo software (Tree Star Inc. Ashland, OR).

IV. Imaging of the Cytoskeleton and Associated Proteins Using Light Microscopy

Giardia is a small motile flagellated cell with ultrastructure that is clearly visible in live and fixed cells using various types of light microscopy [e.g., phase, differential interference contrast (DIC), and epifluorescence] and electron microscopy (EM) (scanning EM, transmission EM, and, more recently, cryoelectron tomography). For live imaging of trophozoites, contrast-enhancing transmitted light microscopy techniques (phase contrast, DIC, and dark field) take advantage of perturbations of the optical light path to produce images of *Giardia* cells without killing, fixing, or staining them, allowing dynamic cytoskeletal processes to be observed with low exposure time (video rate) and high contrast.

A. Live Imaging Using Light Microscopy

To perform live imaging of any cell, appropriate incubation conditions (temperature, humidity, and atmospheric environment) should be provided, and digital imaging hardware and software must be available. Imaging live *Giardia* trophozoites on a microscope is possible if warm (37°C) anoxic conditions are maintained using incubation chambers amenable to high-resolution microscopy. Inverted microscopes equipped with phase contrast, DIC, and epifluorescence optics, as well as high-speed motorized control and image acquisition in the *X, Y* and *Z* planes aid in live imaging of *Giardia*.

Several methods may be used to culture trophozoites for live imaging under physiological conditions. For imaging on slides, trophozoites can be kept in their growth medium or resuspended in 1X HBS. A simple sealed imaging chamber can be created by mounting a coverslip to a standard slide with parallel lines of double-sided tape cut to define the desired imaging area. Cells are loaded into the chamber via capillary action using a wide-bore pipette, and the edges are sealed with melted VALAP (equal parts Vaseline, lanolin, and paraffin). This chamber provides a micro-oxic environment sufficient for short-term experiments (<1 h).

Free-swimming or attached trophozoites can also be cultured and imaged in small, sealed coverglass bottom Petri dishes (or in sealed 48-well plates). In addition experiments can be done using Petri dishes with glass bottoms (Matek P35G-1.5-14-C). Dishes are gassed with a N_2 atmosphere and can be imaged for up to four hours. Cytoskeletal dynamics can be imaged directly in these dishes using the DeltaVision Real-Time (RT) deconvolution microscope with images acquired over several minutes.

Cells are pelleted, resuspended in TYI-S-33 medium, and transferred to the dish. The dish is then placed at 37°C in a sealed chamber gassed with 100% N_2 and the cells are allowed to attach for 1 h. The medium is gently removed by aspiration and replaced with warmed 1X HBS. Any necessary drugs or dyes (e.g., propidium iodide (PI) or Hoechst) are added at this time. To allow for dye incubation in 1X HBS, the dish is incubated anaerobically at 37°C for 15 minutes. The dish is then removed, sealed tightly with parafilm, and imaged on an inverted microscope. A No. 1.5 coverslip can be affixed to a hole drilled in the dish lid to accommodate imaging that requires Koehler alignment (phase, DIC, Hoffman). Alternatively, the Petri dish lid can be replaced with a large coverslip sealed in place with vacuum grease. Live 3D images are generally collected with the DeltaVision System using the 60x/1.4 NA or 100x/1.2 NA objectives and a digital camera. Serial sections are acquired at 0.2 μm intervals, and deconvolved using the softWoRx deconvolution software (Applied Precision). Two-dimensional (2D) projections and 3D models or movies can then be created from the 3D data sets using either softWoRx or Imaris BitPlane (see below).

Embedding trophozoites in low-melt agarose facilitates the imaging of cell division or other processes in single cells by limiting their motility. Cells in glass bottom Petri dishes (see above) are allowed to attach for one hour in warmed 1X HBS containing 0.1% PI. PI is a marker for cell viability; PI staining of the cell's interior indicates that membranes have been compromised and the cells are inviable. The 1X HBS is overlaid with at least 1 ml of 3%, low-melt agarose (Agarose Type IX-A, Sigma A2576) warmed to 37°C. 100% N_2 is introduced under the lid and the dish is sealed immediately with parafilm. Alternatively, the open Petri dish can be placed in a warmed environmental chamber gassed with nitrogen (Pecon Stage Incubator 160-800 046 or Okolab microscope incubators). This method facilitates *Z* plane serial sectioning as the cell is prevented from movement. Embedding trophozoites is particularly useful for fluorescence recovery after photobleaching (FRAP) studies (see below) when cells often detach and swim from the field of view.

B. Live Imaging of Giardial Attachment to Surfaces via the Ventral Disc

Perhaps the most important function of the giardial MT cytoskeleton is to promote the attachment of *Giardia* to the intestinal microvilli. Giardial attachment, or surface "adherence," is most often operationally defined as the number of cells that remain adhered to a given surface after an experimental treatment (Feely and Erlandsen, 1982; Gillin and Reiner, 1982; Magne *et al.*, 1991; Mariante *et al.*, 2005; Perez *et al.*, 2001; Sousa *et al.*, 2001). Trophozoites that remain adhered following a treatment are "attached"; those that are not adhered are "not attached." Most attachment assays have measured the number of attached cells at the population level after long incubation periods (~2–24 h), as opposed to quantifying the attachment dynamics of individual cells (seconds). Thus while trophozoites are known to attach/detach from substrates in less than a second (Feely and Erlandsen, 1981; Hansen and Fletcher, 2008; Hansen *et al.*, 2006), attachment has routinely been quantified over a time scale that is ~7000–80,000 times longer than it actually takes to occur. *Giardia* detaches when dividing (Sagolla *et al.*, 2006; Tumova *et al.*, 2007), when nonviable, or when exposed to oxygen or low temperature (Gillin and Reiner, 1982). Attachment generally has not been correlated with cell viability; thus, previous studies of attachment may have simply quantified the indirect effects of experimental treatments on cell viability.

Giardial attachment to biological or inert surfaces is reversible and has been broadly quantified using three approaches: (1) direct/indirect counts of "attached" and "unattached" cells (Feely and Erlandsen, 1982; Gillin and Reiner, 1982; Magne *et al.*, 1991; Mariante *et al.*, 2005; Perez *et al.*, 2001; Sousa *et al.*, 2001); (2) live imaging (Feely and Erlandsen, 1981, 1985; Hansen and Fletcher, 2008; Holberton, 1973, 1974; Narcisi *et al.*, 1994); and (3) a novel centrifuge assay of normal attachment force (Hansen *et al.*, 2006). The widely varying conditions used in these different attachment assays therefore make it difficult to assess attachment quantitatively or to rectify conflicting findings. Live imaging, quantitation of attachment forces, and measurement of the area of surface contacts [as in focal adhesion assays (Décavé *et al.*, 2002)] are preferable to other attachment assays. 3D live imaging of attachment dynamics informs real-time functional assays that may also aid in evaluating the efficacy of novel anti-giardial drugs. Several assays that incorporate live quantitative imaging of giardial attachment over a short time scales are available.

1. Shear Forces (Flow-cell) Assay of Attachment

Using a syringe pump to create laminar flow, the effect of shear forces on live trophozoites attached to a glass substrate can be imaged. Live imaging of attachment is facilitated by the use of a closed, temperature-controlled parallel plate flow cell chamber with a stage adapter (RC-31 Flow Chamber, Harvard Apparatus) that is mounted on a conventional inverted microscope. The syringe pump is attached to the flow chamber via PE-90 tubing and a three-way stopcock for introduction of cells. Trophozoites in growth medium are loaded into the syringe pump chamber through the stopcock using an epoxy (flat) blunt 18 gauge needle and ½ inch plastic PE-90 tubing. The syringe pump is attached to the flow cell chamber along with two other syringes, one containing a control solution (plain medium) and the other containing experimental additives, if needed. *Giardia* trophozoites are flowed into the flow cell chamber and allowed to attach for 1–5 min. The percentage of cells detaching in a given field is imaged and quantified after flowing in medium at various rates. The rate of flow/time (generally >3 ml/min) is then correlated with the numbers of detached mutant cells, normalized to wild-type cells and can be converted to a "shear" force. Cell counting is automated using image analysis software (such as Metamorph or ImageJ) and the fraction of cells that maintain attachment over a range of shear forces is determined.

2. Normal Force (Centrifuge) Assay of attachment

Defects in normal forces of attachment can be assayed at the population level, using a physical attachment assay (Hansen *et al.*, 2006). Briefly, trophozoites are cultured, detached from culture tubes by chilling at 4°C, and then transferred to custom sample holders capped with thick, custom circular glass slides. The cells are incubated at 37°C for 1 h to allow attachment to the glass slides. The sample holders are then centrifuged at 37°C in a hanging bucket centrifuge rotor at speeds from 500 to 13,000 rpm (4.0 pN–2.7 nN normal force). Controls are prepared in the same manner and incubated at 37°C, but not centrifuged. Immediately after centrifugation, the glass slides are removed from the chamber and images of ~5 areas (~5000 cells) are captured with a digital camera. Cells that maintain attachment are counted using Metamorph or ImageJ software and normalized to the noncentrifuged control. The centrifuge parameters can be converted to a normal force exerted on the cells during centrifugation.

C. Imaging GFP-Tagged Cytoskeletal Proteins in Live and Fixed Cells

GFP-tagged proteins are used extensively in other experimental systems (Sullivan *et al.*, 1999) and have revolutionized the study of MT-associated proteins, including plus end-tracking proteins and kinesins (Schuyler and Pellman, 2001). The advantage is that GFP-tagged proteins can be used as subcellular markers to study cytoskeletal dynamics in live cells, avoiding potential fixation artifacts and allowing the dynamic visualization of cytoskeletal processes such as cell division or attachment. Because many aspects of the giardial MT cytoskeleton are dynamic, it is crucial to investigate the behavior and kinetics of MT-associated proteins in living trophozoites using live imaging. Kinesins (Dawson *et al.*, 2007; Hoeng *et al.*, 2008), flagellar-associated proteins (Hoeng *et al.*, 2008), centromeres (Dawson *et al.*, 2007, and the plus end tip-tracking protein EB1 (Dawson *et al.*, 2007a, b) have each been GFP tagged and visualized in live and fixed trophozoites.

Artifacts may arise with all methods of protein localization. Many artifacts arise from the fixation process; thus, there is no way to rule out protein mislocalization if a single localization method is used. Epitope tags, and several amino acid tags such as AU1 and HA, are small protein tags that have been used extensively in *Giardia*, yet these require that alcohol-based fixation (e.g., methanol) is used to preserve their antigenicity. Alcohol fixation is known to cause severe cytoskeletal artifacts because it causes proteins to precipitate, producing cytoskeletal defects. Aldehyde fixatives such as paraformaldehyde (PFA) crosslink proteins and better preserve cytoskeletal architecture (Oka *et al.*, 1994). When antibodies are used for protein localization, they must first be shown to be specific for the protein of interest; there are many cases in *Giardia* where “specific” antibodies have been found to localize to different structures depending upon the fixation method or heterologous antibody used. A further complication with antibodies is that some animals have been previously exposed to *Giardia*, so anti-giardial antibodies may be present even in preimmune serum.

While GFP is a larger protein tag than epitope tags like AU1, it is used in many organisms and its larger size rarely affects protein localization. GFP mislocalization tends to be cytoplasmic, rather than to specific cytoskeletal structures. Alternative tags, such as SNAP (Covalys Biosciences AG, Witterswil, Switzerland) tags, are also useful for live imaging in *Giardia*. The advantage of the SNAP tag strains is the ability to anaerobically image *Giardia* in TYI-S-33 medium (Regoes and Hehl, 2005). SNAP-tagged proteins can also use diverse fluorophore substrates with various emission spectra; this is useful for pulse/chase type experiments. Similar-sized tags, such as the tandem affinity purification tag, could be used successfully to tag components of the giardial cytoskeleton and to affinity purify interacting proteins.

Although development of methods for live cell imaging of *Giardia* is ongoing, *Giardia* can currently be imaged on sealed slides, in sealed coverglass bottom Petri dishes with a N_2 atmosphere, or embedded in ~1% low-melt agarose. To assess expression and protein localization in live GFP-tagged strains, trophozoites are chilled on ice for 15 min and pelleted at 900×*g*. They are then resuspended in 1 ml cold 1X HBS plus glucose (1 g/l) and incubated on ice for 10 min. TYI-S-33 medium has a high protein content and is generally autofluorescent; thus imaging in HBS allows the signal of the GFP-tagged protein to be readily visible against the background. A wet mount is prepared with 10 μl of cell suspension and GFP localization is examined using the epifluorescent microscope. The slide is then warmed to 37°C for 5 min; if necessary, slides can be sealed with VALAP to prevent evaporation. The cells will attach and begin to express the GFP-tagged protein within 5–10 min (see Fig. 3).

To visualize GFP-expression in fixed cells, trophozoites that have been resuspended in 1–2 ml of 1X HBS are incubated on coverslips for 15–30 min at 37°C. The attached cells are then fixed in 1% PFA (EMS 15714-S). The coverslips should be examined with an inverted microscope throughout the protocol to ensure that the cells remain attached. The fixed cells are washed three times in 1X HBS; the supernatant should be saved for analysis of unattached cells. For PI counterstaining, the cells are first incubated with 5 μg/ml RNase for 30 min at 37°C, washed three times in 1X HBS, and stained with 2–10 μg/ml PI for 50 min at 37°C. The coverslips are then washed three times in 1X HBS and mounted on slides using ProLong Gold antifade mounting medium (Invitrogen P36930). They can then be readily visualized using an epifluorescent microscope. Cytoskeletal buffers and gentle formaldehyde fixation are used for the specific purpose of preserving native MT structure in *Giardia* as seen for median body associated α7.3-giardin (Fig. 3).

D. Fluorescence Recovery After Photobleaching

GFP fusion proteins are ideal for use in FRAP studies because they can be bleached without detectable damage to trophozoites. Laser fluorescence photobleaching of specific regions is used to measure the movement and steady-state turnover of proteins in *Giardia* (Fig. 4), as in other organisms (Sullivan *et al.*, 1999). GFP-expressing trophozoites are embedded in ~1% low-melt agarose (see above), and imaged using an Olympus FV1000 scanning laser confocal microscope equipped with a four channel PMT (or similar). A prebleach image of the cell is acquired using the 488 nm laser. A short, powerful 405 nm laser pulse follows this prebleach imaging to photobleach a specific region (15% 405 nm laser power for 500 ms). To assess recovery, cells are imaged every 1–5 s after photobleaching, with a 488 nm laser. Initially, protein turnover is assessed once every minute for up to 10 min, and successive observations are

been put in place. The slides are allowed to harden at least 2 h before viewing with the epifluorescent microscope. GFP-fusions have the potential to mislocalize, thus the localization of cytoskeletal proteins should be confirmed by constructing epitope-tagged strains. High-resolution localization to specific cytoskeletal elements can also be confirmed by anti-GFP immunoEM.

F. 3D Deconvolution Light Microscopy and Image Analysis

Giardia's compact and complex molecular architecture requires 3D imaging for proper analysis of protein localization. 3D imaging with computational enhancement (removing out of focus information with deconvolution algorithms) has permitted successful visualization of giardial centromeres, spindles, and intraflagellar raft particles (Dawson *et al.*, 2007a, b; Hoeng *et al.*, 2008; Sagolla *et al.*, 2006). The Applied Precision DeltaVision RT deconvolution microscope system equipped with a Quantitative Laser Module for photokinetics at 406 and 488 nm wavelengths is one tool for achieving 3D imaging. 3D images of fixed cells are collected with the DeltaVision image acquisition software softWoRx (Applied Precision) using an Olympus IX70 wide-field inverted fluorescence microscope (or similar), an Olympus UPlanApo 100X, NA 1.35, oil immersion objective, and a Photometrics CCD CH350 camera cooled to –35°C (Roper Scientific). This system allows optical sectioning of live or fixed trophozoites in less than 300 ms. Serial sections are obtained at 0.2 μm intervals and are deconvolved using the softWoRx deconvolution software. For presentation purposes, 2D (maximum intensity) projections can be created from the 3D stacks using softWoRx. This strategy has been useful for calculating flagellar length from 3D image stacks of anti-tubulin immunostained axonemes using softWoRx image analysis tools. The BitPlane AG Imaris 5.0.3 software package can be used to track particles and do quantitative colocalization analysis on images acquired using the DeltaVision or confocal microscopes. Alternatively, a Laser Scanning confocal microscope with a photokinetic scan head (405 nm laser line) and two spectral scan emission detectors can be used [e.g., an Olympus FV1000 inverted confocal microscope (or similar), an UPlanApo 60x, 1.42NA, oil immersion objective, and a Photometrics CCD CH350 camera cooled to –35°C (Roper Scientific)].

V. EM of Trophozoites and Cysts

A. Transmission Electron Microscopy

Optimal methods for embedding and thick sectioning *Giardia* that give good contrast for transmission electron microscopy (TEM) or electron tomography (Hoeng *et al.*, 2008; Poxleitner *et al.*, 2008; Sagolla *et al.*, 2006) have been developed and used extensively. Fixation methods compatible with antibody localization at the EM level have been adapted from studies of mitotic spindle ultrastructure in yeasts (Ding *et al.*, 1993; Winey *et al.*, 1995). Trophozoites are best preserved using high-pressure freezing (HPF) for fixation using a Balzer high-pressure freezer, followed by freeze substitution and fixation in PFA /glutaraldehyde, prior to embedding in Lowicryl or LR White. Specific trophozoites can be identified in plastic using light microscopy. These can then be excised and mounted prior to serial sectioning (Sagolla *et al.*, 2006).

Specifically, trophozoites are prepared for TEM by first allowing them to attach to clean sapphire discs in a 37°C incubation chamber under anaerobic conditions. Sapphire discs or tabs with attached cells are then dipped into TYI-S-33 medium containing 20% BSA and placed in a specimen holder. The discs are then high pressure frozen using standard methods (Sawaguchi *et al.*, 2003). Trophozoites can also be high pressure frozen using a Baltech high pressure freezer, and freeze substituted using a Leica freeze substitution apparatus. Prior to serial sectioning on an Ultracut E, trophozoites are embedded with Epon resin. Ultrathin sections (50–65 nm) are stained with uranyl acetate in 70% methanol and lead citrate prior to acquiring images on an electron microscope such as the JEOL 1200 TEM.

B. Scanning Electron Microscopy of Trophozoites

Trophozoites are first allowed to attach to either Aclar or track membrane filters. Then these substrates with attached *Giardia* are fixed for 1 h in 2% glutaraldehyde in cacodylate buffer. Trophozoites are postfixed with 2% osmium tetroxide, dehydrated in ethanol, critically pointed dried, and coated with iridium using a MED 020 Baltec High Vacuum Coating System. Scanning electron images are acquired using a Hitachi S5000 FESEM at 10 Kv or similar electron microscope.

VI. Other Cytoskeletal Methods

A. Use of MT Depolymerizing and Stabilizing Drugs to Assess Dynamics

MT dynamics are important for chromosome segregation (Dawson *et al.*, 2007; Sagolla *et al.*, 2006), flagellar duplication (Nohynkova *et al.*, 2006), disc division (Tumova *et al.*, 2007), and likely encystation and excystation (Feely *et al.*, 1990). Polymerization dynamics can not only rapidly reorganize the cytoskeleton but also generate pushing and pulling forces (Hunter and Wordeman, 2000) on a scale of seconds to minutes (Mitchison and Kirschner, 1984). Both interphase and mitotic cytoskeletal arrays in *Giardia* are sensitive to MT depolymerizing and stabilizing compounds. Flagella are generally insensitive to MT stabilizing or destabilizing drugs (Dentler and Adams, 1992; Tilney and Gibbins, 1968). Giardial flagella, however, are sensitive to such agents, with the MT destabilizing drug nocodazole causing a significant (20–40%) decrease in external flagellar length of all eight flagella, as well as a shortening of median body MTs (Dawson *et al.*, 2007). Conversely, incubation with the MT stabilizing drug Taxol for less than one cell cycle results in increased length of all flagella (+30–60%) and the median body MTs (+50%). Short (15 min to 1 h) incubations with Taxol and nocodazole affect MT dynamics in dividing cells, causing broken spindles and chromosome segregation defects (Sagolla *et al.*, 2006). MT dynamics in the giardial median body are likely to be critical throughout the life cycle (Kabnick and Peattie, 1990), yet there is no evidence that the ventral disc or MTs undergo dynamic growth and shrinkage during attachment.

Taxol or nocodazole (resuspended in DMSO) can be diluted directly into culture tubes containing medium to a final concentration of either 10 μM or 20 μM. DMSO controls are also useful to compare to treated trophozoites. To assess MT depolymerization in *Giardia*, nocodazole (Sigma M1404) is added at a final concentration of 10 μm to trophozoites, which are then incubated at 37°C for 5 h, prior to tubulin immunostaining. To assess the effect of MT stabilization, Taxol (paclitaxel, Sigma T7191) is added at a final concentration of 20 μM, and the trophozoites are incubated for 1 h at 37°C. It should be noted that short incubations (1–5 h, i.e., less than one cell cycle) are preferable to long incubations of 24–72 h; after prolonged incubation, dead cells are prevalent (Mariante *et al.*, 2005).

B. Detergent Extraction of the Giardial Cytoskeleton

A goal in the isolation of giardial cytoskeletal structures such as the ventral disc or axonemes for proteomic or high-resolution image analysis is the maintenance of MT-associated proteins, by removing radicals and metal ions and by stabilizing MT depolymerization using drugs like Taxol. A protocol modified from Holberton and Ward (1981) can be used to isolate disc and flagellar cytoskeletons from *Giardia*. First, TYI-S-33 medium is decanted from one confluent 12 ml culture tube of trophozoites. Cytoskeletons are extracted by adding 1 ml of 1X PHEM plus Taxol (60 mM PIPES, 25 mM HEPES, 10 mM EGTA, 1 mM $MgCl_2$, pH 7.4, 1 mM DTT, 10 μM Taxol) and 1% Triton X-100. Protease inhibitors (Roche 11206893001) are also added to the preparation to prevent proteolysis. The tubes are then inverted several times to detach cells before demembranating the trophozoites by vortexing them for 3 min at the maximum setting. The demembranated cells and debris (about 1 ml) are transferred to a clean, sterile Eppendorf tube. Cytoskeletons are then pelleted by centrifugation at 16,000×*g* (maximum setting) for 5 min. After pelleting, the supernatant is decanted and the cytoskeletal pellet is washed four times in 1X PHEM plus Taxol (lacking the Triton X-100 detergent). After washing, the pellet is resuspended in 250 μl 1X PHEM and stored at 4°C. Extraction of cytoskeletons can be confirmed by wet mount using phase-contrast or DIC microscopy.

VII. Perspectives

Giardia is not known to produce toxins or specific virulence factors; it is the functioning of the giardial MT cytoskeleton that is essentially the etiologic agent of giardiasis. Perhaps the most important function of the giardial MT cytoskeleton is to promote the attachment of *Giardia* to the intestinal microvilli via the ventral disc. Beyond its role in pathogenesis, the cytoskeleton is critical for motility, intracellular transport, and cell division. MT dynamics are important during interphase, during cell division, and during encystation/excystation. The pivotal role that the MT cytoskeleton plays in pathogenesis highlights the need for a better understanding of giardial cytoskeletal biology (Elmendorf *et al.*, 2003). Future studies on the giardial MT cytoskeleton are sure to include the high-resolution imaging of complex MT structures such as the ventral disc, the identification of proteins and complexes comprising those novel structures, and the analysis of their function in *Giardia*'s complex biology during interphase and during cell division.

Conventional TEM-based analyses of the cytoskeletal architecture in *Giardia* have formed the basis of our understanding of *Giardia*'s complex ultrastructure, yet these analyses were done before modern methods of cryopreservation such as HPF and 3D cryoET without fixation were developed (Baumeister, 2005; McIntosh *et al.*, 2005; Subramaniam, 2005). Ultrastructural information derived from these older methods may have common TEM artifacts associated with specimen preparation, choice of fixation chemicals, and the process of thin sectioning itself. One new method of 3D imaging—cryoelectron tomography or cryoET—allows the imaging of native hydrated structures at <5 nm resolution in vitreous ice using liquid ethane. When combined with volume averaging (PEET software) of regular repetitive structures (Baumeister and Steven, 2000), this strategy has permitted the unprecedented visualization of hydrated molecular structures at high resolution, including individual tubulin subunits within a MT. 3D cryoET not only offers better resolution of structures but also minimizes fixation artifacts that could lead to misinterpretations of giardial cytoskeletal architecture.

In terms of the molecular identification of components of the cytoskeleton, only about 100 of the roughly 6000 ORFs identified in the 12 Mb giardial genome (Adam, 2000) are known to encode proteins that associate with the MT cytoskeleton (see Table I). The *Giardia* MT cytoskeleton has several novel structures—the funis, the median body, and the ventral disc (Dawson, 2010), and the *Giardia* genome contains many novel proteins. Thus, there could exist many novel MT-associated proteins in this important and widespread parasitic protozoan. Based on the complexity of cytoskeletal structures in *Giardia* like the ventral disc (Dawson, 2010), one would assume a similar complexity of protein composition. For example, over 250 proteins are known to comprise the flagellar axoneme (Ostrowski *et al.*, 2002; Pazour *et al.*, 2005), yet only 10 proteins are thus far known to localize to the complex MT spiral arrays and associated structure of *Giardia*'s ventral disc (Bauer *et al.*, 1999; Peattie, 1990; Weiland *et al.*, 2003; Weiland *et al.*, 2005; Nohria *et al.*, 1992; Weiland and McArthur, 2005) (Davids *et al.*, 2008). As binucleate diploid diplomonads are not amenable to classical forward genetic analyses, the use of morpholino-based knockdowns and the overexpression of cytoskeletal proteins with dominant-negative mutations should aid in assessing the function of novel MT-associated proteins in dynamic cytoskeletal movements.

Finally, one can easily argue that by far the majority of eukaryotic cytoskeletal diversity lies in the protists, reflecting their diverse evolutionary status. Fundamental research in cytoskeletal biology is often initiated in protists. "Non-model" once obscure microbial eukaryotes, such as *Tetrahymena* and *Chlamydomonas*, have been developed into robust experimental systems that have made profound contributions toward our understanding of cytoskeletal functioning and dynamics (Cande and McDonald, 1985; Gibbons and Rowe, 1965; Greider and Blackburn, 1987; Salisbury *et al.*, 1988) and flagellar biology. While *Giardia* has complex and novel cytoskeletal arrays, other protists mirror this complexity. Thus, the analysis of the giardial MT cytoskeleton should continue to inform basic studies of MTs in other eukaryotes.

Acknowledgments

We acknowledge members of the Cande Laboratory (UC Berkeley): Meredith Sagolla, Joel Mancuso, Meredith Carpenter, Alex Paredez, and Stephan Gourgechon, and Zac Cande; the Fletcher Laboratory (UC Berkeley): Wendy Hansen, Dan Fletcher; the Boulder Laboratory for the 3D Electron Microscopy of Cells: Cindi Schwartz and Andreas Hoenger; the Wang Laboratory (UCSF): Lei Li and C.C. Wang; as well as members of the Dawson Laboratory (UC Davis): David Woessner, Jonathan Pham, Michael Cipriano, and Moises de la Torre for development of many of the cytoskeletal procedures and protocols in *Giardia*. We also acknowledge Kari Hagen (UC Davis) for critical review of this chapter. S.C.D. is funded through NIH 5R01AI077571-02.

References

Adam, R. D. (2000). The Giardia lamblia genome. *Int. J. Parasitol.* **30**(4), 475–484.

Adam, R. D. (2001). Biology of Giardia lamblia. *Clin. Microbiol. Rev.* **14**(3), 447–475.

Bauer, B., Engelbrecht, S., *et al.* (1999). Functional identification of alpha 1-giardin as an annexin of Giardia lamblia. *FEMS Microbiol. Lett.* **173**(1), 147–153.

Baumeister, W. (2005). From proteomic inventory to architecture. *FEBS Lett.* **579**(4), 933–937.

Baumeister, W., and Steven, A. C. (2000). Macromolecular electron microscopy in the era of structural genomics. *Trends Biochem. Sci.* **25**(12), 624–631.

Benchimol, M., Piva, B., *et al.* (2004). Visualization of the funis of Giardia lamblia by high-resolution field emission scanning electron microscopy–new insights. *J. Struct. Biol.* **147**(2), 102–115.

Boleti, H., Karsenti, E., *et al.* (2001). The use of dominant negative mutants to study the function of mitotic motors in the in vitro spindle assembly assay in Xenopus egg extracts. *Methods Mol. Biol.* **164**, 173–189.

Brown, C. L., Maier, K. C., *et al.* (2005). Kinesin-2 is a motor for late endosomes and lysosomes. *Traffic* **6**(12), 1114–1124.

Brugerolle, G. (1975). Contribution à l′ étude cytologique et phylétique des diplozoaires (Zoomastigophorea, diplozoa, dangeard 1910). V. Nouvelle interpretation de l'organisation cellulaire de giardie. *Protistologica* **11**, 99–109.

Campanati, L., Holloschi, A., *et al.* (2002). Video-microscopy observations of fast dynamic processes in the protozoon Giardia lamblia. *Cell Motil. Cytoskeleton* **51**(4), 213–224.

Canada, P. H. A. O. (2010). Giardia lamblia—material safety data sheets (MSDS).http://www.phac-aspc.gc.ca/msds-ftss/msds71e-eng.php

Cande, W. Z., and McDonald, K. L. (1985). In vitro reactivation of anaphase spindle elongation using isolated diatom spindles. *Nature* **316**(6024), 168–170.

Carpenter, M. L., and Cande, W. Z. (2009). Using morpholinos for gene knockdown in Giardia intestinalis. *Eukaryot Cell* **8**(6), 916–919.

Crossley, R., and Holberton, D. V. (1983). Selective extraction with sarkosyl and repolymerization in vitro of cytoskeleton proteins from Giardia. *J. Cell Sci.* **62**, 419–438.

Crossley, R., and Holberton, D. (1985). Assembly of 2.5 nm filaments from giardin, a protein associated with cytoskeletal microtubules in Giardia. *J. Cell Sci.* **78**, 205–231.

Davids, B. J., Williams, S., *et al.* (2008). Giardia lamblia aurora kinase: A regulator of mitosis in a binucleate parasite. *Int. J. Parasitol.* **38**(3–4), 353–369.

Davis-Hayman, S. R., and Nash, T. E. (2002). Genetic manipulation of Giardia lamblia. *Mol. Biochem. Parasitol.* **122**(1), 1–7.

Dawson, S. C. (2010). An insider's guide to the microtubule cytoskeleton of Giardia. *Cell. Microbiol. Biol.* **12**(5), 588–598.

Dawson, S. C., Sagolla, M. S., *et al.* (2007). The cenH3 histone variant defines centromeres in Giardia intestinalis. *Chromosoma* **116**(2), 175–184.

Dawson, S. C., Sagolla, M. S., *et al.* (2007). Kinesin-13 regulates flagellar, interphase, and mitotic microtubule dynamics in Giardia intestinalis. *Eukaryot Cell* **6**(12), 2354–2364.

Décavé, E., Garrivier, D., *et al.* (2002). Shear flow-induced detachment kinetics of Dictyostelium discoideum cells from solid substrate. *Biophys. J.* **82**, 2383–2395.

Dentler, W. L., and Adams, C. (1992). Flagellar microtubule dynamics in Chlamydomonas: Cytochalasin D induces periods of microtubule shortening and elongation; and colchicine induces disassembly of the distal, but not proximal, half of the flagellum. *J. Cell Biol.* **117**(6), 1289–1298.

Ding, R., McDonald, K. L., *et al.* (1993). Three-dimensional reconstruction and analysis of mitotic spindles from the yeast, Schizosaccharomyces pombe. *J. Cell Biol.* **120**(1), 141–151.

Elmendorf, H. G., Dawson, S. C., *et al.* (2003). The cytoskeleton of Giardia lamblia. *Int. J. Parasitol.* **33**(1), 3–28.

Feely, D. E., and Erlandsen, S. L. (1981). Isolation and purification of Giardia trophozoites from rat intestine. *J. Parasitol.* **67**(1), 59–64.

Feely, D. E., and Erlandsen, S. L. (1982). Effect of cytochalasin-B, low Ca^{++} concentration, iodoacetic acid, and quinacrine-HCl on the attachment of Giardia trophozoites in vitro. *J. Parasitol.* **68**(5), 869–873.

Feely, D. E., and Erlandsen, S. L. (1985). Morphology of Giardia agilis: Observation by scanning electron microscopy and interference reflexion microscopy. *J. Protozool.* **32**(4), 691–693.

Feely, D. E. H., Holberton, D. V., and Erlandsen, S. L. (1990). The biology of Giardia. *In "Giardiasis"* (E. A. Mayer, ed),**Vol. 3**, pp. 11–49. Elsevier, Amsterdam and New York.

Feely, D. E., Schollmeyer, J. V., *et al.* (1982). Giardia spp.: Distribution of contractile proteins in the attachment organelle. *Exp. Parasitol.* **53**(1), 145–154.

Gaechter, V., Schraner, E., *et al.* (2008). The single dynamin family protein in the primitive protozoan Giardia lamblia is essential for stage conversion and endocytic transport. *Traffic* **9**(1), 57–71.

Gelfand, V. I., Le Bot, N., *et al.* (2001). A dominant negative approach for functional studies of the kinesin II complex. *Methods Mol. Biol.* **164**, 191–204.

Gibbons, I. R., and Rowe, A. J. (1965). Dynein: A protein with adenosine triphosphatase activity from cilia. *Science* **149**(3682), 424–426.

Gillin, F. D., and Reiner, D. S. (1982). Attachment of the flagellate Giardia lamblia: Role of reducing agents, serum, temperature, and ionic composition. *Mol. Cell. Biol.* **2**(4), 369–377.

Gillin, F. D., Reiner, D. S., *et al.* (1996). Cell biology of the primitive eukaryote Giardia lamblia. *Annu. Rev. Microbiol.* **50**, 679–705.

Greider, C. W., and Blackburn, E. H. (1987). The telomere terminal transferase of Tetrahymena is a ribonucleoprotein enzyme with two kinds of primer specificity. *Cell* **51**(6), 887–898.

Hansen, W. R., and Fletcher, D. A. (2008). Tonic shock induces detachment of Giardia lamblia. *PLoS Negl. Trop. Dis.* **2**(2), e169.

Hansen, W. R., Tulyathan, O., *et al.* (2006). Giardia lamblia attachment force is insensitive to surface treatments. *Eukaryot Cell* **5**(4), 781–783.

Hoeng, J. C., Dawson, S. C., *et al.* (2008). High resolution crystal structure and in vivo function of a kinesin-2 homolog in Giardia intestinalis. *Mol. Biol. Cell.* **19**(7), 3124–37.

Holberton, D. V. (1973). Fine structure of the ventral disk apparatus and the mechanism of attachment in the flagellate Giardia muris. *J. Cell Sci.* **13**(1), 11–41.

Holberton, D. V. (1973). Mechanism of attachment of Giardia to the wall of the small intestine.. *Trans. R. Soc. Trop. Med. Hyg.* **67**(1), 29–30.

Holberton, D. V. (1974). Attachment of Giardia-a hydrodynamic model based on flagellar activity. *J. Exp. Biol.* **60**(1), 207–221.

Holberton, D. V. (1981). Arrangement of subunits in microribbons from Giardia. *J. Cell Sci.* **47**, 167–185.

Holberton, D. V., and Ward, A. P. (1981). Isolation of the cytoskeleton from Giardia. Tubulin and a low-molecular-weight protein associated with microribbon structures. *J. Cell Sci.* **47**, 139–166.

Hunter, A. W., and Wordeman, L. (2000). How motor proteins influence microtubule polymerization dynamics. *J. Cell Sci.* **113**(Pt 24), 4379–4389.

Kabnick, K. S., and Peattie, D. A. (1990). In situ analyses reveal that the two nuclei of Giardia lamblia are equivalent. *J. Cell Sci.* **95**(Pt 3), 353–360.

Keister, D. B. (1983). Axenic culture of Giardia lamblia in TYI-S-33 medium supplemented with bile. *Trans. R. Soc. Trop. Med. Hyg.* **77**, 487–488.

Kline-Smith, S. L., Khodjakov, A., *et al.* (2004). Depletion of centromeric MCAK leads to chromosome congression and segregation defects due to improper kinetochore attachments. *Mol. Biol. Cell* **15**(3), 1146–1159.

Kulda, J., and Nohynkova, E. (1995). Giardia in humans and animals. In "Parasitic Protozoa" (J. P. Kreier, ed.), **Vol. 10**, pp. 225–423. Academic Press, Inc, San Diego.

Landy, A., (1989). Dynamic, structural, and regulatory aspects of lambda site-specific recombination. *Ann. Rev. Biochem.* **58**, 913–949.

Lin-Jones, J., Parker, E., *et al.* (2003). Disruption of kinesin II function using a dominant negative-acting transgene in Xenopus laevis rods results in photoreceptor degeneration. *Invest. Ophthalmol. Vis. Sci.* **44**(8), 3614–3621.

Lippincott-Schwartz, J., Snapp, E., *et al.* (2001). Studying protein dynamics in living cells.. *Nat. Rev. Mol. Cell Biol.* **2**(6), 444–456.

Magne, D., Favennec, L., *et al.* (1991). Role of cytoskeleton and surface lectins in Giardia duodenalis attachment to caco2 cells. *Parasitol. Res.* **77**(8), 659–662.

Mariante, R. M., Vancini, R. G., *et al.* (2005). Giardia lamblia: Evaluation of the in vitro effects of nocodazole and colchicine on trophozoites. *Exp. Parasitol.* **110**(1), 62–72.

McIntosh, R., Nicastro, D., *et al.* (2005). New views of cells in 3D: An introduction to electron tomography. *Trends Cell Biol.* **15**(1), 43–51.

Mitchison, T., and Kirschner, M. (1984). Dynamic instability of microtubule growth. *Nature* **312**(5991), 237–242.

Morrison, H. G., McArthur, A. G., *et al.* (2007). Genomic minimalism in the early diverging intestinal parasite Giardia lamblia. *Science* **317**(5846), 1921–1926.

Narcisi, E. M., Paulin, J. J., *et al.* (1994). Presence and localization of vinculin in Giardia. *J. Parasitol.* **80**(3), 468–473.

Nohria, A., Alonso, R. A., *et al.* (1992). Identification and characterization of gamma-giardin and the gamma-giardin gene from Giardia lamblia. *Mol. Biochem. Parasitol.* **56**(1), 27–37.

Nohynkova, E., Tumova, P., *et al.* (2006). Cell division of Giardia intestinalis: Flagellar developmental cycle involves transformation and exchange of flagella between mastigonts of a diplomonad cell. *Eukaryot Cell* **5**(4), 753–761.

Oka, M. T., Arai, T., *et al.* (1994). Different reactivity with monoclonal anti-tubulin antibodies between native and fixed mitotic microtubules in sea urchin eggs. *Cell Motil. Cytoskeleton* **29**(3), 241–249.

Ortega-Pierres, G., Smith, H. V., *et al.* (2009). New tools provide further insights into Giardia and Cryptosporidium biology. *Trends Parasitol.* **25**(9), 410–416.

Ostrowski, L. E., Blackburn, K., *et al.* (2002). A proteomic analysis of human cilia: Identification of novel components. *Mol. Cell Proteomics* **1**(6), 451–465.

Palm, D., Weiland, M., *et al.* (2005). Developmental changes in the adhesive disk during Giardia differentiation. *Mol. Biochem. Parasitol.* **141**(2), 199–207.

Pazour, G. J., Agrin, N., *et al.* (2005). Proteomic analysis of a eukaryotic cilium. *J. Cell Biol.* **170**(1), 103–113.

Peattie, D. A. (1990). The giardins of Giardia lamblia: Genes and proteins with promise. *Parasitol. Today* **6**(2), 52–56.

Perez, P. F., Minnaard, J., *et al.* (2001). Inhibition of Giardia intestinalis by extracellular factors from lactobacilli: An in vitro study. *Appl. Environ. Microbiol.* **67**(11), 5037–5042.

Piva, B., and Benchimol, M. (2004). The median body of Giardia lamblia: An ultrastructural study. *Biol. Cell.* **96**(9), 735–746.

Poxleitner, M. K., Carpenter, M. L., *et al.* (2008). Evidence for karyogamy and exchange of genetic material in the binucleate intestinal parasite Giardia intestinalis. *Science* **319**(5869), 1530–1533.

Raikov, I. B. (1994). The diversity of forms of mitosis in protozoa: A comparative review. *Eur. J. Protistol.* **30**(3), 252–269.

Regoes, A., and Hehl, A. B. (2005). SNAP-tag mediated live cell labeling as an alternative to GFP in anaerobic organisms. *Biotechniques* **39**(6), 809–810, 812.

Roxstrom-Lindquist, K., Palm, D., *et al.* (2006). Giardia immunity–an update. *Trends Parasitol.* **22**(1), 26–31.

Sagolla, M. S., Dawson, S. C., *et al.* (2006). Three-dimensional analysis of mitosis and cytokinesis in the binucleate parasite Giardia intestinalis. *J. Cell Sci.* **119**(Pt 23), 4889–4900.

Salisbury, J. L., Baron, A. T., *et al.* (1988). The centrin-based cytoskeleton of Chlamydomonas reinhardtii: Distribution in interphase and mitotic cells. *J. Cell Biol.* **107**(2), 635–641.

Savioli, L., Smith, H., *et al.* (2006). Giardia and Cryptosporidium join the "neglected diseases initiative." *Trends Parasitol.* **22**(5), 203.

Sawaguchi, A., Yao, X., *et al.* (2003). Direct attachment of cell suspensions to high-pressure freezing specimen planchettes. *J. Microsc.* **212**(Pt 1), 13–20.

Schuyler, S. C., and Pellman, D. (2001). Search, capture and signal: Games microtubules and centrosomes play. *J. Cell Sci.* **114**(Pt 2), 247–255.

Sousa, M. C., Goncalves, C. A., *et al.* (2001). Adherence of Giardia lamblia trophozoites to int-407 human intestinal cells. *Clin. Diagn. Lab. Immunol.* **8**(2), 258–265.

Subramaniam, S. (2005). Bridging the imaging gap: Visualizing subcellular architecture with electron tomography. *Curr. Opin. Microbiol.* **8**(3), 316–322.

Sullivan, K. F., Kay, S. A., *et al.* (1999). *Green Fluorescent Proteins.* Academic Press, San Diego, CA; London.

Sun, C. H., Su, L. H., *et al.* (2005). Influence of 5′ sequences on expression of the Tet repressor in Giardia lamblia. *Mol. Biochem. Parasitol.* **142**(1), 1–11.

Sun, C. H., and Tai, J. H. (2000). Development of a tetracycline controlled gene expression system in the parasitic protozoan Giardia lamblia. *Mol. Biochem. Parasitol.* **105**(1), 51–60.

Tilney, L. G., and Gibbins, J. R. (1968). Differential effects of antimitotic agents on the stability and behavior of cytoplasmic and ciliary microtubules. *Protoplasma* **65**(1), 167–179.

Touz, M. C., Conrad, J. T., *et al.* (2005). A novel palmitoyl acyl transferase controls surface protein palmitoylation and cytotoxicity in Giardia lamblia. *Mol. Microbiol.* **58**(4), 999–1011.

Tumova, P., Kulda, J., *et al.* (2007). Cell division of Giardia intestinalis: Assembly and disassembly of the adhesive disc, and the cytokinesis. *Cell Motil. Cytoskeleton* **64**(4), 288–298.

Weiland, M. E., McArthur, A. G., *et al.* (2005). Annexin-like alpha giardins: A new cytoskeletal gene family in Giardia lamblia. *Int. J. Parasitol.* **35**(6), 617–626.

Weiland, M. E., Palm, J. E., *et al.* (2003). Characterisation of alpha-1 giardin: An immunodominant Giardia lamblia annexin with glycosaminoglycan-binding activity. *Int. J. Parasitol.* **33**(12), 1341–1351.

Winey, M., Mamay, C. L., *et al.* (1995). Three-dimensional ultrastructural analysis of the Saccharomyces cerevisiae mitotic spindle. *J. Cell Biol.* **129**(6), 1601–1615.

Woods, A., Sherwin, T., *et al.* (1989). Definition of individual components within the cytoskeleton of Trypanosoma brucei by a library of monoclonal antibodies.. *J. Cell Sci.* **93**(Pt 3), 491–500.

CHAPTER 18

Live Cell-Imaging Techniques for Analyses of Microtubules in *Dictyostelium*

Matthias Samereier[*], **Irene Meyer**[*], **Michael P. Koonce**[†], *and* **Ralph Gräf**[*]

[*]Department of Cell Biology, Institut for Biochemistry and Biology, University of Potsdam, Potsdam-Golm 27708, Germany D-14476

[†]Division of Translational Medicine, Wadsworth Center, Albany, New York 12201-0509

Abstract

Dictyostelium amoebae provide a popular model system for analyses of cell and cytoskeletal dynamics. Yet, the sensitivity of *Dictyostelium* cells to phototoxic effects, their rapid cell movement, and the extraordinary motility of their microtubule system are specific challenges for live cell imaging. The protocols outlined in this chapter are optimized to minimize these challenges, using *Dictyostelium* cells expressing green

978-0-12-381349-7
DOI: 10.1016/S0091-679X(10)97018-0

fluorescent tubulin or microtubule plus-end markers such as TACC. We describe suitable specimen preparations, treatments with microtubule-depolymerizing drugs, and applicable settings on wide-field and confocal microscopy systems for four-dimensional time-lapse and fluorescence recovery after photobleaching analyses of microtubule dynamics.

I. Introduction

Dictyostelium discoideum cells are free-living soil amoebae that feed on other microorganisms. Although positioned at the evolutionary base of today's metazoans and only distantly related to vertebrates, they possess many similarities to amoeboid mammalian cells such as neutrophils or macrophages. Furthermore, upon starvation chemotactic cAMP signaling initiates a developmental program leading to a multicellular fruiting body with two types of differentiated cells. Axenic laboratory strains can easily and cost-effectively be grown to high cell densities in liquid media and are well accessible for biochemical, genetic, and microscopy studies. Due to this unique combination of properties, *Dictyostelium* amoebae constitute a well-established model system for the analyses of phagocytosis, development, chemotaxis, signaling pathways, cell migration, and cytoskeletal functions (Kessin, 2001).

While the actin system plays the major role in cell motility, the microtubule system is crucial for vesicle transport, organelle positioning, mitotic spindle organization, and cytokinesis. The actin and microtubule systems cross talk with one another, especially at the cell cortex where microtubule plus-end-associated proteins appear to interact with the cortical actin network (Hestermann *et al.*, 2002). The microtubule system of *Dictyostelium* cells is strikingly similar to those in amoeboid mammalian cells. It is organized by a centrosome, which remains firmly attached to the nucleus. The *Dictyostelium* centrosome contains no centrioles but instead is composed of a layered core structure surrounded by a so-called electron-dense corona. The corona contains dense nodules that harbor γ-tubulin nucleation complexes (Euteneuer *et al.*, 1998). Unlike centrosomes in higher organisms, the *Dictyostelium* centrosome duplicates at the G2/M transition and organizes an intranuclear spindle for a closed mitosis (reviewed by Gräf, 2009). *Dictyostelium* expresses one each α-, β-, and γ-tubulin, whose primary sequences diverge considerably from corresponding tubulins in higher organisms. Microtubule function further involves a number of associated proteins (MAPs). In addition to 13 kinesin and one dynein motors (Kollmar and Glöckner, 2003; Koonce *et al.*, 1994; Tikhonenko *et al.*, 2009), homologs for MAPs such as EB1, DdCP224, Lis1, and TACC have been characterized at the molecular level (Gräf *et al.*, 2000; Koch *et al.*, 2006; Rehberg and Gräf, 2002; Rehberg *et al.*, 2005). The dynamics of these interesting tubulins and MAPs can so far only be studied *in vivo* in *Dictyostelium*, since it has not yet been practical to isolate polymerization-competent α/β-tubulin from these cells (e.g., see Koonce and Khodjakov, 2002).

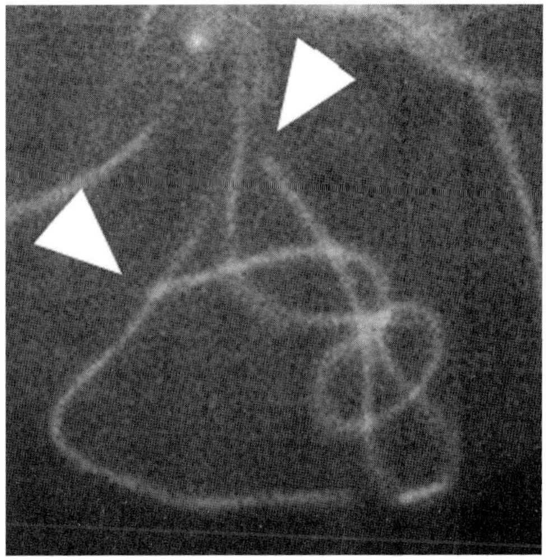

Fig. 1 Microtubule flexibility. Still frame from a movie sequence showing the degree by which microtubules can be bent or twisted *in situ*. Arrowheads mark the two ends of this single microtubule. Robust dynein and kinesin motor activities anchored on organelles or the cell cortex are likely responsible for the bending and lateral microtubule movements.

Live cell imaging of GFP-α-tubulin-labeled microtubules (GFP = green fluorescent protein) and cells expressing GFP-labeled MAPs has demonstrated that *Dictyostelium* microtubules show prominent lateral bending movements and are relatively stable in length (Brito *et al.*, 2005; Kimble *et al.*, 2000; Koonce and Khodjakov, 2002). Indeed, one interesting feature of *Dictyostelium* microtubules is their tremendous flexibility (Fig. 1). Although microtubules exhibit alterations in length at the cell periphery close to the cortex, there are almost no complete catastrophes and rescues during interphase as we illustrate in this chapter (Movie 2). However, microtubule dynamics do change dramatically during mitosis [reviewed by (Gräf, 2009)]. In early prophase, the corona is shed from the duplicating centrosome and is accompanied by the depolymerization of all microtubules (Fig. 2). By prometaphase the centrosomal core structure has split into two pole bodies that remain connected through a nascent central spindle. In metaphase, cytoplasmic astral microtubules begin to grow and elongate continuously until telophase. *Dictyostelium* microtubules are also unusual in regard to their sensitivity against microtubule-depolymerizing drugs such as nocodazole or thiabendazole (TBZ). Although these drugs are quite effective in suppressing spindle formation during mitosis, they are relatively ineffective during interphase where even high concentrations (200 μM for 3 h) cause only shortening, but not complete depolymerization of microtubules.

Compared to many types of vertebrate cells, the interphase *Dictyostelium* centrosome nucleates a relatively small number (30–70) of microtubules (Kuriyama *et al.*, 1982). This feature readily enhances the ability to follow activities of individual microtubules. However, the analysis of fluorescently labeled microtubules in *Dictyostelium* does

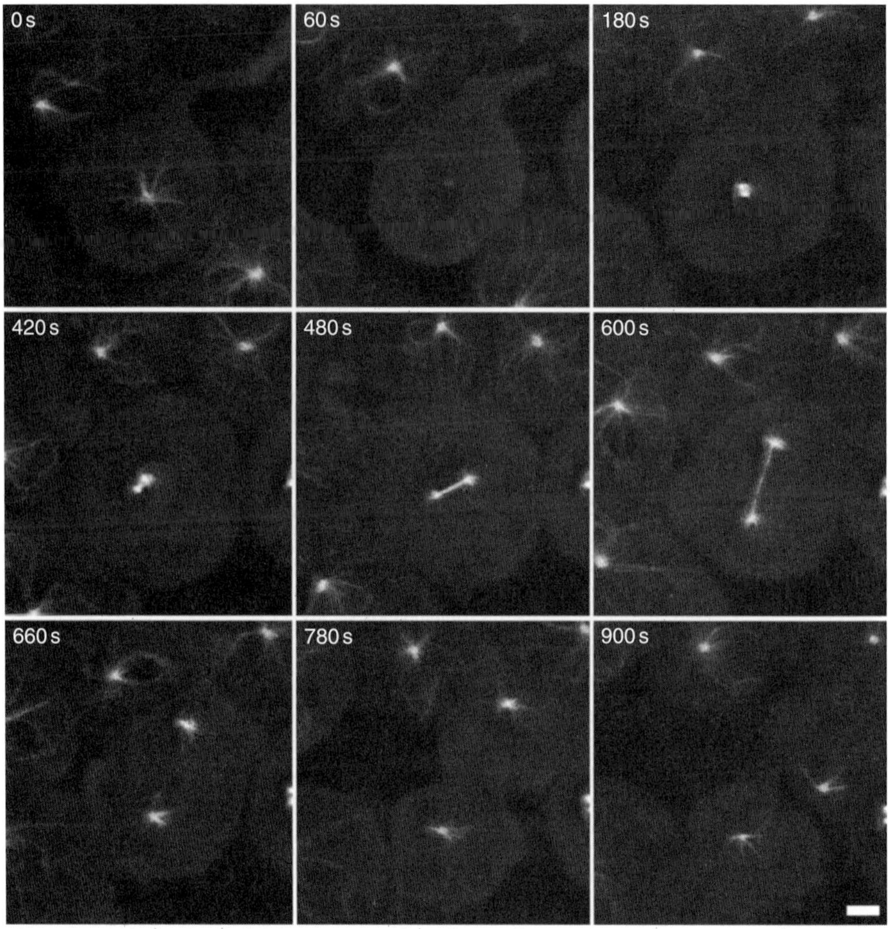

Fig. 2 Microtubule behavior during mitosis by wide-field fluorescence deconvolution microscopy. The figure shows a montage of selected time points from Movie 1. GFP-α-tubulin cells were viewed under agar overlay in a Zeiss CellObserver HS system as described in the text. Live cell imaging of z-stacks with five focus layers occurred at 5 frames/s. Time-lapse was 15 s. Image stacks were deconvolved using the fast iterative deconvolution of Axiovision 4.6 and projected into one plane by maximum intensity projection. Note that the microtubule cytoskeleton depolymerizes completely in prophase (time point 60 s). Growth of astral microtubules occurs after spindle elongation has started (time point 480 s). Bar = 2 μm.

present a few challenges. First, *Dictyostelium* amoebae are not flat, and individual microtubules often curve in and out of the focal plane. Second, *Dictyostelium* amoebae are highly motile and tend to crawl out of the field of view or defined region of interest (ROI). Third, the rapid, bending movements of microtubules render it difficult to follow individual microtubules over time, especially at their plus-ends. Fourth, *Dictyostelium* cells are particularly sensitive to the blue excitation light used to elicit GFP fluorescence. Without special precautions, there are indications of phototoxic effects such as mitotic arrest and cell rounding/blebbing, long before bleaching of GFP becomes evident. Due to these facts, analyses of microtubule dynamics in *Dictyostelium* require microscopes with highly sensitive detectors, capable of fast image acquisition and three-dimensional (3D) stacking. While the examination of individual microtubule dynamics is often difficult using wide-field and conventional laser-scanning microscopes, the measurement of the behavior of microtubule arrays within a defined ROI using the fluorescence recovery after photobleaching (FRAP) can be easier to realize and very informative.

II. Rationale

Dictyostelium amoebae offer a robust, highly dynamic experimental system that integrates cell motility with molecular genetic approaches for detailed analyses of microtubule activities. In this chapter, we provide protocols for the optimization of live cell-imaging conditions and the study of their microtubule behavior using wide-field and laser-scanning confocal microscopes.

III. Specimen Preparation for Live Cell Imaging of *Dictyostelium* Amoebae

This protocol for live cell imaging of mitotic and interphase *Dictyostelium* cells is suitable for inverted microscopes.

A. Materials and Media

- Glass-bottom dishes: 35 mm Petri dishes with 0.17 mm optical quality uncoated glass bottom (e.g., Fluorodish cell culture dishes, World Precision Instruments, Sarasota, FL, USA).
- HL5c Medium (Formedium, Hunstanton, UK).

  Ready to use; contains 5 g peptone, 5 g yeast extract, 5 g tryptone, 1.2 g KH_2PO_4, 0.35 g Na_2HPO_4 per liter. Autoclave for 20 min.

- Low fluorescence (LoFlo) Medium (Liu *et al.*, 2002) (Formedium).

  Ready to use; contains 11 g glucose, 0.68 g KH_2PO_4, 5 g casein peptone, 26.8 mg NH_4Cl, 37.1 mg $MgCl_2$, 1.1 mg $CaCl_2$, 8.11 mg $FeCl_3$, 4.84 Na_2-EDTA, 2.3 mg $ZnSO_4$, 1.11 mg H_3BO_3, 0.51 mg $MnCl_2{\cdot}4H_2O$, 0.17 mg $CoCl_2$, 0.15 mg $CuSO_4{\cdot}5H_2O$, 0.1 mg $(NH_4)_6Mo_7O_{24}{\cdot}4H_2O$ per liter. Adjust pH to 6.5 and autoclave for 20 min.

- Phosphate buffer: 14.6 mM KH_2PO_4, 2 mM Na_2HPO_4.
- Vitamin C = sodium L(+) ascorbate, 200 mg/ml.
- TBZ: 200 mM, dissolved in dimethyl sulfoxide (DMSO). Store at –20° C.

B. *Dictyostelium* Cell Preparation

All media should be prewarmed to room temperature before use.

- Grow *Dictyostelium* cells [GFP-α-tubulin strain (Rehberg and Gräf, 2002)] at 21° C in HL5c medium in adherent culture in tissue culture flasks to about 50% confluency.
- At the day before use, or at least 3 h before use, resuspend cells, count cells, and dilute to a density of 10^5 cells/ml in LoFlo medium.

C. Microscopic Specimen Preparation

- Add 1 ml of cell suspension at the center of the glass-bottom dish and let the cells settle for 10 min. Remove the medium and replace with 2 ml of fresh LoFlo medium.
- Add L-(+)-ascorbate as a radical scavenger to reduce phototoxic effects to a final concentration of 2 mg/ml. This step turned out to be very important for cell survival under fluorescent imaging conditions.
- Incubate overnight or for at least 3 h at 21° C.

Optional: Agar overlay (Fukui *et al.*, 1987)

- Dissolve 2% agarose in phosphate buffer by heating in a microwave oven.
- Place two coverslips at the narrow edges of a microscopic slide, pipet ~250 μl of molten agarose onto the center of the slide and put a second slide on top by pressing on the two narrow edges. The coverslips serve as spacers to create an agar sheet of the same thickness. After 10 min at room temperature carefully lift off the upper slide and cut the agar sheet in pieces of approximately 1 × 2 cm.
- Lift off the agar sheet with the edge of a coverslip and gently slide it into the dish without dislodging the attached cells. Remove the LoFlo Medium with a pipette from the side so that the agar sheet gently settles onto the cells. Remove excess liquid with a Kimwipe. The extent of liquid removal determines the degree of cell flattening. Put a rolled piece of wetted Kimwipe at the side of the dish and close the lid to create a humid chamber. The wetted Kimwipe is necessary to keep the agarose sheet from drying completely, and it must be positioned so that it does not touch the agar. Under these conditions, cells can be imaged for 3–4 h.

Optional: TBZ treatment under agar overlay.

If the agar overlay should be combined with TBZ treatment for microtubule depolymerization, the agar sheet must contain the same concentration of drug as the medium. Since microtubule-depolymerizing drugs are temperature sensitive, they cannot be added to the hot, liquid agarose. Therefore, 3 h before use TBZ at a final concentration of 200 μM is added to the glass-bottom dish containing the cells, LoFlo medium, and the agarose sheet. The time should be sufficient to equilibrate the agar sheet with the drug-containing medium. Of course, preincubation of the agar sheet with the TBZ -containing medium should be omitted, if cells should be released from TBZ treatment at the time of agar overlay.

Notes: The standard HL-5 growth media is autofluorescent. Imaging cells directly in this media results in a high level of background fluorescence, with particular accumulation in endosomes and vacuoles. Thus, it is important to wash cells in LoFlo medium before use. We have also had excellent imaging results with cells attached to coverslips in the presence of phosphate buffer. Cells remain viable and mitotically active in phosphate buffer for many hours, but since this buffer does not contain any growth nutrients, it should only be used 1–2 h before imaging cells.

Rose chambers can also be used in place of the glass-bottom dishes.

IV. Setup and Settings for Live Cell Fluorescence Microscopy of *Dictyostelium* Microtubules

General considerations:

- As with any microscopic recording system, there is always a trade-off between image quality and acquisition rate. One has to find a compromise where both image quality and acquisition speed are sufficient to yield a dataset that can be evaluated.
- *Dictyostelium* cells are sensitive to phototoxic effects. Therefore, imaging conditions must be optimized to reduce light exposure to the minimum required to obtain a desired image quality.
- Due to the thickness of *Dictyostelium* amoebae, extension of microtubule arrays and centrosomal movements in the axial direction are considerable. Thus, in order to capture the whole volume of interest in the cell, recording of image stacks, i.e., consecutive focal planes, is required. To alleviate this problem, cells can be flattened with agarose (Fukui *et al.*, 1987) to considerably reduce the number of image planes required to capture the whole microtubule cytoskeleton. This relatively gentle procedure has the additional advantage of reducing the mobility of amoeboid cells. Depending on the degree of flattening, agar overlaid *Dictyostelium* cells are roughly 3–6 μm thick in the axial direction. Following recording, the images of one focus stack are projected onto a single plane using a maximum intensity projection algorithm (as provided for example by ImageJ). To minimize the movement of fluorescent structures and to avoid artifacts, recording of individual stacks requires high frame rates with the minimum number of images collected that contain objects of interest.
- If mitosis is of special interest, cells should be cultivated overnight in LoFlo medium in glass-bottom observation dishes as described above. Under these conditions, cells divide as normal and a sufficient number of mitotic cells can readily be found during the image acquisition period. This approach is preferred over cell synchronization treatments such as cold treatment or TBZ treatments (Maeda, 1986; Rehberg and Gräf, 2002). First, these treatments are not very effective in *Dictyostelium*. Second, although many dividing cells can be found during a temporal time peak following treatment release (up to 20% of all cells), this wave usually ceases during the acquisition time of the first movie so that there are almost no mitotic cells left for a second trial. If cytokinesis should be observed, cells must not be made too flat by the agar overlay procedure.

Precautions:

- Search for desired regions on the coverslip in bright field or phase contrast modes and not with fluorescent light. If fluorescent light is necessary, close the aperture stop of the reflected light beam path and dim the light source (if possible) to an extent that is just sufficient to see the cells.
- Provide a constant temperature of 21° C. This could be achieved either by placing the whole microscope setup in an air-conditioned room or by using temperture-controlled incubation boxes covering the objectives and the microscope stage. Alternatively temperature-controlled stages and objective collars may be used. Shared microscopes that are also used at 37° C for mammalian cell work must be allowed to cool to room temperature before use with *Dictyostelium*.

A. Wide-Field Microscopy

We use a Zeiss Cell Observer HS system on a Zeiss Axiovert 200 M inverted microscope. This setup includes a Sutter DG-4 wavelength changer, an ASI piezo focusing stage, a Uniblitz shutter in the transmitted light beam path, and a Axiocam MRm Rev. 3 or Axiocam HSm CCD camera. Focusing steps and light source management are triggered by a camera signal to increase acquisition speed.

Optimization of the microscope:

- Use a highly sensitive CCD camera (e.g., Axiocam MRm Rev. 3, Roper Coolsnap HQ, Hamamatsu Orca ER or similar types).
- Use objectives with the highest level of light transmission. These are not necessarily ones with the highest numerical aperture. Objectives that do not correct spherical aberration usually contain less lenses and thus have a higher transmission. On standard setups with a 1× camera adaptor, any of the above-mentioned cameras covers only the central part of the field of view and thus spheric aberration is usually not a big concern. For example, a Zeiss Fluar 40×/1.3 NA has a clearly higher transmission than the EC Plan-Neofluar 40×/1.3 NA and both are superior to the Plan-Apochromat 63×/1.4 NA in this regard. Optimal lenses for live cell imaging of *Dictyostelium* cells are Zeiss Fluar 40×/1.3 and Zeiss LCI Plan-Neofluar 63×/1.3 or equivalents from other companies.
- Avoid HBO lamps. These lamps have strong UV emission lines, which are not completely blocked by most excitation filters. UV light is even more harmful to cells than visible blue light. For imaging of GFP or mRFP, halogen lamps are sufficiently bright and emit no UV light. In the case of Xenon lamps connected with a liquid light guide, choose a light guide with a low UV transmission. These are also better suited for imaging of red fluorescent proteins. For example, the Lumatec liquid light guide series 380 is preferred over series 300 (Lumatec, Deisenhofen, Germany).
- Speed up z-stacking by using a piezo focusing unit, e.g., a piezo stepper or a piezo insert in the microscopic stage.

Optimization of the acquisition program:

- Choose the correct filter sets for excitation and emission, respectively.
- Use program settings that ensure the excitation light is turned on only during image acquisition (i.e., no excitation light during focusing, wavelength change, and time-lapse). This can be accomplished by appropriate settings for the Uniblitz shutter, Sutter Lambda DG-4 wavelength changer, or similar equipment.
- Depending on the task, 2 × 2 binning may be appropriate since it reduces exposure time (but also pixel resolution) by a factor of 4 and increases acquisition speed.
- In cells flattened by agar overlay (see above), set z-stack settings to five frames per stack at a frame spacing of 0.8 μm and select the center frame at the main focal point of interest.
- Choose an exposure time of 100–200 ms (or less if possible) for centrosomes and microtubules. Excitation light intensity and exposure time may be reduced until there is still 5% coverage of a 12-bit gray-level scale (i.e., ~200 gray levels). Under these conditions, image quality remains good after output level adjustment owing to the low noise of modern CCD cameras. Generally avoid conditions that saturate pixels in the areas of interest.

An example for wide-field imaging of mitotic GFP-α-tubulin cells under agar overlay is shown in Fig. 2 (Movie 1). Note that single microtubules are clearly discernible, there is no considerable bleaching during the acquisition time of 18.5 min and that cytokinesis proceeds normally.

B. Confocal Microscopy

The most useful setups for confocal microscopy are multibeam systems such as the spinning disk or array scanning systems (Gräf *et al.*, 2005). However, the more commonly available single beam laser-scanning confocal systems are also suitable for live cell observation of microtubules and centrosomes. Single beam systems are also well suited for FRAP experiments, whereas multibeam systems require an additional independent scanner for bleaching regions of interest. The following protocols are aimed at conventional single beam laser-scanning microscopes such as a Zeiss LSM710, which was used for the confocal images shown in Fig. 3, Movie 2. For camera-based multibeam confocal systems the protocol for wide-field microscopy is applicable as well.

Optimization of the microscope:

- In theory the highest possible optical resolution depends on the numerical aperture of the objective. Yet, under practical conditions in live cell-imaging experiments resolution can strongly be affected by noise or artifacts due to movements of the object of interest. Thus, as in wide-field microscopy an objective with the highest optical transmission may provide better results than a fully corrected high NA-lens such as a Plan-Apochromate 63×/1.4. Higher transmission allows lower detector voltage settings at photomultiplier-based systems and thus produces less image

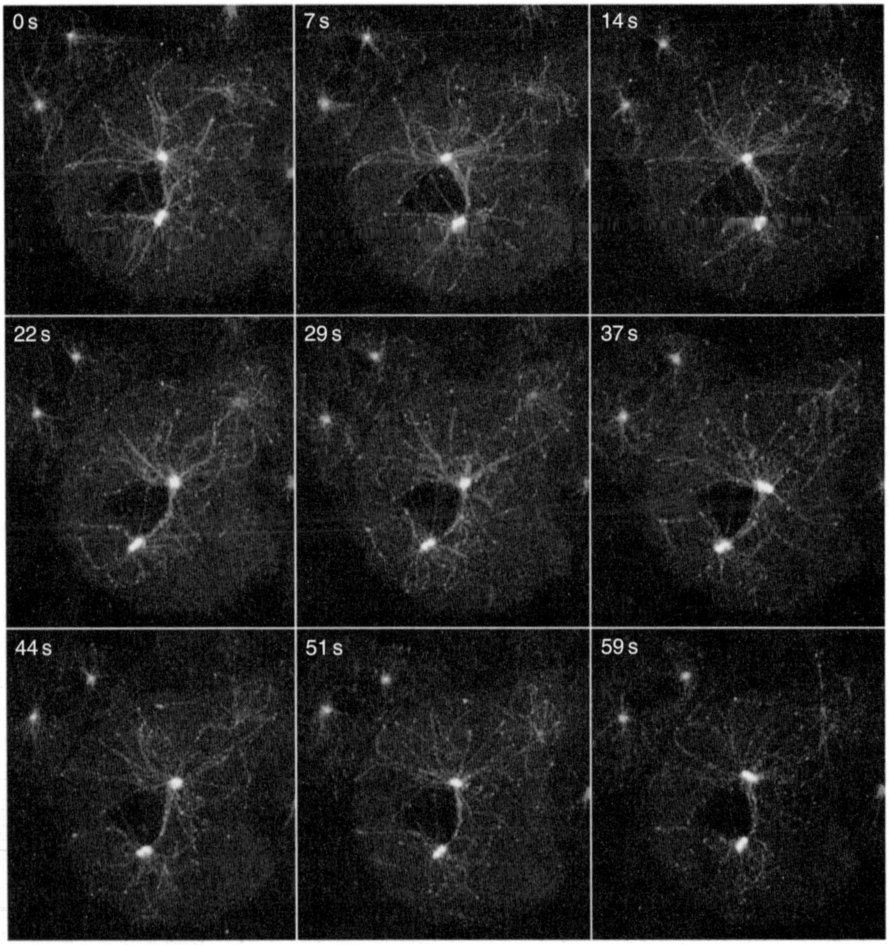

Fig. 3 Dynamics of individual microtubule plus ends can be resolved by confocal laser scanning microscopy. The figure presents a montage of selected time points from Movie 2. Cells expressing GFP-TACC-domain (green) and mRFP-α-tubulin (red) were viewed under agar overlay at a Zeiss LSM710 system as described in the text. Live cell imaging of z-stacks with 5 focus layers occurred at 1.5 frames/s without time-lapse. Image stacks were projected into one plane by maximum intensity projection. (See Plate no. 11 in the Color Plate Section.)

noise. Reduced noise often abrogates a need for frame averaging and thus increases acquisition rates.

- Avoid phototoxic stress in your specimen while searching for a ROI to scan. Thus, follow the same rules regarding fluorescence excitation and light sources as stated above for wide-field systems.
- Speed up z-stacking by using a piezo focusing unit, e.g., a piezo stepper or a piezo insert in the microscopic stage.

Optimization of the acquisition settings:

- Choose the appropriate laser line for excitation (e.g., 488 nm for GFP, 561 nm for mRFP) and the appropriate emission filter or, on emission filter-free systems, the optimal detection window for the chosen fluorophore.
- Set the pinhole size to 1 airy unit (AU). Pinhole size may be increased up to 2 AU in case of weak signals, instead of increasing the excitation intensity.
- Set the zoom factor to a value not higher than 4 (63×/1.4 objective) or 6 (40×/1.3 objective). The higher the zoom factor, the higher the light dose to individual cells. Photobleaching increases with the square of the zoom factor (Centonze and Pawley, 2006).
- Try to meet the Nyquist criterion when setting the pixel dimensions of the image; pixel dimensions should be half the size of the lateral optical resolution. Thus, the required pixel dimensions depend on the chosen zoom factor. Smaller pixel dimensions may be appropriate if the frame rate has to be increased.
- Adjust the frame rate to 100–200 ms (or less if possible) for centrosomes and microtubules. There are four possibilities to increase the frame rate:
- Scan speed: this should be set to the maximum or almost to the maximum.
- Line step: A line step of "2" means that every second line is scanned and the data in-between is interpolated. This increases the acquisition rate by a factor of two with only minor effects on image quality.
- Bidirectional scan: this speeds up frame rate by a factor of two. However, the shift between lines scanned in forth and back direction, respectively, has to be corrected by the scan software. This correction must be repeated whenever scan speed or scan field are altered in any direction. Therefore, this option to increase scan speed is awkward and therefore usually the last resort to increase scan speed.
- Reduce scan field in *y*-direction if possible. For example, acquisition of a 512 × 512 pixel image requires twice the time of a 512 × 256 pixel image.
- Settings of laser power, transmission by the acousto optical tunable filters (AOTF), detector gain, and digital gain depend on the fluorophore and the confocal microscope. Generally, to reduce phototoxic effects and bleaching, excitation intensity should be as low as possible to still allow recording of an image with a moderately low noise. At the Zeiss LSM710 and with GFP-α-tubulin, we usually use 50% laser power of the 30 mW Ar laser, 1% AOTF transmission of the 488 nm laser line, 500 V detector gain, and maximal digital gain (at this instrument digital gain does not increase noise). In contrast to camera-based systems, the 12-bit gray-level scale should be covered at least by 50% due to the usually higher noise of the PMT-based systems. Here a great difference between the highest recorded gray level and the highest output gray level (usually 4095 = white) results in an unpleasant appearance of the image after output level adjustment. Generally avoid saturated pixels in areas of interest.
- For agar-overlayed cells (see above), set z-stack settings to five frames per stack at a frame spacing of 0.8 μm and set the center frame to the main focal plane of interest.

Figure 3 shows an example of a confocal time-lapse recording of a *Dictyostelium* strain-expressing red fluorescent marsRFP-α-tubulin (Müller-Taubenberger *et al.*, 2006) and the GFP–TACC domain as a marker for microtubule plus-ends (Koch *et al.*, 2006). Growth and shrinkage events of individual microtubules in the cell periphery are clearly discernible.

V. Analysis of Microtubule Dynamics by FRAP

FRAP is an elegant way to study protein dynamics *in vivo* (Axelrod *et al.*, 1976; Rabut and Ellenberg, 2005). This section describes the design of a FRAP experiment to study microtubule and centrosome dynamics in *Dictyostelium*. In this method, a defined ROI among the fluorescent structures is photobleached, followed by monitoring fluorescence recovery within the bleached region. Since recovery can only occur through transport or diffusion of nonbleached fluorescent molecules into the bleached region, it is used as measures for protein mobility and macromolecular assembly. FRAP and the related inverse FRAP (iFRAP) or fluorescence loss in photobleaching experiments can be performed with any single beam laser-scanning confocal microscope or camera-based system equipped with a laser-scanning device for bleaching.

In a FRAP experiment, ROIs for bleaching and measurement have to be selected in a 3D time series recording. Three such regions are necessary (Fig. 4): (1) a ROI for bleaching and measurement of recovery (ROI), (2) a background ROI in a region with no cells (BG), (3) a control ROI for measurement of constitutive bleaching during image acquisition (so-called acquisition bleaching; CO). Background measurement is not necessary if the background is set to zero using the offset control of the confocal microscope. Measurement of acquisition bleaching should occur in a second cell to avoid fluorescence loss due to movement of fluorophores into the bleached ROI.

In conventional confocal laser-scanning microscopes, settings of excitation light intensity and ROIs are interpreted by the AOTF. This device has switching times in the

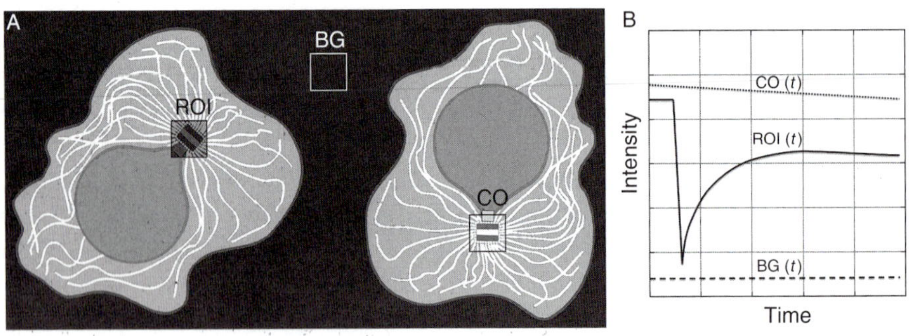

Fig. 4 A. Schematic drawing of selected regions for a FRAP experiment. ROI = bleached region of interest, BG = background region of interest, CO = control region of interest for acquisition bleaching. B. Idealized graphic display of bleaching and recovery processes, plotting fluorescent intensity versus time.

microsecond range and turns the excitation light on once the scanner arrives at the defined ROI and turns it off again as soon as it leaves the ROI. Furthermore, it controls the percentage of laser light transmission. The imaging phase of a FRAP experiment typically requires similar principal settings as in the 3D time series experiment described above (see Section III. B.) with relatively low excitation intensities to avoid phototoxic effects and acquisition bleaching of the whole cell. However, the bleaching phase should be as fast and as complete as possible to avoid perturbing molecules that were outside the ROI at the beginning of the bleach but migrated into the ROI during the photoexposure. Thus, the AOTF is set to 100% transmission during the bleach and to very low transmission during the imaging phase (1% or lower for GFP-α-tubulin).

A. Optimal FRAP settings:

FRAP settings have to be optimized for each type of experiment. The following gives just a general guideline:

- Turn the output power of the excitation laser to the maximum.
- Follow the protocol in Section III. B. for the imaging phase. Reduce AOTF transmission accordingly, if laser output power is reduced to 50% in standard time series imaging to preserve the laser.
- If acquisition bleaching is an issue, it may make sense to open the pinhole, since this allows the reduction of excitation intensity without affecting brightness of the signal.
- Set the FRAP conditions in the bleaching dialog of your scanning software.
- Choose the appropriate laser line for the bleach (e.g., 488 nm for GFP, 561 nm for mRFP).
- Set AOTF transmission to 100% for the bleaching process.
- The scan speed for the bleaching process depends on the size of the ROI and the kinetics of recovery. The bleaching effect increases with increased pixel dwell times.
- Set the number of iterations for the bleach (higher numbers mean a higher light dose but longer bleaching times). Use the lowest number of iterations that leads to complete or almost complete bleaching. In case of *Dictyostelium* GFP-α-tubulin we use 10 iterations at the same, fast scan speed as used for imaging.
- Set the number of prebleach time points (3–10).
- Set an excessive number of time points in the time series dialog to avoid ending the time series before recovery is completed.

B. Data evaluation:

Data evaluation is described here for the free software package ImageJ (Wayne Rasband, http://rsb.info.nih.gov/ij/). This software offers intensity measurements in defined ROIs and several plugins that may be useful for data evaluation such as "FRAP Profiler" and "intensity versus time measurement." However, like other standard software packages for FRAP experiments, these plugins are not suitable to evaluate recovery events of moving objects, unless the individual images are aligned. While a free alignment tool for ImageJ, "TurboReg," is available (Thevenaz *et al.*, 1998), this tool is also not suitable for amoeboid cells with laterally motile, fluorescent microtubules.

Thus, data evaluation of GFP-α-tubulin FRAP experiments in *Dictyostelium* has to be performed manually as follows:

- Make a maximum intensity projection of the image stack using the "grouped z-projector" (Plugins/Stacks-analysis menu).
- Define a ROI for measurement with a suitable selection tool (usually a "rectangular selection").
- Measure intensity using the "Measure" command in the "Analyze" menu.
- Proceed with the next image. Reposition the measurement ROI with the cursor keys, if necessary. All measurements are recorded in a single table, which can be saved as a text file and is compatible with standard calculation programs such as Microsoft Excel.
- Measure a control ROI in the same way to calculate acquisition bleaching.
- Define a background region to measure the background intensity that has to be subtracted from each value.
- Each measured FRAP value has to be corrected for background and acquisition bleaching to yield corrected fluorescence intensities. Within the relevant time period fluorescence decay according to acquisition bleaching is considered as linear.

$$\frac{\mathrm{ROI}\ (t) - \mathrm{BG}}{\mathrm{ROI}\ (t = 0) - \mathrm{BG}} \times (1 - t \times m_{\mathrm{ab}}) = \mathrm{FRAP\ (normalized,\ corrected)}$$

where m_{ab} (has a negative value) is the slope of the regression line of acquisition bleaching and is derived from the calculation program used to plot the acquisition bleach values (e.g., Microsoft Excel).

Another possibility to evaluate FRAP data is to use the individual values of ROI(t) and CO(t) to calculate corrected fluorescence intensity values (Rabut and Ellenberg, 2005). However, in this method, fluctuations of acquisition bleaching intensity values along the regression line increase intensity fluctuations in FRAP (corrected):

$$\frac{\mathrm{ROI}\ (t) - \mathrm{BG}}{\mathrm{CO}\ (t) - \mathrm{BG}} = \mathrm{FRAP\ (corrected)}$$

Values may be normalized according to

$$\frac{\mathrm{ROI}\ (t) - \mathrm{BG}}{\mathrm{CO}\ (t) - \mathrm{BG}} \times \frac{\mathrm{CO}\ (t = 0) - \mathrm{BG}}{\mathrm{ROI}\ (t = 0) - \mathrm{BG}} = \mathrm{FRAP\ (normalized,\ corrected)}$$

The bleaching process reduces the total number of fluorescent molecules within the cell and recovery is fueled by the remaining fluorophores outside the bleached ROI. Therefore, recovery of fluorescence can never reach the prebleach value (replenishment of

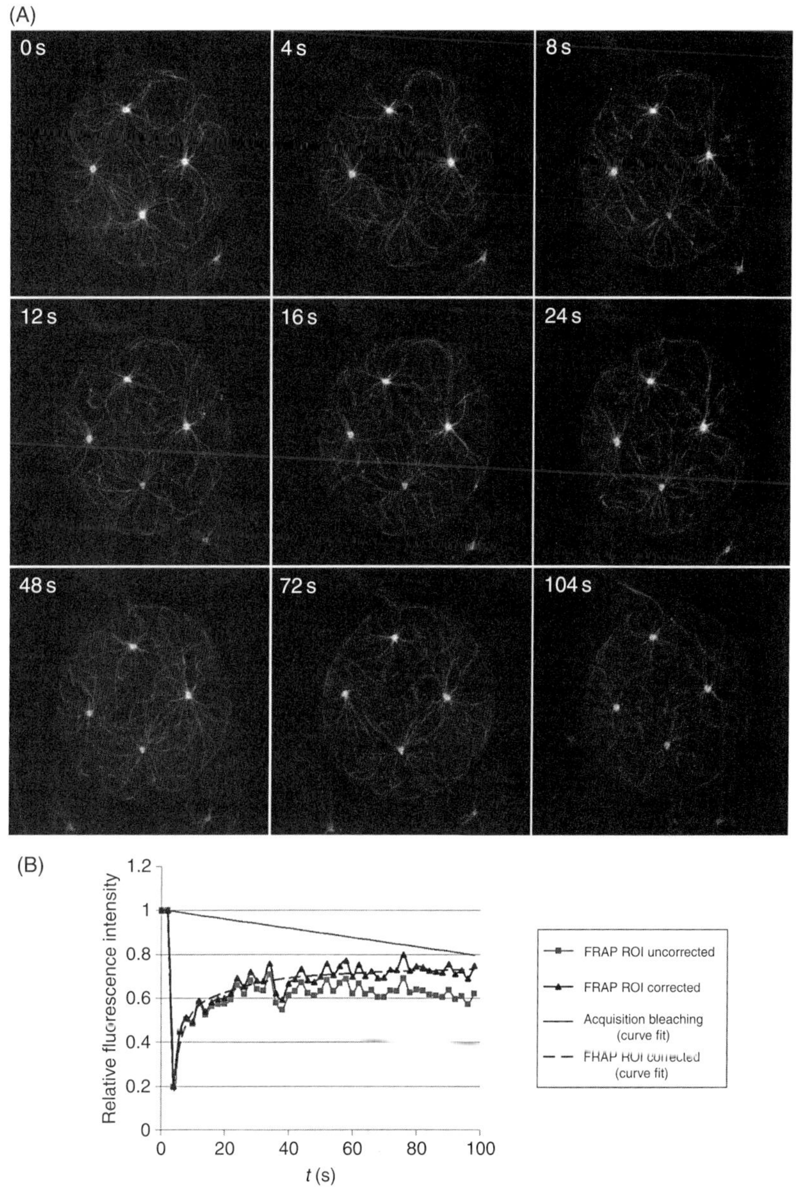

Fig. 5 FRAP experiment of GFP-α-tubulin reveals a rapid recovery of the GFP-α-tubulin signal at the centrosome. Panels in (A) present a montage of selected time points from Movie 3. Cells expressing GFP-α-tubulin were viewed under agar overlay at a Zeiss LSM710 system as described in the text. Live cell imaging of z-stacks with five focus layers occurred at 2.5 frames/s without time-lapse. Image stacks were projected into one plane by maximum intensity projection. A tetranucleated cell with four centrosomes is shown. Here a reference centrosome can be measured in the same cell. The bleached ROI is marked with a white rectangle at time point 0 s. (B) shows corrected and uncorrected FRAP curves together with the acquisition bleaching regression line which was measured at the uppermost centrosome.

fluorophores by protein biosynthesis of GFP fusion proteins can usually be neglected). Thus, in case of large bleach ROIs it may be more useful to define the whole cell as a reference ROI and to measure the total fluorescence (TOT). TOT can then be used instead of CO in the formulas given above. Here the FRAP curve is corrected both against constitutive bleaching during image acquisition and loss of total number of fluorophores due to the bleaching process. Thus, the normalized intensity value at the last time point should reach the prebleach level, provided that there is no immobile fraction of fluorescent molecules.

An example for FRAP analysis of GFP-α-tubulin dynamics at the centrosome is shown in Fig. 5 (Movie 3). This experiment revealed an unexpectedly high turnover of GFP-α-tubulin at the centrosome. A closer look reveals that fast recovery occurs only at the centrosome itself but not at the bleached microtubules in the pericentrosomal area. This interesting result suggests that the centrosome itself harbors a pool of dynamic tubulin dimers, which are not rapidly incorporated into microtubules. In contrast, the microtubules emanating from the centrosome appear to be rather stable, a result consistent with other dynamic measurements and with microtubule-based drug studies.

Acknowledgments

Work in the authors' laboratories is supported by the Deutsche Forschungsgemeinschaft (DFG GR1642/3-1 to R.G.) and the National Science Foundation (MCB-0542731 to M.P.K.). Figure 1 was produced in collaboration with Drs. Daniela Brito and Alexey Khodjakov.

References

Axelrod, D., Koppel, D. E., Schlessinger, J., Elson, E., and Webb, W. W. (1976). Mobility measurement by analysis of fluorescence photobleaching recovery kinetics. *Biophys.* J. **16**, 1055–1069.

Brito, D. A., Strauss, J., Magidson, V., Tikhonenko, I., Khodjakov, A., and Koonce, M. P. (2005). Pushing forces drive the comet-like motility of microtubule arrays in *Dictyostelium. Mol. Biol. Cell* **16**, 3334–3340.

Centonze, V., and Pawley, J. (2006). Tutorial on practical confocal microscopy and use of the confocal test specimen. *In* "Handbook of Biological Confocal Microscopy" (J., Pawley ed.), pp. 549–570. Springer, New York.

Euteneuer, U., Gräf, R., Kube-Granderath, E., and Schliwa, M. (1998). *Dictyostelium* gamma-tubulin: Molecular characterization and ultrastructural localization. *J. Cell Sci.* **111**, 405–412.

Fukui, Y., Yumura, S., and Yumura, T.K. (1987). Agar-overlay immunofluorescence: High resolution studies of cytoskeletal components and their changes during chemotaxis. *Methods Cell Biol.* **28**, 347–356.

Gräf, R. (2009). Microtubule organization in *Dictyostelium. In* "Encyclopedia of Life Sciences (ELS)" (Zheng, Y. ed.), John Wiley & Sons, Ltd, Chichester. DOI: 10.1002/9780470015902.a0021852.

Gräf, R., Daunderer, C., and Schliwa, M. (2000). Dictyostelium DdCP224 is a microtubule-associated protein and a permanent centrosomal resident involved in centrosome duplication. *J. Cell Sci.* **113**, 1747–1758.

Gräf, R., Rietdorf, J., and Zimmermann, T. (2005). Live cell spinning disc microscopy. *Adv. Biochem. Eng. Biotechnol.* **95**, 57–75.

Hestermann, A., Rehberg, M., and Gräf, R. (2002). Centrosomal microtubule plus end tracking proteins and their role in *Dictyostelium* cell dynamics. *J. Muscle Res. Cell Motil.* **23**, 621–630.

Kessin, R. H. (2001). "*Dictyostelium*: Evolution, Cell Biology, and the Development of Multicellularity." Cambridge University Press, Cambridge.

Kimble, M., Kuzmiak, C., McGovern, K. N., and de Hostos, E. L. (2000). Microtubule organization and the effects of GFP-tubulin expression in *Dictyostelium discoideum*. *Cell Motil. Cytoskeleton* **47**, 48–62.

Koch, K. V., Reinders, Y., Ho, T. H., Sickmann, A., and Gräf, R. (2006). Identification and isolation of *Dictyostelium* microtubule-associated protein interactors by tandem affinity purification. *Eur. J. Cell Biol.* **85**, 1079–1090.

Kollmar, M., and Glöckner, G. (2003). Identification and phylogenetic analysis of *Dictyostelium discoideum* kinesin proteins. *BMC Genomics*. **4**, 47.

Koonce, M. P., Grissom, P.M., Lyon, M., Pope, T., and McIntosh, J. R. (1994). Molecular characterization of a cytoplasmic dynein from *Dictyostelium*. *J. Eukaryot. Microbiol.* **41**, 645–651.

Koonce, M. P., and Khodjakov, A. (2002). Dynamic microtubules in *Dictyostelium*. *J. Muscle Res. Cell Motil.* 23, 613–619.

Kuriyama, R., Sato, C., Fukui, Y., and Nishibayashi, S. (1982). In vitro nucleation of microtubules from microtubule-organizing center prepared from cellular slime mold. *Cell Motil.* **2**, 257–272.

Liu, T., Mirschberger, C., Chooback, L., Arana, Q., Dal Sacco, Z., MacWilliams, H., and Clarke, M. (2002). Altered expression of the 100 kDa subunit of the Dictyostelium vacuolar proton pump impairs enzyme assembly, endocytic function and cytosolic pH regulation. *J. Cell Sci.* **115**, 1907–1918.

Maeda, Y. (1986). A new method for inducing synchronous growth of *Dictyostelium discoideum* cell using temperature shifts. *J. Gen. Microbiol.* **132**, 1189–1196.

Müller-Taubenberger, A., Vos, M. J., Bottger, A., Lasi, M., Lai, F. P., Fischer, M., and Rottner, K. (2006). Monomeric red fluorescent protein variants used for imaging studies in different species. *Eur. J. Cell Biol.* **85**, 1119–1129.

Rabut, G., and Ellenberg, J. (2005). Photobleaching techniques to study mobility and molecular dynamics of proteins in live cells: FRAP, iFRAP, and FLIP. *In* "Live Cell Imaging: A Laboratory Manual" (R. D. Goldman and D. L. Spector, eds.), pp. 101–143. Cold Spring Harbor Laboratory Press, Cold Spring Harbor.

Rehberg, M., and Gräf, R. (2002). *Dictyostelium* EB1 is a genuine centrosomal component required for proper spindle formation. *Mol. Biol. Cell* **13**, 2301–2310.

Rehberg, M., Kleylein-Sohn, J., Faix, J., Ho, T. H., Schulz, I., and Gräf, R. (2005). *Dictyostelium* LIS1 is a centrosomal protein required for microtubule/cell cortex interactions, nucleus/centrosome linkage, and actin dynamics. *Mol. Biol. Cell* **16**, 2759–2771.

Thevenaz, P., Ruttimann, U. E., and Unser, M. (1998). A pyramid approach to subpixel registration based on intensity. *IEEE Trans. Image Process* **7**, 27–41.

Tikhonenko, I., Nag, D. K., Robinson, D. N., and Koonce, M. P. (2009). Microtubule-nucleus interactions in *Dictyostelium discoideum* mediated by central motor kinesins. *Eukaryot. Cell* **8**, 723–731.

CHAPTER 19

Imaging of Mitotic Spindle Dynamics in *Caenorhabditis elegans* Embryos

Mika Toya*, Yumi Iida*, *and* Asako Sugimoto*,†

*Laboratory for Developmental Genomics, RIKEN Center for Developmental Biology, Kobe 650-0047, Japan

†Laboratory of Developmental Dynamics, Graduate School of Life Sciences, Tohoku University, Sendai 980-8577, Japan

978-0-12-381349-7
DOI: 10.1016/S0091-679X(10)97019-2

Abstract

Development of the nematode *Caenorhabditis elegans* is highly reproducible, and the cell division patterns are virtually invariant. Transparency of the eggshell and cells enables the observation of intracellular events with a high temporal and spatial resolution. These unique features, along with the sophisticated genetic techniques, make this organism one of the most attractive model systems for dissecting regulatory mechanisms of dynamic cellular behaviors, such as mitosis, at an organismal level. In this chapter, we describe immunofluorescence and live imaging methods for analyzing mitotic spindle regulation. In particular, we present the use of double- or triple-labeled fluorescent strains for high-resolution two-dimensional and three-dimensional live imaging to analyze dynamic behaviors of mitotic spindles.

I. Introduction

A. Early *Caenorhabditis elegans* Embryos as a Model System to Study Mitotic Spindle Dynamics *In Vivo*

The cell lineage of *Caenorhabditis elegans* is invariant, and the timing and orientation of cell division is highly reproducible (Sulston *et al.*, 1983). The cell cycle in early embryos is very rapid, and cell divisions take place every 15–20 min (Sulston *et al.*, 1983); thus, mitotic spindle formation and positioning has to be strictly regulated for proper development of this animal. The small size of the embryos (roughly 50 μm long and 30 μm diameter) and transparency of the eggshell and cells allows the high-magnification imaging of the whole embryo using light microscopes. In addition to these anatomical features, sophisticated genetic techniques including transgenesis and RNAi can be used for this organism. Comparison of the genome sequences revealed that over 60% of the human genes have homologs in *C. elegans* (The *C. elegans* Sequencing Consortium, 1998), thus dissecting conserved phenomena (such as, mitotic spindle dynamics) in this organism would reveal common regulatory mechanisms. Indeed, many of the known components involved in mitosis/mitotic spindle formation are evolutionarily conserved between *C. elegans* and humans.

To further understand the regulatory mechanisms of mitotic spindle formation and positioning, it is important to visualize dynamic behaviors of the mitotic apparatus *in vivo* and to analyze how molecular components interact with each other within a cell.

B. Imaging of Mitosis: Immunofluorescence Versus Fluorescent-Tagged Proteins

Immunofluorescence and live imaging using a fluorescent fusion protein, such as GFP, are the two major visualization techniques of protein components within a cell or an organism.

Although the live imaging with fluorescent fusion proteins is an excellent method to analyze dynamic behaviors of microtubules *in vivo*, some cautions should be taken. First, a protein tag might alter the structure of the target protein, which would cause loss or alteration of the protein function. Second, the expression level or timing of the tagged proteins might be different from the ones of the authentic proteins. Therefore, it is preferable to confirm the functionality of the tagged proteins whenever possible. This can be achieved by testing whether the tagged protein can rescue the phenotype of the mutant of the gene of interest. Alternatively, transgenes can be designed using synthetic oligonucleotides in which multiple silence mutations are incorporated so that the transgene become resistant to RNAi for the authentic gene sequence; in this strain, the functionality of the transgene can be examined by removing authentic protein by RNAi (Green *et al.*, 2008).

Since different expression levels or structural alteration by the protein tag might affect subcellular localization, it is also preferred to confirm the consistency of the localization of fluorescent-tagged protein and that of the authentic protein examined by immunofluorescence.

An advantage of immunofluorescence over live imaging of fluorescent proteins is that a large number of samples can be handled simultaneously and stored for certain time. Also, spatial resolution of immunofluorescence (fixed, "still" images) can be higher than live imaging, especially for dynamic events including behaviors of microtubules/mitotic spindles dependent on the exposure time to detect the fluorescent signal. Therefore, to analyze microtubule-related phenomena, we initially use immunofluorescence to examine a relatively large number of samples to obtain general ideas about the phenotypes and to examine the reproducibility. Live imaging is then used for more detailed analysis of the dynamics.

We describe below the methods for analyzing microtubules and mitotic spindles in *C. elegans* early embryos by immunofluorescence and live imaging using fluorescent-tagged proteins.

II. Immunofluorescence Staining for Microtubule Observation in *C. elegans* Embryos

There are several choices for fixatives including methanol–acetone and aldehydes, but in our experience formaldehyde in PIPES-based buffer (see below; Kentaro Nakano, personal communication) gives the best results in preserving microtubule fibers, which is described below.

A. Reagents

1. 4×PEM buffer pH 6.9 (adjust with KOH): 400 mM PIPES, 20 mM EDTA, 20 mM $MgCl_2$
2. Fixative solution: 3.2% Paraformaldehyde (dilute Polyscience, Inc. 16% solution CAT#18814), 0.24 M Sorbitol, prepare in 1×PEM

3. PEMT solution: 0.1% Triton X-100 in 1×PEM
4. PEMBT solution: 0.1% Triton X-100, 0.1% Bovine Serum Albumin, prepare in 1×PEM

B. Sample Preparation

1. Mark a 1.5 cm diameter circle on the top side of a MAS-coated slide glass (Matsunami Glass Ind., Ltd., Japan) with a liquid-repellent slide marker pen (PAP pen; Daido Sangyo Co. Ltd.) to mark the location of the specimen.
2. Add 10 μl of milli-Q water to the top side of the slide in the marked circle. Pick 5–10 gravid worms into the mounted milli-Q water and cover with a size 18 coverslip (Matsunami Glass Ind., Ltd., Japan).
3. Under the dissecting microscope, suck excess fluid from the edge of the coverslip with a small piece of paper towel to press the worms until eggs are released from the vulva. Stop sucking when the outline of the nuclei in four-cell stage embryos become visible under the dissecting microscope.
4. Place the slide horizontally on a dry ice block over 10 min. The frozen slides can be kept at –80°C for a couple of days. In that case, the slides should be placed on a dry ice block before going to the next step (C).

C. Paraformaldehyde Fixation and Antibody Staining

1. To permeabilize the embryos for chemical fixation, the eggshell is cracked by removing the coverslip of frozen samples (made in B.) using a razor blade while the slide is still frozen (Miller and Shakes, 1995).
2. Once the coverslip is removed, immediately place the slide into –20°C methanol for 15 s.
3. Fix in fixative solution for 30 min at room temperature.
4. Wash the slides three times with PEMT solution.
5. Incubate the slides for 30 min in PEMBT solution to block nonspecific binding.
6. Apply the primary antibody diluted in PEMBT solution.
7. Incubate overnight at 4°C.
8. Wash the slides three times with PEMT solution.
9. Apply the secondary antibody diluted in PEMBT solution.
10. Incubate for 2 h at room temperature.
11. Wash the slides three times with PEMT solution.
12. Mount the coverslips onto the slides with the mounting solution containing DAPI (VECTASHIELD; Vector Lab. Inc.) and seal with nail polish.
13. View under a microscope.

D. Microscopy and Image Acquisition

Specimens are observed with an UPlanApo 100× 1.40NA oil immersion lens by using CSU-X1 spinning-disc confocal system (Yokogawa Electric Corp.) mounted on a IX71 inverted microscope (Olympus).

The specimens are illuminated with a 408 nm (for DAPI), 488 nm (for FITC), or 561 nm (for Alexa Fluor 568) diode pumped solid-state laser (CVI Melles Griot). Images are acquired with an Orca-R2 12bit/16-bit cooled CCD camera (Hamamatsu Photonics) under the control of MetaMorph software (Molecular Devices, Inc.). Fluorescence images are processed with MetaMorph software and Adobe Photoshop (Adobe systems).

In general, images of 10–20 serial *Z*-axis sections with 0.5 or 1 μm step size are taken. For a standard observation with our setting, images are taken with 200 ms exposure for FITC signal (488 nm laser power: 50%), 200 ms exposure for Alexa 568 signal (561 nm laser power: 50%), and 300 ms for DAPI signal (408 nm laser power: 100%).

Figure 1 shows the example of the immunofluorescence images stained with the above procedure. Microtubules are stained with mouse monoclonal anti-α-tubulin-FITC antibody (Sigma F2168). Centrosomal protein SPD-5 (Fig. 1A) or AIR-1 (Fig. 1B) are stained with rabbit polyclonal antibodies made in our laboratory with Alexa Fluor 568 goat anti-rabbit IgG (Invitrogen) as the secondary antibody.

III. Live Imaging of Fluorescent-Tagged Proteins in *C. elegans* Embryos

A. General Consideration for Transformation Methods for Constructing Fluorescent-Tagged Strains

Three methods can be used to construct fluorescent-tagged transgenic worms for the analysis of early embryos: microinjection, microparticle bombardment, and Mos1 transposon insertion.

Microinjection of transgenes into the adult germline syncytium results in extrachromosomal arrays that contain hundreds of copies of the injected plasmids. Expression of a gene in such a repetitive array is generally silenced in the germline and early embryos. This problem can be circumvented by coinjection of genomic DNA to increase the complexity of the array (Strome *et al.*, 2001), although the efficiency is variable.

In microparticle bombardment, transgenes are integrated with a low copy number into a random locus in the genome (Praitis *et al.*, 2001). This method produces a higher ratio of transgenic lines expressing transgenes in the germline and early embryos than that by microinjection. Thus, it has been widely used to construct fluorescent-tagged strains for the analysis of early embryos. Notably, cobombardment of multiple plasmids can cause cointegration of these plasmids into a single chromosomal locus; thus, cobombardment can be applied to construct multilabeled strains without crossing (see below). A disadvantage of microparticle bombardment is that the efficiency of the transformation is very low, thus a large amount of worms needs to be prepared and the screening process for the transgenic lines tends to be time-consuming (4–6 weeks).

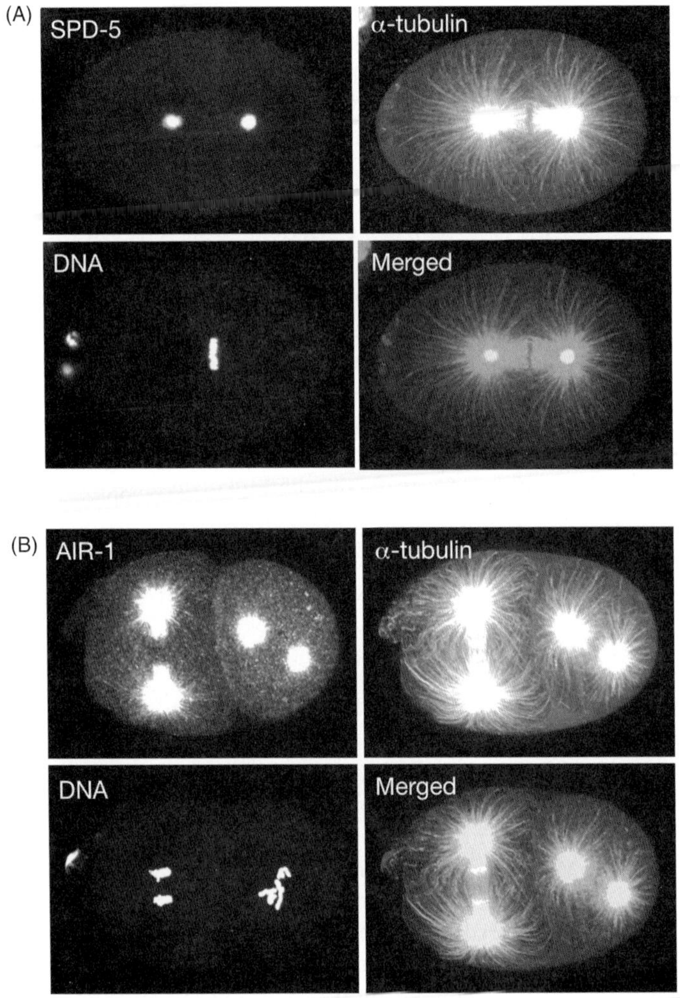

Fig. 1 Immunofluorescence images of the first (A) and the second (B) mitosis in a *C. elegans* embryo. Merged images of the 12 serial Z-sections with 0.5 μm step size are shown. (A) and (B) Microtubules are visualized with anti-α-tubulin antibody. Chromosomes are stained with DAPI. (A) Centrosomes are visualized with anti-SPD-5 antibody. (B) AIR-1 was stained with anti-AIR-1 antibody.

More recently, a Mos1 transposon-mediated single-copy transgene insertion (MosSCI) technique was established (Frokjaer-Jensen *et al.*, 2008). This method inserts transgenes at a defined chromosomal locus as a single copy, thus suitable for expression in germline and early embryos. Currently two chromosomal loci (Chromosome II and IV) are available for transgene insertion. Since this is a relatively new technique, further improvement will be expected including additional loci for insertion and various types of insertion vectors.

We describe below construction of multifluorescent marker strains by microparticle bombardment and its use for the analysis of formation and behaviors of mitotic spindles. For a detailed description of the microparticle bombardment technique, please see the recent excellent article (Green *et al.*, 2008).

B. Plasmid Vectors for Bombardment

To express fluorescent-tagged proteins in early embryos, promoters of germline specific genes, such as *pie-1* and *mex-5*, are used to drive transcription of the transgene in the adult gonad.

We constructed a series plasmid vectors (pMTN series) for microparticle bombardment by modifying the plasmid pID2.02 (Praitis *et al.*, 2001) that contain *pie-1* promotor, Gateway destination assette B, *pie-1* 3′-UTR, and *unc-119* rescuing fragment. First, at the upstream of the Gateway cassette B of pID2.02, an *AscI* restriction enzyme site was created. This *AscI* site was used to insert the coding sequence of GFP or mCherry (either as a single copy or tandemly arranged multiple copies) followed by a linker sequence. The coding DNA fragment of GFP was obtained from pID3.01B (Praitis *et al.*, 2001). The coding DNA fragment of mCherry was obtained from pAA64, whose codon usage was optimized for *C. elegans* (gift from Anjon Audhya and Karen Oegema, Ludwig Institute for Cancer Research UCSD, USA). The linker peptide sequence, GAGAGAGAGAFSV, was based on the yeast GFP expression vector plasmid, pMS-2GK (Sato *et al.*, 2009). An open reading frame of the gene of interest can be inserted using the Gateway recombination cloning system (Invitrogen). The resultant plasmid will produce a protein whose N-terminus is tagged with GFP or mCherry, in the *C. elegans* germline and early embryos.

C. Construction of Multilabeled Strains by Cobombardment and Crossing

Strains in which multiple proteins are fluorescently labeled are highly useful for analyzing dynamic cellular behaviors. Such multilabeled strains can be constructed by crossing multiple single-labeled strains or by cobombardment of multiple fluorescent transgenes (Audhya *et al.*, 2005). Multicolor (for example, GFP and mCherry) strains are useful for analyzing two proteins whose localization might overlap. Single-color, multilabeled strains can also be valuable when the proteins of interest localize to distinct regions within a cell. Since a single color imaging can shorten the exposure time compared to the multicolor imaging that uses multiple lights with distinct wavelength, it can be beneficial for imaging for high time resolution (shorter time interval) or for long-period recording (less bleaching). Some examples for construction of multi-labeled strains are described in Section III. E.

D. Sample Preparation for Live Imaging

1. Fluorescent-tagged strain is grown on NGM plate (Brenner, 1974) at appropriate temperature. (Some fluorescent-tagged strains should be grown at 24~25°C to avoid silencing.)

2. Young adult worms having around 10 eggs are suitable for the observation of early developmental stages. Such gravid worms are dissected in egg buffer (25 mM Hepes pH 7.3, 118 mM NaCl, 48 mM KCl, 2 mM $MgCl_2$, 2 mM $CaCl_2$) or M9 buffer (3 g KH_2PO_4, 6 g Na_2HPO_4, and 5 g NaCl are dissolved in 1 L water. Add 1 ml of 1 M $MgSO_4$ after autoclave treatment) with a 23-gauge needle (Terumo, Japan) on to coverslips, which are then inverted onto agarose pads [2% Agarose (molecular biology grade) in Egg buffer] made on a slide glass. Alternatively, gravid worms are dissected in M9 buffer by a needle or a scalpel on a depression slide; the early embryos are then transferred onto the poly-lysine-coated or MAS-coated slide glass (Matsunami Glass Ind., Ltd., Japan) by mouth pipetting using glass capillaries pulled to a slightly larger diameter than embryos. The stages of mounted embryos can be checked under the dissecting microscope before filming with a confocal microscope.

E. Live Imaging Using Multilabeled Strains

Microscope setting and experimental conditions should be optimized for each experiment. We describe below three experimental conditions using multilabeled strains.

1. Mitotic Spindle Microtubule Assembly in *C. elegans* Embryo

a. ***Strain Construction: Two-Color, Triple-Labeled Strain (Microtubule, Centrosomes, and Chromosomes).*** First, two-color, double-labeled strain (SA240; GFP::β-tubulin and 2×mCherry::γ-tubulin) was constructed by cobombardment of plasmids pMTN1G_*tbb*-2 and pMTN2R_*tbg-1*. SA240 was crossed with the strain SA245 that express mCherry::Histone H2B. The line in which both insertion loci are homozygous (i.e., express three fluorescent markers) were selected and established as the strain SA250. See Table I for detailed genotypes of the strains.

b. ***Image Acquisition.*** The same settings as described in Section II. D. were used for the two-color time-lapse imaging. Images are taken every 7–15 s with 1–3 serial sections along the *Z*-axis with 500–1000 ms exposure for the GFP signal and 100–300 ms exposure for the mCherry signal. 50% of the laser power is used for each laser. During the interval of filming, focal planes were manually adjusted using the fine-focus knob of the microscope.

c. ***Observation.*** The strain SA250 described above microtubules are visualized in green, chromosomes in red, and centrosome in yellow (as merged signal of β- and γ-tubulin at the centrosome) simultaneously as shown in Fig. 2. In this strain, the correlation between the progress of mitotic stages and dynamic microtubule behaviors can be easily observed. Astral microtubules that emanate from centrosomes show dynamic behavior at the cell cortex as previously described (Srayko *et al.*, 2005). Figure 1 also shows that centrosome rotation starts during prophase then ends at metaphase. As anaphase proceeds, the number of astral microtubule tips that reach to the cell cortex was increased, suggesting that astral microtubules became more stable as anaphase proceeds. After the completion of

Table I
Fluorescent Strains Used in This Chapter

Strain name	Marker	Genotype	Cross	Reference
OD58	GFP::PH$^{PLC1\delta1}$ [Cell membrane]	*unc-119(ed3);ltIs38*[pAA1; *pie-1 promoter-GFP:: PH$^{PLC1\delta1}$; unc-119^{+}*]		Audhya *et al.* (2005)
TH32	GFP::Histone H2B; GFP::γ-tubulin	*unc-119(ed3); ruIs32*[pAZ132; *pie-1 promoter-GFP::histone H2B*] III; *ddIs6* [*pie-1 promoter-GFP::tbg-1; unc-119^{+}*]		Cheeseman *et al.* (2004)
SA164	GFP::PH$^{PLC1\delta1}$; GFP:: γ-tubulin	ltIs38[pAA1; *pie-1 promoter-GFP:: PH$^{PLC1\delta1}$; unc-119^{+}*]; *ddIs6* [*pie-1 promoter-GFP::tbg-1; unc-119^{+}*]	OD58 × TH32	This chapter
AZ235	GFP::tubulin	*unc-119(ed3) III; ruIs48[pie-1 promoter -tubulin::GFP; unc-119^{+}]*.		Praitis *et al.* (2001)
SA442	mCherry::AIR-1	*unc-119(ed3); tjIs222[pie-1 promoter-mCherry::air-1; unc-119^{+}]*		This chapter
SA449	GFP:: tubulin; mCherry:: AIR-1	*ruIs48[pie-1 promoter -tubulin::GFP; unc-119^{+}]; tjIs222 [pie-1 promoter-mCherry::air-1r; unc-119^{+}]*	AZ235 × SA442	This chapter
SA240	GFP::β-tubulin; 2×mCherry:: γ-tubulin	*unc-119(ed3); tjIs54[pie 1 promoter-gfp::tbb-2; pie-1promoter-2×mCherry::tbg-1; unc-119^{+}]*		This chapter
SA245	mCherry::Histone H2B	*unc-119(ed3); tjIs57[pie-1 promoter-mCherry::his-48; unc-119^{+}]*		This chapter
SA250	GFP:: β-tubulin2×mCherry:: γ-tubulin; mCherry:: Histone H2B	*tjIs54[pie-1 promoter-gfp::tbb-2; pie-1 promoter-2xmCherry:: tbg-1; unc-119^{+}]; tjIs57[pie-1 promoter-mCherry::his-48; unc-119^{+}]*	SA240 × SA245	This chapter

mitosis, centrosomes in the posterior cell become flattened (Keating and White, 1998), and eventually dispersed, which appears to be caused by pulling of the astral microtubules from the cell cortex.

2. Localization of AIR-1 During the First Mitosis

a. ***Strain Construction: Two-Color, Double-Labeled Strain (Microtubule and Aurora A kinase (AIR-1).*** The strain SA449 that expresses GFP::β-tubulin and mCherry:: AIR-1 was constructed by crossing the strains of AZ235 expressing GFP::β-tubulin (Praitis *et al.*, 2001) and SA442 that express mCherry::AIR-1 (bombarded plasmid: pMTN1R_*air-1*).

b. ***Image Acquisition.*** The same settings as described in Section II. D. were used for the two-color time-lapse imaging (see below). Images are taken every 7–15 s within a single focal plane with 1000 ms exposure for the GFP signal and 300 ms exposure for the mCherry signal. 50% of the laser power is used for each laser.

c. ***Observation.*** Fixed images showed that AIR-1 localizes to a donut-shaped region peripheral to centrosomes and along the base of microtubules (Hannak *et al.*, 2001). The amount of AIR-1 near centrosomes increases toward late anaphase (Motegi *et al.*, 2006). To know how the AIR-1 localization changes as mitosis proceeds, the

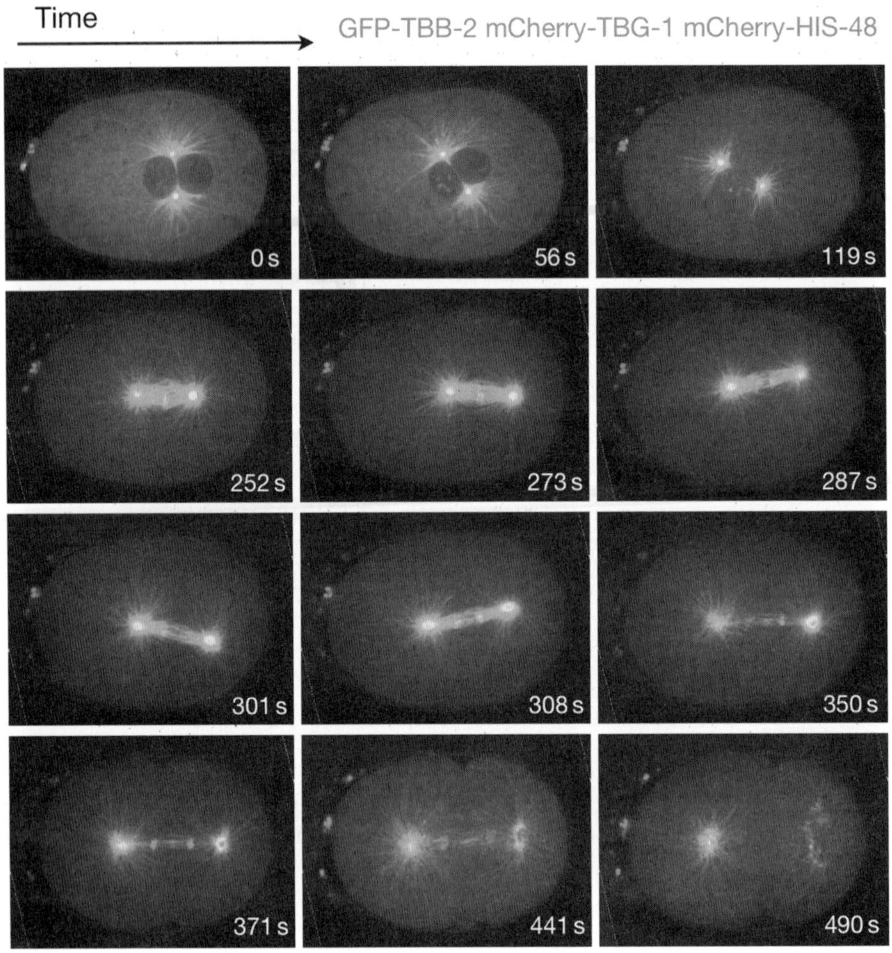

Fig. 2 Mitotic spindle microtubule assembly of the first mitosis in a *C. elegans* embryo. Microtubules, centrosome, and chromosome are visualized with GFP::β-tubulin, mCherry::γ-tubulin, and mCherry::histone H2B, respectively. Images were recorded every 7 s. Selected images of the time-lapse observation are shown. After the meeting of male and female pronuclei, the centrosome–nuclear complex rotates to align with the A-P axis of the embryo during prophase (0 s–119 s). Metaphase spindle with chromosomes aligned in line (252 s). The mitotic spindle oscillates during anaphase (273 s–308 s). The spindle mid-zone is clearly observed while cytokinesis proceeds in telophase (350 s–411 s). γ-tubulin at the centrosome in the posterior cell dispersed soon after the mitosis is completed (490 s).

strain SA449 was observed under the confocal microscope. The obtained images confirmed that the localization of AIR-1 at the centrosomes and along the base of microtubules (Fig. 3). The time-lapse observations clearly showed that the AIR-1 localization along the base of microtubules gradually expands toward the cell cortex as mitosis proceeds (Fig. 3). In addition, the live imaging showed that, during

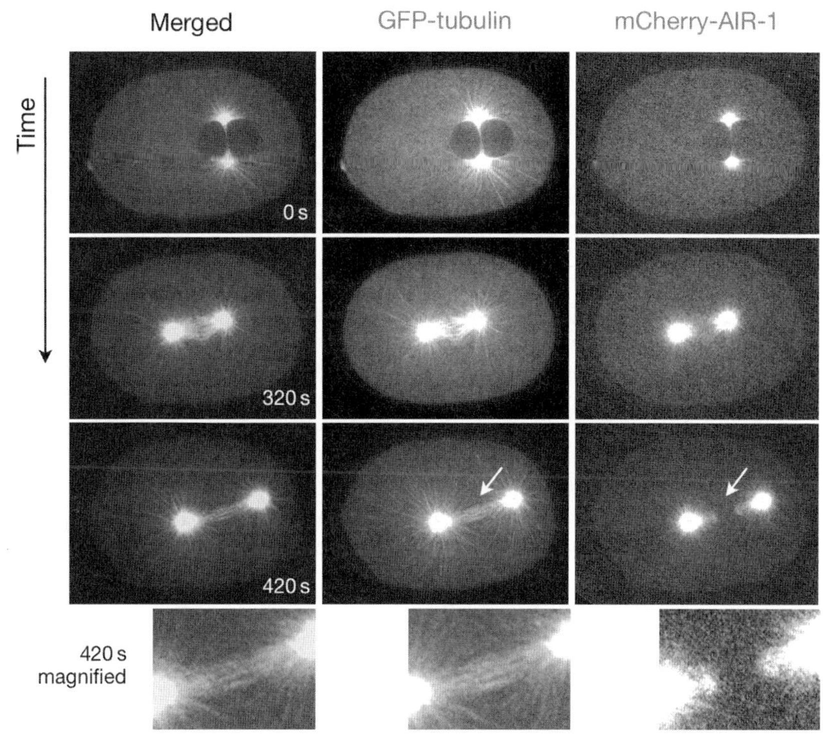

Fig. 3 GFP::tubulin and mCherry::AIR-1 during the first mitosis in a *C. elegans* embryo. Images were taken every 10 s. Selected images of the time-lapse observation are shown. AIR-1 does not localize spindle to the spindle mid-zone (arrows and magnified images).

chromosome separation, AIR-1 localized along microtubules between the centrosome and the chromosomes, but not along the microtubules between two separating chromosomes. The observations may suggest that AIR-1 localizes along kinetochore microtubules, but not along the pole-to-pole microtubules in the mitotic spindle microtubules (Fig. 3, magnified view).

3. Three-Dimensional Time-Lapse (4D) Imaging for Centrosome Positioning

a. ***Strain Construction: Single-Color, Double-Labeled Strain (Centrosome and Plasma Membrane).*** The strain SA164 in which centrosomes and plasma membrane are both labeled by GFP was constructed by crossing the strains OD58 that expresses GFP::PH$^{PLC1\delta1}$ (Audhya *et al.*, 2005) and TH32 that expresses GFP::TBG-1(Cheeseman *et al.*, 2004) (see Table I for genotypes).
b. ***Image Acquisition.*** For the single-color 3D time-lapse (4D) imaging, the following settings were used. Specimens are observed with an UPlan S Apo 60× 1.20NA water immersion lens by using CSU21 spinning-disk confocal system (Yokogawa

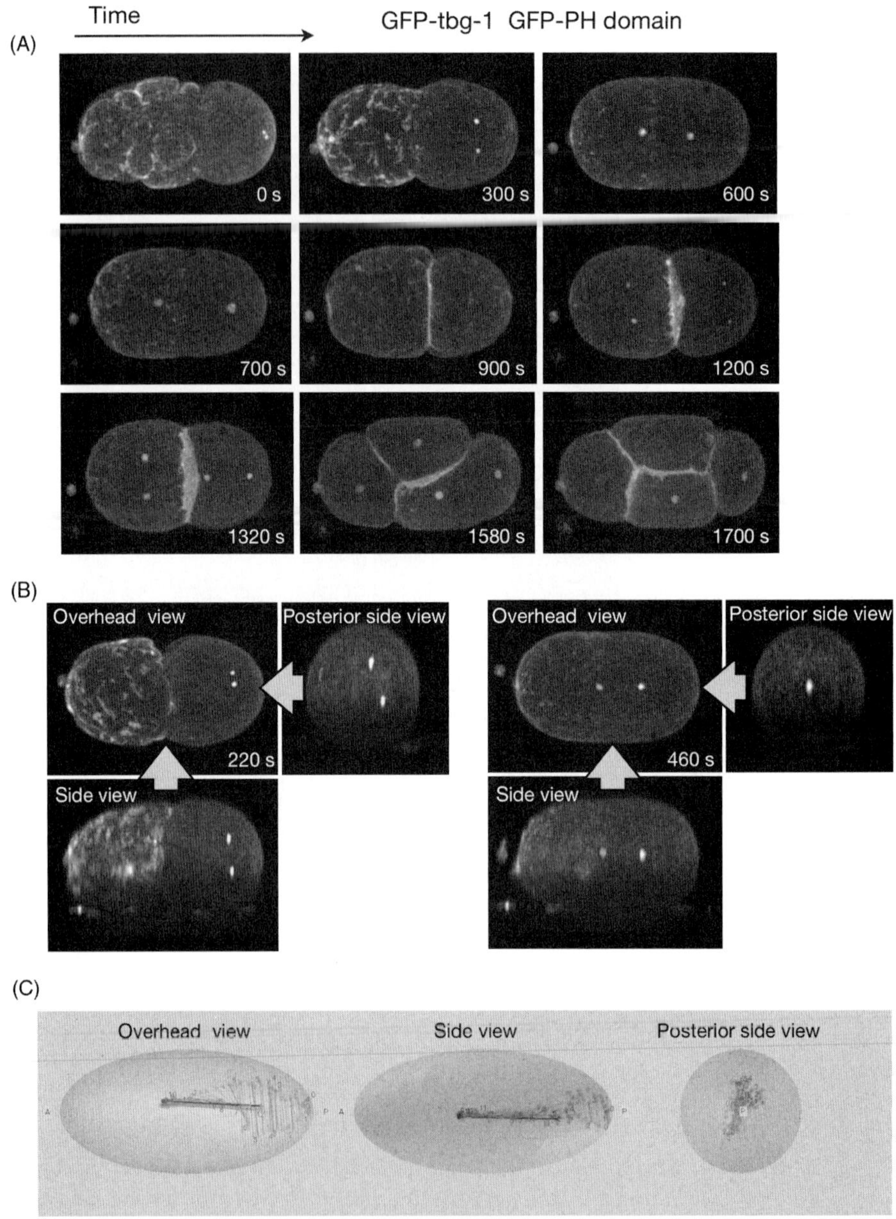

Fig. 4 4D imaging of GFP::γ-tubulin (centrosomes) and GFP::PH $^{PLC1\delta1}$ (cell membrane) during the first two cell divisions in a *C. elegans* embryo. (A) Projected images of 35 Z-sections with 1 μm step size. Centrosome pairs of the male pronucleus move toward the center (300 s) and rotate (600 s). In the second division, only posterior centrosomal pairs rotate (1200 s–1320 s). (B) 3D reconstruction of a whole embryo. (C) Schematic visualization of centrosome movement in the one-cell embryo based on the 4D images.

Electric Corp.) mounted on a BX51 upright microscope (Olympus). The specimens are illuminated with a Sapphire 488 semiconductor laser (20 mW; Coherent, Inc.). Images are acquired with an EM-CCD camera (Andor), and the acquisition system is controlled by RT3D and iXON software (Andor).

To cover a whole embryo, images of 35 serial *Z*-axis sections (150 ms exposure for each frame) with 1 μm step size were acquired. The *Z*-axis position was controlled by a Piezo actuator. The Z-stack images were collected at 10 s intervals for 200 time points, which covered the first two cell divisions. 4D fluorescence images are processed with Imaris software (Bitplane AG).

c. ***Observation.*** Position and orientation of mitotic spindles is strictly regulated and highly reproducible in the development of *C. elegans.* Recent advance in fluorescent proteins and improvement of sensitivity of CCD cameras, enabled the 3D time-lapse (4D) imaging of a developing whole embryo. In this example, plasma membrane and centrosomes are concurrently visualized using GFP::PH$^{PLC1\delta1}$ and GFP::TBG-1. Figure 4A shows the 3D-reconstructed images at different time points using Imaris software (Bitplane AG). These 3D images can be freely rotated in the software (Fig. 4B) and used for quantitative analysis of centrosome movement (Fig. 4C).

With these 4D images of the double-labeled strain, centrosome movement can be easily correlated with cell shape and cell division cycle. For example, during the polarity establishment of a zygote, cortical ruffling is evident with the GFP::PH$^{PLC1\delta1}$ signal (Fig. 4A, 0 s, 300 s). The centrosome pair (attached to the male pronucleus) originally located at the posterior end of the embryo (0 s) starts moving toward the cell center before the psuedocleavage ceases (300 s) and is fully rotated 90° by the time of complete disappearance of the psuedocleavage (600 s).

IV. Summary

The early embryos of *C. elegans* were an excellent model system for studying microtubule behaviors *in vivo*. Recent developments of diverse fluorescent proteins and improvements in the methodology for constructing low-copy transgenic worms dramatically expanded the possibilities of live imaging in *C. elegans* embryos. Specifically, multifluorescent-labeled strains are highly useful for tracing dynamic protein behaviors within a developing embryo. Immunofluorescence using paraformaldehyde-based fixation described here is suitable for analyzing microtubules at a high spatial resolution. These visualization techniques combined with gene knockdown by RNAi should provide better understanding of how microtubules are regulated *in vivo*. In this chapter we limited our focus to early cell divisions, but the same techniques can also be applied for studying other microtubule-mediated phenomena, such as meiosis and cell shape changes during morphogenesis.

Acknowledgments

We thank Geraldine Seydoux, Anjon Audhya, and Karen Oegema for plasmids, Kentaro Nakano for sharing the fixation protocol. We are grateful for Masamitsu Sato and members of the Sugimoto laboratory for discussion, M.T. is a RIKEN Spetial Postdoctoral Researcher and supported by JSPS KAKENHI 21570209. A.S. is supported by MEXT KAKENHI 17017038 and JSPS KAKENHI 19671003.

References

Audhya, A., Hyndman, F., McLeod, I. X., Maddox, A. S., Yates, 3rd, J. R., Desai, A., and Oegema, K. (2005). A complex containing the Sm protein CAR-1 and the RNA helicase CGH-1 is required for embryonic cytokinesis in *Caenorhabditis elegans*. *J. Cell Biol.* **171**, 267–279.

Brenner, S. (1974). The genetics of *Caenorhabditis elegans*. *Genetics* **77**, 71–94.

Cheeseman, I. M., Niessen, S., Anderson, S., Hyndman, F., Yates, 3rd, J. R., Oegema, K., and Desai, A. (2004). A conserved protein network controls assembly of the outer kinetochore and its ability to sustain tension. *Genes Dev.* **18**, 2255–2268.

Frokjaer-Jensen, C., Davis, M. W., Hopkins, C. E., Newman, B. J., Thummel, J. M., Olesen, S. P., Grunnet, M., and Jorgensen, E.M. (2008). Single-copy insertion of transgenes in *Caenorhabditis elegans*. *Nat. Genet.* **40**, 1375–1383.

Green, R. A., Audhya, A., Pozniakovsky, A., Dammermann, A., Pemble, H., Monen, J., Portier, N., Hyman, A., Desai, A., and Oegema, K. (2008). Expression and imaging of fluorescent proteins in the *C. elegans* gonad and early embryo. *Methods Cell Biol.* **85**, 179–218.

Hannak, E., Kirkham, M., Hyman, A. A., and Oegema, K. (2001). Aurora-A kinase is required for centrosome maturation in *Caenorhabditis elegans*. *J. Cell Biol.* **155**, 1109–1116.

Keating, H. H., and White, J. G. (1998). Centrosome dynamics in early embryos of *Caenorhabditis elegans*. *J. Cell Sci.* **111**(Pt 20), 3027–3033.

Miller, D. M., and Shakes, D. C. (1995). Immunofluorescence microscopy. *Methods Cell Biol.* **48**, 365–394.

Motegi, F., Velarde, N. V., Piano, F., and Sugimoto, A. (2006). Two phases of astral microtubule activity during cytokinesis in *C. elegans* embryos. *Dev. Cell* **10**, 509–520.

Praitis, V., Casey, E., Collar, D., and Austin, J. (2001). Creation of low-copy integrated transgenic lines in *Caenorhabditis elegans*. *Genetics* **157**, 1217–1226.

Sato, M., Toya, M., and Toda, T. (2009). Visualization of fluorescence-tagged proteins in fission yeast: The analysis of mitotic spindle dynamics using GFP-tubulin under the native promoter. *Methods Mol. Biol.* **545**, 185–203.

Srayko, M., Kaya, A., Stamford, J., and Hyman, A. A. (2005). Identification and characterization of factors required for microtubule growth and nucleation in the early *C. elegans* embryo. *Dev. Cell* **9**, 223–236.

Strome, S., Powers, J., Dunn, M., Reese, K., Malone, C. J., White, J., Seydoux, G., and Saxton, W. (2001). Spindle dynamics and the role of gamma-tubulin in early *Caenorhabditis elegans* embryos. *Mol. Biol. Cell* **12**, 1751–1764.

Sulston, J. E., Schierenberg, E., White, J. G., and Thomson, J. N. (1983). The embryonic cell lineage of the nematode *Caenorhabditis elegans*. *Dev. Biol.* **100**, 64–119.

The *C. elegans* Sequencing Consortium (1998). Genome sequence of the nematode *C. elegans*: A platform for investigating biology. *Science* **282**, 2012–2018.

CHAPTER 20

Microtubule Dynamics in Plant Cells

Henrik Buschmann, Adrian Sambade, Edouard Pesquet, Grant Calder, *and* Clive W. Lloyd

Department of Cell and Developmental Biology, John Innes Centre, Norwich NR4 7UH, United Kingdom

Henrik Buschmann, Adrian Sambade and Edouard Pesquet contributed equally to the work

Abstract

This chapter describes some of the choices and unavoidable compromises to be made when studying microtubule dynamics in plant cells. The choice of species still depends very much on the ability to produce transgenic plants and most work has been

978-0-12-381349-7
DOI: 10.1016/S0091-679X(10)97020-9

done in the relatively small cells of *Arabidopsis* plants or in tobacco BY-2 suspension cells. Fluorescence-tagged microtubule proteins have been used to label entire microtubules, or their plus ends, but there are still few minus-end markers for these acentrosomal cells. Pragmatic decisions have to be made about probes, balancing the efficacy of microtubule labeling against a tendency to overstabilize and bundle the microtubules and even induce helical plant growth. A key limitation in visualizing plant microtubules is the ability to keep plants alive for long periods under the microscope and we describe a biochamber that allows for plant cell growth and development while allowing gas exchange and reducing evaporation. Another major difficulty is the limited fluorescence lifetime and we describe imaging strategies to reduce photobleaching in long-term imaging. We also discuss methods of measuring microtubule dynamics, with emphasis on the behavior of plant-specific microtubule arrays.

I. Introduction

Despite their immobility higher plant cells contain highly dynamic microtubules. Microtubules made *in vitro* from purified plant tubulin have a higher intrinsic dynamicity than animal microtubules, undergoing more frequent catastrophes with a greater shortening velocity (Moore *et al.*, 1997). Microtubules also behave differently inside the plant, with fluorescently labeled microtubules recovering from photobleaching more rapidly than animal microtubules (Hush *et al.*, 1994; Shaw *et al.*, 2003). The minority of animal microtubules that are not attached to the centrosome undergo treadmilling where polymerization at the fast-growing end can be matched by depolymerization at the slow-growing end. By contrast, plant cortical microtubules are dispersed, unanchored to a common nucleation site, and microtubules move along the plasma membrane by a hybrid form of sustained treadmilling in which plus-end assembly outweighs loss from the opposite end (Shaw *et al.*, 2003). It is from the interaction of these treadmilling microtubules that the interphase cortical array is organized. Gliding toward one another at a shallow angle, microtubules tend to coalign (Dixit and Cyr, 2004) whereas at steeper angles microtubules collide and this can result in depolymerization (Dixit and Cyr, 2004), crossover, or severing at the crossover point (Chan *et al.*, 2009; Wightman and Turner, 2007). The sum of such interactions is a mobile cortical array in which microtubules tend to come together in parallel bundles. However, bundling is not the only outcome since new microtubules branch away from nucleation sites upon mother microtubules (Murata *et al.*, 2005). A proportion of newly nucleated microtubules does not deviate but grows along the mother microtubule in the direction of its plus end (Chan *et al.*, 2009). Cortical microtubules therefore grow, shrink, interact, branch (or not) and over time these separate activities add up to describe the global behavior of the interphase array. At this level, emergent behaviors appear that could not be guessed from the behavior of single microtubules. For example, while microtubules move serially along tracks, groups of similarly polarized tracks form domains that slowly rotate around the surface in clockwise or anticlockwise directions (Chan *et al.*, 2007). Cellulose synthases have been seen to move along lines corresponding to the underlying microtubules (Paredez *et al.*, 2006), and so microtubule reorientations (and corresponding shifts in cellulose alignment) are part of the process by which these immobile cells readjust their direction of growth.

In addition to the complexities of interphase dynamics a further kind of dynamicity is seen as the plant cell goes through the division cycle. The future division plane is foretold by the way that the cortical microtubules of interphase give way to the narrow preprophase band of cortical microtubules, and this is reported to involve a change in microtubule dynamics (Dhonukshe and Gadella, 2003; Vos *et al.*, 2004). As the cortical microtubules give way to the microtubules of the spindle and phragmoplast, dynamic methods allow the subtleties of these key transitions to be followed in detail. In our own laboratory it has been possible to follow the two different ways that the spindle forms, depending on the presence or absence of the preprophase band (Chan *et al.*, 2005); it has also been possible to see how the phragmoplast grows out to the site anticipated by the preprophase band (Buschmann *et al.*, 2006).

Most of these cited investigations have been based on the use of the expression of tubulin or microtubule-associated proteins linked to jellyfish fluorescent protein (Service, 2008), and these have revealed features that simply do not exist in fixed cells. Indeed, over the last 12 years (Marc *et al.*, 1998) studies based on fluorescent proteins (the green fluorescent protein, GFP, or its derivatives) have revolutionized our view of plant microtubules *in vivo*. The main purpose of this review is to introduce strategies we have used for imaging and analyzing microtubule dynamics in living plant cells.

II. Rationale

Once the researcher has identified an interesting question relating to microtubule dynamics he or she has to make a number of decisions relating to the experimental setup, as outlined in Fig. 1. Each decision has direct implications for the quality of the data obtained, and this review aims to provide help in making these decisions. Because this review focuses upon microtubule dynamics in living cells—as opposed to implying changes from fixed cells—much of the discussion will be based on acquiring and analyzing time-lapse movies. "Dynamics" is therefore taken to mean either the short-term behavior of individual microtubules or the longer term changes to microtubule arrays. This chapter focuses on techniques based on expression of jellyfish fluorescent proteins and visualization of microtubules with the confocal microscope.

III. Methods

A. Choice of Organism and Transformation Technique

To date, most analyses of live cell microtubule dynamics in multicellular plants have been performed in *Arabidopsis* or *Nicotiana*, although researchers have recently begun to analyze microtubule dynamics in the moss *Physcomitrella* (Hiwatashi *et al.*, 2008;

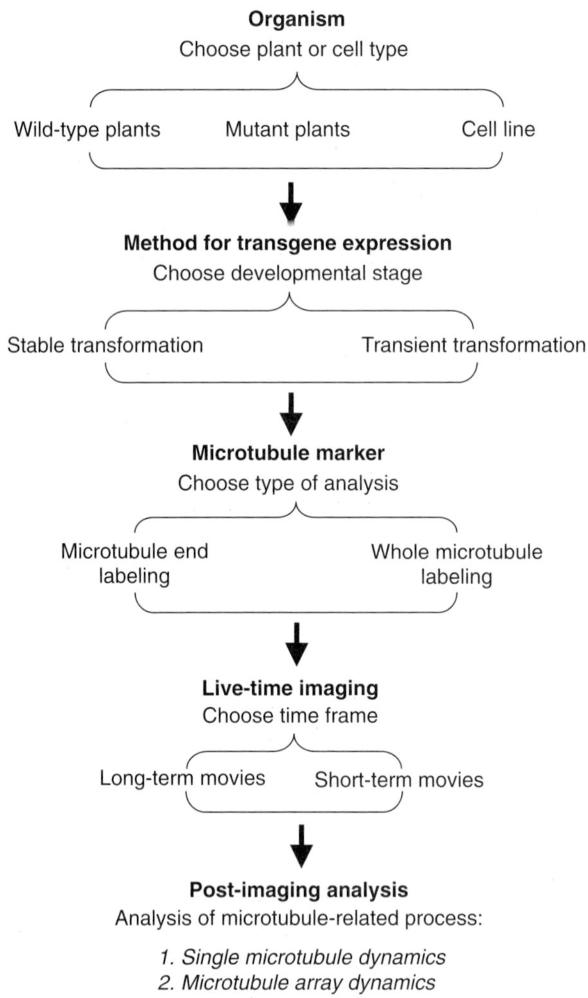

Fig. 1 Flow chart of the working steps and decisions to be made when analyzing plant microtubules. The chart gives an overview of the methods and analyses discussed in the chapter.

Oda *et al.*, 2009). As this restricted list implies, the choice of organism is heavily dependent on whether methods for transformation and transgene expression are available for that organism. *Agrobacterium*-mediated transformation is technically straightforward; however, ballistic transformation may be applicable to a wider range of plant species. In some cases the analysis of cell suspension lines (e.g., *Arabidopsis* and tobacco BY-2) is advantageous: it is generally less time consuming to produce stably transformed cell lines than plants, and in some cases cell suspensions are better suited for microscopy (e.g., larger cells without impervious cuticles and without autofluorescent chloroplasts). Transient expression experiments are even faster. Transient expression of microtubule markers has been used in leaves of *Nicotiana benthamiana* (Brandner *et al.*, 2008) and in *Arabidopsis* suspension cells (Chan *et al.*, 2005). The following citations give details on well-established transformation methods: stable transformation of *Arabidopsis* plants by the "floral dip method" (Clough and Bent, 1998); stable transformation of *Arabidopsis* suspension cells (Pesquet *et al.*, 2010); transient transformation of *Arabidopsis* suspension cells (Mathur *et al.*, 1998); stable transformation to produce *Nicotiana* plants (Horsch *et al.*, 1985); stable transformation of tobacco BY-2 (Koroleva *et al.*, 2006); ballistic transformation of tobacco BY-2 (Vetter *et al.*, 2004); and transient transformation of tobacco leaves (Sparkes *et al.*, 2006).

B. Choice of Microtubule Marker

The ideal GFP-based microtubule marker is expressed in every cell type at every stage of the life cycle and at sufficiently high levels to allow even high sampling rate imaging. In addition, expression of this marker should not produce any phenotype in the plant. Probably no marker described so far reaches this ideal. At the moment, most general microtubule markers are expressed from the constitutive CaMV 35S promoter. One major concern is that high expression levels of certain markers produce side effects (see below). Although it seems reasonable to use endogenous promoters to drive GFP fusions in a complemented mutant background (of the same gene) whenever possible, it is not always possible to apply this counsel of perfection: the level of expression, for example, may be too weak to follow by time lapse without photobleaching. Furthermore, the genetic background could become too complicated to use the marker to analyze additional mutations. When using constitutive expression from the CaMV 35S promoter, one pragmatic way of dealing with these difficulties is to use independent transgenic lines that exhibit differing expression levels and then critically ask whether the process in question is affected by the expression level of the marker (Ambrose *et al.*, 2007). Another approach is to use two different markers (e.g., α-tubulin and EB1 or MAP4-MBD) and compare the results obtained (Buschmann *et al.*, 2006; Chan *et al.*, 2009). We now present a number of markers that have been used in plants to analyze microtubule dynamics.

1. Plus-End Markers EB1 and SPR1

The microtubule-end-binding protein EB1 has a higher affinity for the growing plus end and a lower affinity for the sidewall of the microtubule, thus forming a comet-like gradient at the leading end of dynamic microtubules (Fig. 2) (Mimori-Kiyosue *et al.*, 2000). In *Arabidopsis*, EB1 has three family members (Chan *et al.*, 2003; Mathur *et al.*, 2003). AtEB1a and AtEB1b fused to fluorescent proteins can be used as plus-end markers in interphase cells and cell division (Chan *et al.*, 2003, 2005), whereas EB1c localizes to nuclei at interphase but appears to have specific plus-end-related functions in mitosis and cytokinesis (Komaki *et al.*, 2010). AtEB1a was successfully used to

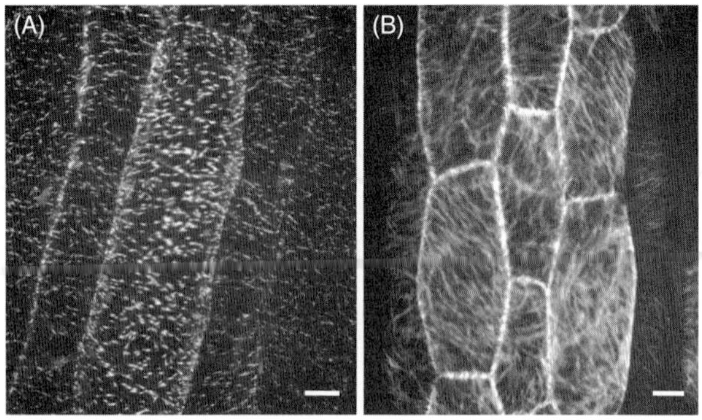

Fig. 2 Transgenic *Arabidopsis* plants showing labeled microtubules. Microtubules in hypocotyl epidermal cells are highlighted by (A) EB1a-GFP and (B) GFP-tubulin β6. Both figures are based on projected z-stacks each acquired at a single time point. The low time resolution of the EB1 (A) results in a dotted pattern of plus ends while the tubulin probe (B) labels whole microtubules. Bars (A, B) 10 μm.

characterize either single microtubules growing in short-term movies or the behavior of the microtubule array in long-term movies (Chan *et al.*, 2007). The advantage of labeling microtubules with AtEB1a over probes like α-tubulin is that it allows the polarity of the microtubule to be easily identified. The EB1 comet can also be visualized moving along coaligned bundles of microtubules whereas this is virtually impossible when all microtubules are labeled evenly with GFP-tubulin. On the other hand, EB1 does not label stable, nondynamic microtubules. In addition, EB1 falls off microtubules undergoing catastrophe and so cannot be used to measure catastrophe rate.

An alternative probe for the microtubule plus-end is the *Arabidopsis* microtubule-localized SPIRAL1 protein (Nakajima *et al.*, 2004; Sedbrook *et al.*, 2004). It has been used to verify data obtained with EB1, although microtubule labeling was less clear with a higher cytoplasmic background (Chan *et al.*, 2009).

2. Universal Microtubule Markers Based on Tagging α- or β-tubulin

GFP fusions with plant α- and β-tubulin genes are being used widely, and in these cases the fluorescent protein is always attached N-terminally, as the C-terminal fusions appear to be nonfunctional (Abe and Hashimoto, 2005; Ueda *et al.*, 1999). GFP–tubulin fusions highlight the entire tubulin polymer, which allows for the appreciation of the entire microtubule network in simple stills without the need to project time courses (Fig. 2). Fluorescence recovery after photobleaching analyses based on GFP-tubulin revealed that cortical microtubule mainly translocate by treadmilling and not by sliding (Shaw *et al.*, 2003). Furthermore, GFP–tubulin fusions can be used to analyze growth as well as catastrophe of the leading end, and it has been used to characterize lagging end dynamics (Nakamura *et al.*, 2004; Shaw *et al.*, 2003). Classical parameters of microtubule dynamics are therefore more completely covered by tagged tubulin, when compared to EB1. In cases where microtubules are highly bundled, as for example in the preprophase band, the GFP-tubulin marker may be of limited use when trying to analyze leading or lagging end dynamics. In this case it is advisable to use plus-tip markers like EB1 or even mammalian CLIP170 (Dhonukshe and Gadella, 2003). In terms of probe selection, GFP–tubulin fusions have been reported to produce weak right-handed organ twisting in *Arabidopsis* (Hashimoto, 2002). This is not seen under all growth conditions and it is possible that the extent of twisting depends on the specific transfer DNA (T-DNA) insertion line used. One report claims that twisting was seen only in fusions with α-tubulin and not in fusions with β-tubulin (Abe and Hashimoto, 2005). This will need confirmation by rigorous testing but indicates the need to check that a particular probe does not have an effect on important phenotypes, such as organ twisting.

3. NEDD1 as an Emerging Minus-End Marker

NEDD1 (*Arabidopsis* homologue to neural precursor cell expressed, developmentally down-regulated 1) has recently been used to study the minus ends of plant microtubules, providing an effective minus-end marker in interphase and division (Motose *et al.*, 2008; Zeng *et al.*, 2009). It was possible to colocalize NEDD1 with sites of apparent microtubule nucleation and these foci were often colabeled with EB1 (Chan *et al.*, 2009). These studies were based on transient expression of NEDD1-mRFP in *N. benthamiana*; however, it would be highly desirable to analyze this marker further in stably transformed *Arabidopsis thaliana* plants.

4. Using Non-plant Proteins for Microtubule Labeling in Planta

The microtubule-binding domain (MBD) of mammalian MAP4 (microtubule-associated protein 4) has also been extensively used to enable visualization of microtubule behavior and dynamics in interphase [e.g., Granger and Cyr (2001) and Marc *et al.* (1998)] and cell division of plants (Granger and Cyr, 2000). Overexpressed MAP4-MBD appears to decorate the entire array of a given cell with excellent signal-to-noise ratios. MAP4-MBD has no preference for microtubule ends but microtubule labeling is fast enough to allow for the study of plus-end dynamics (Vos *et al.*, 2004). The MAP4-MBD construct has been used in a variety of plant species and the initial study described microtubule dynamics in leaf epidermal cells of *Vicia faba*. High expression levels of MAP4-MBD produce phenotypes *in planta*. In our lab, overexpression of MAP4-MBD leads to moderate microtubule bundling and the slower proliferation of *Arabidopsis* cell suspensions (Pesquet, unpublished). It also produces mild organ twisting in *Arabidopsis* seedlings (Hashimoto, 2002).

The animal CLIP170 protein was used as a leading end marker by Dhonukshe and Gadella (2003) who coupled a mammalian sequence of CLIP170 to YFP. The protein's behavior was then analyzed in plant protoplasts and, in a pilot study, on microtubule dynamics in preprophase band formation of tobacco BY-2 cells (Dhonukshe and Gadella, 2003). It appears that obvious CLIP170 homologues are absent from *Arabidopsis* (Kirik *et al.*, 2007).

Interestingly, some viral proteins such as the movement protein of *tobacco mosaic virus* can function as a microtubule-associated protein (MAP) (Ashby *et al.*, 2006). In particular, the deletion mutant C55, lacking the last 55 amino acids of the protein, exhibits an enhanced microtubule-binding activity (Boyko *et al.*, 2000). This could be a useful unrelated microtubule marker to be used in live microtubule imaging.

5. Motor Proteins as Mobile Microtubule Markers

The *Arabidopsis* genome contains at least 61 kinesins (Richardson *et al.*, 2006) and mutational analyses have revealed their functional importance in diffuse growth, tip growth, mitosis, meiosis, and cytokinesis (Buschmann and Lloyd, 2008). It is therefore highly desirable to dynamically visualize kinesins in living cells. Kinesins might have general use as GFP markers because they translocate on microtubules toward their minus or plus ends, depending on the type of motor they possess. Kinesins could therefore act as powerful tools in identifying the polarity of a given microtubule (array) in live cell imaging. Few analyses have made use of GFP-labeled kinesins in plants so far (Ambrose *et al.*, 2005; Bannigan *et al.*, 2007). However, one recent study described two GFP-labeled N-terminal kinesins, KINID1a and b from *Physcomitrella* (related to PAKRP2 from *Arabidopsis*) (Lee *et al.*, 2001), which localize to the phragmoplast midline, suggesting they localize to microtubule plus-ends (Hiwatashi *et al.*, 2008). Vanstraelen *et al.* (2006) have utilized the kinesin KCA1 as a (negative) marker for the cortical division site, although this kinesin preferentially binds to membranes.

C. Movies—the Basis of Modern Dynamics

The experimental setup needed to acquire images for characterizing a specific aspect of microtubule dynamics is described in the general scheme presented in Fig. 3. This contains a flow chart of the steps for analyzing a particular aspect of microtubule behavior—all of which are developed in the paragraphs below. Obviously, in live imaging the sample will change with time and time-course recordings can provide important clues for understanding cellular and tissue behavior. To study a dynamic event, such as the polymerization of a single microtubule, it is necessary to capture consecutive images at a frequency that acquires sufficient information about the speed of change. In principle, the simplest approach for following a dynamic event would be to obtain a series of pictures over time (t) in a single focal plane (xy). However, plant cells are not as flat as animal tissue culture cells, the cytoplasm is wrapped around a large central vacuole, and the cell surface is often domed. Due to the three-dimensional complexity of plant microtubule arrays it is often necessary to take a series of optical slices down the z axis (z-stacks) at points throughout the time course (Fig. 3A). Each z-stack can be condensed or "projected" to form a single image. Then by replaying a succession of projected z-stacks it is possible to see how the microtubule array changes over time, as if it were a movie of a single focal plane (Fig. 3A). As we will see, z-stacks are also useful for keeping the

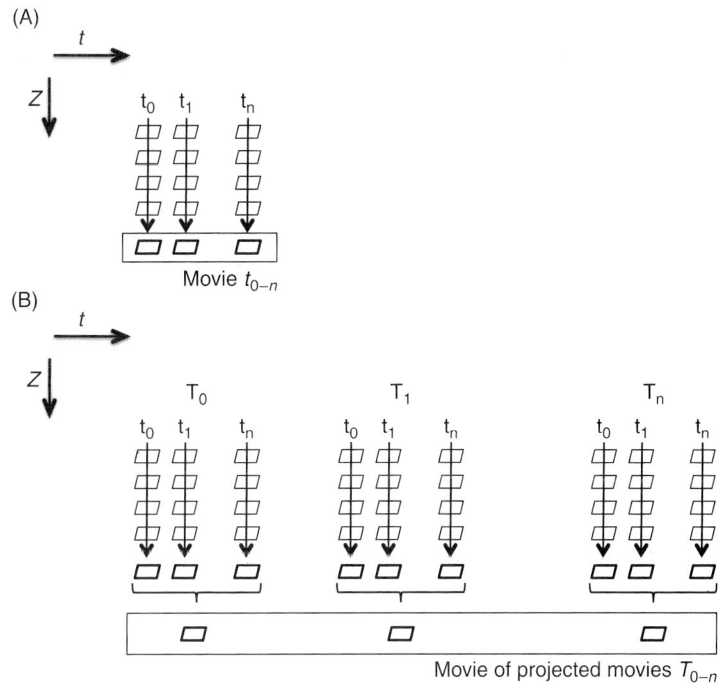

Fig. 3 Imaging strategies for time-lapse studies. (A) As plant cells and tissues often exhibit complex 3D shapes it is usually necessary to acquire images from several z-levels (producing z-stacks) in order to trace microtubules over several minutes (z = depth through the z axis; t = time). Each of the images obtained at the intervals t_0–t_n is then projected and the sequence of projections played back as a movie. (B) To reduce photobleaching during the long-term imaging of microtubules, while still acquiring some dynamic data at high temporal resolution, we often combine short-term with long-term acquisitions (Chan *et al.*, 2007). Basically, short movies are made from z-stack projections as in (A) then the process is repeated after an interval designed to reduce photobleaching. The projections obtained at each phase of sampling are played back as a combined movie. The movie of projected movies provides information on long-term behavior while high-resolution microtubule dynamics can be inspected at the time points T_0, T_1, or T_n.

cortex in focus because long time courses are difficult to make when using a single focal plane due to machine drift and continued cell expansion. Z-stacks are also the starting point for making 3D reconstructions.

1. Short-Term Movies—High Sampling Rate

Several things need to be borne in mind when acquiring data for a time course. First, the size of the sample and the resolution (number of pixels per micrometer) required will directly influence lens selection (for microtubule observations, oil or water immersion lens ranging from ×40 to ×100 are usually used). Once lens and zooming setups are selected, the next step is to decide the number of z slices and the number of time points required to follow the dynamic event in question. Obviously, fast events will require fast

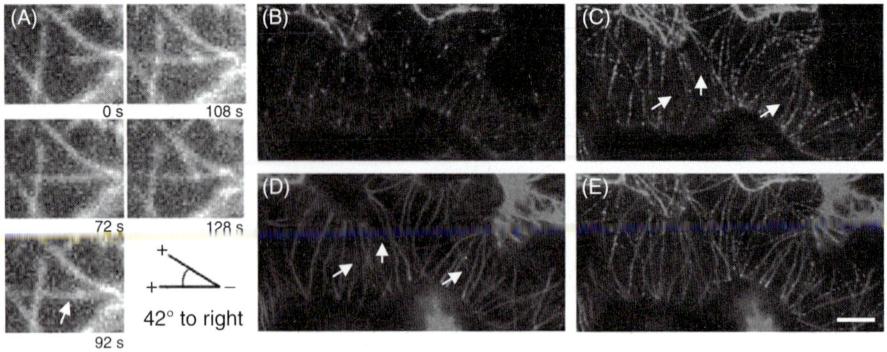

Fig. 4 Results from high sampling rate movies using microtubule markers. (A) Movie of *Arabidopsis* leaf epidermal cells stably transformed with GFP-tubulin β6. The movie is based on *z*-projections acquired at 4 s intervals and shows a branching event (arrow). Because the life history of the mother and daughter microtubules are known it is possible to deduce plus-end polarity and, therefore, the geometry of the branching event (i.e., whether the daughter microtubule is angled toward or away from the mother microtubule's plus end). In this case the daughter microtubule branches with 42 degrees to the right. (B–E) Transient coexpression of microtubule markers EB1a-GFP and mRFP-labeled MAP4-MBD in leaf epidermal cells of *N. benthamiana*. The movie is based on *z*-projections acquired every 20 s. (B) Single time point shows a dotted pattern of EB1a-GFP comets. (C) The movie was then projected along the time axis to produce a more continuous pattern of the EB1 trajectories. (D)The cortical microtubule network of the same cell is labeled with mRFP-tagged MAP4-MBD. (E) Overlay of (C) and (D). The overlay reveals that a subset of microtubules is not labeled with EB1 (arrows). These microtubules appear to be nondynamic and stabilized. Bar (in E) 10 μm.

acquisition, reducing the interval between time points. In general, to follow the dynamics of a single microtubule polymerization in detail, acquisitions are made every 4–30 s. Some fast events, like catastrophes (fast microtubule depolymerization from the microtubule's plus-end), which occur at rates of about 10 μm per minute, would require shorter acquisition rates of 1–4 s. Thus, the speed of the dynamic object under observation can represent a direct limitation for the acquisition setup (i.e., setting up of *xy* and *z* axis). Alternative fluorescence microscopy processes can, to some extent, circumvent these limitations, as we will discuss. The dynamic sequence seen in Fig. 4A is an excerpt of a movie obtained by initiating a *z*-stack of five with a step-size of 0.25 μm every 4 s for 4 min using a spinning disc confocal microscope.

Another key limiting factor is the bleaching of the fluorescent tag. Strong excitation light would extinguish faster the fluorescent properties of the tag. During multi-color acquisitions (e.g., using both green and red fluorescent proteins), one of the fluorescent tags usually decays before the other and so each channel requires different and specific settings to ensure optimal survival of the probes. Although bleaching may not become problematical for some time into the time course it is eventually likely to impact negatively upon the length over which images can be acquired and their quality. The settings will therefore need to be modified to reduce the amount of excitation light on the sample, compromising the quality of the full acquisition. The projected time course in Fig. 4B–E, which is based on EB1 and MAP4-MBD two-color labeling, show the results obtained by carefully balancing z-sectioning with acquisition frequency. Because sampling was reiterated every 20 s in order to reduce bleaching, EB1-GFP comets do not line up to form continuous tracks but appear like pearls on a string. However, mRFP-labeled MAP4-MBD, which labels the entire microtubule, clearly highlights the microtubule bundle along which EB1 traveled.

2. Long-Term Observations Using Biochambers—Low Sampling Rate

Compared to the analysis and characterization of microtubule dynamics, which require short-term movies with a very high sampling rate, long-term movies with a lower sampling rate are used when focusing on the behavior of the entire microtubule array throughout developmental processes such as cell growth, cell division, or cell differentiation. While short-term movies have an overall time frame of a few minutes, long-term movies can cover up to several days (Chan *et al.*, 2007; Pesquet *et al.*, 2010). In order to follow cellular events over long time periods, one can either increase the number of time points (balancing this against the bleaching of probes), or increase the time between each time point acquisition. However, either of these possibilities will affect the kind of data obtained.

In the first case (by adding more time points), the data will show clearly the dynamics but the resulting bleaching will limit the final extent of the acquisition. Furthermore, the size of the data files will represent a limiting factor. For example, a time course with acquisition time points every 30 s and four z slices for each time point would produce a file with 480 pictures per hour. Such an image series will result in huge files for long-term time courses and, additionally, would create extensive photobleaching. On the other hand, increasing the time intervals to 20 min will give a general picture of cortical array behavior, although the obtained data will lack temporal resolution and will not allow the analysis of single microtubule dynamics. Nevertheless, in many cases this approach is sufficient and it has been successfully used to follow the evolution of microtubule orientation during root cell elongation in *A. thaliana* plants expressing GFP-labeled MAP4-MBD (Granger and Cyr, 2001).

An alternative way of sampling microtubule dynamics at larger intervals over many hours or even days is represented in Fig. 3B. In this combined method, first a short-term movie is made that can be projected to make a single image. This is repeated at larger intervals, e.g., every 20 min. The movie made at a each time point provides detailed information about microtubule dynamics, but its projection can be played with other projections taken at successive intervals to make a "movie of projected movies" that provides longer term information about array behavior as well. For example, this combination of short- and long-term observations was used by Chan *et al.* (2007) to follow the growth of microtubules in *A. thaliana* plants labeled with the plus-end probe AtEB1a-GFP. By following this in short movies (roughly 5 min in length) it was possible to obtain information on the polarity of single microtubules and to see that microtubules tend to move successively along the same tracks and that groups ("domains") of adjacent tracks tend to share the same polarity. When projected, the movement of these plus-end dots was summed to form lines. Confocal microscopy was then stopped for 15 min to prevent photobleaching, before another short-term movie was initiated and compressed into another single projection. By playing these projections made every 20 min it was possible to see that the EB1 trajectories rotated around the cell and this could be followed for periods of at least 24 h. Acquiring data in a way that allows the making of "movies of projected movies" has two important advantages; (1) the simultaneous acquisition of long term dynamics and short-term dynamics while reducing photobleaching and (2) providing sufficient sample density to project the EB1 plus-end trace into a line. For such data acquisition we use the MetaMorph software to drive a spinning disc confocal microscope. Iterations of time-lapse observations are made possible by using the "Loop a journal" function.

Material used for long-term imaging has included both whole plants (Chan *et al.*, 2007) and plant cell cultures (Buschmann *et al.*, 2006; Pesquet *et al.*, 2010). Both rely on one central technical aspect: the ability to maintain plant or cell viability throughout the observation process. This was achieved with the biochamber. This custom-made device provides a microenvironment for long-term imaging. Figure 5 illustrates the making of the biochamber (also consult Section IV). Specimens are placed between the coverslip and a gas-permeable bioFOLIE membrane. This is taped to a frame and placed on the microscope for imaging. Incubation conditions during imaging therefore need to be optimized for the sample and include the following:

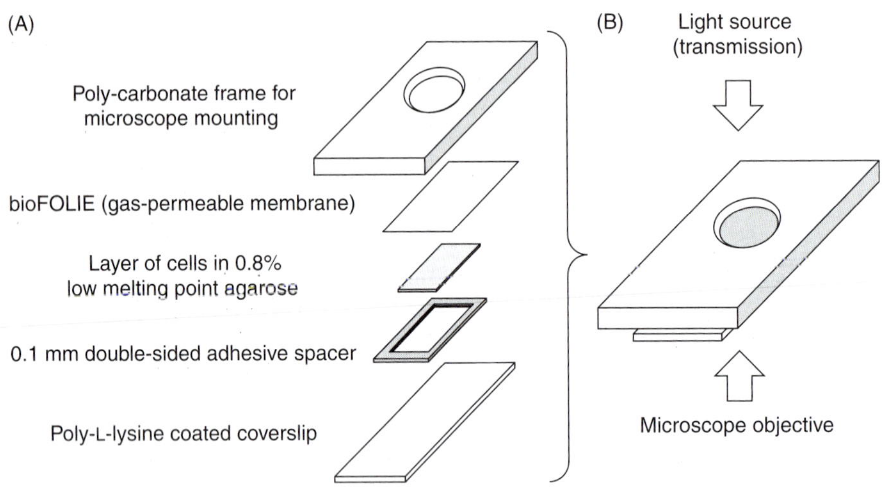

Fig. 5 The biochamber: an imaging device for plant cells. The illustration presents the setup used for suspension cells, although we have used a simpler version without agarose or poly-L-lysine coating for whole *Arabidopsis* seedlings. (A) Exploded view of biochamber. (B) Assembled biochamber. The chamber can be used on inverted (shown) as well as normal microscopes. When the gas-permeable bioFOLIE is omitted the specimen can be treated with drugs during observation (see description in Section III). In this case the chamber must be used with an inverted microscope.

a. Maintenance of Sample Viability. It is important to establish conditions that minimize the effects of incubation on cellular functions. This includes preventing hypoxia (the gas-permeable film allows seedlings to grow in the biochamber for many hours), water loss (biochambers can be placed in a humid environment), heat-stress (the imaging system has to be maintained in a temperature-controlled room), and physiological status (samples need to be kept in optimal culture conditions, which can include illumination of plants on the microscope stage or feeding with culture medium). The fact that the biochamber allows efficient gas exchange was recently confirmed by Koroleva *et al.* (Koroleva *et al.*, 2009) who performed reversible hypoxia experiments by occluding the membrane with an additional gating coverslip.

b. Reduction of Sample Movement in the Chamber. It is, of course, essential to keep the tissue being imaged both in focus and within the field of view in order to perform successful long-term imaging. Lower magnification allows a growing specimen to be tracked for longer. The inadvertent displacement of suspension cells during focal sectioning can be minimized by placing them in a semi-solid mounting medium (0.8% low-melting point agarose). When movement cannot be avoided for other technical reasons, tiled imaging frames (overlapping fields of view) can be made, ensuring that the specimen is kept under observation for as long as possible.

c. Optimization of the Imaging Condition. Clearly, the illumination time for each time point should be reduced as far as possible to avoid heating, photobleaching, and phototoxicity (UV-excited probes may prove problematic).

D. How to Measure Single Microtubule Dynamics

As described above, high sampling rate movies can be used to quantify microtubule dynamics. In particular, the expression of fluorescently tagged tubulins has been successfully used to estimate dynamicity and total amount of polymer gained or lost per time (Ishida *et al.*, 2007; Shaw *et al.*, 2003), while EB1 has been used to describe microtubule nucleation and branching (Chan *et al.*, 2009).

One way of determining the rate of microtubule polymerization is to compare the growth of the microtubule plus end between at least two successively recorded images. However, the growth rate of a microtubule leading end varies considerably over time and different microtubules grow at different rates. Furthermore, microtubules often do not grow along straight paths. It is therefore advisable to make measurements based on high sampling rate movies. Growth measurements are then made on the basis of time–space plot images, also known as kymographs (Fig. 6). Kymographs can be created using free software such as ImageJ (http://reb.info.nih.gov/ij/) and are analyzed as follows. In brief, identify the desired microtubule in the movie, draw a line (straight, or segmented for a curved growth path) along the growing track, use the "reslice" tool to produce an image (2D) by stacking pixels underneath this line for every frame of the movie. In this new image, the y axis represents time and the x axis distance. The pattern obtained shows the apparent movement along the track and the slope is an expression

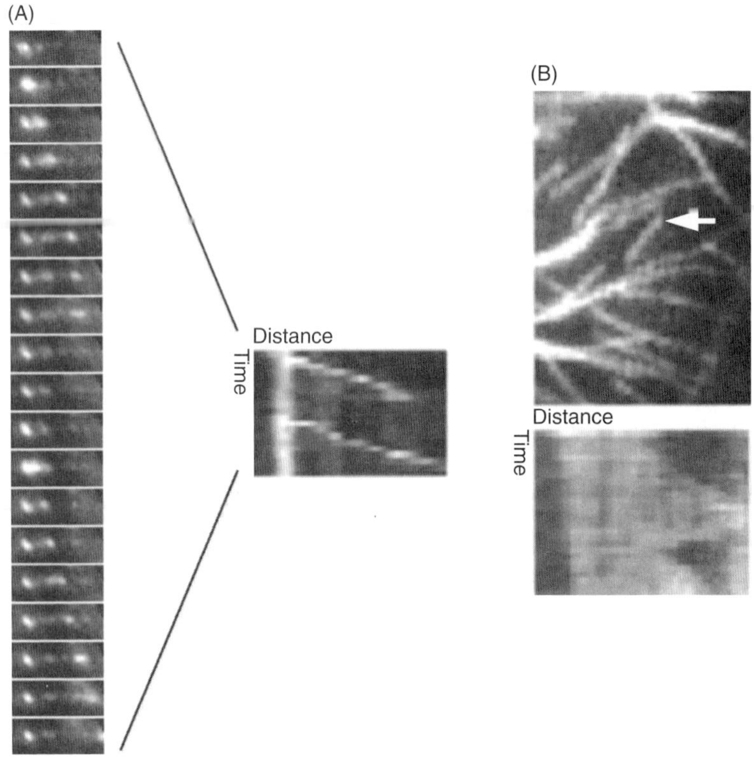

Fig. 6 Movie-based kymographs reveal microtubule behavior at the pixel level. (A) Analysis based on a movie obtained from expressing EB1a-GFP acquired at 10 s time intervals. On the left: excerpts from the movie. On the right of A is the resulting kymograph. An outgrowing microtubule is labeled at the plus-tip. The microtubule then undergoes catastrophe and vanishes. The microtubule is not labeled during catastrophe. A new microtubule is subsequently initiated from the same site resulting in two growth excursions, which yields a saw-toothed edge to the kymograph. (B) Analysis based on a movie obtained from a cell expressing GFP-tubulin β6. The upper panel shows an excerpt from the underlying movie, which was acquired at 4 s time intervals (arrow indicates the plus end of the microtubule under study). The kymograph in the panel below shows the complex behavior of this microtubule. The plus end (toward the right) grows out and undergoes catastrophe after colliding with another microtubule. The microtubule is then recued and grows out again. Throughout this time the minus end slowly shrinks (as indicated by the slightly sloping minus end to the left), suggesting treadmilling.

of the leading end growth rate (Fig. 6A and B). Bouts of growth and shrinkage produce a saw-toothed edge.

The major advantage of kymographic representation is that it allows dynamic events to be appreciated at the pixel level; kymographs also allow different events (e.g., microtubule branching) to be discriminated as they occur along the same track. Where microtubules exhibit treadmilling (opposite end assembly/disassembly), kymographs have been applied to quantify the movement of plus and minus ends of a single microtubule along the same path (Shaw *et al.*, 2003). Important parameters of microtubule dynamics can be obtained by using movies and kymographs as tools. Such parameters include rescue and catastrophe frequencies, the relative time spent in growth or pause or shrinkage, and microtubule dynamicity (total polymer length gained or lost; usually expressed in μm/min) (Abe and Hashimoto, 2005; Buschmann *et al.*, 2009; Shaw *et al.*, 2003). Kymographs further reveal the full complexity of microtubule end dynamics, with all possible transitions between growth, pause, and shrinkage occurring with specific frequencies (Shaw *et al.*, 2003). Multicolor labeling of tubulin and MAPs applied to kymographs allows the analysis of separate molecule classes that are bound to the same track (Varga *et al.*, 2009).

Recent research has shown that microtubule branching has an important impact on cortical array behavior (Nakamura and Hashimoto, 2009; Zeng *et al.*, 2009). Most branching events occur at an angle of nearly 45 degrees (Chan *et al.*, 2009; Murata *et al.*, 2005); however, this can be changed in mutant backgrounds. Once a branching event is located, angle measurements can be easily done using the angle macro in ImageJ. It is only possible to distinguish left-handed and right-handed branching where the polarity of the mother microtubule is known (Chan *et al.*, 2009) (Fig. 4A).

TIP

In live microscopy, over the time course of data acquisition, the tissue usually undergoes small displacements that affect the alignment of cellular structures from picture to picture. This distortion will directly affect data analysis, especially speed calculations. A solution for aligning slices along the stacks is provided by the ImageJ macro, Stackreg (Thevenaz et al.*, 1998). It is equally useful for aligning z-stacks (for example, for 3D projection) and 2D time-lapse movies.*

E. How to Measure Array Dynamics

During cell cycle progression, plant microtubules organize into several array types of varying shape, behavior, and function. The question arises how these microtubule arrays behave on a more global scale and how they interchange. We will now discuss how this behavior can be measured and described.

F. Dynamic Behavior of Cortical Interphase Microtubules

Fast-growing cells often show transverse microtubule arrays, but this alignment is not necessarily static over a long period of time. A change in cortical microtubule alignment can be the response to changing vectors of light or gravity (Matsumoto *et al.*, 2010; Nick *et al.*, 1990; Paredez *et al.*, 2006; Ueda, 2000), and also occurs as a developmental reorientation when cell elongation ceases (Crowell *et al.*, 2009; Granger and Cyr, 2001). In the latter case, in *Arabidopsis* root cells, microtubule alignment mainly switches from transverse to oblique/longitudinal. In contrast, seemingly continuous rotation, i.e., beyond the more limited transverse to longitudinal reorientation, can be seen in slower-growing *Arabidopsis* hypocotyl epidermal cells (Fig. 7A). In this case cortical microtubules (or rather, long-lived microtubule

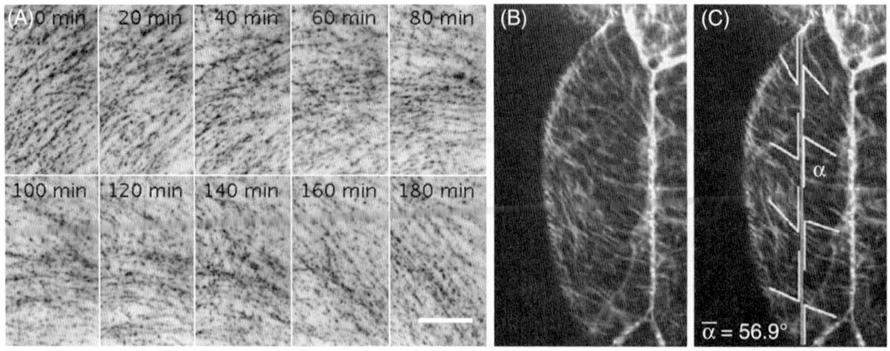

Fig. 7 Alignment of cortical microtubules in relation to cell polarity. (A) Time sequence of projected movies based on EB1a-GFP expression in growing *Arabidopsis* hypocotyls. This long-term observation (3 h) demonstrates the continuous rotation of cortical microtubule arrays. The microtubules start with a right-handed oblique alignment and finish 180 min later with a left-handed oblique array. (B and C) Measurement of cortical microtubule alignment seen in hypocotyls labeled by GFP-tubulin β6. (B) Original *z*-projection. (C). Measurement based on the ImageJ software. Eight measurements were made at regular intervals along the cell's surface. The average alignment is indicated. In this example, values between 0 and 90 degrees indicate a left-handed array, values between 90 and 180 degrees indicate a right-handed array. Bar 10 μm.

tracks or bundles) have been seen to undergo continuous rotations through more than 360 degrees in 200–800 min. This type of continuous reorientation may be associated with the more complex cellulose microfibril patterns seen in aerial plant organs (Chan *et al.*, 2007).

Relatively few studies have addressed the whole-array behavior of cortical microtubules using GFP techniques. The absolute angle of alignment is an important parameter of cortical array behavior that changes during growth and is an important factor in describing microtubule behavior in twisted growth mutants (e.g., *spiral* and *tortifolia* mutants). Microtubule alignment is usually expressed in relation to cell polarity, where transverse equals 90 degrees and longitudinal equals 0 or 180 degrees. It is often too laborious to measure every microtubule of a given cell and different approaches to solve this problem were applied in the past. In many cases orientations were described on a "one cell—one microtubule orientation" basis, and cells were simply classified as being transverse, oblique, or longitudinal by eye. This fast approach lacks resolution and becomes problematic when the cortical array in question shows complex microtubule orientations, for example, radial arrays, and the distinct behavior of separate microtubule domains as described by Chan *et al.* (2007). In this case it is possible to subdivide the cell into smaller areas. A good estimate of orientation is obtained by drawing a line along the axis of a cell and by measuring microtubule orientation at regular intervals using ImageJ (see example in Fig. 7B, C). This approach yields a range of orientation values and it is convenient to present the data in the form of histograms [e.g., Buschmann *et al.* (2004)]. Although this method has been applied successfully in the past it is now necessary to advance the method of data acquisition and employ software to calculate microtubule orientation.

There are further aspects of interest in cortical microtubule array behavior. One question is whether the direction of reorientation is biased (clockwise or anticlockwise rotation). Chan *et al.* (2007) reported left-handed and right-handed rotations in wild-type hypocotyls of *Arabidopsis*. The slower reorientations in roots and the rotation in tracheids appear to show a bias in handedness (Abe *et al.*, 1995; Liang *et al.*, 1996). We have further assessed the rate of reorientation (angular velocity) in rotating cortical arrays of three-day-old *Arabidopsis* hypocotyls and found that these arrays rotate with a speed of 0.76 ± 0.17 degrees/min (Buschmann and Lloyd, unpublished). Another question is whether the acentrosomal cortical array of microtubules is biased (i.e., polar) in regard to its plus-end directionality. Studies suggest that the cortical arrays contains mobile subdomains of coherent polarity (Chan *et al.*, 2007; Dixit *et al.*, 2006).

G. Dynamics of Mitotic and Cytokinetic Arrays

Time-lapse studies and kymographs have been used to analyze the symmetry of prophase and metaphase spindles. The degree of spindle bi-polarity, orientation of the spindle, spindle length, and spindle width have been found to be altered in certain kinesin mutants of *Arabidopsis* (Ambrose and Cyr, 2007; Ambrose *et al.*, 2005; Chen *et al.*, 2002; Marcus *et al.*, 2003). Microtubule dynamics are important in regulating spindle biogenesis and, when perturbed by applying taxol at the preprophase band stage, produces misaligned spindles (Ambrose and Cyr, 2008). Furthermore, the width of the preprophase band appears to be a read-out for subsequent spindle integrity (Chan *et al.*, 2005; Ambrose and Cyr, 2008) (and references therein). For example, in the *clasp1* mutant, preprophase bands are less uniform in width and the subsequently formed spindles are misaligned and shorter than in wild type (Ambrose *et al.*, 2007). Also, in subpopulations of *Arabidopsis* suspension cells that divided without preprophase bands, the spindle is misaligned, mitosis is delayed, and the phragmoplast forms in abnormal division planes (Chan *et al.*, 2005). Another measure is the time required to advance through mitosis. This can be quantified by following cells with GFP-labeled microtubules through mitosis. For example, in the *atk1* mutant the time from prophase to anaphase is longer than in wild type (Marcus *et al.*, 2003).

We provide examples how kymographs can be used to analyze metaphase spindle shape and the rate of phragmoplast growth in tobacco BY-2 cells. The kymograph reveals that in the transition from pro-metaphase to late metaphase the spindle diameter grows and then shrinks (Fig. 8). We have repeated this for movies of further cells and find this is consistent behavior of tobacco BY-2 spindles. Another analysis aimed at measuring the rate of phragmoplast growth in cytokinesis. The slope of the signal in the kymograph of Fig. 9 indicates that the phragmoplast edge approaches the cortical division site at 0.47–0.58 μm/min, which is slightly faster than deduced for the smaller cells of *Arabidopsis* roots (based on) (Cutler and Ehrhardt, 2002).

Preprophase bands forecast the landing of the centrifugally growing cell plate at the cortex. The analysis of preprophase band function therefore requires observing cells throughout mitosis and cytokinesis (Buschmann *et al.*, 2006). In tobacco BY-2 cells this process typically requires between 90 min and 2 h. Good resolution in time and

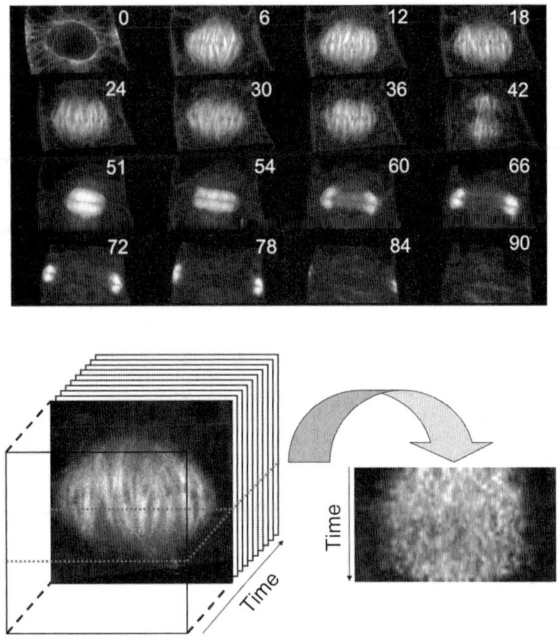

Fig. 8 Gallery of GFP-tubulin in mitosis and cytokinesis of tobacco BY-2 and kymographic analysis of the spindle stage. Upper panel: gallery of single median confocal sections of a dividing cell from preprophase to cytokinesis. Time is indicated in minutes. Below: the stack of time points was analyzed by making a kymograph. The data beneath the central dotted line (on the left) were plotted against time using the ImageJ reslice tool, in order to see what happens to the spindle volume through time. The resulting kymograph (on the right) shows that spindle width is not constant but first widens then contracts.

space is obtained when a z-stack is initiated every 1 or 2 min. In the case of tobacco BY-2, a z-stack should contain 20–30 z-levels (1 μm apart), which allows for data acquisition of conveniently one half the cell's diameter. Such observations produce considerable photobleaching and it is therefore sometimes necessary to restrict sampling frequency or depth—depending on the microscope. Outstanding results are obtained using a spinning disc microscope, because of reduced photobleaching.

H. Functional Analyses

Mechanistic studies of cellular functions are greatly facilitated by specifically interfering with gene function. One promising experimental setup is therefore to monitor fluorescently labeled microtubules in wild-type and mutant background and compare the results. This allows for the deduction of gene function in respect to microtubule dynamics. This is relevant because treatments or conditions that change the dynamics of the microtubule system lead to reproducible phenotypes including cell shape defects, decrease of growth rate, abnormalities in cell division. The plant microtubule

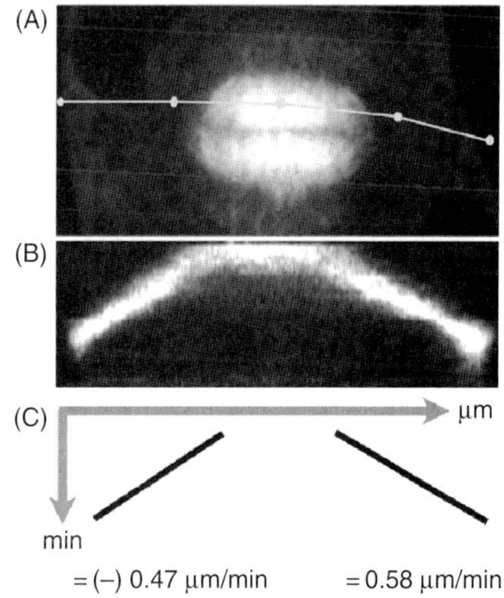

Fig. 9 Analysis of the expansion rate of a tobacco BY 2 phragmoplast. (A) Early phragmoplast stage. The gray line indicates the line which was used to produce the kymograph using ImageJ. In this case a segmented line was used because the direction of phragmoplast growth was curved. (B) Resulting kymograph. (C) Schematic representation of the measurement based on the kymograph. The slope of the growing phragmoplast edge is indicated (black bars). The left and the right edge grew with slightly different speeds.

system is assumed to be composed of several hundred microtubule-associated proteins and mutants are now available for many of these. We have recently published a database of *Arabidopsis* microtubule mutants (Buschmann and Lloyd, 2008). The database features mutant loci with published microtubule-related phenotypes and is updated whenever new microtubule mutants are published. The database assists researchers when designing the experimental setup by providing an overview of available microtubule mutants (http://www.jic.ac.uk/staff/clive-lloyd/henrik-buschmann/microtubulemutants/).

When working with *Arabidopsis* plants, fluorescently labeled markers are normally introduced by crossing. In this case F2 families are selected that carry the desired genotypic combination. Ideally, mutant and wild-type reference lines should be selected from this crossing. In some cases it is necessary to transform wild type and mutant separately with the fluorescent marker construct. This is problematic because T-DNA insertion occasionally produces gene knock-outs in *Arabidopsis*. In this case the direct comparison of wild-type and mutant background is no longer valid, as more than one genetic difference is present. The associated problems can be circumvented by analyzing various lines in parallel, but this can be quite laborious. Another difficulty arises when expression of the fluorescent maker protein changes the phenotype of the mutant under scrutiny. Although this can be turned to advantage in investigating gene function it is advisable to analyze several independent fluorescent markers. The markers listed above can serve as alternatives. Dynamic studies on GFP-labeled microtubules have been used to analyze growth and division in mutant and wild-type backgrounds. Analyses in helical growth mutants showed that certain tubulin mutations disturb almost all dynamic parameters of microtubule behavior (rescue, shrinkage, pause frequencies, rate of growth and shrinkage, and more) (Buschmann *et al.*, 2009; Ishida *et al.*, 2007). The analysis of movies in mutants of the kinesin AtKRP125c (*rsw7*) revealed details of spindle and phragmoplast formation (Bannigan *et al.*, 2007).

Where mutants are not available drugs may serve the same purpose. Many drugs readily enter root cells of *Arabidopsis* but aerial plant organs have a cuticle and this can interfere with the entry of drugs. In leaves of larger plants, like *Nicotiana*, it is possible to infiltrate antimicrotubule drugs using a syringe (Brandner *et al.*, 2008). Generally, although experiments based on drugs are convenient, the specificity of the drug for the protein of interest should be considered. The microtubule-destabilizing drugs oryzalin and propyzamide have been widely used in plants and the effects on microtubule dynamics have been studied (Nakamura *et al.*, 2004). Dynamic studies using oryzalin revealed that microtubules are required for the recruitment but not for the retention of TANGLED-YFP to the cortical division site (Walker *et al.*, 2007). Similar results were obtained for RanGAP1-GFP, whose retention at the preprophase band site persists even after microtubules are disassembled (Xu *et al.*, 2008). Recently a gaseous microtubule-destabilizing drug has been published that might be useful for the treatment of aerial organs (Chaimovitsh *et al.*, 2009). Stabilizing drugs explored in plants are Epothilon B and Paclitaxel; however, both appear to attach to the same binding pocket of β-tubulin (Hause *et al.*, 2005). Drugs that potentially target plant microtubule-associated proteins rather than tubulin itself are Morlin (DeBolt *et al.*, 2007) and CIPC (Clayton and Lloyd, 1984).

TIP

Drugs can be applied to cells during the imaging process providing an on-off effect. We have used a modified setup based on the biochamber for such experiments. Basically, the chamber is assembled normally but without the bioFOLIE. Cells should be embedded in 0.8% low-melting point agarose prior to the experiment. The cells together with the agar will comprise a volume of no more than 50 μl. The chamber is then placed on an inverted microscope. To avoid drying of the cells they are overlaid with 1 ml of liquid growth medium which is pipetted into the cavity of the frame (Fig. 5). A cell is selected and imaging is initiated. At the required point in time the medium is carefully replaced with medium containing the drug (e.g., 20 μM oryzalin). Imaging of the same cell can continue and effects are usually seen within minutes.

IV. Materials

A. Material Required for Plant Transformation

The generation of transgenic plants assumes access to standard molecular biology laboratory equipment, including a sterile laminar flow hood and growth chambers.

Several publications present fluorescent protein vector sets for plant transformation [e.g., Curtis and Grossniklaus (2003), Goodin *et al.* (2007), Karimi *et al.* (2002), and Lee and Gelvin (2008)]. Vectors were either requested from researchers or purchased, e.g., from Ghent University (http://www.psb.ugent.be/gateway/) or ABRC (http://www.arabidopsis.org/abrc/catalog/vector_1.html). Plants recommended for microtubule dynamic studies are *A. thaliana* ecotypes Col-0 or Landsberg *erecta, N. benthamiana, Arabidopsis* cell culture Col-0, and tobacco BY-2 cell culture. *Agrobacterium* strains recommended are hypervirulent LBA 4404 or classical GV3101; however, *Agrobacterium* strain EHA105 is at times superior when transforming tobacco BY-2.

B. Material Required for the Biochamber

The biochamber (Chan *et al.*, 2005) is a custom-made device that keeps suspension cells alive and whole seedlings growing during the imaging process. The setup enables gas exchange while minimizing evaporation of water. This becomes possible by applying a gas-permeable membrane to the microscope slide. The specimen is mounted between the membrane (bioFOLIE; VivaScience, Göttingen, Germany) and the coverslip (#1.5 VWR International, Lutterworth, UK), usually supplied with a drop of growth medium or water. A spacer (Grace Bio-Labs, JTR13R-0.5) can be used to provide additional room for the sample. Membrane, specimen, and coverslip are taped together to a polycarbonate frame. This frame has the shape of a normal microscope slide and must have a hole in its center enabling gas exchange (Fig. 5). For long-term observations of cell suspension it is necessary to immobilize cells. We have adopted a twofold strategy. First, we use coverslips coated with 0.1% (w/v) poly-L-lysine solution (Sigma). Cells are applied to this coverslip and allowed to settle. Second, liquid is then replaced with medium containing 0.8% (w/v) low-melting point agarose using a pipette. The bioFOLIE is then applied and the chamber is assembled as usual.

C. Software

In general, confocal microscopes run under the control of specific software, for example, Leica with the LAS software (http://www.leica-microsystems.com/) and Zeiss with the LSM (http://www.zeiss.com/), providing control of the device as well as some analytical tools. However, in most cases there are royalty limitations for their use in additional computers and alternative free access versions of the same software are functionally limited. We use the MetaMorph® software for custom integration with various automated microscope devices. The software also provides a wide range of analysis tools. An alternative solution for the opening and analysis of microscopy data is provided by the open source ImageJ software (http://rsbweb.nih.gov/ij/) from the National Institutes of Health, USA. This software is capable of working on many types of computer operating systems (e.g., Windows, Apple, Linux) and can open most imaging format files using the LOCI Bio-formats library (http://www.loci.wisc.edu/software/bio-formats). ImageJ is capable of handling multidimensional data, and its functionality is expandable through plug-ins/macros. Most of the analyses used to support this work were made using this software. Other useful and freely available software, which can help to analyze and interpret 3D data are Voxx (http://www.nephrology.iupui.edu/imaging/voxx/), a multiplatform program from Indiana University (USA), which performs real-time rendering of large multichannel 3D/4D microscopy data sets (Clendenon *et al.*, 2002), and Osirix (http://www.osirix-viewer.com/), an open-source program devised for MacOS X. Although these programs are tailored to medical applications rather than cell biology they are simpler than expensive commercial volume-rendering programs. Used in conjunction with the tools provided by ImageJ, it is possible to obtain a reliable analysis of 3D data. Amira (http://www.amira.com/) is a powerful commercially available 3D visualization and analysis software package; it has extensive analysis tools but in our experience is not as easy to use. Other potentially useful free software includes CellProfiler for image analysis (http://www.cellprofiler.org/) and ImageSurfer for 3D visualization and analysis (http://imagesurfer.cs.unc.edu/). However, we have not extensively tested these programs.

D. Microscopes

Certain imaging systems are better adapted to particular applications (i.e., short- versus long-term imaging) and their acquisition characteristics mainly vary in their versatility, acquisition rate, and resolution. As analysis of microtubule dynamics depends on live-time imaging, and visualization of microtubules depends on fluorescent detection of the gene fusion, it is necessary to use either a standard epifluorescence microscope or a confocal laser scanning microscope. The higher resolution and removal of out-of-focus light provided by confocal microscopy is generally preferred.

Type of confocal microscopy

- *Point-scanning* represents the most general and flexible type of confocal microscopy, with adjustable pinholes to match any objective and full control over the scanning beam. We are using Leica SP2 and SP5 microscopes (freely adjustable emission range selection) and the Zeiss LSM510meta that, in addition to its classical filter sets, presents a new system for emission spectrum separation and combination. Using biochambers, point-scanning confocal microscopy can be used for short-term and long-term imaging but bleaching of the fluorescent tag is sometimes problematic. Newly developed point-scanning confocal microscopes (e.g., the resonant scanning system of the Leica SP5) have a much reduced acquisition time and can provide a solution for minimizing bleaching. One of the limiting factors of point scanning confocal microscope is its sensitivity, which is a key factor for both speed of acquisition and imaging of low signal intensity. The Leica SP5 AOBS (Acoustic Optical Beam Splitter) represents an advance in microscope design. It allows for efficient light throughput, suppressing more reflected laser light than standard dichroic mirrors. This allows emitted light to be collected closer to the excitation laser wavelength, thus harvesting more of the probe's emission spectra, resulting in increased brightness. There have also been advances in detector-noise suppression. Noise is a particular problem with photomultiplier tubes (PMT) in point scanners; thermal noise is perhaps the major problem, explaining why spinning disc microscopy uses a cooled camera. Leica perform an averaging step at the PMT collection level so that the signal from a single captured pixel is sampled several times then averaged. The Zeiss LSM780 has a new design giving its GaSaP QUASAR detector much higher quantum efficiency (sensitivity) than a standard PMT, making it better suited for live cell imaging.

– *Spinning disc* microscopy allows probes with a low fluorescent yield to be imaged with reduced bleaching. This represents the best alternative for probes possessing a range of fluorescent yields as the time of acquisition can be specifically adjusted. We use a Visitech QLC-100 assembly comprised of a Yokogawa CSU10 confocal scan head mounted on a Nikon E800 upright microscope with a Hamamatsu Orca ER cooled charge-coupled device (CCD) detector. The microscope is run by Metamorph® software. New advances in manufacture of filters (Semrock 3625 Buffalo Road, Suite 6 Rochester, NY 14624 USA) and new camera designs, such as electron multiplying CCD cameras, have produced enhancements in spinning disc confocal microscopy. However, although the spinning disc microscope is superior in sensitivity the obtained images are of somewhat poorer quality than a point scanner because the pinholes on the spinning disc are of a fixed size. While 50 μm pinholes are optimal for ×100 objectives they will produce an effectively opened pinhole at lower magnification, allowing out-of-focus light to contaminate the image, thereby lowering the contrast.

Images from epifluorescence or point scanning/spinning disc confocal microscopes can be improved using deconvolution—an image restoration technique. This calculates the optical distortion and creates a new, mathematically corrected image. This is an intensive postprocessing step and to work effectively requires data with a high spatial (*z*) resolution. There are several commercially available packages. We use AutoQuant X AutDeblur by Media Cybernetic (http://www.mediacy.com/index.aspx?page=AutoDeblurVisualize) that employs an automatic blind deconvolution algorithm based on the inputting of data concerning the type of lens, microscope, scale, and emission wavelength of the probe.

V. Outlook

We are beginning to understand how the classical microtubule arrays seen in plants are formed through microtubule dynamics, and how they function (Ambrose and Cyr, 2008; Ambrose *et al.*, 2007; Ehrhardt and Shaw, 2006; Lloyd and Chan, 2008; Müller *et al.*, 2009). There is still much work to be done on understanding nonclassical arrays like the filiform apparatus of synergid cells, the phycoplast, and astral spindles seen in lower plants. In some cases it will be necessary to employ other types of microscope, such as two-photon confocal microscopes, when it is necessary to assess cells deep inside tissues. Multiphoton microscopes use infrared instead of visible light to give better optical penetration and lower phototoxicity. However, they currently have slow point scanning and their lasers are expensive. To study dynamics with higher temporal and spatial resolution it will be necessary to use other approaches. Total internal reflection fluorescence microscopes (TIRFM) offer these advantages as well as very thin optical sectioning with an excellent signal-to-noise ratio. The technique, which has a lower phototoxicity than spinning disk confocal microscopy, depends upon achieving a critical angle of incident illumination at the interface between two media with different refractive indices, generating an evanescent wave that excites fluorophores within 100 nm of the interface. The outer epidermal wall might be thought to frustrate the application of this technique to plant tissues since the domed outer surface does not easily stick to the coverslip and the wall can be thicker than the wave of excitation. However, Uchida *et al.* (2010) have visualized microtubule dynamics with a 1 s resolution in walled Neurospora hyphae expressing β-tubulin-GFP and so it may be possible to adapt this method for walled higher plant cells and not just for cortical microtubules in protoplasts (Uchida *et al.*, 2010). By modifying the TIRFM imaging system, Konopka and Bednarek (2008) have shown that subcritical angles of laser illumination can be obtained to produce variable angle epifluorescence microscopy (VAEM). Using plant epidermal cells expressing GFP-tagged MAP4-MBD, this technique was capable of visualizing cortical microtubules with a high signal-to-noise ratio (Konopka and Bednarek, 2008). The improved time resolution of VAEM should be useful in capturing the highly dynamic behavior of plant cortical microtubules. Continuing developments such as these should enable the finest details of cortical microtubule dynamics to be studied in plants.

Acknowledgments

We are grateful to Kim Findlay and Peter Shaw for their support and innovative development of the John Innes Bioimaging Facilities. The research was funded by Biotechnology and Biological Sciences Research Council research grants to CWL and by a grant-in-aid to the John Innes Centre.

References

Abe, H., Funada, R., Imaizumi, H., Ohtani, J., and Fukazawa, K. (1995). Dynamic changes in the arrangement of cortical microtubules in conifer tracheids during differentiation. *Planta* **197**, 418–421.

Abe, T., and Hashimoto, T. (2005). Altered microtubule dynamics by expression of modified alpha-tubulin protein causes right-handed helical growth in transgenic *Arabidopsis* plants. *Plant J.* **43**, 191–204.

Ambrose, J. C., and Cyr, R. (2007). The kinesin ATK5 functions in early spindle assembly in *Arabidopsis*. *Plant Cell* **19**, 226–236.

Ambrose, J. C., and Cyr, R. (2008). Mitotic spindle organization by the preprophase band. *Mol. Plant* **1**, 950–960.

Ambrose, J. C., Li, W., Marcus, A., Ma, H., and Cyr, R. (2005). A minus-end-directed kinesin with plus-end tracking protein activity is involved in spindle morphogenesis. *Mol. Biol. Cell* **16**, 1584–1592.

Ambrose, J. C., Shoji, T., Kotzer, A. M., Pighin, J. A., and Wasteneys, G.O. (2007). The *Arabidopsis* CLASP gene encodes a microtubule-associated protein involved in cell expansion and division. *Plant Cell* **19**, 2763–2775.

Ashby, J., Boutant, E., Seemanpillai, M., Groner, A., Sambade, A., Ritzenthaler, C., and Heinlein, M. (2006). Tobacco mosaic virus movement protein functions as a structural microtubule-associated protein. *J. Virol.* **80**, 8329–8344.

Bannigan, A., Scheible, W. R., Lukowitz, W., Fagerstrom, C., Wadsworth, P., Somerville, C., and Baskin, T. I. (2007). A conserved role for kinesin-5 in plant mitosis. *J. Cell Sci.* **120**, 2819–2827.

Boyko, V., van der Laak, J., Ferralli, J., Suslova, E., Kwon, M. O., and Heinlein, M. (2000). Cellular targets of functional and dysfunctional mutants of tobacco mosaic virus movement protein fused to green fluorescent protein. *J. Virol.* **74**, 11339–11346.

Brandner, K., Sambade, A., Boutant, E., Didier, P., Mely, Y., Ritzenthaler, C., and Heinlein, M. (2008). Tobacco mosaic virus movement protein interacts with green fluorescent protein-tagged microtubule end-binding protein 1. *Plant Physiol.* **147**, 611–623.

Buschmann, H., Chan, J., Sanchez-Pulido, L., Andrade-Navarro, M. A., Doonan, J.H., and Lloyd, C.W. (2006). Microtubule-associated AIR9 recognizes the cortical division site at preprophase and cell-plate insertion. *Curr. Biol.* **16**, 1938–1943.

Buschmann, H., Fabri, C. O., Hauptmann, M., Hutzler, P., Laux, T., Lloyd, C. W., and Schäffner, A. R. (2004). Helical growth of the *Arabidopsis* mutant *tortifolia1* reveals a plant-specific microtubule-associated protein. *Curr. Biol.* **14**, 1515–1521.

Buschmann, H., Hauptmann, M., Niessing, D., Lloyd, C. W., and Schäffner, A.R. (2009). Helical growth of the *Arabidopsis mutant tortifolia2* does not depend on cell division patterns but involves handed twisting of isolated cells. *Plant Cell* **21**, 2090–2106.

Buschmann, H., and Lloyd, C. W. (2008). *Arabidopsis* mutants and the network of microtubule-associated functions. *Mol. Plant* **1**, 888–898.

Chaimovitsh, D., Abu-Abied, M., Belausov, E., Rubin, B., Dudai, N., and Sadot, E. (2009). Microtubules are an intracellular target of the plant terpene citral. *Plant J.* **61**, 399–408.

Chan, J., Calder, G. M., Doonan, J.H., and Lloyd, C. W. (2003). EB1 reveals mobile microtubule nucleation sites in *Arabidopsis*. *Nat. Cell Biol.* **5**, 967–971.

Chan, J., Calder, G., Fox, S., and Lloyd, C. (2005). Localization of the microtubule end binding protein EB1 reveals alternative pathways of spindle development in *Arabidopsis* suspension cells. *Plant Cell* **17**, 1737–1748.

Chan, J., Calder, G., Fox, S., and Lloyd, C. (2007). Cortical microtubule arrays undergo rotary movements in *Arabidopsis* hypocotyl epidermal cells. *Nat. Cell Biol.* **9**, 171–175.

Chan, J., Sambade, A., Calder, G., and Lloyd, C. (2009). *Arabidopsis* cortical microtubules are initiated along, as well as branching from, existing microtubules. *Plant Cell* **21**, 2298–2306.

Chen, C., Marcus, A., Li, W., Hu, Y., Calzada, J. P., Grossniklaus, U., Cyr, R. J., and Ma, H. (2002). The *Arabidopsis* ATK1 gene is required for spindle morphogenesis in male meiosis. *Development* **129**, 2401–2409.

Clayton, L., and Lloyd, C.W. (1984). The relationship between the division plane and spindle geometry in Allium cells treated with CIPC and griseofulvin: An anti-tubulin study. *Eur. J. Cell Biol.* **34**, 248–253.

Clendenon, J.L., Phillips, C.L., Sandoval, R.M., Fang, S., and Dunn, K.W. (2002). Voxx: A PC-based, near real-time volume rendering system for biological microscopy. *Am. J. Physiol. Cell Physiol.* **282**, C213–C218.

Clough, S. J., and Bent, A. F. (1998). Floral dip: A simplified method for *Agrobacterium*-mediated transformation of *Arabidopsis thaliana*. *Plant J.* **16**, 735–743.

Crowell, E. F., Bischoff, V., Desprez, T., Rolland, A., Stierhof, Y. D., Schumacher, K., Gonneau, M., Hofte, H., and Vernhettes, S. (2009). Pausing of Golgi bodies on microtubules regulates secretion of cellulose synthase complexes in *Arabidopsis*. *Plant Cell* **21**, 1141–1154.

Curtis, M. D., and Grossniklaus, U. (2003). A gateway cloning vector set for high-throughput functional analysis of genes in planta. *Plant Physiol.* **133**, 462–469.

Cutler, S. R., and Ehrhardt, D.W. (2002). Polarized cytokinesis in vacuolate cells of *Arabidopsis*. *Proc. Natl. Acad. Sci. U.S.A.* **99**, 2812–2817.

DeBolt, S., Gutierrez, R., Ehrhardt, D. W., Melo, C.V., Ross, L., Cutler, S. R., Somerville, C., and Bonetta, D. (2007). Morlin, an inhibitor of cortical microtubule dynamics and cellulose synthase movement. *Proc. Natl. Acad. Sci. U.S.A.* **104**, 5854–5859.

Dhonukshe, P., and Gadella, T.W. Jr. (2003). Alteration of microtubule dynamic instability during preprophase band formation revealed by yellow fluorescent protein-CLIP170 microtubule plus-end labeling. *Plant Cell* **15**, 597–611.

Dixit, R., Chang, E., and Cyr, R. (2006). Establishment of polarity during organization of the acentrosomal plant cortical microtubule array. *Mol. Biol. Cell* **17**, 1298–1305.

Dixit, R., and Cyr, R. (2004). Encounters between dynamic cortical microtubules promote ordering of the cortical array through angle-dependent modifications of microtubule behavior. *Plant Cell* **16**, 3274–3284.

Ehrhardt, D. W., and Shaw, S. L. (2006). Microtubule dynamics and organization in the plant cortical array. *Annu. Rev. Plant Biol.* **57**, 859–875.

Goodin, M. M., Chakrabarty, R., Banerjee, R., Yelton, S., and Debolt, S. (2007). New gateways to discovery. *Plant Physiol.* **145**, 1100–1109.

Granger, C. L., and Cyr, R. J. (2000). Microtubule reorganization in tobacco BY-2 cells stably expressing GFP-MBD. *Planta* **210**, 502–509.

Granger, C. L., and Cyr, R. J. (2001). Spatiotemporal relationships between growth and microtubule orientation as revealed in living root cells of *Arabidopsis thaliana* transformed with green-fluorescent-protein gene construct GFP-MBD. *Protoplasma* **216**, 201–214.

Hashimoto, T. (2002). Molecular genetic analysis of left-right handedness in plants. *Philos. Trans. R. Soc. Lond. B Biol. Sci.* **357**, 799–808.

Hause, G., Lischewski, S., Wessjohann, L. A., and Hause, B. (2005). Epothilone D affects cell cycle and microtubular pattern in plant cells. *J. Exp. Bot.* **56**, 2131–2137.

Hiwatashi, Y., Obara, M., Sato, Y., Fujita, T., Murata, T., and Hasebe, M. (2008). Kinesins are indispensable for interdigitation of phragmoplast microtubules in the moss *Physcomitrella patens*. *Plant Cell***20**, 3094–3106.

Horsch, R. B., Rogers, S. G., and Fraley, R. T. (1985). Transgenic plants. *Cold Spring Harb. Symp. Quant. Biol.* **50**, 433–437.

Hush, J. M., Wadsworth, P., Callaham, D. A., and Hepler, P. K. (1994). Quantification of microtubule dynamics in living plant cells using fluorescence redistribution after photobleaching. *J. Cell Sci.* **107(Pt 4)**, 775–784.

Ishida, T., Kaneko, Y., Iwano, M., and Hashimoto, T. (2007). Helical microtubule arrays in a collection of twisting tubulin mutants of *Arabidopsis thaliana*. *Proc. Natl. Acad. Sci. U.S.A.* **104**, 8544–8549.

Karimi, M., Inze, D., and Depicker, A. (2002). GATEWAY vectors for *Agrobacterium*-mediated plant transformation. *Trends Plant Sci.* **7**, 193–195.

Kirik, V., Herrmann, U., Parupalli, C., Sedbrook, J. C., Ehrhardt, D. W., and Hulskamp, M. (2007). CLASP localizes in two discrete patterns on cortical microtubules and is required for cell morphogenesis and cell division in *Arabidopsis*. *J. Cell Sci.* **120**, 4416–4425.

Komaki, S., Abe, T., Coutuer, S., Inze, D., Russinova, E., and Hashimoto, T. (2010). Nuclear-localized subtype of end-binding 1 protein regulates spindle organization in *Arabidopsis*. *J. Cell Sci.* **123**, 451–459.

Konopka, C. A., and Bednarek, S. Y. (2008). Variable-angle epifluorescence microscopy: A new way to look at protein dynamics in the plant cell cortex. *Plant J.* **53**, 186–196.

Koroleva, O. A., Calder, G., Pendle, A. F., Kim, S. H., Lewandowska, D., Simpson, C. G., Jones, I. M., Brown, J. W., and Shaw, P. J. (2009). Dynamic behavior of *Arabidopsis* eIF4A-III, putative core protein of exon junction complex: Fast relocation to nucleolus and splicing speckles under hypoxia. *Plant Cell* **21**, 1592–1606.

Koroleva, O. A., Roberts, G. R., Tomlinson, M. L., and Doonan, J. H. (2006). Novel approaches for cell cycle analysis in BY-2. *In* "Tobacco BY-2 Cells: From Cellular Dynamics to Omics" (T.Nagata, K. Matsuoka, and D. Inze, eds.), **Vol. 58**, pp. 3–21. Springer, Berlin, Heidelberg, New York.

Lee, L. Y., and Gelvin, S. B. (2008). T-DNA binary vectors and systems. *Plant Physiol.* **146**, 325–332.

Lee, Y. R., Giang, H. M., and Liu, B. (2001). A novel plant kinesin-related protein specifically associates with the phragmoplast organelles. *Plant Cell* **13**, 2427–2439.

Liang, B. M., Dennings, A. M., Sharp, R.E., and Baskin, T. I. (1996). Consistent handedness of microtubule arrays in maize and *Arabidopsis* primary roots. *Protoplasma* **190**, 8–15.

Lloyd, C., and Chan, J. (2008). The parallel lives of microtubules and cellulose microfibrils. *Curr. Opin. Plant Biol.* **11**, 641–646.

Marc, J., Granger, C. L., Brincat, J., Fisher, D. D., Kao, T., McCubbin, A. G., and Cyr, R. J. (1998). A GFP-MAP4 reporter gene for visualizing cortical microtubule rearrangements in living epidermal cells. *Plant Cell* **10**, 1927–1940.

Marcus, A. I., Li, W., Ma, H., and Cyr, R.J. (2003). A kinesin mutant with an atypical bipolar spindle undergoes normal mitosis. *Mol. Biol. Cell* **14**, 1717–1726.

Mathur, J., Mathur, N., Kernebeck, B., Srinivas, B. P., and Hulskamp, M. (2003). A novel localization pattern for an EB1-like protein links microtubule dynamics to endomembrane organization. *Curr. Biol.* **13**, 1991–1997.

Mathur, J., Szabados, L., Schaefer, S., Grunenberg, B., Lossow, A., Jonas-Straube, E., Schell, J., Koncz, C., and Koncz-Kalman, Z. (1998). Gene identification with sequenced T-DNA tags generated by transformation of *Arabidopsis* cell suspension. *Plant J.* **13**, 707–716.

Matsumoto, S., Kumasaki, S., Soga, K., Wakabayashi, K., Hashimoto, T., and Hoson, T. (2010). Gravity-induced modifications to development in hypocotyls of *Arabidopsis* tubulin mutants. *Plant Physiol.* **152**, 918–926.

Mimori-Kiyosue, Y., Shiina, N., and Tsukita, S. (2000). The dynamic behavior of the APC-binding protein EB1 on the distal ends of microtubules. *Curr. Biol.* **10**, 865–868.

Moore, R. C., Zhang, M., Cassimeris, L., and Cyr, R. J. (1997). In vitro assembled plant microtubules exhibit a high state of dynamic instability. *Cell Motil. Cytoskeleton* **38**, 278–286.

Motose, H., Tominaga, R., Wada, T., Sugiyama, M., and Watanabe, Y. (2008). A NIMA-related protein kinase suppresses ectopic outgrowth of epidermal cells through its kinase activity and the association with microtubules. *Plant J.* **54**, 829–844.

Müller, S., Wright, A. J., and Smith, L. G. (2009). Division plane control in plants: New players in the band. *Trends Cell Biol.* **19**, 180–188.

Murata, T., Sonobe, S., Baskin, T. I., Hyodo, S., Hasezawa, S., Nagata, T., Horio, T., and Hasebe, M. (2005). Microtubule-dependent microtubule nucleation based on recruitment of gamma-tubulin in higher plants. *Nat. Cell Biol.* **7**, 961–968.

Nakajima, K., Furutani, I., Tachimoto, H., Matsubara, H., and Hashimoto, T. (2004). SPIRAL1 encodes a plant-specific microtubule-localized protein required for directional control of rapidly expanding *Arabidopsis* cells. *Plant Cell* **16**, 1178–1190.

Nakamura, M., and Hashimoto, T. (2009). A mutation in the *Arabidopsis* γ-tubulin-containing complex causes helical growth and abnormal microtubule branching. *J. Cell Sci.* **122**, 2208–2217.

Nakamura, M., Naoi, K., Shoji, T., and Hashimoto, T. (2004). Low concentrations of propyzamide and oryzalin alter microtubule dynamics in *Arabidopsis* epidermal cells. *Plant Cell Physiol.* **45**, 1330–1334.

Nick, P., Bergfeld, R., Schafer, E., and Schopfer, P. (1990). Unilateral reorientation of microtubules at the outer epidermal wall during photo- and gravitropic curvature of maize coleoptiles and sunflower hypocotyls. *Planta* **181**, 162–168.

Oda, Y., Hirata, A., Sano, T., Fujita, T., Hiwatashi, Y., Sato, Y., Kadota, A., Hasebe, M., and Hasezawa, S. (2009). Microtubules regulate dynamic organization of vacuoles in Physcomitrella patens. *Plant Cell Physiol.* **50**, 855–868.

Paredez, A., Somerville, C., and Ehrhardt, D. (2006). Visualization of cellulose synthase demonstrates functional association with microtubules. *Science* **312**, 1491–1495.

Pesquet, E., Calder, G., Korolev, A. V., and Lloyd, C. W. (2010). The microtubule-associated protein AtMAP70-5 regulates secondary wall patterning in *Arabidopsis* wood cells. *Curr. Biol.* **20**, 744–749.

Richardson, D. N., Simmons, M. P., and Reddy, A. S. (2006). Comprehensive comparative analysis of kinesins in photosynthetic eukaryotes. *BMC Genomics* **7**, 18.

Sedbrook, J. C., Ehrhardt, D. W., Fisher, S. E., Scheible, W. R., and Somerville, C. R. (2004). The *Arabidopsis* SKU6/SPIRAL1 gene encodes a plus end-localized microtubule-interacting protein involved in directional cell expansion. *Plant Cell* **16**, 1506–1520.

Service, R. F. (2008). Nobel Prize in chemistry. Three scientists bask in prize's fluorescent glow. *Science* **322**, 361.

Shaw, S. L., Kamyar, R., and Ehrhardt, D. W. (2003). Sustained microtubule treadmilling in *Arabidopsis* cortical arrays. *Science* **300**, 1715–1718.

Sparkes, I. A., Runions, J., Kearns, A., and Hawes, C. (2006). Rapid, transient expression of fluorescent fusion proteins in tobacco plants and generation of stably transformed plants. *Nat. Protoc.* **1**, 2019–2025.

Thevenaz, P., Ruttimann, U.E., and Unser, M. (1998). A pyramid approach to subpixel registration based on intensity. *IEEE Trans. Image Process.* **7**, 27–41.

Uchida, M., Mourino-Perez, R. R., and Roberson, R. W. (2010). Live-cell imaging of microtubule dynamics in hyphae of *neurospora crassa*. *In* "Methods in Molecular Biology" (A.Sharon, ed.), **Vol. 638**, pp. 259–268. Springer, Heidelberg, New York.

Ueda, K., and Matsuyama, T. (2000). Rearrangement of cortical microtubules from transverse to oblique or longitudinal in living cells of transgenic *Arabidopsis thaliana*. *Protoplasma* **213**, 28–38.

Ueda, K., Matsuyama, T., and Hashimoto, T. (1999). Visualization of microtubules in living cells of transgenic *Arabidopsis thaliana*. *Protoplasma* **206**, 201–206.

Vanstraelen, M., Van Damme, D., De Rycke, R., Mylle, E., Inze, D., and Geelen, D. (2006). Cell cycle-dependent targeting of a kinesin at the plasma membrane demarcates the division site in plant cells. *Curr. Biol.* **16**, 308–314.

Varga, V., Leduc, C., Bormuth, V., Diez, S., and Howard, J. (2009). Kinesin-8 motors act cooperatively to mediate length-dependent microtubule depolymerization. *Cell* **138**, 1174–1183.

Vetter, G., Hily, J. M., Klein, E., Schmidlin, L., Haas, M., Merkle, T., and Gilmer, D. (2004). Nucleo-cytoplasmic shuttling of the beet necrotic yellow vein virus RNA-3-encoded p25 protein. *J. Gen. Virol.* **85**, 2459–2469.

Vos, J., Dogterom, M., and Emons, A. (2004). Microtubules become more dynamic but not shorter during preprophase band formation: A possible "search-and-capture" mechanism for microtubule translocation. *Cell Motil. Cytoskeleton* **57**, 246–258.

Walker, K. L., Muller, S., Moss, D., Ehrhardt, D. W., and Smith, L. G. (2007). *Arabidopsis* TANGLED identifies the division plane throughout mitosis and cytokinesis. *Curr. Biol.* **17**, 1827–1836.

Wightman, R., and Turner, S. R. (2007). Severing at sites of microtubule crossover contributes to microtubule alignment in cortical arrays. *Plant J.* **52**, 742–751.

Xu, X. M., Zhao, Q., Rodrigo-Peiris, T., Brkljacic, J., He, C. S., Muller, S., and Meier, I. (2008). RanGAP1 is a continuous marker of the *Arabidopsis* cell division plane. *Proc. Natl. Acad. Sci. U.S.A.* **105**, 18637–18642.

Zeng, C. J., Lee, Y. R., and Liu, B. (2009). The WD40 repeat protein NEDD1 functions in microtubule organization during cell division in *Arabidopsis thaliana*. *Plant Cell* **21**, 1129–1140.

CHAPTER 21

Melanophores for Microtubule Dynamics and Motility Assays

Kazuho Ikeda, Irina Semenova, Olga Zhapparova, *and* Vladimir Rodionov
Department of Cell Biology, R. D. Berlin Center for Cell Analysis and Modeling, University of Connecticut Health Center, Farmington, Connecticut 06032-1507

Abstract

Microtubules (MTs) are cytoskeletal structures essential for cell division, locomotion, intracellular transport, and spatial organization of the cytoplasm. In most interphase cells, MTs are organized into a polarized radial array with minus-ends clustered at the centrosome and plus-ends extended to the cell periphery. This array directs transport of organelles driven by MT-based motor proteins that specifically move either to plus- or to minus-ends. Along with using MTs as tracks for cargo, motor proteins can organize MTs into a radial array in the absence of the centrosome. Transport of organelles and motor-dependent radial organization of MTs require MT dynamics, continuous addition and loss of tubulin subunits at minus- and plus-ends. A unique experimental system for studying the role of MT dynamics in these processes is the melanophore,

978-0-12-381349-7
DOI: 10.1016/S0091-679X(10)97021-0

which provides a useful tool for imaging of both dynamic MTs and moving membrane organelles. Melanophores are filled with pigment granules that are synchronously transported by motor proteins in response to hormonal stimuli. The flat shape of the cell and the radial organization of MTs facilitate imaging of dynamic MT plus-ends and monitoring of their interaction with membrane organelles. Microsurgically produced cytoplasmic fragments of melanophores are used to study the centrosome-independent rearrangement of MTs into a radial array. Here we describe the experimental approaches to study the role of MT dynamics in intracellular transport and centrosome-independent MT organization in melanophores. We focus on the preparation of cell cultures, microsurgery and microinjection, fluorescence labeling, and live imaging of MTs.

I. Introduction

Microtubules (MTs) are highly dynamic structures that continuously grow and shorten by addition and loss of tubulin subunits (Cassimeris *et al.*, 1987; Desai and Mitchison, 1997; Howard and Hyman, 2009). Polarized radial array of MTs growing from the centrosome defines spatial organization of the cytoplasm and supports transport of organelles driven by MT-based motor proteins (Cole and Lippincott-Schwartz, 1995; Gross, 2004; Lane and Allan, 1998; Welte, 2004). In the absence of the centrosome, motor proteins establish radial organization of MTs (Borisy and Rodionov, 1999; Compton, 1998; Hyman and Karsenti, 1996; Sharp *et al.*, 2000). MT dynamics are critical for both transport of membrane organelles and motor-dependent organization of a radial MT array. The major experimental approach to study the role of MT dynamics in these processes is live imaging of cells with fluorescently labeled MTs. The cells used for these assays should be suitable for both the observation of dynamic MTs and the moving organelles; these features are combined in melanophores—pigment cells of lower vertebrates.

The main function of melanophores is fast and synchronous redistribution of numerous pigment granules, which aggregate at the cell center or disperse uniformly throughout the cytoplasm in response to hormones (Nascimento *et al.*, 2003). The rapid and highly coordinated redistribution of pigment granules changes the color of animal skin and helps to elude predators (Nascimento *et al.*, 2003). Pigment granules are transported along MTs by motor proteins—kinesins move them to the MT plus-ends during dispersion and cytoplasmic dynein to the minus-ends during aggregation (Nilsson and Wallin, 1997; Rodionov *et al.*, 1991; Tuma *et al.*, 1998). We have used *Xenopus* melanophores to test whether MT dynamics facilitates the interaction of pigment granules with MTs (Lomakin *et al.*, 2009). Live cell imaging of melanophores with fluorescently labeled MTs revealed that the initiation of minus-end transport involved capturing of pigment granules by the growing MT plus-ends. We also found that stabilization of MTs with taxol dramatically inhibited pigment aggregation. These data demonstrate that dynamic MTs are required for the initiation of minus-end transport of pigment granules (Lomakin *et al.*, 2009).

MTs in melanophores are organized into a polarized radial array as a result of nucleation and anchoring at the centrosome. Remarkably, cytoplasmic fragments of melanophores form a radial MT array in the absence of the centrosome after the activation of dynein motors bound to pigment granules (McNiven *et al.*, 1984; Rodionov and Borisy, 1997). Radial organization of MTs might result from minus-end-directed transport of preassembled MTs by dynein motors, seen in mitotic cell extracts (Hyman and Karsenti, 1996; Verde *et al.*, 1991) or from disassembly and reassembly of MTs. To determine the mechanism of radial MT organization, we injected cells with fluorescent tubulin taken at a low concentration (Waterman-Storer and Salmon, 1998). The labeled tubulin unevenly incorporated along the MTs and the resulting bright speckles allowed us to follow the position of reference points on individual MTs and determine whether they were moved in the cytoplasm by motor proteins and discriminate between the two mechanisms of MT reorganization. We found that fluorescent speckles did not move after the activation of dynein motors bound to pigment granules (Vorobjev *et al.*, 2001), which indicated that MTs remained immotile. These data demonstrate that in centrosome-free fragments of melanophores, radial organization of MTs results from disassembly and reassembly of MTs, rather than motor-driven movement in the cytoplasm. Therefore, MT dynamics are important for the formation of a radial array in the absence of the centrosome.

Here we report the methods for studying the role of dynamic MTs in melanophores and describe the details of cell culture preparation, microsurgery and microinjection, fluorescence labeling, and live imaging of MTs.

II. Experimental Procedures

A. Cultivation of Melanophores

Melanophores are obtained either from fish (*Gymnocorymbus ternetzi*) scales or from frog (*Xenopus laevis*) tadpoles. Primary cultures of fish melanophores are prepared from black tetra scales prior to each experiment. Cells are separated from the scales by collagenase treatment, plated on carbon-coated coverslips, and incubated overnight in tissue culture medium. Permanent cell lines of *X. laevis* melanophores are obtained according to the modified method of Daniolos *et al.* (1990) that includes the following steps: trituration of tadpoles, purification of melanophores by Percoll density centrifugation, and generation of stable cell lines by limiting dilution. *Xenopus* melanophores become spontaneously immortalized during the first 2–3 weeks of growth in primary culture, and therefore melanophore cell lines can be continually maintained in culture. *Xenopus* melanophores containing pigment do not survive freezing, and therefore cells should be depleted of melanin by treatment with PTU (*N*-phenylthiocarbamide) that inhibits tyrosinase, an enzyme responsible for melanin synthesis.

1. Primary Culture of Fish Melanophores

1. Remove 10–20 scales from black tetra (*G. ternetzi*) under a dissection microscope using Dumond #5 or #7 tweezers.
2. Place scales in fish Ringer's solution (5 mM Tris–HCl, 103 mM NaCl, 1.8 mM KCl, 0.8 mM NaHCO3, 2 mM $CaCl_2$, pH 7.2) supplemented with 5 mg/ml BSA and 1 mg/ml collagenase (Worthington, type 3).
3. Incubate scales at 30°C for about 1 h.
4. Under a dissection microscope, hold down scales (pin them) using tweezers and detach cells by pipetting using a 2–20 μl automatic pipettor.
5. Wash off the excess collagenase by transferring individual cells with a pipettor from one 35 mm cell culture dish containing Ringer's solution into another dish; repeat three to five times. To prevent attachment of cells to the bottom of the dish, treat dishes with 1% BSA solution in PBS for 10–15 min, and rinse with water prior to use.
6. Place cell suspension onto 22 × 22 mm carbon-coated coverslips. Carbon coating is applied using the DV-502 evaporator (Denton Vacuum, Inc.) and coverslips are mounted with silicon vacuum grease over a hole drilled in a 35 mm cell culture dish. Instead of carbon, coverslips can be coated with laminin (40 μg/ml) or poly-L-lysine (0.1 mg/ml), although melanophores spread more readily on carbon.
7. Add 3 ml of tissue culture medium into each dish (DMEM supplemented with 20 mM HEPES, 20% fetal bovine serum, 200 μg/ml streptomycin, and 200 units/ml penicillin) and incubate overnight at 30°C for the complete spreading of melanophores.

2. Generation of Immortalized Cell Lines of *X. laevis Melanophores*

1. Rinse 20 *X. laevis* tadpoles (stage 30–35; Nieuwkoop and Faber, 1967) three times with amphibian Ringer's solution (9 mM HEPES, 115 mM NaCl, 3 mM KCl, 2 mM $CaCl_2$, pH 7.3).
2. Rinse with 70% ethanol for 5 s.
3. Rinse three times with amphibian Ringer's solution.
4. Grind tadpoles in 10 ml of *Xenopus* tissue culture medium (XTCM; 0.7× L-15 medium (Sigma-Aldrich, St. Louis, MO), supplemented with 10% heat-inactivated fetal bovine serum (Gibco, Invitrogen Corp., Carlsbad, CA), 200 units/ml penicillin, 200 μg/ml streptomycin, and 5 μm insulin, pH 7.3–7.4).
5. Plate the resulting suspension of cells onto three 60 mm cell culture dishes.
6. Incubate at 27°C for 1 week.
7. Rinse with XTCM to remove unattached cells.
8. Cultivate for 1–2 months changing XTCM twice a week until black colonies of melanophores appear.
9. Detach cells by trypsinization (briefly incubate in trypsin solution containing 2 mg/ml trypsin, 0.2 mg/ml EDTA in PBS) and collect by centrifugation.
10. Purify melanophores by centrifugation through Percoll cushion. Resuspend cell pellet in 20% Percoll solution, load onto 12 ml of 33% Percoll solution (both

Percoll solutions prepared on XTCM), and pellet cells by centrifugation at $500 \times g$ at for 5 min at a room temperature.

11. Resuspend the pellet, which contains 80–90% melanophores, in 5 ml of XTCM, plate into a 25 cm^2 flask, and cultivate at 27°C. Change the medium twice a week.
12. When melanophores become confluent, repeat steps 9–11 until unpigmented cells are completely removed. Usually three rounds of Percoll density gradient purification yield pure primary culture of melanophores.
13. Prepare feeder layer of *Xenopus* fibroblasts by plating them into a 48-well plate and cultivate until cells form a monolayer. Treat cells with 10 μg/ml mitomycin C (M0440, Sigma-Aldrich) for 3 h at 27°C to arrest cell growth. Wash three times with XTCM to remove mitomycin C.
14. Detach melanophores by trypsinization, resuspend in 10 ml of XTCM, count using a hemocytometer (Bright-line Counting Chamber, #3100 Haussen Scientific, Horsham, PA), and dilute the suspension to concentration of one cell per 400 μl.
15. Add 200 μl of cell suspension to each well of the 48-well plate containing a monolayer of *Xenopus* fibroblasts (step 13).
16. Cultivate for about 1 month; colonies of melanophores will appear in the wells. Be sure that wells containing two or more colonies are not used for further steps. Change the medium twice a week.
17. Select colonies of cells that grow fast and respond well to hormones by completely aggregating or dispersing pigment granules within 20 min after the treatment with 10 nM melatonin or 10 nM melanocyte stimulating hormone (MSH), respectively.
18. Grow cells in the selected wells until they form a monolayer, detach by trypsinization, and transfer into a 24-well plate.
19. Repeat step 18, and replate cells consecutively into 12- and 6-well plates, and, finally, onto the 25 cm^2 flask. Always maintain melanophores in a semi-confluent culture since at low density cells stop dividing and responding to hormones.
20. If required, freeze melanophores as follows. Incubate cells in the XTCM supplemented with 1 mM PTU (P7629, Sigma) for at least 3 weeks to deplete melanophores of melanin. Trypsinize cells and freeze cell suspension in 20% DMSO in XTCM. After thawing in PTU-free XTCM melanophores produce pigment within 5–6 days.

B. Fluorescent Labeling of Tubulin

Purified tubulin is obtained from porcine brain by a standard procedure involving two cycles of MT temperature-dependent assembly-disassembly (Borisy *et al.*, 1975). MT-associated proteins (MAPs) are removed by an additional cycle of disassembly-reassembly of MTs in a high-salt buffer (HS buffer). Conjugation with Cy3 involves incubation of assembled MTs with the dye followed by the two cycles of MT assembly-disassembly and purification of labeled MTs by centrifugation through a glycerol cushion.

1. Experimental Procedure

1. Thaw 1 ml aliquot of twice cycled tubulin (8–10 mg/ml) in PM buffer (0.1 M PIPES, 1.0 mM $MgCl_2$, pH 6.9) in water bath at 37°C.
2. Add guanosine triphosphatase (GTP) to a final concentration of 1 mM.
3. Polymerize MTs by incubating at 37°C for 10 min. The solution should become viscous and turbid.
4. Pellet MTs at 100,000 xg (50,000 rpm, Beckman TLA 100.3 rotor) at 37°C for 5 min; discard the supernatant.
5. Measure MT pellet volume and resuspend it in equal volume of HS buffer (0.5 M PIPES, 2 mM $MgCl_2$, 1 mM EGTA, pH 6.9).
6. Add DMSO to a final concentration 10% and GTP to 1 mM.
7. Polymerize MTs and pellet them by centrifugation as described in steps 2–3. At this step, the most MAPs are removed.
8. Resuspend the MT pellet in a twofold excess volume of cold PEM buffer (0.1 M PIPES, 5 mM EGTA, 2 mM $MgCl_2$, pH 6.9).
9. Depolymerize MTs by incubating on ice for 10 min.
10. Add DMSO to a final concentration of 10% and GTP to 1 mM.
11. Polymerize MTs by incubating at 37°C for 10 min.
12. Dissolve one vial of commercial aliquot of Cy3 bis-reactive dye (GE Healthcare, PA23000) in 20 μl of anhydrous DMSO. Each commercial vial of Cy3 dye is sufficient for conjugation with 8–10 mg of tubulin.
13. Add the Cy3 solution to polymerized MTs and vortex immediately. Incubate at 37°C for 30 min.
14. Pellet MTs at 100,000 xg at 37°C for 5 min; discard the supernatant.
15. Resuspend the MT pellet in a fivefold excess volume of cold PEM buffer.
16. Add GTP to a final concentration of 1 mM.
17. Depolymerize MTs by incubating on ice for 10 min.
18. Centrifuge tubulin solution at 100,000×*g* at 4°C for 5 min. Transfer the supernatant into a new cold tube.
19. Measure the volume of supernatant and adjust DMSO and GTP concentrations to 10% and 1 mM, respectively.
20. Polymerize and centrifuge MTs as in steps 2–3.
21. Repeat steps 14–18.
22. Polymerize MTs by incubating at 37°C for 10 min.
23. Load MTs onto 900 μl warm 33% glycerol (37°C) cushion prepared on PEM buffer containing 1 mM GTP. Pellet MTs by centrifugation in TLS55 swinging bucket rotor (Beckman, Fullerton, CA; #343778 tubes) at 135,000×*g* (40,000 rpm) at 37°C for 20 min. Aspirate the supernatant and glycerol cushion and resuspend MT pellet in 200 μl of PEM buffer with 1 mM GTP.
24. Depolymerize MTs on ice for 10 min. Centrifuge tubulin solution at 100,000×*g* at 4°C for 5 min. Transfer the supernatant into a new cold tube. The concentration of tubulin solution will be about 6–8 mg/ml.
25. Freeze 10 μl aliquots of supernatant containing Cy3-labeled tubulin and store in liquid nitrogen.

C. Microinjection and Microsurgery

For fluorescent labeling of cytoplasmic MTs in melanophores, cells are injected with Cy3-labeled tubulin. Immediately before microinjection, Cy3-labeled tubulin is clarified by high-speed centrifugation to avoid clogging of microneedles. Injected cells are incubated for about 1 h to allow for the incorporation of Cy3-labeled tubulin into MTs. To produce fluorescent speckles along MTs, cells are injected with Cy3-tubulin solution at a low (0.5 mg/ml) needle concentration. Centrosome-free fragments of melanophores with fluorescently labeled MTs are obtained by microsurgical dissection of the microinjected cells with a sharp glass microneedle.

1. Microinjection

1. Plate cells onto photo-etched carbon-coated coverslips (24 × 24 mm; Bellco Biotechnology, Vineland, NJ) mounted over a hole drilled in a 35 mm cell culture dish; photo-etched coverslips have grids that enable relocation of the microinjected cells.
2. Thaw a 10 μl aliquot of Cy3-labeled tubulin and clarify by centrifugation at 135,000×*g* (40,000 rpm) for 5 min at 4°C.
3. Transfer 1 μl of Cy3-labeled tubulin solution into a glass microneedle with tip diameter of about 0.1 μm produced using micropipette puller, such as Narishige PB-7 (Narishige, East Meadow, NY), from thin-wall borosilicate glass capillaries with filament (an outer capillary diameter is 1.5 mm and inner diameter—1.12 mm; WPI Inc., Sarasota, FL).
4. Place the glass microneedle into a micromanipulator with pipette holder and inject cells with Cy3-labeled tubulin under a pressure of 40 hPa applied by a microinjector (model 5242, Eppendorf, Hauppauge, NY).
5. Incubate for at least 1 h at 27°C for *Xenopus* or at 30°C for fish melanophores to allow the incorporation of labeled tubulin into MTs.

2. Microsurgery

1. Plate fish melanophores on photo-etched carbon-coated coverslips (24 × 24 mm; Bellco Biotechnology) and grow in tissue culture medium overnight.
2. Prepare microneedles as described above.
3. For the dissection of a cytoplasmic fragment, place the tip of an empty glass microneedle close to the cell surface; the fragment is separated by capillary suction. Try to select cells with processes and perform dissection at the base of the process.

D. Live Cell Imaging and Data Analysis

Cells injected with Cy3-tubulin are observed on an inverted fluorescence microscope. Images are acquired with a sensitive cooled charge-coupled device (CCD) camera. To prevent photodamage of cells caused by light exposure during the observation, the oxygen scavenger Oxyrase is added to the medium and mineral oil is overlaid on the culture medium to retard gas exchange.

1. Take a dish with microinjected cells and add Oxyrase (Oxyrase, Ashland, OH) and its substrate lactic acid (L4263 Sigma) to final concentrations of 2% and 20 mM, respectively; check the pH and adjust, if necessary, to 7.3–7.4.
2. Overlay 2 ml of mineral oil (purchased in a local drug store) to retard gas exchange and incubate for 10 min at a room temperature for oxygen depletion.
3. Acquire fluorescence images of cells with an inverted fluorescence microscope, such as a Diaphot 300 (Nikon, Melville, NY) equipped with a Plan ×100 1.25 numerical aperture objective lens and 100 W mercury arc lamp, and a narrow band rhodamine filter set, which is compatible with the excitation of Cy3 fluorescence. For image acquisition, use a CCD camera, such as Andor iXon EM-CCD sensor (Andor, South Windsor, CT), Photometrics series 300 cooled CCD camera (Photometrics, Tucson, AZ), or other cameras which have high quantum efficiency at the emission wavelengths of Cy3 and rhodamine. Collect time series of images of labeled MTs with the exposure time 100–500 ms and 2–3 s intervals between the frames using Metamorph software (MDC, Downingtown, PA).
4. Determine parameters of MT dynamics by tracking the positions of MT plus-ends and recording *X, Y* coordinates using Metamorph. Analyze the data with an appropriate computer software such as the program based on the Multiscale Trend Analysis algorithm (Zaliapin *et al.*, 2005) to determine the following parameters: the duration, length and velocity of MT growth and shortening, and the frequency of catastrophes (transition from growth or pause to shortening) and rescues (the transition from shortening or pause to growth). Alternatively, determine these parameters manually by decomposing life history (distance vs time) plots of MTs into periods of growth and shortening and pauses assuming the changes of MT length over 0.5 μm as growth or shortening events, and others changes as pauses. The parameters of MT dynamics in control and Taxol-treated melanophores with stabilized MTs are represented in Table I.

Table I
Parameters of MT Dynamic Instability in Control and Taxol-treated *Xenopus* Melanophores

	Control	Taxol
Growth distance (μm)	2.96 ± 0.16	0.23 ± 0.01
Growth rate (μm/s)	0.17 ± 0.03	0.06 ± 0.01
Shortening distance (μm)	3.17 ± 0.18	0.23 ± 0.01
Shortening rate (μm/s)	0.18 ± 0.04	0.06 ± 0.001
Catastrophe frequency (s^{-1})	0.026 ± 0.001	0.026 ± 0.002
Rescue frequency (s^{-1})	0.029 ± 0.001	0.042 ± 0.002
Duration of pauses (s)	3.99 ± 0.31	11.51 ± 0.57
Number of analyzed MTs	30	30
Number of analyzed cells	5	6

E. Quantification of Aggregation and Dispersion of Pigment Granules

Responses of melanophores to hormones that induce aggregation or dispersion of pigment granules are quantified by estimating the fractions of cells with aggregated, partially aggregated, or dispersed pigment granules at a fixed time interval after induction. Cells in each category are manually counted using phase contrast microscopy. Alternatively, the kinetics of aggregation or dispersion of pigment granules are determined by measuring pixel gray values within cell outlines using Metamorph region measurement tool. This parameter reflects the degree of homogenous pigment distribution throughout the cytoplasm.

1. Quantification of Cells with Aggregated, Partially Aggregated, or Dispersed Pigment Granules

1. Induce aggregation or dispersion of pigment granules. In the case of fish melanophores, treat cells with adrenaline taken at a final concentration 500 nM to induce aggregation. Induce dispersion by the addition of 5 mM caffeine. To induce pigment aggregation or dispersion in *Xenopus* melanophores, place cells in a serum-free medium, incubate for 1 h at 27°C, add melatonin or MSH, respectively, to the final concentration 10 nM.
2. Fix cells with 4% formaldehyde solution 5 or 10 min after the application of adrenaline or caffeine (fish melanophores) or 10 or 20 min after the stimulation with melatonin or MSH (*Xenopus* melanophores).
3. Count the cells with aggregated, partially aggregated, dispersed pigment granules by counting cells in each category using phase contrast microscopy. Completely aggregated cells have a compact aggregate of pigment granules in the center and no granules at the periphery. Fully dispersed cells contain granules randomly distributed throughout the cytoplasm. Partially aggregated cells have no granules at the margins and less compact pigment aggregate in the center.

2. Quantification of Kinetics of Aggregation and Dispersion of Pigment Granules in *Xenopus* Melanophores

1. Induce aggregation or dispersion of pigment granules in fish or *Xenopus* melanophores as described above.
2. Acquire time series of bright-field images of melanophores with 10 s time intervals using Metamorph time-lapse acquisition mode.
3. Measure the gray levels as integrated pixel values within cell outlines in each of the acquired images in the time-lapse series using Metamorph region measurement tool. The values in the fully dispersed state are taken as 100%. Percentage of gray levels is calculated for each image using the following equation:

$$A = (I_b - I_t)/(I_b - I_d) \times 100$$

where I_b is averaged background levels measured outside cell outlines, I_t is integrated pixel value within a cell outline at a given moment t, and I_d is integrated pixel value within a cell outline in the fully dispersed state. Percentages of gray levels for each time point are averaged across the recorded cells and plotted as a function of time (Lomakin *et al.*, 2009).

III. Discussion

We have described the experimental approaches used to study the role of MT dynamics in intracellular transport and radial organization of MTs in melanophores. Fish or *Xenopus* melanophores provide a unique experimental system for imaging both dynamic MTs and moving membrane organelles. Dynamic MT plus-ends are easily followed in large and flat lamellae of melanophores. Synchronous movement of pigment granules driven by motor proteins in response to distinct hormonal stimuli can be observed using conventional light microscopy (Fig. 1). Large and flat fish

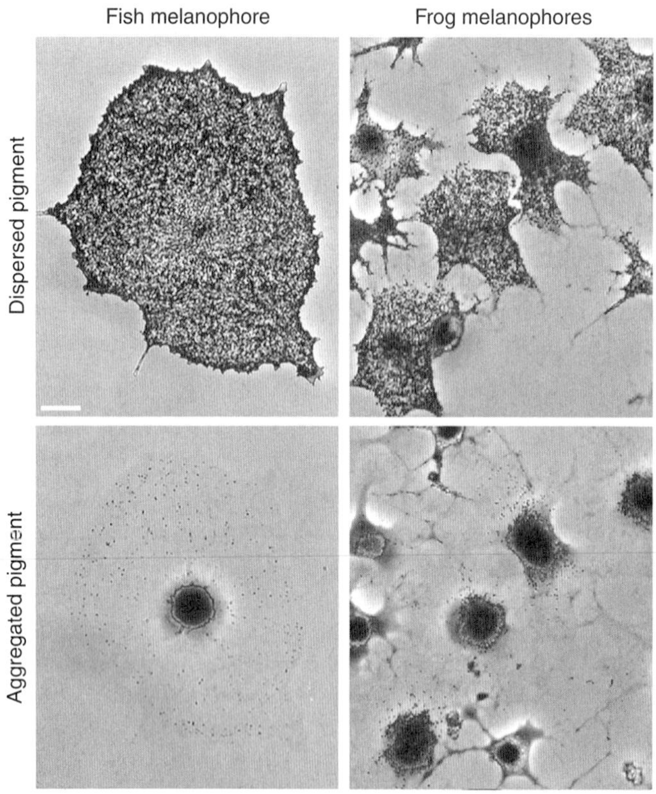

Fig. 1 Phase contrast images of melanophores with dispersed or aggregated pigment granules. Fish (*G. ternetzi*, left images) and frog (*X. laevis*, right images) melanophores with dispersed (top) or aggregated pigment granules (bottom). Bar, 25 μm.

melanophores are particularly good for micromanipulation, including microsurgical dissection of cytoplasmic fragments. However, they cannot be grown in large quantities, and each experiment requires a laborious procedure of cell isolation. Therefore immortalized *Xenopus* melanophores are recommended for the experiments that do not involve microsurgery. Another advantage of *Xenopus* cells is that they are easy to transfect and are suitable for various biochemical experiments.

For live cell imaging of MTs, melanophores are microinjected with fluorescently labeled tubulin. An alternative way of MT labeling involves the expression of tubulin fused with a fluorescent protein. This method is less laborious but involves transfection that is inefficient in cells grown in primary culture, such as fish melanophores.

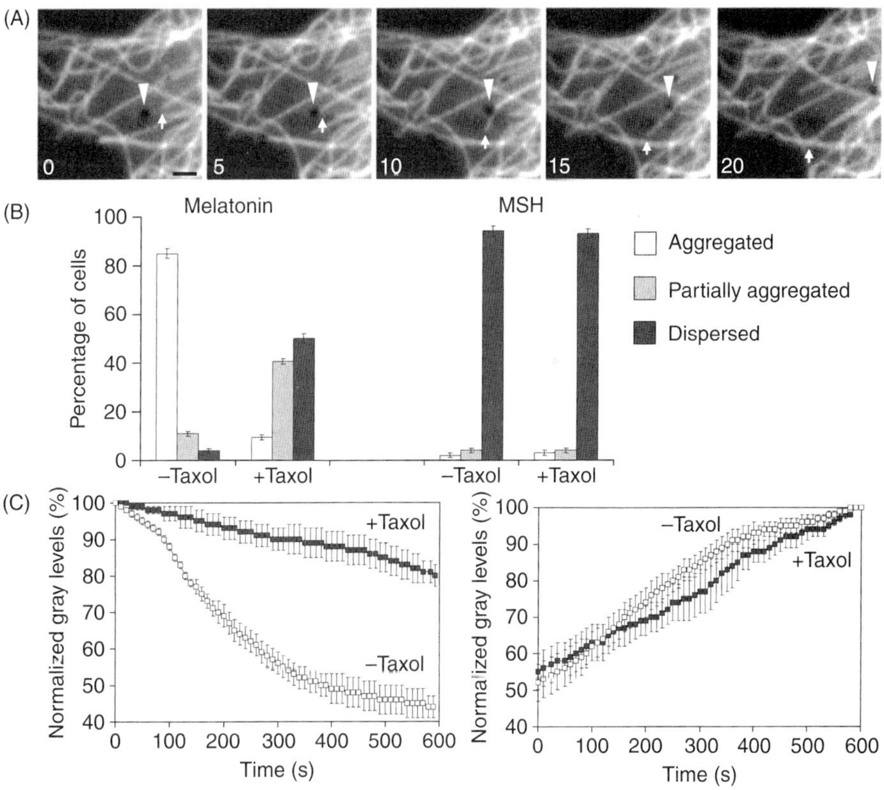

Fig. 2 MT dynamics are required for aggregation of pigment granules in *Xenopus* melanophores. (A) Time series of images of fluorescently labeled MTs during pigment granule aggregation. The arrow and arrowhead indicate a growing MT tip and a pigment granule, respectively. Numbers indicate time in seconds. Bar, 2 μm. (B) Stabilization of MTs reduces pigment granule transport. Responses to aggregation (melatonin, 10 min) or dispersion (MSH, 15 min) in control melanophores or melanophores treated with taxol (1 μM) are quantified. The data are expressed as the percentages of cells with aggregated (white bars), partially dispersed (gray bars), or dispersed (black bars) pigment granules. (C) Kinetics of pigment granule aggregation (left) or dispersion (right) in control (white squares) or taxol-treated (black squares) melanophores.

Besides, the level of exogenous tubulin expression varies from cell to cell being sometimes insufficient for real-time observation, or oppositely, excessive expression generates high background. Therefore, microinjection is a preferable approach for MT labeling in melanophores. The amount of fluorescent tubulin introduced into the cytoplasm by microinjection can be adjusted according to the aim of an experiment. High tubulin concentration results in even labeling of the entire MT and enables observation of dynamic plus-ends, whereas low concentration generates fluorescent speckles that allow studying MT movement and distinguishing between two types of translocation—growth and shortening at the ends or transport of the entire MT (Vorobjev *et al.*, 2001).

The described methods were used to define the role of MT dynamics in intracellular transport and in MT organization, driven by motor proteins. We hypothesized that constantly growing and shortening plus-ends searched the cytoplasm for cargo and that such dynamic behavior increased the probability of organelle capturing. Live cell imaging of melanophores stimulated to aggregate pigment granules revealed the events of granule capturing by the growing plus-ends leading to the initiation of transport (Fig. 2). We further confirmed the role of MT dynamics in this process by treating cells with a MT-stabilizing agent and demonstrated that aggregation of pigment granules was inhibited under these conditions (Lomakin *et al.*, 2009).

In the absence of the centrosome MTs are organized by motor proteins (Borisy and Rodionov, 1999; Compton, 1998; Hyman and Karsenti, 1996; Sharp *et al.*, 2000). We suggested that MT dynamics were important for this organization. The rearrangement could result either from the transport of previously assembled MTs by motor proteins or from the nucleation of new MTs and their rapid reorganization due to active growth and shortening. Imaging of MTs labeled with fluorescent speckles (Fig. 3) revealed

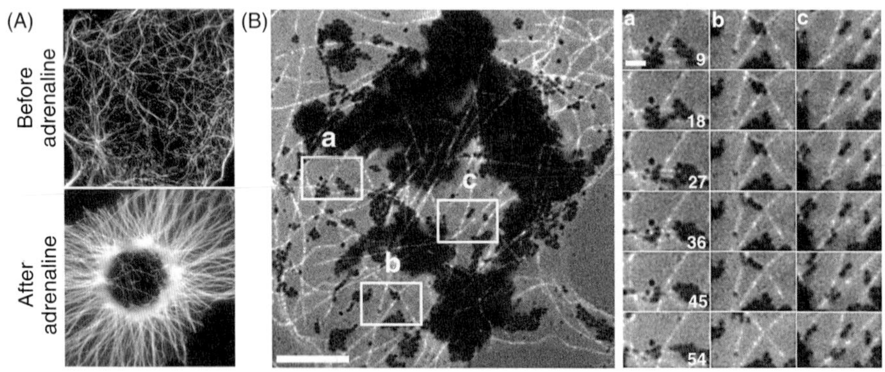

Fig. 3 Organization of a polarized radial array of MTs in cytoplasmic fragments of fish melanophores. (A) Fluorescence images of MTs in a centrosome-free fragment of fish melanophore before (top) and after (bottom) stimulation of aggregation with adrenaline. (B) Fluorescence image of MTs with speckles at low magnification (left) and time sequences of speckles on MTs in the regions indicated on the left panel (right). Numbers indicate the time in seconds. Scale bars, 10 μm (left) and 2 μm (right).

that MTs were not transported in the cytoplasm but were nucleated de novo, formed small local asters, which merged into a single array as a result of MT growth and shortening (Rodionov and Borisy, 1997). Therefore, live cell imaging of MTs in melanophores elucidated the role of dynamics in intracellular transport and centrosome-independent organization of MTs.

Acknowledgments

This work was supported by NIH grant GM62290 to V.I.R.

References

Borisy, G. G., Marcum, J. M., Olmsted, J. B., Murphy, D. B., and Johnson, K. A. (1975). Purification of tubulin and associated high molecular weight proteins from porcine brain and characterization of microtubule assembly *in vitro*. *Ann. N. Y. Acad. Sci.* **253**, 107–132.

Borisy, G. G., and Rodionov, V. I. (1999). Lessons from the melanophore. *FASEB J.* **13**(Suppl. 2), S221–S224.

Cassimeris, L. U., Walker, R. A., Pryer, N. K., and Salmon, E. D. (1987). Dynamic instability of microtubules. *Bioessays* **7**, 149–154.

Cole, N. B., and Lippincott-Schwartz, J. (1995). Organization of organelles and membrane traffic by microtubules. *Curr. Opin. Cell Biol.* **7**, 55–64.

Compton, D. A. (1998). Focusing on spindle poles. *J. Cell Sci.* **111**(Pt 11), 1477–1481.

Daniolos, A., Lerner, A. B., and Lerner, M. R. (1990). Action of light on frog pigment cells in culture. *Pigment Cell Res.* **3**, 38–43.

Desai, A., and Mitchison, T. J. (1997). Microtubule polymerization dynamics. *Annu. Rev. Cell Dev. Biol.* **13**, 83–117.

Gross, S. P. (2004). Hither and yon: A review of bi-directional microtubule-based transport. *Phys. Biol.* **1**, R1–R11.

Howard, J., and Hyman, A. A. (2009). Growth, fluctuation and switching at microtubule plus ends. *Nat. Rev. Mol. Cell Biol.* **10**, 569–574.

Hyman, A. A., and Karsenti, E. (1996). Morphogenetic properties of microtubules and mitotic spindle assembly. *Cell* **84**, 401–410.

Lane, J., and Allan, V. (1998). Microtubule-based membrane movement. *Biochim. Biophys. Acta* **1376**, 27–55.

Lomakin, A. J., Semenova, I., Zaliapin, I., Kraikivski, P., Nadezhdina, E., Slepchenko, B. M., Akhmanova, A., and Rodionov, V. (2009). CLIP-170-dependent capture of membrane organelles by microtubules initiates minus-end directed transport. *Dev. Cell* **17**, 323–333.

McNiven, M. A., Wang, M., and Porter, K. R. (1984). Microtubule polarity and the direction of pigment transport reverse simultaneously in surgically severed melanophore arms. *Cell* **37**, 753–765.

Nascimento, A. A., Roland, J. T., and Gelfand, V. I. (2003). Pigment cells: A model for the study of organelle transport. *Annu. Rev. Cell Dev. Biol.* **19**, 469–491.

Nieuwkoop, P. D., and Faber, J. (1967). "A Normal Table of *Xenopus laevis* (Daudin)." North Holland Publishing Co, Amsterdam.

Nilsson, H., and Wallin, M. (1997). Evidence for several roles of dynein in pigment transport in melanophores. *Cell Motil. Cytoskeleton* **38**, 397–409.

Rodionov, V. I., and Borisy, G. G. (1997). Self-centring activity of cytoplasm. *Nature* **386**, 170–173.

Rodionov, V. I., Gyoeva, F. K., and Gelfand, V. I. (1991). Kinesin is responsible for centrifugal movement of pigment granules in melanophores. *Proc. Natl. Acad. Sci. U. S.A.* **88**, 4956–4960.

Sharp, D. J., Rogers, G. C., and Scholey, J. M. (2000). Roles of motor proteins in building microtubule-based structures: A basic principle of cellular design. *Biochim. Biophys. Acta* **1496**, 128–141.

Tuma, M. C., Zill, A., Le Bot, N., Vernos, I., and Gelfand, V. (1998). Heterotrimeric kinesin II is the microtubule motor protein responsible for pigment dispersion in *Xenopus* melanophores. *J. Cell Biol.* **143**, 1547–1558.

Verde, F., Berrez, J. M., Antony, C., and Karsenti, E. (1991). Taxol-induced microtubule asters in mitotic extracts of xenopus eggs: Requirement for phosphorylated factors and cytoplasmic dynein. *J. Cell Biol.* **112**, 1177–1187.

Vorobjev, I., Malikov, V., and Rodionov, V. (2001). Self-organization of a radial microtubule array by dynein-dependent nucleation of microtubules. *Proc. Natl. Acad. Sci. U.S.A.* **98**, 10160–10165.

Waterman-Storer, C. M., and Salmon, E. D. (1998). How microtubules get fluorescent speckles. *Biophys. J.* **75**, 2059–2069.

Welte, M. A. (2004). Bidirectional transport along microtubules. *Curr. Biol.* **14**, R525–R537.

Zaliapin, I., Semenova, I., Kashina, A., and Rodionov, V. (2005). Multiscale trend analysis of microtubule transport in melanophores. *Biophys. J.* **88**, 4008–4016.

CHAPTER 22

Imaging Cilia in Zebrafish

Kimberly M. Jaffe[*], Stephan Y. Thiberge[†], Margaret E. Bisher[*], *and* Rebecca D. Burdine[*]

[*]Department of Molecular Biology, Princeton University, Princeton, New Jersey 08544

[†]Lewis-Sigler Institute for Integrative Genomics, Princeton University, Princeton, New Jersey 08544

Abstract

Research focused on cilia as extremely important cellular organelles has flourished in recent years. A thorough understanding of cilia regulation and function is critical, as disruptions of cilia structure and/or function have been linked to numerous human diseases and disorders. The tropical freshwater zebrafish is an excellent model organism in which to study cilia structure and function. We can readily image cilia and their motility in embryonic structures including Kupffer's vesicle during somite stages and the pronephros from 1 day postfertilization onward. Here, we describe how to image cilia by whole-mount immunofluorescence, transverse cryosection/immunohistochemistry, and transmission electron microscopy. We also describe how to obtain videos of cilia motility in living embryos.

978-0-12-381349-7
DOI: 10.1016/S0091-679X(10)97022-2

I. Introduction

Cilia are microtubule-based organelles that protrude from the cell while still maintaining a connection to the cell membrane. Most nondividing cells produce a single cilium, referred to as a monocilium or primary cilium. However, specialized cells, such as those in the respiratory tract, can be multiciliated. Traditionally, cilia were thought to consist of nine outer microtubule doublets with an inner microtubule doublet. When the inner doublet is present, the cilia are in the "9 + 2" conformation. However, other microtubule configurations can occur including cilia with nine outer doublets, but missing the inner doublet. When the inner doublet is missing, the cilia are said to be in the "9 + 0" conformation. It was previously believed that all "9 + 0" cilia were immotile and all "9 + 2" cilia were motile; however, recent data challenge this strict categorization [for reviews, see Berbari *et al.* (2009), Cardenas-Rodriguez and Badano (2009), and Gerdes *et al.* (2009)]. For the purposes of this chapter, we will focus on imaging motile cilia in zebrafish.

Cilia have recently come to the forefront of research as an extremely important cellular organelle. When motile cilia are not formed correctly or functioning properly, several different diseases and disorders can arise. These are loosely termed "ciliopathies" [for a review, see Baker and Beales (2009)]. Recent investigation into the cilium's role during early vertebrate development has linked cilia to multiple signaling pathways and many developmental processes—from left-right patterning to kidney cystogenesis [for a review, see Gerdes *et al.* (2009)].

The zebrafish embryo is an excellent model organism in which motile cilia can be studied. In this chapter, we focus on two areas of the embryo in which motile cilia of an organism can be readily visualized: Kupffer's vesicle (KV) (Fig. 1A) and the embryonic kidney, or pronephros (Fig. 1B). The motile cilia in these areas are of the "9 + 2" conformation (Kramer-Zucker *et al.*, 2005).

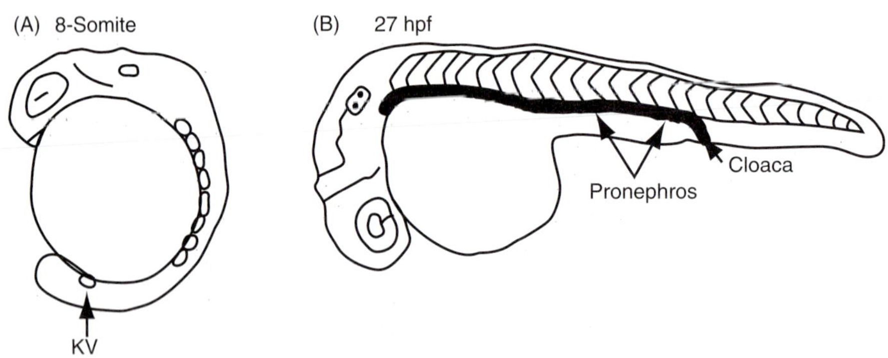

Fig. 1 Hand-drawn images of zebrafish embryos, highlighting two stages of embryonic development when motile cilia can be readily imaged. (A) At the 8 somite stage, Kupffer's vesicle (arrow) is a ciliated organ at the base of the notochord. (B) At 27 h postfertilization, the pronephros, and cloaca are indicated by the arrows and thicker black line.

A. Rationale

We describe the following ways in which cilia can be visualized and imaged in zebrafish embryos: (1) immunofluorescence (IF) on whole embryos, (2) immunohistochemistry (IHC) on transverse cryosections of the embryo, (3) transmission electron microscopy (TEM) on sectioned embryos, and (4) video microscopy of live embryos. Our protocols were developed specifically for cilia in KV and the pronephros, but they can be successfully utilized to image cilia in other regions of the embryo including the neural tube.

Each technique provides different information to the researcher. IF is useful to determine length, location, and organization of cilia within the whole embryo. If desired, you can make quantitative length measurements of the cilia from these images to reveal subtle phenotypes and to facilitate direct comparisons among different embryos.

IHC on cryosectioned embryos is better for less-efficient antibodies that fail to penetrate the embryo in whole-mount IF. Cryosections also allow you to visualize structures in transverse view.

TEM from sectioned embryos provides visualization of cilia ultrastructure. It can tell you whether the microtubule doublets and associated proteins within the cilia are intact and whether the cilia of interest are in the "9 + 0," "9 + 2," or "other" conformation.

The above methods visualize cilia within fixed embryos and provide information about cilia presence and structure. Live video microscopy methodology provides information about cilia motility. Live microscopy allows you to assay cilia function and calculate beat frequencies from videos.

II. Methods

A. Imaging Cilia by IF on Whole Embryos

Imaging cilia in the zebrafish embryo can be done at a variety of stages. For cilia in KV, embryos should be imaged during somite stages, approximately 12–16 h post-fertilization (hpf; Fig. 1A). Cilia in KV are longer than those in surrounding tissues and easy to detect from ~8 somites onward. For cilia in the pronephros (and the neural tube), excellent images can be obtained from embryos at 27 hpf (Fig. 1B). At later time points, the kidney becomes convoluted and the tubules narrow significantly, making good whole-mount confocal images of the cilia more difficult to acquire.

Imaging cilia by IF in whole zebrafish embryos takes roughly 3 days. This protocol is performed in an Eppendorf tube, with no more than 15–20 embryos per tube. Unless otherwise stated, all washes are done with approximately 1 ml volume each and performed at room temperature, sitting on the benchtop.

1. Day 1

Fix zebrafish embryos in 4% paraformaldehyde (PFA)/phosphate-buffered saline (PBS) for 1–5 h at room temperature (this can also be done overnight at 4° C). Rinse the embryos in 1 × PBS + 0.1% Tween 20 (PBST) and dechorionate embryos with forceps. Wash embryos with four 5 min PBST washes. While this is not recommended for the IF protocol because it can lead to reduction in signal, embryos can be rinsed into methanol (MeOH) for long-term storage at –20° C.

The next step is to permeabilize the embryos so that antibodies may enter and bind to the appropriate tissue. (If the embryos are in MeOH at this point, they must first be rehydrated into PBST. Do this stepwise with 5 min washes: 75% MeOH/25% PBST wash, 50% MeOH/50% PBST wash, 25% MeOH/75% PBST wash, followed by five PBST washes.) Permeabilize embryos by washing into ddH_2O for 5 min, prechilled acetone (kept at –20°C) for 7 min, and ddH_2O again for 5 min. Wash the embryos into PBDT (PBST with 1% DMSO) for 5 min to equilibrate them into the IF buffer.

The final step of day 1 includes blocking and adding the primary antibody. Block embryos in PBDT + 10% normal animal serum for 2 h on a rocker. The type of serum you use will depend on the animal that the secondary antibody was produced in; for example, if you plan to use a goat antimouse secondary, you should use normal goat serum (NGS) in the block. It is also important *not* to use sera from the animal in which the primary antibody was derived. For example, if the primary antibody is a rabbit polyclonal, rabbit sera should not be used in the block.

Add the primary antibody at an appropriate dilution in PBDT + 1% NGS and let sit overnight at 4° C (200 μl volume per tube is sufficient). We typically use monoclonal antiacetylated tubulin (clone 6-11B-1 from mouse ascites fluid) at 1:400 to detect cilia. This antibody labels stable tubulin structures and thus detects stable cilia as well as neuronal processes. Antibodies to other modifications found on tubulin can also be utilized [see Pathak *et al.* (2007) and Wloga *et al.* (2009)]. The isotype of the 6-11B-1 primary is IgG_{2b}. Thus, one can use mouse antibodies of different isotypes for colabeling their sample followed by detection with isotype-specific secondary antibodies (see below).

2. Day 2

Wash excess unbound antibody from the sample (all washes done on rocker): one 1 min wash in PBDT + 1% NGS + 0.1 M NaCl, five 30 min washes in PBDT + 1% NGS + 0.1 M NaCl, followed by one 30 min wash in PBDT + 1% NGS.

Add 200 μl of secondary antibody, diluted in PBDT, and let incubate over night at 4° C. We typically use goat antimouse IgG_{2b}–FITC at 1:500.

From this point on, keep samples covered in foil to protect them from light.

3. Day 3

We recommend that you include a nuclear staining step before washing, although it is not necessary: Wash the embryos in PBDT quickly and incubate embryos with 200 μl of nuclear dye for 15 min on a rocker (protected from light). We typically use Hoechst dye at 0.8 μl/ml.

Wash excess unbound antibody from sample (all washes done on rocker, protected from light): one 1 min wash in PBDT + 1% NGS + 0.1 M NaCl, five 30 min washes in PBDT + 1% NGS + 0.1 M NaCl, followed by one 30 min wash in PBDT. The embryos are now ready to be imaged. Until they are imaged, store samples protected from light and at 4° C. Samples can be stored in appropriate antifade reagents if desired.

4. Mounting and Imaging for KV

We typically use Aqua-Poly/Mount, Dumont #55 forceps, and razor blades for mounting. We have found that the most consistent way to image KV is to flat-mount the embryos between two coverslips, dorsal side up.

Place one embryo into Aqua-Poly/Mount on a normal glass slide. While looking under a dissecting microscope, slice through only the yolk of the embryo with forceps or a razor blade—taking care not to slice all the way through the actual embryo. Gently remove as much of the yolk as possible with forceps without disrupting the orientation of the embryo (ventral up, dorsal down). (If you lose track of the orientation, you will simply have to check both sides of the embryo when you image.) Add some Aqua-Poly/Mount to a coverslip and carefully transfer your embryo with forceps to the coverslip so that the embryonic tissue lies flat. We transfer deyolked embryos to a new coverslip to limit the amount of free yolk particles present during imaging. Add another coverslip on top to flatten the embryo between the coverslips. It is possible to flat-mount more than one embryo between one set of coverslips, but we recommend mounting no more than four at a time so that the embryos are not too crowded for imaging. With a permanent marker, draw a circle on the coverslip around each embryo to help you find them during imaging. Keep the slides protected from light at 4° C until imaging.

KV cilia can be nicely imaged with a 40× water objective. Find the field to image by locating the circle you drew around the embryo. Search for the embryo in this field using the nuclear dye under the UV laser. Once you have located the embryo, switch to brightfield to find the notochord, which is a narrow stripe that runs along most of the anterior–posterior length of the embryo. Follow the notochord away from the head and eyes to the posterior end and switch again to the UV laser to visualize the nuclear dye. The easiest way to identify KV is to first focus on the nuclei at the base of the notochord. Then adjust the focus while imaging for the acetylated tubulin signal at a 2× zoom. In wild-type embryos, KV is a relatively small spherical structure, and the cilia within KV are longer than those in surrounding tissues (Fig. 2A). The spherical nature of KV is typically obvious when utilizing both a nuclear stain and the acetylated tubulin stain.

5. Mounting and Imaging for Kidney

We have found that the most consistent way to image the kidney is to mount the embryos between two coverslips, lateral side up.

Lay the embryo on a glass slide in a small volume of PBDT. While looking under a dissecting microscope, slice the embryo's head and yolk off with forceps or a syringe

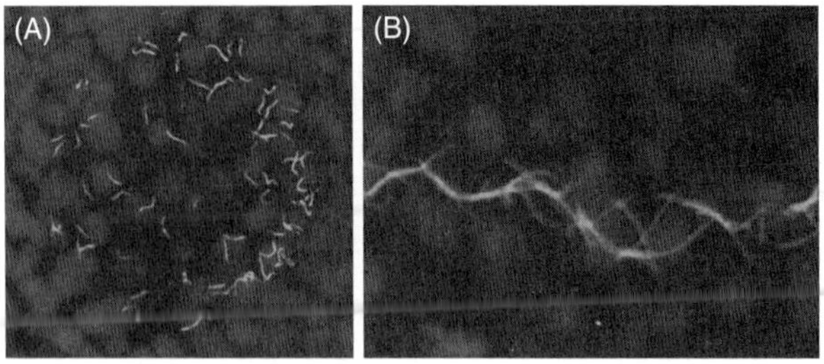

Fig. 2 Stacked 40 × confocal images of primary cilia visualized by IF performed on whole embryos. (A) Cilia in Kupffer's vesicle at the 8 somite stage. The spherical structure of KV is obvious from the cilia arrangement. Cilia are present on cells outside of KV, but they are significantly shorter and not evident in this image. (B) Cilia in the anterior region of the kidney in a 27 h postfertilization embryo. Green: acetylated tubulin; Blue: nuclei. (See Plate no. 12 in the Color Plate Section.)

needle. One method that works well for us is to squeeze the embryo with one pair of forceps at the exact point where you want to cut and use the other pair to scrape away the excess tissue across the edge this has created, like a scalpel. The pronephros is located posterior to the ear (Fig. 1B), so we typically cut the embryo behind the ear and make sure to clear away as much of the yolk ball as possible (do not remove the yolk extension as it will not interfere with mounting and removal typically destroys the embryo area you want to image). Transfer the clean tail to a coverslip. Add a drop of Aqua-Poly/Mount to the embryo and reposition it as needed before adding another coverslip to flatten the sample. It is possible to mount more than one embryo in one drop, but we recommend mounting no more than four at a time so that the embryos are not too crowded for imaging. With a permanent marker, draw a circle on the coverslip around each embryo to help you find them during imaging. Keep the slides protected from light at 4° C until imaging.

Kidney cilia can be nicely imaged with a 40× water objective. Find the field to image by locating the circle you drew around the embryo. Search for the embryo using the nuclear dye under the UV laser. Zoom to 2× and focus first on the area where the yolk extension ends—this is where the most posterior region of the kidney, the cloaca, is located—before switching to the antiacetylated tubulin-specific laser. The cilia should be obvious in this area, forming a line from the cloaca to the anterior region of the pronephros (Fig. 2B). If you use antiacetylated tubulin to visualize the cilia, other structures such as the neural circuitry dorsal to the kidney will also be labeled.

6. Measuring Cilia Length

You can quantify cilia length from your stacked confocal images. We typically use the ImageJ program (a Java-based image processing program from the National Institutes of Health); download this software for free at http://rsbweb.nih.gov/ij/. In ImageJ, open your stacked confocal image. Before you can begin measuring, you need to update the microscope parameters used for your specific image: Go to the Image dropdown menu, click Properties, and enter the pixel size (open the same image within your standard confocal software to find the original pixel sizes).

To measure individual cilium in ImageJ use the Line tool to draw a line from the beginning to the end of a cilium, go to the Analyze dropdown menu, and click Measure to get a length measurement. ImageJ will conveniently record measurements as you go and save them together in a Microsoft Excel file. To create statistically significant values, we typically measure at least 20 cilia per embryo and 3 embryos per condition before calculating a final average value for each condition. When moving to a new stacked image, you must reenter the corresponding pixel sizes so that you can legitimately make direct comparisons among your samples. To minimize user variability in marking the start and end points of cilia, we typically have the same individual to complete all of the marking and measuring for an experiment. In wild-type embryos, cilia length is approximately 3–4 μm in KV (Kramer-Zucker *et al.*, 2005; Schottenfeld *et al.*, 2007; Serluca *et al.*, 2009) and approximately 7–11 μm in the pronephros (Kramer-Zucker *et al.*, 2005; Wilkinson *et al.*, 2009). We typically see 15–50 cilia per KV (Schottenfeld *et al.*, 2007; Serluca *et al.*, 2009).

a. Materials

Eppendorf tubes
Transfer pipettes…VWR International: Catalog# 14670-200
PBS
ddH_2O
Micropipettes/pipette tips
Graduated cylinders
Glass bottles
Tween 20 (polyoxyethylene sorbitan monolaurate)….Sigma-Aldrich: Catalog# P-9416
Dumont #55 forceps…Fine Science Tools: Catalog# 11255-20
4% PFA (diluted in 1× PBS)…Electron Microscopy Sciences: Catalog# 15713-S
DMSO (dimethyl sulfoxide)…Sigma-Aldrich: Catalog# D8418
Acetone (HPLC grade)…Fisher Scientific: Catalog# A949-1
Benchtop rocker
NaCl (sodium chloride)
Monoclonal antiacetylated tubulin, clone 6-11B-1 from mouse ascites fluid… Sigma-Aldrich: Catalog# T6793
NGS …Jackson ImmunoResearch Laboratories, Inc.:
Catalog# 005-000-121
Goat antimouse IgG_{2b}–FITC…Southern Biotechnology Associates, Inc.:
Catalog# 1090-02
Foil—to cover embryos during incubations and washing
Hoechst 33342, trihydrochloride, trihydrate-10 mg/ml solution in water…Invitrogen: Catalog# H3570
Dissecting microscope
Razor blades

Syringe needles
Permanent marker
Glass slides
Glass cover slips…VWR International: Catalog# 48393 172
Aqua-Poly/Mount…Polysciences, Inc.: Catalog# 18606
Confocal microscope/software…Zeiss LSM 510
40× water immersion lens
Computer /ImageJ software

B. Imaging Cilia by IHC on Transverse Cryosections of the Embryo

This protocol is similar to the one above, except that serial 10–12 μm transverse sections of the embryo are taken before IHC is performed. This protocol is ideal for visualization of stain in cross-section and for less efficient antibodies that cannot be used in whole-mount.

Imaging cilia by IHC on cryosections takes 3–4 days.

1. Day 1

Steps before cryosectioning are done in an Eppendorf tube. Unless otherwise stated, all washes are done with approximately 1 ml volumes and performed at room temperature, on the benchtop.

Fix embryos with 2% PFA/PBS at 4° C. Suggested fixation times for 5 somite stage embryos is 40 min, 24 hpf is 90 min, and 48–72 hpf is 2 h.

Wash in ice-cold PBS once quickly, followed by two 10 min washes. Equilibrate the embryos into sucrose solutions: rinse first in an ice-cold 10% sucrose/PBS wash (tap the vial so that the samples sink to the bottom, or for smaller embryos, keep them in solution for 20 min), followed by an ice-cold 25% sucrose/PBS wash (wait again until the embryos sink to the bottom before proceeding).

Embed samples: Remove solution and embed the embryos into a mixture of 3:1 Tissue-Tek® O.C.T. Compound: 30% sucrose in Tissue-Tek® Cryomolds®. Use forceps or a toothpick to orient the embryo. For transverse kidney sections, orient the embryo so that its axis is perpendicular to the cutting surface. Freeze on dry ice and store at –80° C until cryosectioning is performed.

2. Day 2/3

Use standard cryosection techniques to generate sections of the tissue of interest. We typically use a Leica® Cryostat CM3050 S, kept at –22° C, to cut 10–12 μm sections with a low-profile Leica® 819 Microtome Blade. Once a section is obtained, place it on a Superfrost® Plus glass slide using forceps. Let the sections dry for approximately 2–3 h. The samples can either be processed now or kept in a slide holder overnight.

All washes are done with the slides submerged in the appropriate buffer in glass staining dishes unless otherwise noted. The slide rack mentioned here will hold up to

20 slides. If you are filling all 20 slots in the rack, you should make sure that the sections on the slides are not facing each other or touching adjacent slides.

To process the sections, put the slides into a glass slide rack and submerge the rack into 0.005% saponin/PBS to permeabilize the sections for two 10 min washes at room temperature.

Blocking and addition of primary antibody are done in a handmade humidified chamber that minimizes evaporation and reagent use: Line the bottom of a lidded plastic box or a large Petri dish with wet paper towels. On top of the paper towel layer, create a layer of disposable plastic pipettes (cut the pipettes so that they can fit the width of the container). The slides will lie on top of these pipettes. Add the slides, sample side up. While incubating, cover the tray or dish so as to trap the moisture inside.

Block in a small volume (150–200 μl per individual section) of 5% Normal Animal Serum + 1% BSA/0.005% saponin/PBS for 1 h at room temperature. The choice of serum used for blocking is discussed in the IF protocol.

Add primary antibody in a small volume (150–200 μl per individual section) over night at 4° C in the humidified chamber. We typically use monoclonal antiacetylated tubulin (clone 6-11B-1 from mouse ascites fluid) at 1:400 in blocking solution. The isotype of the 6-11B-1 primary is IgG_{2b}. Thus, you can use mouse antibodies of different isotypes for colabeling your sample followed by detection with isotype-specific secondary antibodies.

3. Day 3/4

Wash excess unbound antibody from slides in 0.005% saponin/PBS: Wash once quickly, followed by three 10 min washes using the glass staining dishes.

Add secondary antibody in a small volume (150–200 μl per individual section) for 2 h at room temperature in the humidified chamber. We typically use goat antimouse IgG_{2b}–FITC at 1:400, made up in 0.005% saponin/PBS. You may also choose to add Rhodamine phalloidin to the secondary mixture at 1:200. Rhodamin phalloidin will stain F-actin.

From this point on, keep slides covered in foil to protect the samples from light.

Wash excess unbound antibody from samples in 0.005% saponin/PBS: wash once quickly, followed by a 5 min wash using the glass staining dishes.

If desired, you can include a nuclear dye stain at this point. Add dye in a small volume (150–200 μl per individual section) for 15 min in the humidified chamber. We typically use Hoechst dye at 0.8 μl/ml. Wash excess dye in 0.005% saponin/PBS: wash once quickly, once for 5 min, once for 10 min, followed by two 20 min washes.

To prepare the samples for imaging, cover the slides in Aqua-Poly/Mount and a cover slip. Let this set for at least 2 h before imaging.

The sections are now ready to image on the confocal microscope. Until they are imaged, store samples protected from light at 4° C. An image of cilia in the neural tube obtained from a cryosection can be seen in Fig. 3.

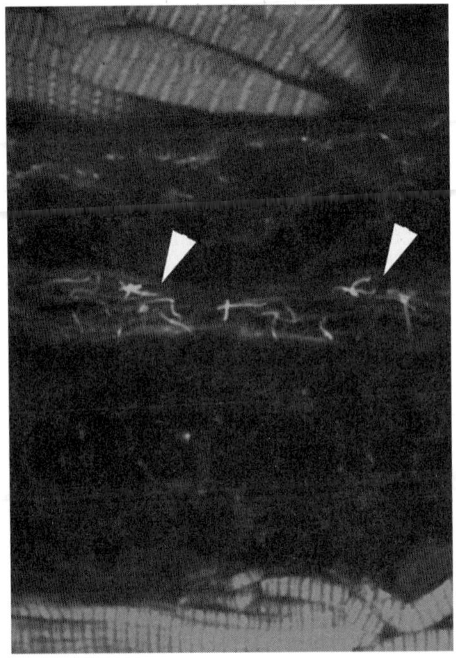

Fig. 3 Longitudinal cryosection of the neural tube in a 2 days postfertilization embryo. Cilia (green; arrowheads) are visualized by IHC performed on the sectioned embryo with an antibody to acetylated tubulin. F-actin in the somites and underlying neural tube is labeled with rhodamine phalloidin (red). (See Plate no. 13 in the Color Plate Section.)

a. Materials

Eppendorf tubes
Transfer pipettes… VWR International: Catalog# 14670-200
Micropipettes/pipette tips
Graduated cylinders
Glass bottles
Pipettes
PBS
Dumont #55 forceps…Fine Science Tools: Catalog# 11255-20
Toothpicks
Dry ice
Foil
Plastic box/large Petri dish
Paper towels
Disposable pipettes
ddH$_2$O
2% PFA (made in 1× PBS)…Electron Microscopy Sciences: Catalog# 15713-S
Sucrose…Roche 100 168

Tissue-Tek® Cryomold®; Sakua Finetek...Electron Microscopy Sciences: Catalog# 62534-10
Cryostat... Leica® CM3050 S
Superfrost® Plus glass slides...Fisher Scientific: Catalog# 12-550-15
Tissue-Tek® O.C.T. Compound...Electron Microscopy Sciences: Catalog# 62550-01
Leica® 819 Microtome Blade, Low Profile...Electron Microscopy Sciences: Catalog# 63065-LP
Glass staining dishes...Electron Microscopy Sciences: Catalog# 70312-23
Glass slide rack...Electron Microscopy Sciences: Catalog# 70312-24
Saponin...Sigma-Aldrich: Catalog# S4521
Bovine serum albumin (BSA)...Sigma-Aldrich: Catalog# A3059
NGS...Jackson ImmunoResearch Laboratories, Inc.: Catalog# 005-000-121
Monoclonal antiacetylated tubulin, clone 6-11B-1 from mouse ascites fluid...Sigma-Aldrich: Catalog# T6793
Goat antimouse IgG_{2b}–FITC...Southern Biotechnology Associates, Inc.: Catalog# 1090-02
Rhodamine phalloidin...Invitrogen. Catalog# R415
Hoechst 33342, trihydrochloride, trihydrate-10 mg/ml solution in water...Invitrogen: Catalog# H3570
Glass cover slips...VWR International: Catalog# 48393 172
Aqua-Poly/Mount...Polysciences, Inc.: Catalog# 18606
Confocal microscope/software...Zeiss LSM 510

C. Imaging Cilia by TEM

While visualizing movement and structure of cilia on a gross-anatomical level is extremely useful and informative, it may be necessary to visualize the ultrastructure of cilia through TEM. As mentioned previously, a cilium is made up of nine outer microtubule doublets and can vary in inner doublet composition. Below, we describe methodology for visualizing KV cilia in 8–10 somite stage embryos and kidney cilia in 27 hpf embryos.

Imaging cilia by TEM takes approximately 4 days.

1. Day 1

Fix embryos overnight in a solution of 1.5% glutaraldehyde, 1% PFA, 3% sucrose, 70 mM $NaPO_4$, pH 7.2. It is critical that this solution be made fresh just prior to use (we recommend using individual 10 ml vials of glutaraldehyde and PFA). 0.1% tannic acid can be added to the fixative if desired. The tannic acid can enhance the final contrast of the microtubules in the images.

For KV-staged embryos, dechorionate the embryos and add them to a glass scintillation vial full of fresh fixative and let them sit overnight at 4° C. Limit the number of embryos per vial to 5–10.

For 27 hpf embryos, dechorionate the embryos, and place them into a glass dish full of fixative. Using dissection scissors, immediately cut off the tail posterior to the cloaca, and then take off the head (aim for right behind the eye, trying not to disturb the yolk). This will allow penetration of the fixative to the pronephric area rapidly. In our experience this is critical for obtaining useful TEM images. Once the embryos are trimmed, place them into a glass scintillation vial full of fresh fixative and let them sit overnight at 4° C. Limit the number of embryos per vial to 5–10.

Embryos will remain in the same glass scintillation vials until further noted.

2. Day 2

The following washes are done at room temperature. Rinse the embryos for three 15 min washes in 0.1 M sodium cacodylate buffer, pH 7.4 (21.40 g of sodium cacodylate, 0.44 g calcium chloride, 40.00 g sucrose in ddH_2O). Post-fix the samples in a 1% osmium tetroxide, 1.5% potassium ferrocyanide (in ddH_2O) solution for 3 h with no agitation (we typically use osmium that comes in a 4% aqueous solution, in sealed vials). Rinse the embryos for three 15 min washes in 0.1 M sodium cacodylate buffer, pH 7.4. After washing, the embryos are taken through a series of EtOH dehydrations. Do this stepwise with six 15 min washes: 30% EtOH/70% ddH_2O, 50% EtOH/50% ddH_2O, 70% EtOH/30% ddH_2O, 95% EtOH/5% ddH_2O, 100% EtOH, 100% EtOH. Rinse embryos through two 15 min washes in 100% propylene oxide (a transitional solvent).

To embed the embryos we use an epoxy-based resin (13 ml of EMBed-812, 8 ml of DDSA, 7 ml of NMA, and 0.47 ml of DMP-30) for ultrathin sectioning. Place these chemicals into a 50 ml conical tube and thoroughly mix on a nutator for 30–45 min. This resin will have the consistency of maple syrup. If you plan to use the resin the next day, it can be stored at room temperature, otherwise store at 4° C (the resin will polymerize at room temperature within 24–48 h).

Place the embryos into a solution of 1 part resin mixture: 1 part propylene oxide. Place the vials on a tissue rotator to provide slow rotation (which helps in the agitation of biological specimens) overnight. Leave the caps off of the vials so as to allow the propylene oxide to evaporate. In the morning, the embryos should be in a more concentrated amount of resin.

3. Day 3

Remove the resin in the scintillation vials and replace with 100% fresh resin. Allow mixing on the tissue rotator for 6 h. During this time you can prepare your labels and embedding molds. We typically type our identification labels using six-point font and place them into the bottom of disposable flat embedding molds. Flat embedding molds are used because the orientation of the sample is very important for sectioning (Fig. 4A). Once the vials have mixed for 6 h, they are ready to be embedded.

Embed samples: Add 100% fresh resin (previously made up during *Day 2*) into the labeled flat embedding molds. Add one embryo to each mold and orient it appropriately (Fig. 4A).

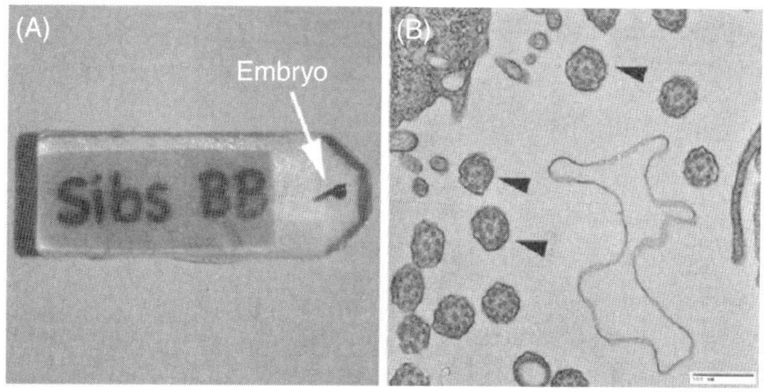

Fig. 4 (A) Example of an embedded 27 h postfertilization embryo [with head removed. This embryo was embedded in a disposable flat embedding mold in the correct orientation [anterior end as far to the cutting edge as possible (right) and posterior end extending to the left, toward the paper label (Sibs BB)]. The paper label is printed with six-point font and embedded into the mold along with the embryo. (B) TEM image of kidney cilia ultrastructure. Image taken at magnification of 12,500 ×. The cilia in this image are in the "9 + 2" conformation (three have been indicated with arrowheads). This image is of a zebrafish mutant that results in pronephric cysts, so the tubule in enlarged and cilia are separated from one another. In wild-type embryos, the tubule is narrower and cilia are often packed tightly together. Scale bar located at the bottom right corner of the image represents 500 nm.

For KV, move the embryo with a toothpick as close as possible to edge of the mold that will be cut. Here, the orientation does not matter.

For the kidney, move the embryo as close as possible to the edge of the mold that will be cut, with the anterior side situated closest to the cutting edge (Fig. 4A).

Place the samples into a 60° C oven and allow the resin to polymerize for 24 h.

4. Day 4

Using standard techniques cut and collect 70 nm sections onto 300 mesh copper grids. We use a Leica UC6 Ultramicrotome with a Diatome diamond knife.

The samples are now ready to be imaged. We observe our samples at 80 kV on a Zeiss912AB Transmission Electron Microscope equipped with an Omega Energy Filter. (Note that because of the Omega Filter on our TEM, post-staining with uranyl acetate (UA) and lead citrate is not needed. If you are using a standard TEM, you must first stain with UA and lead citrate following standard protocols.) We capture micrographs using a digital camera from Advanced Microscopy Techniques and save the images as tiff files onto a computer. For an example of a TEM image of a kidney cilium, see Fig. 4B.

a. Materials

Disposable pipettes
Glass dishes—60 mm^2
Glass scintillation vials
Dissecting scissors
ddH_2O

10% glutaraldehyde, individual 10 ml ampoules…Electron Microscopy Sciences: Catalog #16100
10% PFA, individual 10 ml ampoules…Electron Microscopy Sciences: Catalog # 15712
Tannic acid…Sigma-Aldrich: Catalog #403040
Sodium Cacodylate…Electron Microscopy Sciences: Catalog #12300
Calcium Chloride…Baker: Catalog #1311-01
Sucrose…Sigma-Aldrich: Catalog #S-9378
4% osmium tetroxide, individual 5 ml ampoules…Electron Microscopy Sciences: Catalog # 19160
Potassium ferrocyanide…Fisher Scientific: Catalog #P236-500
Ethyl alcohol, absolute, 200 proof, 99.5% A.C.S. reagent…Acros Chemicals: Catalog #61509-5000
Propylene oxide, EM Grade…Electron Microscopy Sciences: Catalog #20401
Embed-812 (Epon-812)…Electron Microscopy Sciences: Catalog #14900
DDSA (Dodecenyl Succinic Anhydride)…Electron Microscopy Sciences: Catalog #13700
NMA (Nadic Methyl Anhydride)…Electron Microscopy Sciences: Catalog #19000
DMP-30 (2,4,6-9Tri(Dimethylaminoethyl)phenol)…Electron Microscopy Sciences: Catalog #13600
EMS Embedding Molds (Disposable Flat Embedding Mold)…Electron Microscopy Sciences: Catalog #70906-10
Computer/paper/printer
50 ml conical
Tissue rotator
Nutating mixer
300 mesh copper grids…Electron Microscopy Sciences: Catalog #G300-Cu
UC6 Ultramicrotome…Leica
Diatome diamond knife…Diatome
Zeiss912AB Transmission Electron Microscope equipped with an Omega Energy Filter…Zeiss
Digital camera…Advanced Microscopy Techniques

D. Visualizing Cilia Movement by Video Microscopy

Observing cilia motility can be extremely important to describing the action of a particular protein. For example, in *seahorse* mutants, the cilia appear completely normal by IF and TEM but are severely slowed or completely immotile when observed live (Serluca *et al.*, 2009). In the zebrafish, motile cilia can be imaged in areas such as KV, the kidney, the floor plate, and the ear.

In order to directly observe and record cilia motility, you need an upright microscope, an objective that works well for imaging zebrafish embryos, and a good camera. The microscope also needs to have a difference interference contrast (DIC) system. Below, we provide detailed information about choosing such microscope components:

1. Choice of a Camera

Accurate measurement of cilia beating frequency is critically dependent on the frame rate at which images are acquired. The Nyquist–Shannon sampling theorem implies that to measure the frequency F_0 of cilia, acquisition of images should be performed at a rate more than twice as fast ($F > 2 * F_0$). Not fulfilling this condition can lead to artifact known as aliasing. In practice, however, acquisitions at even faster rate are highly recommended (typically, $F > 3–4 * F_0$). The higher acquisition rate compensates for the fact that the movement cannot be exactly periodic and the measurement is of finite duration, overall facilitating the frequency measurement. The choice of a higher frame rate also delivers movies of better quality. Previous measurements have already shown that cilia movement in KV is approximately 30 periods per second, whereas cilia movement in the 2–3 dpf kidney is approximately 40 Hz [KV: (Kramer-Zucker *et al.*, 2005; Okabe *et al.*, 2008); kidney: (Sullivan-Brown *et al.*, 2008)]. Therefore we would recommend working with a frame rate exceeding 120 Hz.

We have also utilized an additional method to measure cilia frequency relying on light scattering (Sullivan-Brown *et al.*, 2008). This approach presents the advantage of a very high sampling frequency (several thousands of points per seconds), therefore avoiding any risk of aliasing. This method has confirmed validity of the results obtained by video microscopy at lower sampling rates (120 Hz and above).

Several cost-effective video cameras (e.g., Dragonfly Express from PointGrey) provide the frame rate required and can therefore be selected for this application. Many laboratories have adopted Electron Multiplication CCDs (EMCCDs) for their fluorescence imaging applications, and these cameras can also be used for fast recording, given only a small portion of the CCD ship is read. EMCCDs come with a feature called “frame transfer” that increases the rate at which images can be acquired. The frame transfer is a particularity of the CCD ship that permits the acquisition of a new image while the previous one is being read. We typically use a Luca EMCCD camera (Andor) in “frame transfer” mode to record images at a rate of 160 Hz on a small portion of the chip of size 100 by 200 pixels.

2. Choice of Objective

A long working distance 60 × objective is well adapted to visualize cilia in zebrafish. A water immersion objective can be used by directly deepening the lens within the medium the fish is held in. Attention should be paid to the numerical aperture (NA) of the objective, as it will directly influence the resolution: higher NA will give better resolution. We typically use a 60 × long working distance, water immersion objective of NA 0.90 from Olympus.

3. Difference Interference Contrast

The components for DIC imaging are a polarizer and Wollaston prism below the specimen and another Wollaston prism and polarizer above the specimen. Commercial microscope suppliers offer different prism pairs specific to the magnification used and the thickness of the sample observed. Although the zebrafish embryo is a thick sample, the DIC system adapted to thick samples will not necessarily give the best contrast and resolution. If available different DIC systems should be tested. We typically use thick specimen specialized DIC prisms on our upright microscope (BX51, Olympus).

4. Setting Up Nomarski DIC

To obtain DIC imaging of a good quality, it is critical to properly implement Köhler illumination. After focusing on the specimen, the high and lateral position of the condenser should be adjusted such that the image of the field iris is well focused and centered in the field of view. It should be opened to the edge of the field of view visible in the eyepiece. For best contrast, the NA of the condenser should match that of the objective used. This is achieved by adjusting the condenser aperture. If glare is visible in the sample, the condenser aperture should be reduced. Once the prism is put in place within the condenser, both the analyzer and the polarizer should be inserted. The contrast should then be adjusted by rotating the polarizer.

Once you have your microscope system set up, you can proceed with imaging.

We provide below detailed instructions for mounting both KV cilia (8 somite stage) and kidney cilia (starting at 27 hpf onward).

5. Mounting for Imaging KV Cilia

Mounting embryos to visualize motile cilia in KV can be difficult, as the embryos are extremely fragile at this stage of development. Because we use an upright microscope with the camera recording from the top of the microscope, the embryo must be mounted so that KV is positioned up with the head of the embryo facing the bottom of a normal glass slide. To obtain this position, we use 1% low melting point (LMP) agarose/PBS as the mounting medium. First, melt the 1% LMP agarose at 70° C and keep it at this temperature until just prior to use.

When the embryos are approximately at the 6–7 somite stage, dechorionate them carefully and allow them to continue developing in an appropriate incubator. It is a good idea to dechorionate more embryos than you plan to image to compensate for any losses during the mounting process.

Once the embryos are at the 8 somite stage, take the melted agarose from the 70° C incubator/water bath and let it cool at room temperature (this will not take long). The agarose should be slightly warm when tested against your inside wrist or face, but not hot. Once the LMP agarose is at the appropriate temperature, rinse the embryos in ddH_2O to get rid of excess salts and put a few embryos (no more than four) onto the slide. Take off excess liquid (if there is too much liquid, the agarose will be too dilute and not solidify correctly). Add two to four drops of agarose to the embryos and gently swirl the embryos and agarose around to mix in any additional liquid left on the slide. Quickly position the embryos with forceps so that KV is sticking straight up. The hardening process may take a few minutes and it can be difficult to get all of the embryos to cooperate in such a small area of hardening agarose (in the beginning, we

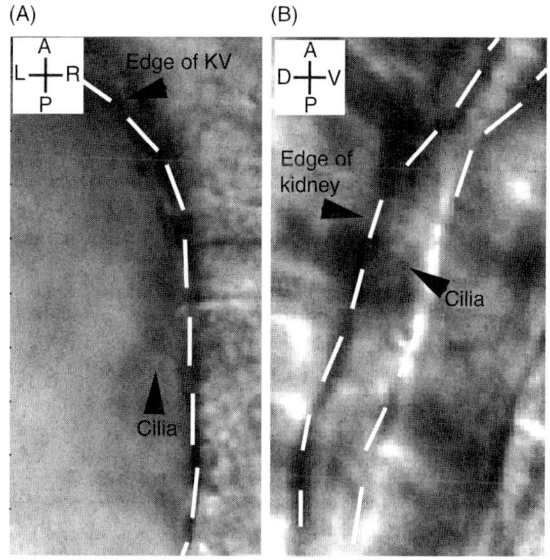

Fig. 5 Stills taken from movies made of (A) KV and (B) 2 days postfertilization kidney. In (A), the cilium in this frame is against the right side of KV (denoted by the dotted white line). In (B), the edge of kidney cells is shown by the dotted white line. Cilia lie within these two lines.

suggest trying two embryos at a time instead of four). Once the agarose is solidified, the embryos are ready to be imaged. The slide can be kept in a Petri dish filled with 1 × E3 Buffer (5 mM NaCl, 0.2 mM KCl, 0.3 mM $CaCl_2$, 0.3 mM $MgSO_4$, ddH_2O—pH 6.5) to prevent the embryos from drying out until you are ready to image. This holding step is not necessary if the embryos are to be imaged immediately.

Once at the microscope, mount the slide and add E3 buffer to the top of the slide. Find KV at the base of the notochord and focus the condenser to obtain proper DIC quality images. While watching synchronous visualization on the computer, search through the focal planes at the edge of KV until you observe cilia. For an example of what this should look like, see Fig. 5A. KV is a spherical organ with cilia projecting from all of the cells lining the fluid filled space (Okabe *et al.*, 2008), so cilia are typically visible from multiple different angles.

6. Mounting for Imaging Motile Kidney Cilia

Motile cilia can be visualized in the kidney as early as 27 hpf, but these cilia will not yet be bundled. From 2 to 3 dpf, the motile cilia will bundle together and the tubule will still be fairly wide, which allows for easier imaging. From 4 dpf on, the tubules become narrow and convoluted, but imaging is still possible. We have been able to achieve quality videos up to 6 dpf (we have not tried later time points). Sullivan-Brown *et al.* provides a good detailed analysis of where to look for cilia motility at the different stages of development (Sullivan-Brown *et al.*, 2008). Motile cilia are also visible in the floor plate (Kramer-Zucker *et al.*, 2005; Serluca *et al.*, 2009) at all of the above-mentioned stages as well as in the ear from approximately 18–24 hpf (Riley *et al.*, 1997).

To image embryos after 1 dpf it is important to first prevent pigment formation so that the kidney is not obscured. To do this, place embryos in 0.003% weight/volume 1-phenyl-2-thiourea (PTU)/embryo medium from 28 hpf onward (they can be put into this solution at any time during somitogenesis). PTU is difficult to dissolve, so make it up the night before. Add 0.03 g of PTU to 1 l of appropriate embryo medium. Stir overnight, protected from light. The following morning stores the PTU solution at the desired temperature and let it equilibrate. Once the PTU solution is at the correct temperature, change embryos from normal embryo medium to the PTU/embryo medium. Keep the solution and embryos away from light. Change the solution each day to help encourage proper development. We typically keep the PTU solution for 1 month before making it fresh.

When ready to image, wash embryos in ddH_2O to remove any residual salts. Prepare a glass depression slide by filling the depression with 3% methylcellulose (a viscous solution that helps immobilize the embryo) and two drops of 0.4% Tricaine (a light anesthetic).

The thickness of the glass depression slide should be carefully selected to match the working distance of the condenser lens. Add enough Tricaine so that the heartbeat slows considerably and blood is barely flowing. While blood flow is not a problem for visualizing cilia in the floor plate or ear, blood movement near the kidney can make cilia observation difficult.

Orient embryos along their lateral side by adding the embryo to the methylcellulose solution and pushing it down until it is at the very bottom, in the middle of the depression.

Put the slide on the microscope and find the cloaca, which is located just above the posterior end of the yolk extension. Focus the condenser to achieve proper DIC images.

7. Localizing Cilia

Because of the fast cilia movement in the pronephros, it is not possible to identify them by eye. Localizing beating cilia is made easier by viewing images taken with the camera using a frame integration time set to a minimum (typically 10 ms or less), thus minimizing the blur due to the beating movement. For the sole purpose of finding cilia, the frame rate does not need to be fast. Typically we use a frame rate of 20–40 Hz to permit synchronous visualization of the images on the computer screen.

While watching synchronous visualization on the computer, search through the focal planes to find the pronephric tubule connecting to the cloaca. Continue to move along the pronephros from the posterior toward the anterior, searching for motile cilia. We provide a snapshot from a movie of a 2 dpf kidney in Fig. 5B as an example of what you should observe. Note the cells that form the wall of the kidney. The blood flow is directly dorsal to these cells.

Keep in mind that some areas of the kidney are easier to image than others, and the orientation of the given embryo can influence how easily the cilia are observed. So, do not become discouraged if cilia are not immediately obvious.

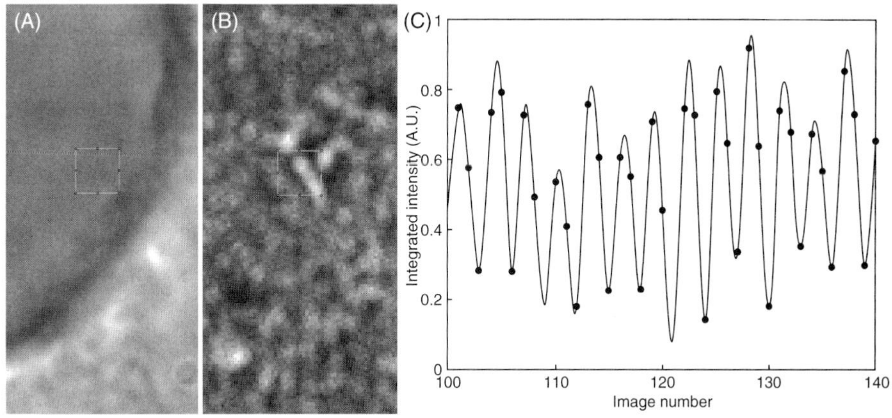

Fig. 6 (A) DIC image of a cilium in KV. (B) Calculated standard deviation of a movie containing about 100 images. The region with the strongest variations in intensity (e.g., where the cilium is) appears brighter. The frame represents the region where the intensity is integrated. (C) The plot of the integrated intensity function of the frame number shows oscillations originating from the cilia moving in and out of the frame. Cilia frequency is then extracted from such a curve.

Once you have found a region of motile cilia, you are ready to record a movie. The camera is set for fast recording, as discussed above, and an acquisition is started.

8. Movie Recording and Cilia Frequency Measurement

We typically record a few hundred images and store them as tiff files.

Any image analysis program capable of measuring pixel intensities should allow the measurement of cilia frequency. We use ImageJ to open our images and measure the cilia frequency. In ImageJ, open the images as a stack. Then select a region in which the cilium moves in and out. For each image, measure the summed or the mean pixel intensity in the region of interest. Plot the values as a function of the image number. The periodic curve observed reflects the passage of the cilia within the region of interest. The frequency of the periodic movement can be measured by counting the number of periods N observed between two distant minima (or maxima) taken at time Tn and Tm. $F_0 = N / (T\text{m} - T\text{n})$.

Figure 6 shows a region of interest (ROI) selected and the average intensity plot as a function of the image number.

a. Materials (see above text for further suggestions)

Upright microscope
60 × objective
Thick specimen specialized DIC prism

Camera
Computer
ImageJ software
Glass beaker/stir bar
ddH_2O
1-phenyl-2-thiourea (PTU)...Sigma-Aldrich: Catalog# P7629
Glass depression slide...Ward's Natural Science
Methylcellulose... Sigma-Aldrich: Catalog# M0387
Tricaine MS-222...Sigma-Aldrich: Catalog# A5040
Glass slides
NaCl (sodium chloride)
KCl (potassium chloride)
$CaCl_2$ (calcium chloride)
$MgSO_4$ (magnesium sulfate)
Agarose, low melt
PBS
Dumont #55 forceps...Fine Science Tools: Catalog# 11255-20

III. Conclusions

Visualizing and obtaining data from motile cilia within the zebrafish embryo is extremely informative. Here we illustrate several different ways in which this can be accomplished. By using these advanced imaging techniques, we can gain a more complete understanding of how cilia function within the developing embryo.

Acknowledgments

The authors would like to thank past and present members of the laboratory for the development of methodology and helpful discussions, particularly Jessica Sullivan-Brown and Noriko Okabe. We would like to especially acknowledge Iain Drummond for his generosity in sharing protocols and suggestions; specifically his suggestions to improve our TEM fixation buffer and informing us of Point Grey's affordable high-speed cameras. We thank Shin-Yi Lin for critical reading of this chapter and Joe Goodhouse for help with confocal microscopy. Methods presented in this report were developed with funding to R.D.B. from the Edward Mallinckrodt Jr. Foundation, the New Jersey Commission on Cancer Research (04-2405-CCR-E0), the Polycystic Kidney Disease Foundation (117b2r), and from the National Institutes of Child Health and Human Development (1RO1HD048584). K.M.J. is supported by postdoctoral grant 0825952D from the American Heart Association. S.Y.T. runs the Lewis-Sigler Institute Imaging Facility, which is supported by grant P50GM071508 from the NIH/NIGMS.

References

Baker, K., and Beales, P. L. (2009). Making sense of cilia in disease: The human ciliopathies. *Am. J. Med. Genet. C Semin. Med. Genet.* **151C**, 281–295.

Berbari, N. F., O'Connor, A. K., Haycraft, C. J., and Yoder, B. K. (2009). The primary cilium as a complex signaling center. *Curr. Biol.* **19**, R526–R535.

Cardenas-Rodriguez, M., and Badano, J. L. (2009). Ciliary biology: Understanding the cellular and genetic basis of human ciliopathies. *Am. J. Med. Genet. C Semin. Med. Genet.* **151C**, 263–280.

Gerdes, J. M., Davis, E. E., and Katsanis, N. (2009). The vertebrate primary cilium in development, homeostasis, and disease. *Cell* **137**, 32–45.

Kramer-Zucker, A. G., Olale, F., Haycraft, C. J., Yoder, B. K., Schier, A. F., and Drummond, I. A. (2005). Cilia-driven fluid flow in the zebrafish pronephros, brain and Kupffer's vesicle is required for normal organogenesis. *Development* **132**, 1907–1921.

Okabe, N., Xu, B., and Burdine, R. D. (2008). Fluid dynamics in zebrafish kupffer's vesicle. *Dev. Dyn.* **237**, 3602–3612.

Pathak, N., Obara, T., Mangos, S., Liu, Y., and Drummond, I. A. (2007). The zebrafish fleer gene encodes an essential regulator of cilia tubulin polyglutamylation. *Mol. Biol. Cell* **18**, 4353–4364.

Riley, B. B., Zhu, C., Janetopoulos, C., and Aufderheide, K. J. (1997). A critical period of ear development controlled by distinct populations of ciliated cells in the zebrafish. *Dev. Biol.* **191**, 191–201.

Schottenfeld, J., Sullivan-Brown, J., and Burdine, R. D. (2007). Zebrafish curly up encodes a pkd2 ortholog that restricts left-side-specific expression of southpaw. *Development* **134**, 1605–1615.

Serluca, F. C., Xu, B., Okabe, N., Baker, K., Lin, S. Y., Sullivan-Brown, J., Konieczkowski, D. J., Jaffe, K. M., Bradner, J. M., Fishman, M. C., and Burdine, R. D. (2009). Mutations in zebrafish leucine-rich repeat-containing six-like affect cilia motility and result in pronephric cysts, but have variable effects on left-right patterning. *Development* **136**, 1621–1631.

Sullivan-Brown, J., Schottenfeld, J., Okabe, N., Hostetter, C. L., Serluca, F. C., Thiberge, S. Y., and Burdine, R.D. (2008). Zebrafish mutations affecting cilia motility share similar cystic phenotypes and suggest a mechanism of cyst formation that differs from pkd2 morphants. *Dev. Biol.* **314**, 261–275.

Wilkinson, C. J., Carl, M., and Harris, W. A. (2009). Cep70 and cep131 contribute to ciliogenesis in zebrafish embryos. *BMC Cell Biol.* **10**, 17.

Wloga, D., Webster, D. M., Rogowski, K., Bre, M. H., Levilliers, N., Jerka-Dziadosz, M., Janke, C., Dougan, S. T., and Gaertig, J. (2009). TTLL3 is a tubulin glycine ligase that regulates the assembly of cilia. *Dev. Cell* **16**, 867–876.

CHAPTER 23

Modeling Microtubule-Mediated Forces and Centrosome Positioning in *Caenorhabditis elegans* Embryos

Akatsuki Kimura* ***and*** **Shuichi Onami†**

*Cell Architecture Laboratory, Center for Frontier Research, National Institute of Genetics, Mishima 411-8540, Japan

†Advanced Computational Sciences Department, RIKEN Advanced Science Institute, Yokohama 230-0045, Japan

Abstract

Microtubules and associated motor proteins are the major generators and mediators of the forces that organize the functional positioning of intracellular structures. The positioning of the centrosomes is a primary target for microtubule-mediated organization. The positioning of the centrosomes further defines the positionings of

978-0-12-381349-7
DOI: 10.1016/S0091-679X(10)97023-4

nucleus, mitotic spindles, and other organelles. Numerical modeling is an effective means by which we can further understand the physical mechanisms underlying microtubule-mediated centrosome positioning. Here, we summarize how we formulated the biophysical properties of microtubules in order to construct a numerical model of centrosome positioning in *Caenorhabditis elegans* embryos. Microtubules elongate and shrink in a stochastic manner, in a process known as "dynamic instability." Upon association with the cell cortex or motor proteins, microtubules mediate pushing and pulling forces. These forces move the centrosome, which is located at the minus-end of the microtubules, to the right place with the right timing. We discuss how the modeling efforts complement experimental knowledge and allow us to evaluate the sufficiency of various candidate hypotheses.

I. Introduction

Inside cells, organelles are positioned at the right place with the right timing. Microtubules play a critical role in organelle movement and positioning in eukaryotic cells (Alberts *et al.*, 2008). When microtubules are disrupted by drug treatment, organelle positioning is disorganized (Wordeman *et al.*, 1986). Microtubules are protein-based, cylindrical structures with a diameter of 25 nm and are located inside the cell. Among the major cytoskeletal filaments (i.e., actin filaments, intermediate filaments, and microtubules), microtubules are the thickest and most rigid (Howard, 2001). Microtubules are polymers consisting of alternating α- and β- tubulin monomers and usually function with various accessory proteins. These accessory proteins modulate nucleation, polymerization, and bundle formation of the microtubules (Alberts *et al.*, 2008). Motor proteins are an important class of accessory proteins that slide along the microtubules and generate forces to move organelles (Howard, 2001). Microtubules dynamically change their length by polymerization and depolymerization, and these processes also generate forces (Howard, 2001).

The intracellular positioning of the centrosomes is a primary target for microtubule-mediated organization. The centrosome is a major microtubule-organizing center in the animal cell (Kellogg *et al.*, 1994). Microtubules grow from and shrink toward the centrosomes with their minus-end associated with the centrosomes. The positioning of the centrosomes is critical in defining the positioning of organelles because many membrane-bound organelles are transported along microtubules (Hirokawa, 1998). The positioning of nucleus and mitotic spindle is defined by the positioning of the centrosomes in a more direct manner (Pearson and Bloom, 2004; Reinsch and Gönczy, 1998). The centrosomes associate with nucleus in most cases, and the centrosomes define the poles of mitotic spindles in animal cells.

In order to understand the mechanical basis of microtubule-mediated positioning of organelles, we first need to estimate the forces that microtubules produce and we then need to evaluate whether the force estimates can explain the positioning of the organelles. Numerical modeling and simulations are effective in evaluating the behaviors of organelles under various hypotheses. Numerical modeling has been extensively used to analyze the mechanical bases of spindle behavior. On the basis of a molecular and biophysical understanding of motors and the dynamic instability of microtubules, the forces acting on the spindle have been calculated and compared with those on the real spindle (Dumont and Mitchison, 2009; Mogilner *et al.*, 2006). Studies of spindles using *Xenopus* egg extracts offer high-quality quantitative data that are suitable for numerical modeling (Burbank *et al.*, 2007). *Drosophila* embryos and cultured cells have also proved to be potent systems for the study of spindles, in that they have provided sufficient quantitative data for numerical modeling (Brust-Mascher *et al.*, 2004; Cytrynbaum *et al.*, 2003; Goshima *et al.*, 2005b). Nédélec and coworkers have developed Cytosim, a computational framework used for simulating cytoskeleton-based movements of objects inside cells (Nédélec, 2002; Nédélec and Foethke, 2007). This sophisticated simulator has been successful in explaining various microtubule-mediated processes, such as microtubule organization, spindle formation, and spindle movements in several systems (Foethke *et al.*, 2009; Goshima *et al.*, 2005a; Janson *et al.*, 2007; Kozlowski *et al.*, 2007). Numerical models have become almost indispensable for studying spindle function (Goshima and Kimura, 2010; Mogilner *et al.*, 2006).

The *Caenorhabditis elegans* embryo is another potent system to study microtubule-based processes. Asymmetric positioning of the mitotic spindle and the accompanying oscillation of the spindle have been extensively studied using numerical modeling. Grill and coworkers have used elegant mathematics and simulations to deduce the force-generating mechanism underlying the oscillatory movements of the spindle (Grill *et al.*, 2001, 2005; Pecreaux *et al.*, 2006). Cytosim has also been applied to the study of spindle oscillation (Kozlowski *et al.*, 2007). Our group has examined related but different microtubule-dependent processes in *C. elegans* embryos, namely pronuclear and centrosome centering (Fig. 1) (Kimura and Onami, 2005), spindle repositioning (Kimura and Onami, 2007), and spindle elongation (Hara and Kimura, 2009).

In general, one of the major purposes of conducting simulations is to test various hypotheses. It is therefore important that we make any model as simple as possible to clarify the cause and effect relationship between the input and the output of the model (Phillips *et al.*, 2009). In this chapter, we describe the numerical model we constructed to simulate microtubule-based forces in *C. elegans* embryos (Hara and Kimura, 2009; Kimura and Onami, 2005, 2007). To examine the commonalities and differences between modeling studies, we compare our assumptions and parameters with those of other published models of microtubule-based processes in *C. elegans* embryos (Kozlowski *et al.*, 2007; Pecreaux *et al.*, 2006).

II. Rationale

Microtubule-mediated forces can be classified into pushing forces and pulling forces (Dogterom *et al.*, 2005; Reinsch and Gönczy, 1998) (Fig. 2). The pushing forces are produced by microtubule polymerization. When the growing tip of a microtubule reaches an object, such as the walls of a microfabricated chamber or the cell cortex, the microtubule can push the object. When the microtubule pushes against an object, the

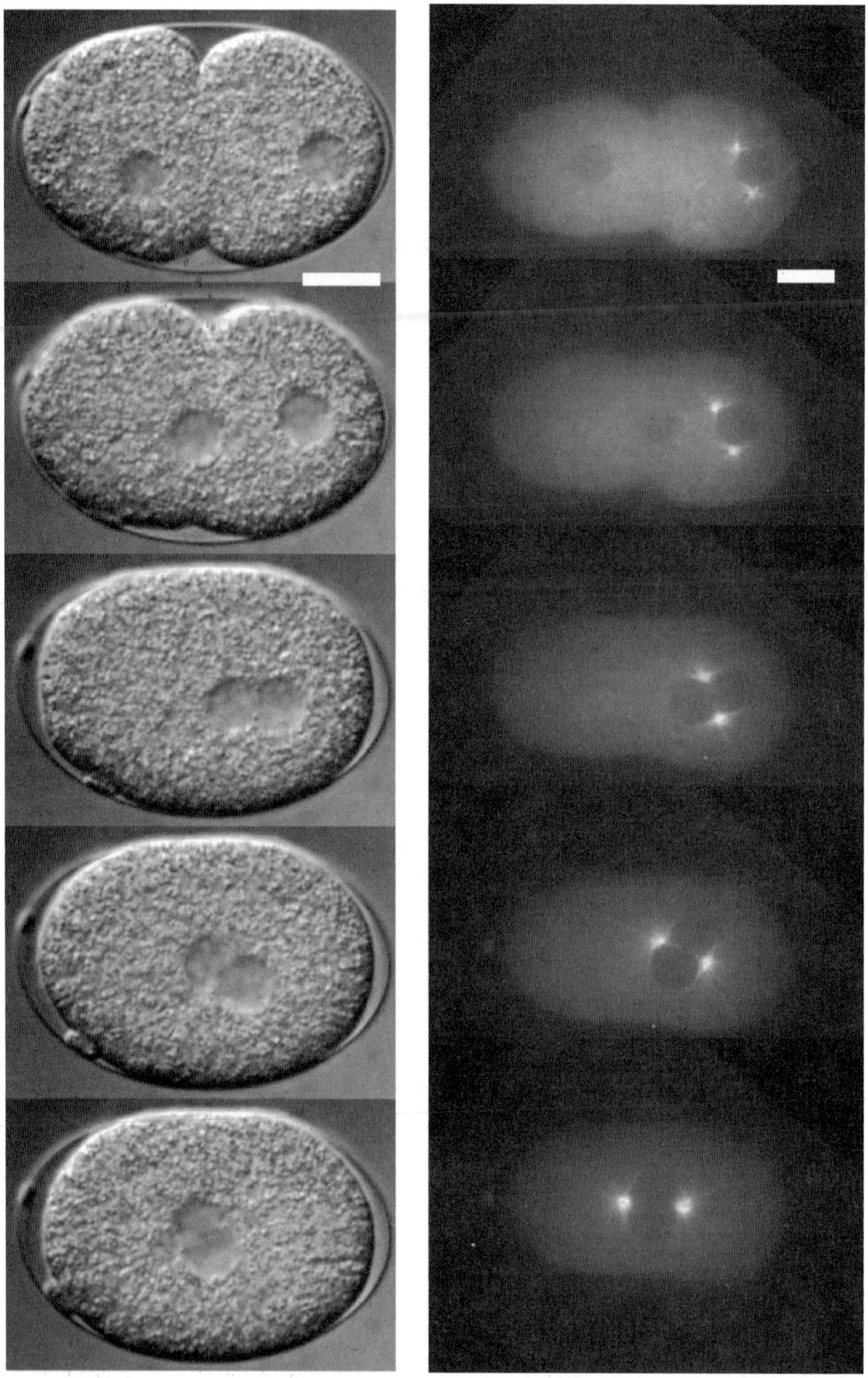

Fig. 1 Microtubule-dependent movement of pronuclei in *C. elegans* embryos. The left panel shows images from Nomarski [differential interference contrast (DIC)] microscopy. The smooth regions in the cytoplasm correspond to the male (right circle) and female (left circle) pronuclei. The right panel shows confocal microscopy images visualizing tubulin-GFP. Filamentous microtubule signals extend outward from the two bright spots of the centrosome. Note that the left and right panels are from different embryos. Bar, 10 μm.

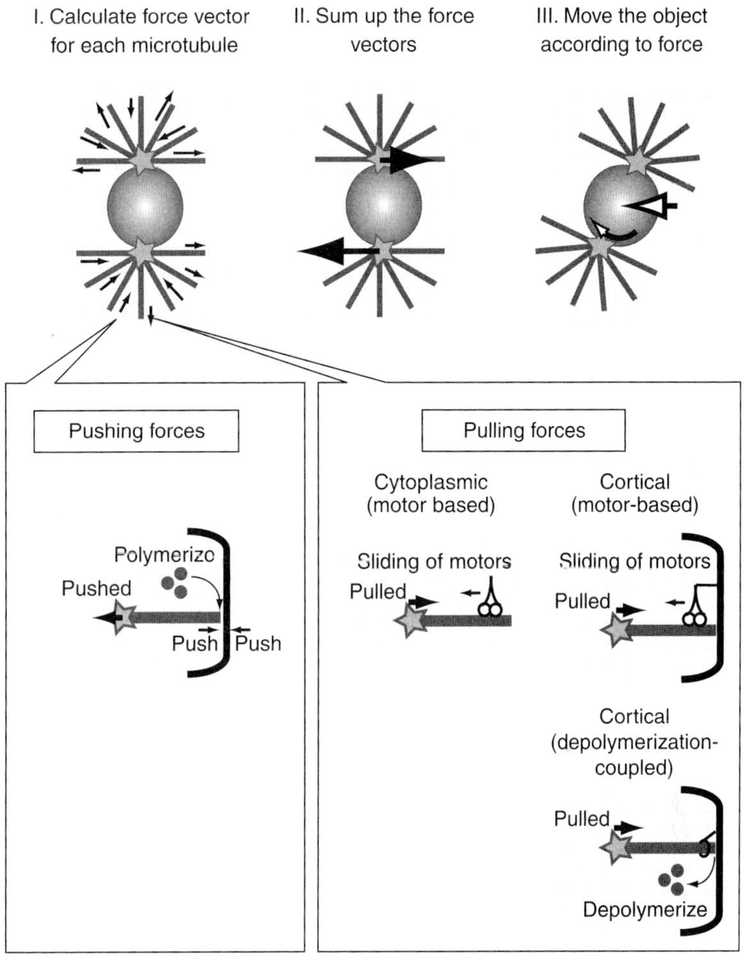

Fig. 2 Microtubule-mediated forces and movement of objects. The upper panel shows the three steps that were used to calculate the movement of an object on the basis of microtubule-mediated forces. The circles at the center of the panels indicate objects to be moved (e.g., pronucleus). The thick lines represent microtubules and the stars represent centrosomes. The solid arrows indicate the direction and magnitude of forces acting on microtubules. The open arrows indicate the resultant translational and rotational movements of the object. It should be noted that, in most cases, the movement of the object and centrosomes affects the forces acting on microtubules, and these three steps are therefore not calculated sequentially but are calculated by solving simultaneous equations. The lower panel is a schema of the force-generating mechanisms. The stars and thick gray lines indicate the centrosomes and microtubules, respectively. The thick black curved lines indicate the cell cortex. The arrows indicate the direction of the forces. The small filled circles indicate α/β tubulin dimers. The two open circles on the microtubules indicate microtubule motor proteins.

counteraction force from the object pushes the microtubule tip back. The force is then transmitted through the microtubule to the other end of the microtubule, the centrosome in animal cells. To estimate the magnitude of pushing forces, two properties of microtubules should be considered: the "force–velocity relationship" (i.e., the relationship between the microtubule pushing force and the velocity of microtubule polymerization) and microtubule buckling (Howard, 2001). The force–velocity relationship means that when the growing end of a microtubule encounters opposing forces, its growth velocity is reduced; the extent of the reduction depends on the magnitude of the opposing forces (Dogterom and Yurke, 1997). Since microtubules are considered to be elastic rods, the pushing forces mediated through them are reduced dramatically when they buckle. Therefore, the buckling force of the microtubules is approximately equal to the maximum pushing force mediated by the microtubules (Dogterom *et al.*, 2005).

The pulling forces are generated at the cell cortex and the cytoplasm (Fig. 2). At the cortex, microtubule-pulling forces can be generated by two mechanisms: motor-based pulling and depolymerization-coupled pulling (Kozlowski *et al.*, 2007; Nguyen-Ngoc *et al.*, 2007). In motor-based pulling, the motor proteins slide along the microtubule but are anchored to a fixed structure. As a consequence, the motor proteins do not move but the microtubule does. In depolymerization-coupled pulling, the microtubule shrinks, but its shrinking end is anchored to a fixed structure. Since the end is fixed, the entire microtubule is pulled toward the fixed position. Microtubule-pulling forces can be generated in the cytoplasm. Although the molecular mechanism is unclear, existence of pulling forces in the cytoplasm is evident from past observations (Hamaguchi and Hiramoto, 1986), and dynein motors exist throughout the cytoplasm (Gönczy *et al.*, 1999). Since force generators in cytoplasm are assumed to be distributed uniformly, the magnitude of cytoplasmic pulling forces mediated by one microtubule should be roughly proportional to the length of the microtubule ("length-dependent pulling force").

In our models of centrosome positioning, we assumed a motor-based mechanism for microtubule-pulling forces at the cortex and the cytoplasm. The motor protein dynein is essential for pronuclear migration and other centrosome positioning processes (Gönczy *et al.*, 1999; Kimura and Onami, 2005). Force generation by a motor protein is involved in asymmetric spindle positioning and oscillation (Nguyen-Ngoc *et al.*, 2007), although depolymerization-coupled force may also be involved in this process (Kozlowski *et al.*, 2007). A motor-based mechanism was also assumed by Pecreaux *et al.* (2006), whereas a depolymerization-coupled mechanism was assumed by Kozlowski *et al.* (2007). Assuming different mechanisms affect how we calculate pulling forces and thus the quantitative behavior of the models. However, the difference between the two mechanisms seems small enough that it does not affect overall conclusions of the modeling in previous studies.

The forces mediated by the microtubules drive the movements of the centrosomes (and associated pronucleus or mitotic spindle) attached to the microtubules (Fig. 2). Inside the cell, viscosity dominates and inertia is negligible—that is, it is a "low Reynolds number" environment (Purcell, 1977). Stokes' law can be used to calculate the translational and rotational movements of spherical objects in cytoplasm (such as organelles), if we assume cytoplasm as a viscous fluid (Berg, 1993). The process of microtubule growth and shrinkage, which is termed dynamic instability (Mitchison and Kirschner, 1984), is modeled as stochastic events, as in a previous model (Mitchison and Kirschner, 1984; Nédélec, 2002). Detailed mechanisms of microtubule polymerization and depolymerization (Karsenti *et al.*, 2006) were not included in our model.

III. Methods

Our model is a three-dimensional force balance model in which the vector sum of the forces mediated by the microtubules is used to calculate the magnitude and direction of the velocity of the centrosomes (and associated pronucleus or mitotic spindle) attached to the microtubules. Throughout this section, bold-faced letters are used to represent vector quantities.

A. Pushing Force

In our "pushing" model, the force acting on the centrosome is the sum of the pushing forces acting on the microtubules attached to the centrosome [Eq. (1)]:

$$\mathbf{F_{total}} = \sum F_{\text{push},i}(-\mathbf{u}_i) \tag{1}$$

where $\mathbf{F_{total}}$ is the force acting on the centrosome, $F_{\text{push},i}$ is the magnitude of the pushing force of the ith microtubule, and $\mathbf{u}_i$, is the direction vector from the centrosome toward the growing tip of the ith microtubule.

A simple assumption regarding the amount of the pushing force exerted on a microtubule ($F_{\text{push},i}$) is that the growing end of the microtubule pushes (and is pushed back by) the cell cortex with a constant amount of force. More realistic assumptions take into account the buckling force and force–velocity relationship.

As described earlier, the force–velocity relationship means that, when a microtubule pushes against an object, the opposing reaction force from the object suppresses polymerization of the microtubule. There is a one-to-one relationship between the pushing force and the velocity of microtubule growth. Since the force exerted by the object equals the force transmitted to the other end of microtubule, we can estimate the force from the growth velocity of the microtubule. The decay in growth velocity of microtubules in response to mechanical force is modeled on the basis of thermodynamic arguments, assuming that the off-rate of tubulin is not affected by the force (Dogterom and Yurke, 1997; Hill, 1987; Howard, 2001; Janson and Dogterom, 2004; Peskin *et al.*, 1993). The force–velocity relationship is thus modeled as Eq. (2):

$$v(F) = V_0(\exp(-F/F_0) - 1) + V_g \tag{2}$$

where $v(F)$ is the growth velocity of a microtubule when encountering an opposing force F, V_0 is a parameter defining the rate of addition of α/β tubulin dimers ("elongation-rate parameter"), F_0 is the characteristic force scale defining the sensitivity of growth to force, and V_g is the growth velocity for freely growing microtubules. Since the plus-end of the microtubule cannot penetrate into the cell cortex, the growth velocity of the microtubule after reaching the cortex (v) is defined by the displacement of the minus-end of the microtubule, and thus can be formulated as Eq. (3):

$$v = -\mathbf{V_{cen}} \cdot \mathbf{u}_i \tag{3}$$

where $\mathbf{V_{cen}}$ is the velocity vector of the centrosome (i.e., the minus-end of the microtubule), $\mathbf{u}_i$ is the direction vector from the centrosome toward the growing tip of the ith microtubule, and $\mathbf{V_{cen}} \cdot \mathbf{u}_i$ represents the inner product of $\mathbf{V_{cen}}$ and $\mathbf{u}_i$. For simplicity, here we assume that the contact point between the growing end of microtubule and the cell cortex is fixed, and that there is no buckling.

The parameter V_0, in theory, equals $k_{on}c\delta$, where k_{on} is the tubulin on-rate, c is the tubulin dimer concentration, and δ is the change in microtubule length per addition of tubulin dimer (0.6 nm). The parameter V_g, in theory, equals $\delta(k_{on}c - k_{off})$ where k_{off} is the tubulin off-rate. These definitions lead to Eq. (4):

$$V_g = V_0 \left(1 - \frac{k_{off}}{k_{on}c} \right) \tag{4}$$

Because quantitative analyses show that $k_{off} << k_{on}c$ (Janson and Dogterom, 2004), V_0 approximately equals V_g. Thus Eq. (2) can be simplified to Eq. (5):

$$v(F) = V_g \exp(-F/F_0) \tag{5}$$

In theory, the parameter F_0 equals $k_B T/\delta = 7$ pN, where k_B is the Boltzmann constant (1.38×10^{-23} J/K) and T is temperature (298 K). When pure tubulin is used to examine the force–velocity relationship *in vitro*, the calculated value of F_0 is 1.9 pN (Dogterom and Yurke, 1997). The F_0 value *in vivo* may be larger than both these values. In yeast cells, when microtubules touch the cell cortex, their growth velocity decreases by about 38% when the opposing force from the cortex is about 15 pN (Tran *et al.*, 2001). A similar result was obtained in another study in yeast by Drummond and Cross (Drummond and Cross, 2000). Substituting these values obtained *in vivo* into Eq. (5), the calculated value of F_0 is 31 pN. This discrepancy between F_0 values might be due to differences in the quality (e.g., flexibility) of the barriers against which the growing microtubules push (i.e., the cell cortex *in vivo* and glass walls *in vitro*) or to the presence or absence of various microtubule-associated proteins that affect assembly and disassembly of microtubules *in vivo* (Kinoshita *et al.*, 2001; Tournebize *et al.*, 2000). In our simulation, we used the F_0 value estimated from *in vivo* results. It should be noted that a recent simulation study assuming F_0 as $\sim$1.7 pN successfully reproduced the general shape of microtubule in *S. pombe* (Foethke *et al.*, 2009).

The buckling force of the microtubule defines the maximum limit of the pushing force transmitted to the other end of microtubule. The buckling force of a microtubule is calculated by Eq. (6):

$$F_{push} = \pi^2 \kappa / L^2 \tag{6}$$

where F_{push} is buckling force, κ is the flexural rigidity of the microtubule, and L is the length of the microtubule (Howard, 2001).

B. Pulling Force

In our "pulling" model, the force acting on the centrosome is the sum of the pulling forces acting on the microtubules attached to the centrosome [Eq. (7)]:

$$\mathbf{F}_{\text{total}} = \sum F_{\text{pull},i} \cdot \mathbf{u}_i \tag{7}$$

where $\mathbf{u}_i$ is the direction vector of the ith microtubule, $F_{\text{pull},i}$ is the magnitude of the pulling force of the microtubule, and $\mathbf{u}_i$ is the direction vector from the centrosome toward the growing tip of the ith microtubule.

1. Number of Force Generators Associated with One Microtubule

The pulling force is generated by force generators and mediated by one microtubule along the direction of the microtubule according to Eq. (8):

$$F_{\text{pull}} = N_{\text{FG}} F_{\text{FG}} \tag{8}$$

where F_{pull} is the pulling force of the microtubule, N_{FG} is the number of active force generators per microtubule, and F_{FG} is the pulling force generated by a single active force generator.

Among the parameters used in the numerical models of microtubule-mediated forces, the parameter N_{FG} is one of the most difficult to estimate from experiments. Therefore, the setting of the parameter N_{FG} is often a key assumption in the model.

In our length-dependent pulling model of centrosome centering (Kimura and Onami, 2005, 2007), N_{FG} is assumed to be proportional to the length of the microtubule (Hamaguchi and Hiramoto, 1986; Reinsch and Gönczy, 1998). In these simulations, the proportionality constants (representing the density of motors on microtubule, D) were 0.1/μm (Kimura and Onami, 2005) and 0.005/μm (Kimura and Onami, 2007); D is a fit parameter, which is dependent on the total number of microtubules.

To model the pulling force generated at the cortex, we make the simple assumption that microtubules associate with a constant number of motors with a constant probability (Hara and Kimura, 2009; Kimura and Onami, 2007). As a more elaborate assumption, the idea of variable probability is introduced. Pecreaux *et al.* (2006) and Kozlowski *et al.* (2007) consider the on- and off-rates for the force generator binding to a microtubule, and they assume that the off-rate increases exponentially with load (Kozlowski *et al.*, 2007; Pecreaux *et al.*, 2006); this phenomenon has been reported for a kinesin motor *in vitro* (Schnitzer *et al.*, 2000). When asymmetry of the pulling force toward the posterior and anterior cortex is modeled, the parameters involved in the numbers of force generators associated with microtubule are changed between the posterior and the anterior cortex (Kimura and Onami, 2007;

Kozlowski *et al.*, 2007), according to the experimental estimation of the number of force generators (Grill *et al.*, 2003).

2. Pulling Force Exerted by a Single Force Generator

In our motor-based pulling model, the pulling force generated by a single motor (F_{FG}) was calculated according to the force–velocity relationship of motors, with the limitation that the maximum force that the motor protein cannot exceed is the stall force of the motor (F_{stall}), which is the force required to stop the movement of the motor. When motor proteins slide along microtubules, their velocity decreases as they encounter larger load forces (force–velocity relationship of motor proteins). Experimental measurements have demonstrated that the force–velocity plots for motor proteins exhibit various shapes, including linear, convex-downward, and convex-upward (Hirakawa *et al.*, 2000; Howard, 2001; Oiwa and Takahashi, 1988; Schnitzer *et al.*, 2000; Visscher *et al.*, 1999). Because of the uncertainty in the shape of the force–velocity plot for dynein motor in *C. elegans* embryos, a simplified relationship for force and velocity is often assumed [Eq. (9)]:

$$v(F) = V_{max}(1 - (F/F_{stall})) \tag{9}$$

where $v(F)$ is the velocity of the motor when exerting a force, V_{max} is the maximum velocity, F is the force exerted by the motor, and F_{stall} is the stall force of the motor. According to this relationship, $v(F)$ decreases in a linear manner from V_{max} when $F = 0$ to a value of zero when $F = F_{stall}$ (Kimura and Onami, 2005, 2007; Pecreaux *et al.*, 2006). The parameter values of F_{stall} and V_{max} used in past simulations are listed in Table I.

In the depolymerization-coupled pulling model of Kozlowski *et al.* (2007), the pulling force generated by the force generator is assumed to be proportional to the stretch required to link the cortex and the tip of the depolymerizing microtubule. The proportionality constant ("FG/cortical rigidity") is set to 220–560 pN/μm, and this constant can be asymmetric between the anterior and the posterior cortex (Kozlowski *et al.*, 2007).

C. Drag Force

When viscosity is high, as in the cytoplasm, inertial terms can be neglected because the viscous force is dominant, a situation known as a low Reynolds number environment (Purcell, 1977). In this case the drag force of the object (i.e., pronucleus) is assumed to be proportional to the velocity of the object, and it can be formulated as Eq. (10):

$$\mathbf{F}_{\mathbf{total}} = \Gamma \mathbf{V}_{\mathbf{trans}} \tag{10}$$

where $\mathbf{F}_{\mathbf{total}}$ is the drag force, Γ is the drag coefficient, and $\mathbf{V}_{\mathbf{trans}}$ is the translational velocity vector.

The pronucleus can be modeled as a sphere, and according to Stokes' law the drag coefficient of a sphere is given by Eq. (11):

$$\Gamma = 6\pi r \eta \qquad (11)$$

where Γ is the drag coefficient of the sphere, r is the Stokes' radius of the sphere, and η is the viscosity (Berg, 1993).

Rotational movement of a sphere can also be formulated according to Stokes' law as Eq. (12):

$$\mathbf{T_{total}} = 8\pi r^3 \eta \mathbf{W_{rot}} \qquad (12)$$

where $\mathbf{T_{total}}$ is the net rotational moment (torque), r is the Stokes' radius of the sphere, η is the viscosity, and $\mathbf{W_{rot}}$ is the rotational vector.

The net torque is calculated as Eq. (13)

$$\mathbf{T_{total}} = \sum F_i(\mathbf{u}_i \times \mathbf{r}_i) \qquad (13)$$

Table I
Examples of Parameter Values Used in Past Numerical Models of *C. elegans* Embryos

			Kimura and Onami (2005, 2007), Hara and Kimura (2009)	Kozlowski *et al.* (2007)	Pecreaux *et al.* (2006)	Source of experimental data[a]
Microtubule (MT) dynamics						
Growth velocity	V_g	μm/s	0.12[b], 0.328[c]	0.51		Srayko *et al.* (2005)
Shrinkage velocity		μm/s	0.288[b], 0.537[c]	0.84		Kozlowski *et al.* (2007)
Catastrophe frequency		s^{-1}	0.014[b], 0.046[c]	0.01 (cyt.)[d] 5 (cor.)[e]		Kozlowski *et al.* (2007)
Rescue frequency		s^{-1}	0.044[b], 0133[c]	0		
Nucleation rate per MT		s^{-1}		0.05		
Max. number of MT per fiber			1	15		
Number of fibers per pole			98[b], 208[c], 224[f]	300		Srayko *et al.* (2005)
Pushing force						
Rigidity of MT	κ	pN μm^2	10[b]	120		Dogterom and Yurke (1997)
Sensitivity of growth to force	F_0	pN	31[b]	1.67		Dogterom and Yurke (1997) and Tran *et al.* (2001)
Pulling force, motor mediated						
Stall force of motor	F_{stall}	pN	1.1[b,c,f]		6.0	Gross *et al.* (2000), Mallik *et al.* (2004), and Toba *et al.* (2006)
Maximum velocity of motor	V_{max}	μm/s	2.0[b,c]		2.0	Gross *et al.* (2000) and Paschal *et al.* (1987)

(*Continued*)

Table I
(*Continued*)

			Kimura and Onami (2005, 2007), Hara and Kimura (2009)	Kozlowski *et al.* (2007)	Pecreaux *et al.* (2006)	Source of experimental data[a]
Pulling force, depolymerization-coupled						
FG/cortical rigidity		pN/μm		220–560[g]		
Pulling force, attachment of FG (cytoplasmic length dependent)						
Density of motors		μm^{-1}	0.1[b], 0.005[c]			
Pulling force, attachment of FG (cortical)						
FG on rate		s^{-1}	0.8[c]	3.2–5[h]	0.6	
FG off rate (without load)		s^{-1}		0.003	0.6	
Sensitivity of FG attachment to force		pN		28	1.5	Grishchuk *et al.* (2005)
Maximum number of motors per MT					28	
Other MT-dependent forces						
Spring constant for centering force		pN/μm			10	
Drag force of pronucleus/ spindle pole						
Drag coefficient	Γ	pN s/μm	188.4		85.8	
Viscosity of cytosol	η	pN s/μm^2	1.0	1.0		

[a] Only relevant references, such as those for measurements in *C. elegans* embryos, are given. For a full list, refer to the main text or the original studies.
[b] Kimura and Onami (2005)
[c] Kimura and Onami (2007)
[d] cyt., cytoplasmic
[e] cort., cortical
[f] Hara and Kimura (2009)
[g] This value can be asymmetric between the anterior and the posterior cortex in a model where the polarity affects FG/cortical rigidity.
[h] This value can be asymmetric between the anterior and the posterior cortex in a model where the polarity affects FG attachment.
Source: Since the modeling framework differs among the various models, parameter values on the same row correspond to parameters that have similar but not necessarily identical functions. For instance, similar parameters for different force-generating mechanisms (i.e., motor-based or depolymerization-coupled models) are displayed in the same row.

where F_i is the magnitude of the pulling force generated on the ith microtubule, $\mathbf{r}_i$ is the direction vector from the center of the object toward the minus-end of the ith microtubule (i.e., the centrosome), and $\mathbf{u}_i$ is the unit direction vector of the ith microtubule.

D. Dynamics of Microtubules

In our model, the dynamic instability of microtubules is modeled as described by Nédélec (2002). The model assumes that the microtubule has a static minus-end and either a growing or shrinking plus-end. The frequency with which a growing microtubule stochastically switches to the shrinking phase is termed the catastrophe frequency, whereas the frequency with which a shrinking microtubule switches to the growing phase is called the rescue frequency. The microtubule catastrophe frequency increases when the growing tip of the microtubule encounters an object *in vitro* (Janson *et al.*, 2003). We assume that the tip of the microtubule does not slip along the cortex and that the direction of the pushing or pulling force is along the axis of the microtubule.

E. Parameter Values

The parameter values used in past modeling studies in *C. elegans* embryos are listed in Table I. It should be noted that similar, but not identical, parameters relating to different force-generating mechanisms (i.e., motor-based or depolymerization-coupled models) are displayed in the same row, for the ease of comparison.

F. Solving the Equations

By solving Eqs. (10) and (12), we obtain translational and rotational velocity vectors, $\mathbf{V}_{\mathbf{trans}}$ and $\mathbf{W}_{\mathbf{rot}}$, respectively. In a simplified version of the modeling, we assumed pushing forces or pulling forces are independent of $\mathbf{V}_{\mathbf{trans}}$ and $\mathbf{W}_{\mathbf{rot}}$ (Hara and Kimura, 2009). In this case, the forces on each microtubule are calculated and summed and are used to move the objects. In more realistic model, we assumed the forces were dependent on $\mathbf{V}_{\mathbf{trans}}$ and $\mathbf{W}_{\mathbf{rot}}$ (Kimura and Onami, 2005, 2007). In such cases, we used the Newton–Raphson method for nonlinear systems of equations (Press *et al.*, 1992). For simplicity, we did not include rotational movement in the calculation when we compared the difference between the pushing and the pulling models (Kimura and Onami, 2005). Since this comparison of the two models indicates that pushing force is not a major force during pronuclear migration in *C. elegans* embryos, we calculate rotational motion by considering pulling forces only in later simulations (Kimura and Onami, 2007). Once the equations are solved and we obtain values for $\mathbf{V}_{\mathbf{trans}}$ and $\mathbf{W}_{\mathbf{rot}}$, we move the object "a little" (for a short period of time, e.g., 0.05 s) according to these values, and we then solve the equations again on the basis of the renewed configuration of the cell, the objects, and microtubules. By repeating this step, we are able to compute the entire movements.

In our case, we solved the equations by programming in C. Specialized software for solving mathematical equations, such as MatLab, can be effective (Pecreaux *et al.*, 2006). Moreover, a simulator, Cytosim, was developed by Nédélec and coworkers to simulate the movement of an object inside cells by microfibers (Kozlowski *et al.*, 2007; Nédélec and Foethke, 2007).

IV. Discussion

Using the above scheme of methods, we were able to address several questions. First, we investigated the contribution of pushing force versus pulling force for centrosome centering in *C. elegans* embryo (Kimura and Onami, 2005). In both the pushing and the pulling models, the simulated male pronucleus migrates toward the cell center and then remains near the center. Interestingly, when we plot the distance of the pronucleus from the posterior pole of the embryo against time, the shape of the distance–time plot differs between the pushing and the pulling models. In the pushing model, the shape of the distance–time plot is convex: the pronucleus rapidly moves away from the cortex at first but then slows down as it approaches the center. In contrast, in the pulling model, the shape of the distance–time plot is sigmoidal: the pronucleus moves slowly at first and then gradually moves faster, and finally slows again when it nears the center. Importantly, by examining a wide range of parameter values, we can demonstrate that the shape of the distance–time plot does not depend on the specific values of the parameters and thus the shape is intrinsic to each model. This difference in the shape of the distance–time plot provided a criterion by which to discriminate between migration driven by the pushing mechanism and that driven by the pulling mechanism. When we examined the shape of the distance–time plots of experimental data derived from *C. elegans* embryos, the curve was sigmoidal. This provides evidence that the pulling mechanism rather than the pushing mechanism is the primary mechanism for male pronuclear migration in *C. elegans* embryos.

Models with the same framework were applied to test various hypotheses relating to switching the position of the centrosome at the center and off-center (Kimura and Onami, 2007) and cell-size-dependent spindle elongation (Hara and Kimura, 2009). Numerical modeling complemented experimental approaches to evaluating the various candidate mechanisms.

V. Summary

In this chapter, we have described the assumptions and parameters used to model microtubule-dependent forces in the positioning of the pronucleus and mitotic spindle in *C. elegans* embryos. In any modeling study, simplification or approximation is inevitable. As described in the book "Physical Biology of the Cell" (Phillips *et al.*, 2009), "The art of model building lies in striking the proper balance between too little detail and too much." Although the importance of numerical modeling in cell biology has increased recently, criteria that discriminate good models from bad models (too little detail, too much) are not fully established. Further transdisciplinary interactions and trial-and-error processes are required to construct good models that contribute to our understanding of the dynamics inside the cell.

Acknowledgments

We thank Y. Azuma, M. Fujita, Y. Hara, H. Hayashi, K. Kimura, H. Koyama, K. Kyoda, T. Maeda, F. Nédélec, R. Niwayama, and J. Takayama for comments.

References

Alberts, B. M., Johnson, A., Lewis, J., Raff, M., Roberts, K., and Walter, P. (2008). "Molecular Biology of the Cell," 5th ed. Garland Science, New York.

Berg, H. C. (1993). "Random Walks in Biology." Princeton University Press, Princeton, NJ.

Brust-Mascher, I., Civelekoglu-Scholey, G., Kwon, M., Mogilner, A., and Scholey, J. M. (2004). Model for anaphase B: Role of three mitotic motors in a switch from poleward flux to spindle elongation. *Proc. Natl. Acad. Sci. U.S.A.* **101**, 15938–15943.

Burbank, K. S., Mitchison, T. J., and Fisher, D. S. (2007). Slide-and-cluster models for spindle assembly. *Curr. Biol.* **17**, 1373–1383.

Cytrynbaum, E. N., Scholey, J. M., and Mogilner, A. (2003). A force balance model of early spindle pole separation in *Drosophila* embryos. *Biophys. J.* **84**, 757–769.

Dogterom, M., Kerssemakers, J. W., Romet-Lemonne, G., and Janson, M. E. (2005). Force generation by dynamic microtubules. *Curr. Opin. Cell Biol.* **17**, 67–74.

Dogterom, M., and Yurke, B. (1997). Measurement of the force-velocity relation for growing microtubules. *Science* **278**, 856–860.

Drummond, D. R., and Cross, R. A. (2000). Dynamics of interphase microtubules in *Schizosaccharomyces pombe*. *Curr. Biol.* **10**, 766–775.

Dumont, S., and Mitchison, T. J. (2009). Force and length in the mitotic spindle. *Curr. Biol.* **19**, R749–R761.

Foethke, D., Makushok, T., Brunner, D., and Nédélec, F. (2009). Force- and length-dependent catastrophe activities explain interphase microtubule organization in fission yeast. *Mol. Syst. Biol.* **5**, 241.

Gönczy, P., Pichler, S., Kirkham, M., and Hyman, A. A. (1999). Cytoplasmic dynein is required for distinct aspects of MTOC positioning, including centrosome separation, in the one cell stage *Caenorhabditis elegans* embryo. *J. Cell Biol.* **147**, 135–150.

Goshima, G., and Kimura, A. (2010). New look inside the spindle: Microtubule-dependent microtubule generation within the spindle. *Curr. Opin. Cell Biol.* **22**, 44–49.

Goshima, G., Nédélec, F., and Vale, R. D. (2005a). Mechanisms for focusing mitotic spindle poles by minus end-directed motor proteins. *J. Cell Biol.* **171**, 229–240.

Goshima, G., Wollman, R., Stuurman, N., Scholey, J. M., and Vale, R. D. (2005b). Length control of the metaphase spindle. *Curr. Biol.* **15**, 1979–1988.

Grill, S. W., Gönczy, P., Stelzer, E. H., and Hyman, A. A. (2001). Polarity controls forces governing asymmetric spindle positioning in the *Caenorhabditis elegans* embryo. *Nature* **409**, 630–633.

Grill, S. W., Howard, J., Schaffer, E., Stelzer, E. H., and Hyman, A. A. (2003). The distribution of active force generators controls mitotic spindle position. *Science* **301**, 518–521.

Grill, S. W., Kruse, K., and Julicher, F. (2005). Theory of mitotic spindle oscillations. *Phys. Rev. Lett.* **94**, 108104.

Grishchuk, E. L., Molodtsov, M. I., Ataullakhanov, F. I., and McIntosh, J. R. (2005). Force production by disassembling microtubules. *Nature* **438**, 384–388.

Gross, S. P., Welte, M. A., Block, S. M., and Wieschaus, E. F. (2000). Dynein-mediated cargo transport in vivo. A switch controls travel distance. *J. Cell Biol.* **148**, 945–956.

Hamaguchi, M. S., and Hiramoto, Y. (1986). Analysis of the role of astral rays in pronuclear migration in sand dollar eggs by the colcemid-UV method. *Dev. Growth Differ.* **28**, 143–156.

Hara, Y., and Kimura, A. (2009). Cell-size-dependent spindle elongation in the *Caenorhabditis elegans* early embryo. *Curr. Biol.* **19**, 1549–1554.

Hill, T. L. (1987). "Linear Aggregation Theory in Cell Biology." Springer-Verlag, New York.

Hirakawa, E., Higuchi, H., and Toyoshima, Y. Y. (2000). Processive movement of single 22S dynein molecules occurs only at low ATP concentrations. *Proc. Natl. Acad. Sci. U.S.A.* **97**, 2533–2537.

Hirokawa, N. (1998). Kinesin and dynein superfamily proteins and the mechanism of organelle transport. *Science* **279**, 519–526.

Howard, J. (2001). "Mechanics of Motor Proteins and the Cytoskeleton." Sinauer Associates, Massachusetts.

Janson, M. E., and Dogterom, M. (2004). Scaling of microtubule force-velocity curves obtained at different tubulin concentrations. *Phys. Rev. Lett.* **92**, 248101.

Janson, M. E., Loughlin, R., Loiodice, I., Fu, C., Brunner, D., Nédélec, F. J., and Tran, P. T. (2007). Crosslinkers and motors organize dynamic microtubules to form stable bipolar arrays in fission yeast. *Cell* **128**, 357–368.

Janson, M. E., de Dood, M. F., and Dogterom, M. (2003). Dynamic instability of microtubules is regulated by force. *J. Cell Biol.* **161**, 1029–1034.

Karsenti, E., Nédélec, F., and Surrey, T. (2006). Modelling microtubule patterns. *Nat. Cell Biol.* **8**, 1204–1211.

Kellogg, D. R., Moritz, M., and Alberts, B. M. (1994). The centrosome and cellular organization. *Annu. Rev. Biochem.* **63**, 639–674.

Kimura, A., and Onami, S. (2005). Computer simulations and image processing reveal length-dependent pulling force as the primary mechanism for *C. elegans* male pronuclear migration. *Dev. Cell* **8**, 765–775.

Kimura, A., and Onami, S. (2007). Local cortical pulling-force repression switches centrosomal centration and posterior displacement in *C. elegans*. *J. Cell Biol.* **179**, 1347–1354.

Kinoshita, K., Arnal, I., Desai, A., Drechsel, D. N., and Hyman, A. A. (2001). Reconstitution of physiological microtubule dynamics using purified components. *Science* **294**, 1340–1343.

Kozlowski, C., Srayko, M., and Nédélec, F. (2007). Cortical microtubule contacts position the spindle in *C. elegans* embryos. *Cell* **129**, 499–510.

Mallik, R., Carter, B. C., Lex, S. A., King, S. J., and Gross, S. P. (2004). Cytoplasmic dynein functions as a gear in response to load. *Nature* **427**, 649–652.

Mitchison, T., and Kirschner, M. (1984). Dynamic instability of microtubule growth. *Nature* **312**, 237–242.

Mogilner, A., Wollman, R., Civelekoglu-Scholey, G., and Scholey, J. (2006). Modeling mitosis. *Trends Cell Biol.* **16**, 88–96.

Nédélec, F. (2002). Computer simulations reveal motor properties generating stable antiparallel microtubule interactions. *J. Cell Biol.* **158**, 1005–1015.

Nédélec, F., and Foethke, D. (2007). Collective Langevin dynamics of flexible cytoskeletal fibers. *New J. Phys.* **9**, 427.

Nguyen-Ngoc, T., Afshar, K., and Gönczy, P. (2007). Coupling of cortical dynein and Galpha proteins mediates spindle positioning in *Caenorhabditis elegans*. *Nat. Cell Biol.* **9**, 1294–1302.

Oiwa, K., and Takahashi, K. (1988). The force-velocity relationship for microtubule sliding in demembranated sperm flagella of the sea urchin. *Cell Struct. Funct.* **13**, 193–205.

Paschal, B. M., Shpetner, H. S., and Vallee, R. B. (1987). MAP 1C is a microtubule-activated ATPase which translocates microtubules in vitro and has dynein-like properties. *J. Cell Biol.* **105**, 1273–1282.

Pearson, C. G., and Bloom, K. (2004). Dynamic microtubules lead the way for spindle positioning. *Nat. Rev. Mol. Cell Biol.* **5**, 481–492.

Pecreaux, J., Roper, J. C., Kruse, K., Julicher, F., Hyman, A. A., Grill, S. W., and Howard, J. (2006). Spindle oscillations during asymmetric cell division require a threshold number of active cortical force generators. *Curr. Biol.* **16**, 2111–2122.

Peskin, C. S., Odell, G. M., and Oster, G. F. (1993). Cellular motions and thermal fluctuations: The Brownian ratchet. *Biophys. J.* **65**, 316–324.

Phillips, R., Kondev, J., and Theriot, J. (2009). "Physical Biology of the Cell." Garland Science, New York.

Press, W. H., Flannery, B. P., Teukolsky, S. A., and Vetterling, W. T. (1992). "Numerical Recipes in C." Cambridge University Press, Cambridge, UK.

Purcell, E. M. (1977). Life at low Reynolds number. *Am. J. Phys.* **45**, 3–11.

Reinsch, S., and Gönczy, P. (1998). Mechanisms of nuclear positioning. *J. Cell. Sci.* **111**, 2283–2295.

Schnitzer, M. J., Visscher, K., and Block, S. M. (2000). Force production by single kinesin motors. *Nat. Cell Biol.* **2**, 718–723.

Srayko, M., Kaya, A., Stamford, J., and Hyman, A. A. (2005). Identification and characterization of factors required for microtubule growth and nucleation in the early *C. elegans* embryo. *Dev. Cell* **9**, 223–236.

Toba, S., Watanabe, T. M., Yamaguchi-Okimoto, L., Toyoshima, Y. Y., and Higuchi, H. (2006). Overlapping hand-over-hand mechanism of single molecular motility of cytoplasmic dynein. *Proc. Natl. Acad. Sci. U.S. A.* **103**, 5741–5745.

Tournebize, R., Popov, A., Kinoshita, K., Ashford, A. J., Rybina, S., Pozniakovsky, A., Mayer, T. U., Walczak, C. E., Karsenti, E., and Hyman, A. A. (2000). Control of microtubule dynamics by the antagonistic activities of XMAP215 and XKCM1 in Xenopus egg extracts. *Nat. Cell Biol.* **2**, 13–19.

Tran, P. T., Marsh, L., Doye, V., Inoue, S., and Chang, F. (2001). A mechanism for nuclear positioning in fission yeast based on microtubule pushing. *J. Cell Biol.* **153**, 397–411.

Visscher, K., Schnitzer, M. J., and Block, S. M. (1999). Single kinesin molecules studied with a molecular force clamp. *Nature* **400**, 184–189.

Wordeman, L., McDonald, K. L., and Cande, W. Z. (1986). The distribution of cytoplasmic microtubules throughout the cell cycle of the centric diatom Stephanopyxis turris: Their role in nuclear migration and positioning the mitotic spindle during cytokinesis. *J. Cell Biol.* **102**, 1688–1698.

CHAPTER 24

Cryo-Electron Tomography of Cellular Microtubules

Roman I. Koning

Department of Molecular Cell Biology, Section Electron Microscopy, Leiden University Medical Center, 2300 RC, Leiden, The Netherlands

Abstract

Microtubules are intrinsically dynamic structures. In the cellular environment many proteins and protein complexes are associated with microtubules that influence or functionalize microtubule dynamics. Therefore, investigation of the structure and dynamics of microtubules with their associated complexes inside the cellular environment lies at the heart of fully understanding their function. Cryo electron microscopy has been essential in structural microtubule research since the atomic structure of tubulin and the structure of microtubules were unraveled using this technique. Furthermore, the specific structures at the microtubule ends linked to the growing or shrinking states were also detected by cryo electron microscopy. Electron microscopy studies on

978-0-12-381349-7
DOI: 10.1016/S0091-679X(10)97024-6

microtubules were mainly performed *in vitro* but microtubules can also be investigated inside cells, using cryo electron tomography. Cryo electron tomography is an important tool in structural biology research because it enables visualization of single and unique protein complexes in a cellular environment and at a molecular resolution. Cryo electron tomography is a three-dimensional (3D) imaging technique in which electron microscopy tomographic imaging is performed on cryogenically cooled, vitrified specimens after which the object is computationally reconstructed. Here, I describe the materials and methods for cryo electron tomography of microtubules and in whole cells, describing cell growth, specimen vitrification, localization of microtubules, cryo electron tomography recording, tomographic image reconstruction, and 3D visualization techniques.

I. Introduction

A. Microtubules

Microtubules (MTs), actin, and intermediate filaments together form the cytoskeleton of the cell, the backbone that gives the cell its shape and strength, and is involved in cellular motion and intracellular transport. MTs play a key role in many cellular processes. They are involved in the segregation of chromosomes during cell division and they mediate the intracellular transport of organelles and vesicles.

MTs are hollow tubes made up of tubulin (Li *et al.*, 2002; Mandelkow *et al.*, 1986). They are highly dynamic biopolymers that can grow and shrink by the addition or removal of tubulin $\alpha\beta$ dimers at their ends. They can grow many micrometers long and extend throughout the cytoplasmic of cells. Tubulin polymerizes in a head-to-tail fashion into so-called protofilaments. *In vivo*, 13 of such protofilaments are positioned side-by-side in a ring forming hollow MTs with a diameter of 25 nm (for reviews see, e.g., Howard and Hyman, 2003; Wade, 2007). In a cell, polymerization mainly occurs at the MT plus end since most minus ends are attached to the centrosome.

The dynamics of MT (de-)polymerization are controlled by guanosine triphosphate (GTP) binding to tubulin. MT polymerization is stimulated when GTP is bound to β-tubulin, while GDP-$\alpha\beta$-tubulin is not able to polymerize into MTs. In $\alpha\beta$-tubulin the GTP that is bound to α-tubulin is not hydrolysable and does not influence MT dynamics. The incorporation of GTP-$\alpha\beta$-tubulin into MTs stimulates hydrolysis of GTP that is bound to tubulin that is already incorporated in the MT lattice. GDP-$\alpha\beta$-tubulin has the tendency to induce depolymerization when it is embedded into the MT lattice and to curl the protofilaments outward. In a MT this is counteracted by lateral binding of protofilaments and by the cap of GTP-$\alpha\beta$-tubulin at protofilament ends. This dynamic behavior results in straightened protofilaments at plus ends of growing MTs and highly outward curving protofilaments at the plus ends of shrinking MTs (Mandelkow *et al.*, 1991; for reviews see, e.g., Nogales and Wang, 2006a, 2006b).

The structure of MTs in complex with various associated proteins can be determined using cryo-electron microscopy (Hoenger and Gross, 2008). For example, the complex of kinesins (e.g., Bodey *et al.*, 2009 and for reviews see Mandelkow and Hoenger,

1999 and Kikkawa, 2008), MAP2c, and tau (Al-Bassam *et al.*, 2002) and ncd80 (Mandelkow and Hoenger, 1999) with MTs were determined by helical averaging techniques *in vitro* and using purified components. However, this type of structure determination depends on proper MT decoration of the associated proteins. Cryo electron tomography can be used to investigate the structure of single MT-associated proteins which are not necessarily repetitively bound to the MT lattice, also inside cells. This is essential in order to relate MT structure in its cellular environment to its dynamics and functions.

B. Cryo Electron Tomography

Cryo electron tomography is a combination of cryogenic techniques for specimen preparation, electron microscopy for data collection, and tomographic reconstruction techniques for visualization in three dimensions. Cryo electron microscopy is electron microscopy that is performed on cryogenically cooled samples which are embedded in an environment of vitreous water. Biological specimens for cryo electron microscopy are prepared by cryo-fixation using ultra-fast cooling of a thin aqueous layer of a protein suspension or parts of a cell that are thinner than roughly a micrometer. When the cooling rate is high enough (>100.000° C/s) the water is not able to form crystals and adopts a glass-like structure that is called vitreous water (for reviews see Costello, 2006; Dubochet *et al.*, 1988). Using vitrification the atomic structures and molecular interactions of complexes are rapidly preserved and devoid of artifacts or distortions. In MT research this is important to ensure proper preservation of the protofilament structures at the end of the highly dynamic MTs. Samples are maintained in cryogenic condition using liquid nitrogen, which limits sublimation of water in the low-pressure environment of the microscope and ensures sample stability. This is a major advantage of cryo electron microscopy since it allows direct observation of the object and consequently the resolution is not limited by the use of staining agents. However, vitrified biological samples are highly sensitive to electrons, resulting in progressive radiation damage at increasing electron dose. As a result the resolution is limited by the electron dose that can be used for imaging before severely damaging the sample.

Cryo electron microscopy images have to be recorded using a minimal dose and therefore have low contrast, while the high-resolution signal is obscured by noise. Image averaging of many similar particles is needed to increase the signal-to-noise ratio and to attain nanometer scale resolution in three dimensions. Several reconstruction techniques are developed to deal with a variety of specimens in order to increase the signal-to-noise ratio and to generate three-dimensional (3D) views: single particle analysis for averaging multiple asymmetric individual particles; icosahedral, helical, and crystallographic reconstruction techniques for averaging specimens that contain several flavors of ordering; and cryo electron tomography, which is capable of visualizing single and unique structures such as organelles and cells.

Tomography is the recording of a projection series from different angles and subsequent computational reconstruction to image the object in three dimensions. Cryo electron tomography is a powerful imaging technique in structural biology that is ultimately suitable for investigation of macromolecular assemblies, organelles *in vitro* but also inside their native environment (Koning and Koster, 2009; Lucic *et al.*, 2005). Biological structures like MTs can be visualized in their native environment in three dimensions at a nanometer scale resolution. The attainable resolution of cellular cryo electron tomograms is limited by both specimen thickness and the total tolerable electron dose that can be deposited on one sample. Nevertheless, the structure of an increasing number of cellular structures is being (re-)investigated using electron tomographic techniques, resulting in unbiased and detailed views that 3D cryo tomograms present. Structures that are studied by cryo electron tomography are for reasons of size and uniqueness often not suitable to study in three dimensions by other structural biological techniques like X-ray crystallography, nuclear magnetic resonance, or light microscopy. However, cryo electron tomography can bridge imaging and resolution gaps between other techniques. It has a suitable resolution range (2–5 nm) to be combined with atomic resolution techniques. Also the possibilities for live imaging and cryo-fluorescent light microscopy techniques (Plitzko *et al.*, 2009; van Driel *et al.*, 2009) will make it possible to study the dynamics of MTs and the structure of MT-associated proteins.

C. The Structure of Cellular MTs by Cryo Electron Tomography

The structure of MTs in the context of their native environment or in intact cells has been studied by cryo electron tomography or cryo-sectioning by several groups. Tomographic investigations of MTs in intact neuronal cells clearly showed that MTs are filled with luminal particles that bind to the MT lattice. These particles were not only abundant in neurons but also present in several other types of cells (Garvalov *et al.*, 2006). Similar particles were observed in MTs in tomograms of cryo-sectioned Chinese hamster ovary cells (Bouchet-Marquis *et al.*, 2007). While in sporozoites the spacing of luminal particles was shown to be abundant, in MTs of mouse embryonic fibroblasts (MEFs) appeared to have less luminal particles than neurons with no apparent spacing pattern (Koning *et al.*, 2008). Although the presence of densities inside MTs was reported earlier in several papers (see references inside Garvalov *et al.*, 2006), only in single slices of the reconstructed tomogram do the shape, size, distribution, and repetitive nature of luminal densities become apparent. Cryo electron tomography shows its power with the visualization of these luminal particles since these are not apparent in two-dimensional (2D) images of cellular MTs.

In two papers the structure of the MT plus ends was reported. In plasmodium sporozoites the MT ends were mostly flared and sometimes capped (Cyrklaff *et al.*, 2007), while in fibroblasts MT plus ends were frayed, sheet-like, or blunt (Koning *et al.*, 2008). Additionally, in fibroblasts it was possible to discern the individual protofilaments at the MT end. In a more recent study (McIntosh *et al.*, 2009) kinesin-like Eg5 was bound to MTs inside detergent-treated 3T3 cells and visualized, which revealed that the lattice of cellular MTs is not cylindrically symmetric and the protofilament lattice has a seam. Additionally it is worth mentioning the doublet MT structure in axonemes, which form the core of flagella and cilia, revealing the interaction of MTs with many proteins that regulate movement (Nicastro *et al.*, 2005, 2006; Sui and Downing, 2006, reviewed in Downing and Sui, 2007). These investigations show a promising future for the structural determination of cellular MTs, including their associated proteins and plus-end structures, using cellular cryo electron tomography.

II. Rationale

The fundamental reason for performing cryo electron tomography on MTs inside cells is to investigate the structure of MTs in a native environment, where all factors and proteins that associate with MTs and influence their behavior are present. Of special interest is the investigation of how MT-associated proteins bind MTs and how they influence the dynamic behavior of a single MT.

Cryo electron tomography is the only technique that can determine 3D structure of individual and unique macromolecular complexes at molecular resolution in its natural environment of the cell. The biological interest ultimately lies in understanding structure–function relationships of proteins and macromolecular complexes, including MTs, inside a cellular environment. This cellular environment is essential, since circumstances *in vitro* are significantly different than *in vivo*. For example, acetylation and detyrosination of tubulin, protein phosphorylation, and cytoplasmic protein complex formation will have its effects on MT functioning.

Here, I describe the materials, methods, and specific issues that were considered while performing cryo electron tomography of MTs in whole cells (Koning *et al.*, 2008). Many good reviews on general aspects of cryo electron tomography (Grunewald and Cyrklaff, 2006; Hoenger and McIntosh, 2009; Lucic *et al.*, 2005; Milne and Subramaniam, 2009) have been published. The focus here will be specifically of cryo-specimen preparation techniques, cryo electron microscopy, tomography data collection, image processing and visualization.

III. Materials and Methods

A. Specimen Preparation

1. Cell Culture

For our cryo electron microscopy experiments on MT plus ends in vitrified cells (Koning *et al.*, 2008), we used primary MEFs that were purified from 13.5-day-old mouse embryos of which the body is cut into pieces and homogenized with collagenase and trypsin. The pieces are cultured in a 1:1 mixture of Dulbecco's modified Eagle's medium (Gibco, Invitrogen, the Netherlands) and Ham's F10 medium (Cambrex, USA) supplemented with 10% fetal calf serum and penicillin and streptomycin antibiotics in a humidified atmosphere with 5% CO_2 at 37° C. Cells that grew out of this culture are a heterogeneous culture of cells with fibroblast-like characteristics. Only the low passage numbers of these primary MEFs were used for cryo electron tomography. MEF cells were cultured in Greiner Cellstar tissue culture flasks and between 10,000 and 30,000 cells were plated into 5 cm diameter Cellstar tissue culture dishes in which EM grids were positioned on the bottom. The amount of cells that are plated should be large enough to ensure proper cell growth but small enough to prevent too high confluence, which can lead to problems during blotting and vitrification and affect cell shape.

For cell growth we only used gold EM grids, since these are chemically stable in medium, not toxic to cells, and commercially available. The mesh size should facilitate high tilts for tomography and provide a stable and strong support layer for cell growth, blotting, and vitrification. Therefore, we mainly used 300 mesh hexagonal or square meshed grids, 300 × 75 Mesh grids, and 100 or 135 Mesh finder grids (AGAR Scientific Essex, England; resp. grid types G2403A, G2300A, G2375A, H6, and HF15). A layer of Formvar was positioned on the support grid for strength and carbon was evaporated onto the formvar support (Emitech K950X, Emitech, the Netherlands). Fiducial markers with a size of 5, 10, or 15 nm in size were applied on the grid by placing it for 1 min on a suspension of colloidal gold stabilized by protein A or BSA, followed by blotting and air drying.

Before cell culture the support layer was made hydrophilic by glow discharging in air at 200 mbar for 2 min at 20–40 mA and was sterilized under UV light for 15 min. Cells were allowed to attach to the grids overnight and occasionally were allowed to grow for a few days before vitrification.

2. Vitrification of Cells

For the vitrification of cells we used both a Vitrobot Mark IV (FEI Company, Eindhoven, the Netherlands) and a custom-built vitrification device, of which both are computer controlled and equipped with a temperature- and humidity-controlled chamber. Electron microscopy grids with cultured cells were transferred as fast as possible from the medium into the climate chamber, which was conditioned at 37° C and 100% humidity. Cells were given time to adapt from the temperature change, excess medium was blotted once between 1 and 2 s from two sides with Whatman filter paper no 4. or filter paper numbers 595 or 597 from Schleicher and Schuell. The blotted grid was plunged without delay into liquid ethane. The liquid ethane was cooled by liquid nitrogen and kept in equilibrium with solid ethane (having a translucent appearance) using a custom-built container which allowed heating of the ethane. After vitrification the grid was repositioned, while keeping the grid under cold nitrogen gas, into a grid storage box. This storage box was kept under liquid nitrogen in a separate container with a lid to shield it from ice contamination. The grid storage boxes were stored in 50 ml polystyrene tubes (Greiner, the Netherlands) in liquid nitrogen Dewars (Cryotech, the Netherlands) until further use.

Important for the structural investigation of MTs is the control of humidity and temperature of the cells during specimen preparation. Cells need to be kept hydrated in medium or in a 100% humidity environment at all times during preparation to prevent dehydration and subsequent loss of morphology. Moreover, even mild dehydration changes protein (e.g., tubulin) concentrations inside the cell resulting into unwanted effects on MT dynamics. Furthermore, MTs are sensitive to temperature changes (Caplow *et al.*, 1988), and for the investigation of structural changes at growing and shrinking MTs (Mandelkow *et al.*, 1991; Zovko *et al.*, 2008) it is important to maintain the temperature at 37° C before cryo-fixation.

3. Cell Thickness

The thickness of cells is a critical parameter in cryo electron tomography using whole cells, because most cells are too thick to image and it is difficult to control cell thickness without the use of sectioning. Consequently, only a limited amount of cell types can be used for whole cell cryo tomography. In practice, therefore, mainly neurons, endothelial and epithelial cells, fibroblasts, Dictyostelium cells, and small (archae-) bacteria are used (for an overview see table I in Koning and Koster, 2009). Usable cells are either intrinsically small or contain thin areas, like an extending cell cortex or processes. In tomography thickness is particularly an issue since during tomogram acquisition the sample is tilted up to 60 or 70 degrees and effective thickness increases, respectively, two- or threefold. A sample thickness up to roughly 300 nm can be used for electron tomography (depending on the accelerating voltage of the microscope).

Thickness can be estimated beforehand using energy electron loss spectroscopy (Shi *et al.*, 1996) but this was not necessary for finding suitable areas. Thinnest MEF cell areas which were suitable for cryo electron tomography were visually selected by their translucency from low-dose digital images taken at low magnifications. Thinnest areas with visibly present MTs were used for imaging. The thickness of recorded cellular area after tomographic reconstruction appeared to range from 50 to 300 nm. This was confirmed by perpendicular sectioning of flat embedded cells.

The limited thickness of MEF cells is important for structural investigations of MTs by electron tomography. First, it avoids sectioning of cells while MTs are always oriented perpendicular to the viewing direction and are not sectioned. Therefore, MTs can be tracked over an extensive distance which additionally allows localization and investigation of MT plus ends. Second, sample thickness is the limiting factor in the attainable resolution. Crowther (Crowther *et al.*, 1970) has shown a formal solution for the resolution in tomograms and shows that this is inversely related to the thickness of the sample. Thin specimens are therefore necessary to visualize individual protofilaments in MTs. Ultimately, in cryo electron tomography the resolution is limited by radiation damage and the total electron dose on a specimen that can be tolerated for imaging is limited. This dose restriction results in low signal-to-noise levels which limits resolution. With growing thickness, the noise levels increase and therefore restrict the attainable resolution.

Several methods and tricks have been reported that deal with cell thickness, apart from cryo-sectioning (Al-Amoudi *et al.*, 2004; Bouchet-Marquis and Fakan, 2009) and Focused Ion Beam milling (Marko *et al.*, 2007; Rigort *et al.*, 2010). Specimen thickness can be reduced by heating the cryo-holder to above the sublimation point of vitreous water −125° C which results in removal of water (supplementary data in Nicastro *et al.*, 2006). Partial lysis of cells using 0.1% Triton detergent in MT-stabilizing buffer for 30 s was used to allow entry of MT binding proteins inside the cell. Although it was not mentioned in the article, it seems that partial cell lysis can have a positive effect on cell thickness for cryo electron tomography (McIntosh *et al.*, 2009). Also rinsing of cells in PBS and extended blotting, once outside the humidity chamber followed by 10 s blotting at 90% humidity, can influence cell thickness and its usability for cryo electron tomography (Berriman *et al.*, 2009).

B. Cryo Electron Tomography

1. MT Localization

Localization of MTs and MT ends in vitrified cells was performed by visual inspection of digital electron images. In order to minimize electron beam damage searching was performed using digital imaging at low magnifications (3000–9000), high spot size number (5–7), and short exposure times (0.05–0.1 s). The first step after insertion of the specimen into the microscope was the assessment of overall quality and usability of the grids at different magnifications. Quality of grids can be insufficient for several reasons, most common being one or several of the following: (1) Excessive thickness of the vitreous water that prevents the beam from penetrating through the sample, (2) limited number of cells, (3) excessive breaking of the support film, (4) crystalline ice due to improper vitrification or handling of vitrified specimens, (5) excessive ice contamination, (6) absence of large thin areas of the cell, (7) cell death, or (8) excessive drying of the cells (Fig. 1). In the second step potentially usable cells are noted and stage positions are stored in a list. MTs were located at low magnifications using systematic manual scanning of cellular areas while plus ends were located by tracking of MTs (Figs. 1 and 4A). Alternatively, systematic scanning of cells can be performed by several programs, including GRACE (Oostergetel *et al.*, 1998), TOM toolbox (Nickell *et al.*, 2005), Leginon (Suloway *et al.*, 2005, 2009), and serial EM (Mastronarde, 2005), which include possibilities stitching the images together. Manually scanned images were stitched together using the photomerge option in Photoshop CS3 (Adobe) after high-pass filtering to remove background density ramps due to variations in cell thickness (Fig. 2).

Fibroblasts that spread have typical shapes in which the location in thin areas of MTs can be deduced from the general cell morphology. MTs usually run from the cell nucleus straight toward the extending regions (Fig. 1G and H in Koning *et al.*, 2008). MTs also run parallel to the cell borders connecting the extending regions, but these regions are relatively thick because of the presence of actin filament bundles and additionally the MTs contain relatively few plus ends.

The direction of the MT with respect to the tilt axis should be taken into account during data collection. Because of anisotropic resolution of single-axis tilt series, structures that lie perpendicular to the tilt axes are not resolved in the final tomogram (Fig. 3). The direction of the MT can be changed by manual rotation of the grid in the

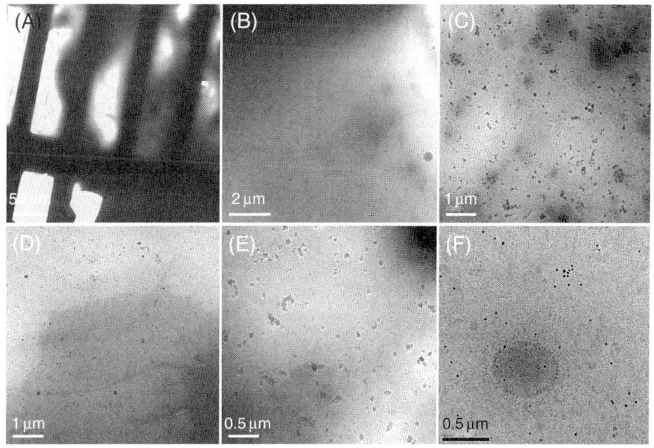

Fig. 1 Cryo electron microscopy example images of different typical views of vitrified MEF cells. (A) Overview of a 300 × 75 mesh EM grid with vitrified cells, showing meshes with broken carbon (lower left), meshes covered with thick ice (lower right), and thick cells (upper right) and meshes with usable cells (top left). Bar is 50 μm. (B) Area of cell that is unsuitable for tomography because of its thickness. On the right round vesicles that resemble mitochondria can be observed. (C) Area of vitrified cell that is excessively covered with ice particles that have precipitated on top of the specimen. (D) Example of a dried cell. In dried cells the macromolecular structures are difficult to recognize and the cell boundaries have high contrast. In hydrated cells the cell boundary is often not clearly visible. The darker area with appendage on lower right is a cellular extension that fell upon the cell. Small black dots are fiducial gold particles that are deposited on the grid for tomographic alignment. (E) Cell showing large amounts of extracellular vesicles and elongated lipid tubular tubes, which are indicative of cell death or bacterial infection. (F) Suitable area for tomographic tilt series acquisition. The area is thin enough to directly observed microtubules (ranging from lower left to top right), actin filaments, and rough microsomes. No ice contamination is visible and the area has suitable amount of gold fiducials for alignment.

cryo-holder, by using a rotational cryo-holder or an electron microscope equipped with a stage capable of flipping the grid. Alternatively, and even better, a double-axis tilt series can be recorded, which improves the anisotropic resolution in the tomogram.

2. Tomographic Data Collection

Data collection was performed on Tecnai F20 or Polara F30 microscopes (FEI Company Eindhoven, the Netherlands) at 200 and 300 keV, respectively. Images were recorded on postcolumn energy filter 2k × 2k CCD cameras (GIF 2002, Gatan GmbH, Germany) in zero-loss mode using a slit width of 20 eV. Grids were mounted in a Gatan 626 high tilt or Gatan 914 cryo electron tomography holder for imaging in the Tecnai F20, while the F30 is equipped with an attached loading device. Tilt series were recorded using Xplore3D software (FEI company) using low-dose mode. To ensure maximum stability of the lenses the spot sizes in all

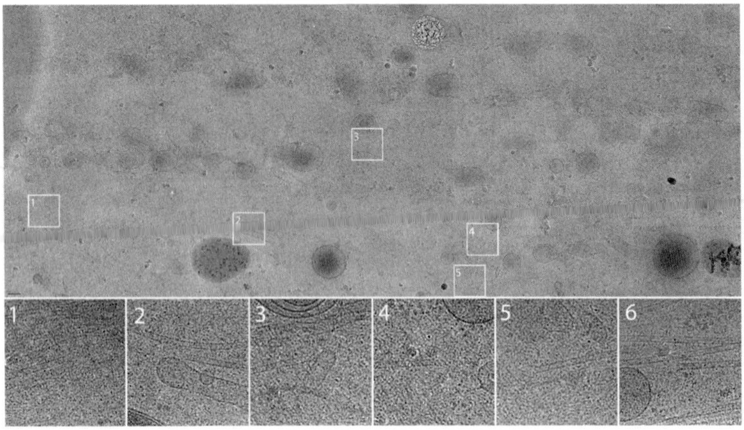

Fig. 2 Cryo electron microscopy overview of a vitrified MEF cell and locations of microtubule plus ends. The overview has an overall area of ~42 μm^2 and is stitched from 24 high-pass filtered images taken at 14,000 × magnification. Small white boxes noted 1–5 outline locations of microtubule plus ends and are shown four times enlarged below. Microtubule end 6 is not shown in the overview. At full magnification in the projected view microtubules are easily tracked. Furthermore, ribosomes, vesicles, rough microsomes, mitochondria, and actin filament bundles can be observed. Scale bar is 1 μm.

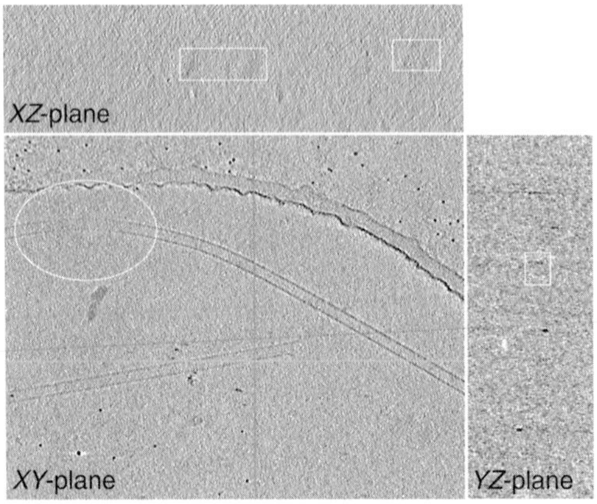

Fig. 3 Anisotropic resolution in cryo electron tomograms. The *XY* plane depicts a 25 nm thick slice through the tomogram including the complete thickness of the microtubules. The tomographic tilt axis lies vertical, parallel to the drak gray line. The *XZ*and*YZ* planes are single slices along the horizontal light gray line and vertical dark gray line, respectively. From the oval area in the *XY* plane, it is clear that the microtubule is almost not resolved in the place where it runs perpendicular to the tilt axis. The contrast of microtubules increases when the microtubules run more parallel to the tilt axis. The same effect was observed with outward curling protofilaments at microtubule plus ends (not shown here).

modes and the magnification of the focus and exposure mode were kept identical. Images were recorded in continuous mode from maximum possible negative to maximum possible positive tilt angles, usually ranging between ± 60° and ±70°, depending on the cryo-holder, local specimen thickness and location on the grid. For recording, a linear tilt scheme was used with 2° intervals or alternatively Saxton schemes with average intervals between 2.5° and 3°. The total electron dose of the tilt series was kept below 200 e/Å^2 and divided over 61–71 images that were recorded with an exposure time of 0.2–0.5 s, depending on the spot size. The exposure time was corrected for increasing angles with a factor of 1.6 between 0° and 60° while the dose was distributed over the whole series. Magnifications varied between 13,500 × and 22,500 × resulting in pixel sizes between, resp., 1.02 and 0.63 nm. The defocus ranged between –4 and –10 μm under focus to optimize contrast transfer, which was calculated using the program CTF explorer (http://www.maxsidorov.com/ctfexplorer/). Focus was corrected every, or every second tilt angle step. Tracking was performed after image acquisition by cross-correlation of filtered images (long wavelength cut-off at 100 nm ± 3 nm and 2 nm ± 0.5 nm low wavelength cut-off, with a taper filter of 8 pixels and the sobel filter and remove X-ray options on).

C. Reconstruction and Visualization

1. Tomographic Reconstruction

Image reconstruction of tomographic tilt series was performed on Dell Precision workstations with two dual core processors (32-bits) with 4 Gb total memory running Microsoft Windows XP professional or on 64-bit workstations running Ubuntu Linux and over 8 Gb of RAM memory. Image processing was performed using the latest stable release of the IMOD software suite (Kremer *et al.*, 1996). Tomograms were generated using Etomo without any exceptional measures. Here the procedural steps are described including some specifics that might apply to cellular cryo electron tomograms, more than for sections or thinner samples.

Hot pixels that were generated by X-rays in the electron microscope were removed using pre-processing, running the CCD-eraser command with the standard values several times until there were no more peaks found. At very low mean gray values ($<$100) in the individual images of the tilt series the difference criterion in the CCD peak eraser had to be reduced. Course alignment was performed by cross-correlation of images. When part of the image was obstructed by dark regions of thick ice, pixel trimming was necessary for proper alignment. Occasionally, consecutive images were manually aligned and saved in Midas, followed by generating the course aligned stack. Fiducial-less alignment of cryo electron tomograms in our hands never generated satisfactory results.

Fiducial model generation was the most time-consuming part of the reconstruction. It is performed by tracking 5–15 nm gold beads as fiducials. The main encountered problem was that not all gold beads were correctly tracked automatically in all images of the tilt series. Several solutions were found to be effective for different cases. (1) Increase the bead diameter; the fiducial diameter is asked during the tomogram setup of a to be processed tomogram, but the actual value of the gold particles might be different than expected and also varies between beads. (2) Slow increase in the amount of seeds; first tracking of the seed model is started with, e.g., 5 beads and after correct tracking beads are increasingly added for tracking over several rounds. (3) Tracking beads on one surface; in cellular samples for cryo electron tomography gold beads are mainly positioned on the carbon surface, while occasionally beads are present at the top of the cells. These latter gold beads often had to be excluded. (4) Subsequent increase in the amount of tilted views to include in the tracking model.

In the fine alignment step several issues were taken into account to improve the quality of the resulting tomogram: (1) Local alignment was performed when there is a sufficient amount of fiducials. (2) Individual (high angle) views with little contrast because of thick ice were removed. (3) As many gold fiducials were used for alignments as possible. (4) The threshold for residual reports was gradually decreased in several rounds to a standard deviation of 2. (5) The fiducials for which it was not possible to visually confirm correct positioning were deleted. (6) Final visualization of the reconstructed alignment of the tomogram was used as a final measure to assess tomogram quality.

During tomogram positioning the sample tomogram thickness was often increased to ~1200 pixels to account for the thickness of cells. If it was not possible to use the sample tomograms due to limited amount of contrast, the whole tomogram was used with binning of 4 to create a boundary model. After the tomogram was generated using back-projection using standard settings, postprocessing included conversion to bytes, minor volume trimming, and swapping y and z dimensions so that the tomograms opens with z positioned in the direction of the beam. When the resulting tomograms exceeded the amount of available memory on the PC, tomograms were opened binned by two in IMOD or converted to 8-bits. Volume squeezing by a factor 2 in all directions was performed for quicker loading for tomogram inspection and generation of images for movies.

2. Visualization

The visualization of cryo electron tomograms is hampered because of the intrinsic low signal-to-noise levels. First, inspection of tomogram slices was carried out by examination of individual 2D slices using the so-called Zap window of the 3dmod display tool from IMOD. To improve the signal-to-noise levels the so-called slicer window was used in which thickness of the tomographic slices can be increased to improve the contrast. Additionally, in the slicer window the orientation of the slice through the tomogram can be controlled by the rotation around all axes. Also median and anisotropic diffusion filters were applied on single slices. Images from all visualization windows in IMOD can be generated by simply saving the views and

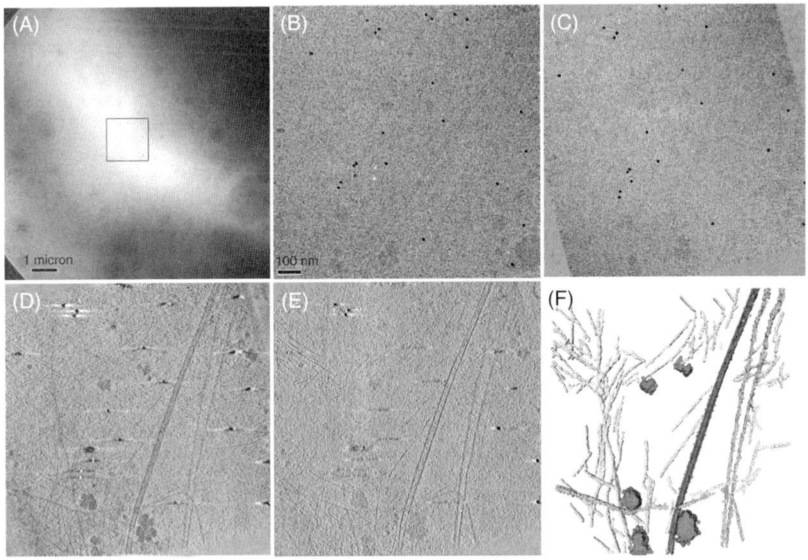

Fig. 4 Typical steps during cellular cryo electron tomography. (A) Low magnification (3000 ×) overview of cellular area that is suitable for cryo electron tomographic recording. Central area is thin and contains vesicles and a microtubule. Thicker surrounding area is indicative for correct vitrification. No ice contamination or crystalline ice is present. Black square denotes area in B–F. Scale bar is 1 μm. (B) Image of area outlined in (A) taken after the tilt series recording. Black dots are fiducial gold markers put on the sample to aid alignment of the tilt series. (C) Single exposure at 0° tilt from aligned tomographic tilt series. (D) Digital slice of 25 nm from the reconstructed tomogram positioned around the microtubule. On the right two intermediate filaments can be seen and on the left an actin network. Dark regions at the bottom are storage granules. (E) Single filtered slice through the tomogram showing more clearly the microtubule, luminal particles, intermediate filaments, actin filaments, storage granules, and gold particles. (F) Surface rendering of tomogram with microtubule (red), intermediate filaments (cyan), actin filaments (yellow), and glycosomes (magenta). (See Plate no. 14 in the Color Plate Section.)

image sequences through the tomogram can be generated for production of movies (Figs. 4 and 7).

3D filtering of tomograms is more computationally intensive and several filters that are present in IMOD, like median filtering, high-pass filtering, and especially nonlinear anisotropic diffusion (Frangakis and Hegerl, 2001), were performed on a Hewlett-Packard XC cluster (Betagraphics NV, Hengelo, the Netherlands) with 56 nodes with a total of 64 Gb RAM running Linux. Nonlinear anisotropic diffusion enhances contrast and preserves the structural elements better than in low-pass and median filtering techniques. Suitable parameters for nonlinear anisotropic diffusion were found by trial and visual inspection of a single slice, which resulted in *k*-values between 5 and 150, while using between 5 and 40 iterations. Increasing the amount of filtering, however, decreases the amount of gray levels and, more important, level of detail and care should be taken with the amount of denoising that is used (Fig. 5).

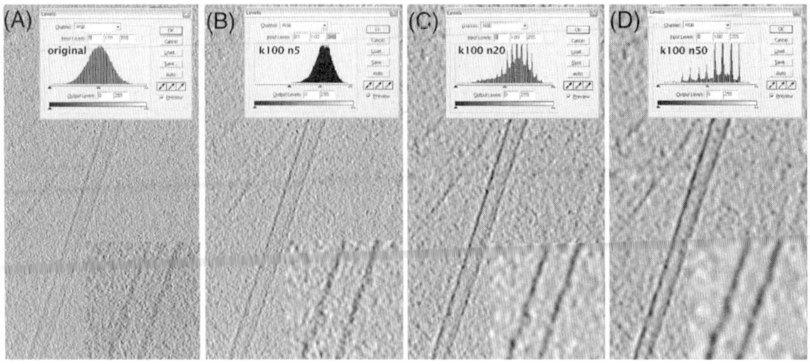

Fig. 5 Effect of nonlinear anisotropic diffusion filtering on details and gray levels of the microtubule lattice. Image shows a single slice of microtubule, two times enlarged area, gray levels, and values for *k* and *n* (iterations). (A) Unfiltered tomogram shows noisy data with smooth gray levels. (B–D) Same slice with increasing amounts of iterative filtering shows that contrast increases but details are lost and the amount of gray levels are reduced. Mild nonlinear anisotropic diffusion filtering is beneficial for contrast and visibility while extended filtering can be utilized for masking and modeling.

The filtered tomograms then were employed for generation of images and 3D models.

3D visualization and surface rendering were performed using Amira Resolve RT version 5.2.0 (Visage Imaging, GmbH, Berlin, Germany) with the electron tomography toolbox plug-in (EM package) for Amira (Pruggnaller *et al.*, 2008) on a HP workstation XW 8200 with GPU capabilities running windows XP 64 bits.

For 3D data visualization we preferentially try to avoid modeling or drawing structures by hand, which are subjective and time-consuming approaches. To avoid user bias as much as possible we try to use a combined approach of mask segmentation and surface rendering in which only the former is subjective in the sense that the user can choose what to include or eliminate from the tomograms.

Image mask segmentation was performed semi-automatically (Fig. 6) on nonlinear anisotropically diffusion-filtered data sets only. The desired regions of interest are hand drawn in the segmentation editor in Amira. MTs can most easily be outlined by circles in the *xz* or *yz* planes and not in the *yx* plane (in which the MTs lay). In these "end-on-views" the MTs are sectioned along their length and appear as circles or ovals. MT outlines can then be selected using the brush in the segmentation editor every 5th or 10th slice followed by interpolation. Additional features like lipid vesicles or mitochondria can best be segmented in the *yx* plane and were added to the segmentations by adding as new material. Within the segmented material highest density pixels are assigned to the material. Afterward the selected material is cleaned up and smoothened by removing small islands and smoothing of labels. This semi-automated masking approach can be time-consuming but produced satisfactory results for surface visualization of tomograms from both stained sections (Knoops *et al.*, 2008) and cryo electron microscopy (Fig. 4F and Fig. 7G).

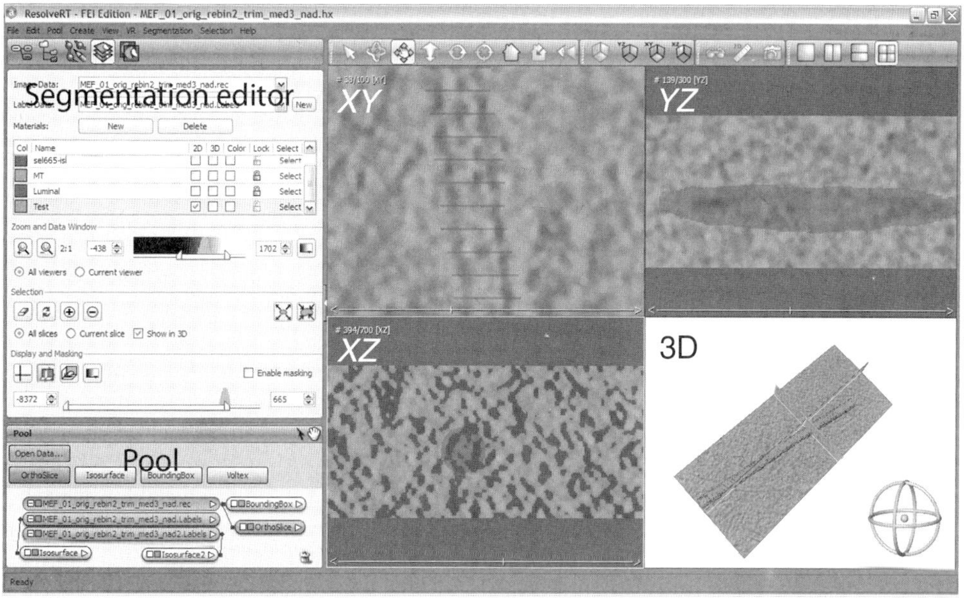

Fig. 6 Workflow for three-dimensional segmentation and rendering of microtubules using Amira ResolveRT. Note that this Amira window screenshot of the segmentation editor is adapted and does not represent a realistic view: the *XY, YZ*, and *XZ* views represent different steps during segmentation and the pool window (bottom left) is normally not present in the segmentation editing window. Segmentation of a microtubule is performed in three steps. First, in the *XZ* view along the microtubule the outline of the complete microtubule is selected using a circular brush every 5 or 10 slices and interpolated to form a tube (*YZ* view). The *XZ* view shows the general outline of the microtubule in red. Additionally the darkest pixels in the tomogram are selected (in purple) by enabling masking to a maximal density (in this case 665). The pixels that are both purple and red are chosen by selecting these pixels using the histogram only for the material that was assigned to the microtubule (not shown). The three-dimensional view of the masked pixels is showed in the lower right window. The pool window (bottom left) shows the original file, in this case two label-fields, one for the microtubule and one for the densities inside the microtubule which are both visualized by an isosurface. (See Plate no. 15 in the Color Plate Section.)

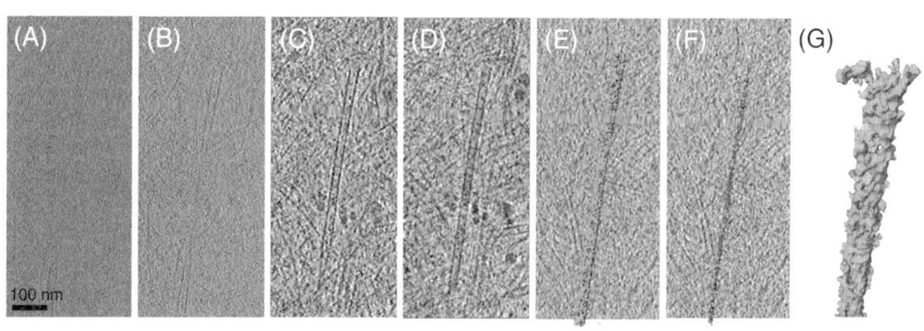

Fig. 7 Different ways of visualization of tomograms before and after automated surface rendering of a microtubule. (A) Single slice from tomogram with the low signal to noise as visualized in Zap window in 3dmod. (B) Sum of four slices from tomogram from slicer window of 3dmod. (C) Single slice from median filtered and anisotropically noise-filtered tomogram. Note that the microtubule is slightly tilted in the slicer window compared to (A), to position a continuous part of the microtubule in the slice. (D) Summed 25 nm thick slice from (C) that includes the whole microtubule. Note that luminal particles are less apparent compared to (B) and (C). (E) Surface-rendered microtubule (green) and luminal particles (red) and (F) luminal particles (red) only. (G) Magnified and slightly rotated view of microtubule plus end from (E). (See Plate no. 16 in the Color Plate Section.)

IV. Summary and Outlook

Here we described methods for nanoscale imaging of MTs in eukaryotic cells using cryo electron tomography. Our goal was to investigate the structure of MTs and MT plus ends in the environment of the cell as a first step in investigating the structure–function relationship of MT-associated proteins and their influence on the dynamic behavior of single MTs.

It was shown that MTs can be imaged in the cortex of vitrified fibroblast cells using cryo electron tomography without the use for chemical fixation, staining, or sectioning. Hereby, the structure of the individual protofilaments at the plus ends of dynamic MTs was optimally preserved and imaged. It was possible to discern straight and curled protofilament conformations, which were shown *in vitro* to be indicative of growing and shrinking MTs, respectively. Therefore, it directly unveiled information on the dynamic state of an individual MT while visualized in its native environment at molecular resolution.

Cryo electron tomography is a perfect method for 3D visualization of macromolecular complexes in the cell, but a visual cellular proteomics approach is hampered since there are no techniques for (1) localization of specific molecules prior to tomography, (2) labeling of structures for identification in tomograms, and (3) dynamic imaging. Room for improvement therefore lies in combining cryo electron tomography with light microscopy so that fluorescent tags can be used for identification and localization of structures. Several papers have already described correlative light and electron microscopy techniques, including imaging of vitrified cells at cryogenic conditions (Plitzko *et al.*, 2009; van Driel *et al.*, 2009). Moreover, it was shown that the tracks of GFP-labeled MTs can be correlated in vitrified cells (Schwartz *et al.*, 2007). Light microscopy is not suitable for identification of individual macromolecules in a tomogram since the resolution difference between light and electron microscopy is two orders of magnitude. Use of super-resolution light microscopy techniques (Huang *et al.*, 2009), however, might prove to be very useful for molecular localization within such a correlative approach.

A second interesting development is the progress that has recently been made to introduce a clonable tag for cryo electron microscopy that can be used in cells to introduce an electron-dense label on an individual protein inside a cell prior to vitrification (Diestra *et al.*, 2009; Mercogliano and DeRosier, 2006; 2007). This potentially enables the labeling of specific MT-associated proteins in cryo electron tomograms of cells.

Inherent to biological electron microscopy is that only fixed materials can be studied and therefore it is unable to resolve the dynamics of processes from its images and structures. A combination of dynamic imaging of MTs using live cell fluorescence imaging in combination with cryo electron tomography of the same vitrified cell and correlation of the MT dynamics would be necessary to directly correlate the structure of the MT plus ends including its associated protein complexes with MT dynamics.

Acknowledgments

I thank Linda van Driel, Christoph Diebolder, and Montserrat Bárcena for critical reading of the chapter and colleague scientist whom I worked with on the investigation of cellular microtubules: Jeffrey van Haren, Niels Galjart (Erasmus Medical Center, Rotterdam, the Netherlands), Gert Oostergetel (University of Groningen, the Netherlands), Sandra Zovko, Kèvin Knoops, Kasia Moscicka, Henk Koerten, Raimond Ravelli, Mieke Mommaas, and Bram Koster. I am also grateful for Dutch Organization of Sciences (NWO) for the Veni grant that enabled to initiate microtubule research.

References

Al-Amoudi, A., Norlen, L. P., and Dubochet, J. (2004). Cryo-electron microscopy of vitreous sections of native biological cells and tissues. *J. Struct. Biol.* **148**(1), 131–135.

Al-Bassam, J., Ozer, R. S., Safer, D., Halpain, S., and Milligan, R. A. (2002). MAP2 and tau bind longitudinally along the outer ridges of microtubule protofilaments. *J. Cell Biol.* **157**(7), 1187–1196.

Berriman, J. A., Li, S., Hewlett, L. J., Wasilewski, S., Kiskin, F. N., Carter, T., Hannah, M.J., and Rosenthal, P.B. (2009). Structural organization of weibel-palade bodies revealed by cryo-EM of vitrified endothelial cells. *Proc. Natl. Acad. Sci. U.S.A.* **106**(41), 17407–17412.

Bodey, A. J., Kikkawa, M., and Moores, C. A. (2009). 9-Angstrom structure of a microtubule-bound mitotic motor. *J. Mol. Biol.* **388**(2), 218–224.

Bouchet-Marquis, C., and Fakan, S. (2009). Cryoelectron microscopy of vitreous sections: A step further towards the native state. *Methods Mol. Biol.* **464**, 425–439.

Bouchet-Marquis, C., Zuber, B., Glynn, A. M., Eltsov, M., Grabenbauer, M., Goldie, K. N., Thomas, D., Frangakis, A. S., Dubochet, J., and Chretien, D. (2007). Visualization of cell microtubules in their native state. *Biol. Cell* **99**(1), 45–53.

Caplow, M., Shanks, J., and Ruhlen, R. L. (1988). Temperature-jump studies of microtubule dynamic instability. *J. Biol. Chem.* **263**(21), 10344–10352.

Costello, M. J. (2006). Cryo-electron microscopy of biological samples. *Ultrastruct. Pathol.* **30**(5), 361–371.

Crowther, R. A., De Rosier, D. J., and Klug, A. (1970). The reconstruction of a three-dimensional structure from projections and its application to electron microscopy. *Proc. R. Soc. Lond.* **317**, 319–340.

Cyrklaff, M., Kudryashev, M., Leis, A., Leonard, K., Baumeister, W., Menard, R., Meissner, M., and Frischknecht, F. (2007). Cryoelectron tomography reveals periodic material at the inner side of subpellicular microtubules in apicomplexan parasites. *J. Exp. Med.* **204**(6), 1281–1287.

Diestra, E., Fontana, J., Guichard, P., Marco, S., and Risco, C. (2009). Visualization of proteins in intact cells with a clonable tag for electron microscopy. *J. Struct. Biol.* **165**(3), 157–168.

Downing, K. H., and Sui, H. (2007). Structural insights into microtubule doublet interactions in axonemes. *Curr. Opin. Struct. Biol.* **17**(2), 253–259.

Dubochet, J., Adrian, M., Chang, J.J., Homo, J. C., Lepault, J., McDowall, A. W., and Schultz, P. (1988). Cryo electron microscopy of vitrified specimens. *Q. Rev. Biophys.* **21**(2), 129–228.

Frangakis, A. S., and Hegerl, R. (2001). Noise reduction in electron tomographic reconstructions using nonlinear anisotropic diffusion. *J. Struct. Biol.* **135**(3), 239–250.

Garvalov, B. K., Zuber, B., Bouchet-Marquis, C., Kudryashev, M., Gruska, M., Beck, M., Leis, A., Frischknecht, F., Bradke, F., Baumeister, W., Dubochet, J., and Cyrklaff, M. (2006). Luminal particles within cellular microtubules. *J. Cell Biol.* **174**(6), 759–765.

Grunewald, K., and Cyrklaff, M. (2006). Structure of complex viruses and virus-infected cells by electron cryo tomography. *Curr. Opin. Microbiol.* **9**(4), 437–442.

Hoenger, A., and Gross, H. (2008). Structural investigations into microtubule-MAP complexes. *Methods Cell Biol.* **84**, 425–444.

Hoenger, A., and McIntosh, J. R. (2009). Probing the macromolecular organization of cells by electron tomography. *Curr. Opin. Cell Biol.* **21**(1), 89–96.

Howard, J., and Hyman, A. A. (2003). Dynamics and mechanics of the microtubule plus end. *Nature* **422**(6933), 753–758.

Huang, B., Bates, M., and Zhuang, X. (2009). Super-resolution fluorescence microscopy. *Annu. Rev. Biochem.* **78**, 993–1016.

Kikkawa, M. (2008). The role of microtubules in processive kinesin movement. *Trends Cell Biol.* **18**(3), 128–135.

Knoops, K., Kikkert, M., Worm, S. H., Zevenhoven-Dobbe, J.C., van der Meer, Y., Koster, A. J., Mommaas, A. M., and Snijder, E.J. (2008). SARS-coronavirus replication is supported by a reticulovesicular network of modified endoplasmic reticulum. *PLoS Biol.* **6**(9), e226.

Koning, R.I., and Koster, A. J. (2009). Cryo-electron tomography in biology and medicine. *Ann. Anat.* **191**(5), 427–445.

Koning, R. I., Zovko, S., Barcena, M., Oostergetel, G. T., Koerten, H. K., Galjart, N., Koster, A. J., and Mommaas, A. M. (2008). Cryo electron tomography of vitrified fibroblasts: Microtubule plus ends in situ. *J. Struct. Biol.* **161**(3), 459–468.

Kremer, J. R., Mastronarde, D. N., and McIntosh, J. R. (1996). Computer visualization of three-dimensional image data using IMOD. *J. Struct. Biol.* **116**(1), 71–76.

Li, H., DeRosier, D. J., Nicholson, W. V., Nogales, E., and Downing, K. H. (2002). Microtubule structure at 8 a resolution. *Structure* **10**(10), 1317–1328.

Lucic, V., Forster, F., and Baumeister, W. (2005). Structural studies by electron tomography: From cells to molecules. *Annu. Rev. Biochem.* **74**, 833–865.

Mandelkow, E., and Hoenger, A. (1999). Structures of kinesin and kinesin-microtubule interactions. *Curr. Opin. Cell Biol.* **11**(1), 34–44.

Mandelkow, E. M., Mandelkow, E., and Milligan, R. A. (1991). Microtubule dynamics and microtubule caps: A time-resolved cryo-electron microscopy study. *J. Cell Biol.* **114**(5), 977–991.

Mandelkow, E.M., Schultheiss, R., Rapp, R., Muller, M., and Mandelkow, E. (1986). On the surface lattice of microtubules: Helix starts, protofilament number, seam, and handedness. *J. Cell Biol.* **102**(3), 1067–1073.

Marko, M., Hsieh, C., Schalek, R., Frank, J., and Mannella, C. (2007). Focused-ion-beam thinning of frozen-hydrated biological specimens for cryo-electron microscopy. *Nat. Methods* **4**(3), 215–217.

Mastronarde, D. N. (2005). Automated electron microscope tomography using robust prediction of specimen movements. *J. Struct. Biol.* **152**(1), 36–51.

McIntosh, J. R., Morphew, M. K., Grissom, P. M., Gilbert, S. P., and Hoenger, A. (2009). Lattice structure of cytoplasmic microtubules in a cultured mammalian cell. *J. Mol. Biol.* **394**(2), 177–182.

Mercogliano, C. P., and DeRosier, D. J. (2006). Gold nanocluster formation using metallothionein: Mass spectrometry and electron microscopy. *J. Mol. Biol.* **355**(2), 211–223.

Mercogliano, C. P., and DeRosier, D. J. (2007). Concatenated metallothionein as a clonable gold label for electron microscopy. *J. Struct. Biol.* **160**(1), 70–82.

Milne, J. L., and Subramaniam, S. (2009). Cryo-electron tomography of bacteria: Progress, challenges and future prospects. *Nat. Rev. Microbiol.* **7**(9), 666–675.

Nicastro, D., McIntosh, J. R., and Baumeister, W. (2005). 3D structure of eukaryotic flagella in a quiescent state revealed by cryo-electron tomography. *Proc. Natl. Acad. Sci. U.S.A.* **102**(44), 15889–15894.

Nicastro, D., Schwartz, C., Pierson, J., Gaudette, R., Porter, M. E., and McIntosh, J. R. (2006). The molecular architecture of axonemes revealed by cryoelectron tomography. *Science* **313**(5789), 944–948.

Nickell, S., Forster, F., Linaroudis, A., Net, W. D., Beck, F., Hegerl, R., Baumeister, W., and Plitzko, J. M. (2005). TOM software toolbox: Acquisition and analysis for electron tomography. *J. Struct. Biol.* **149**(3), 227–234.

Nogales, E., and Wang, H. W. (2006a). Structural intermediates in microtubule assembly and disassembly: How and why? *Curr. Opin. Cell Biol.* **18**(2), 179–184.

Nogales, E., and Wang, H. W. (2006b). Structural mechanisms underlying nucleotide-dependent self-assembly of tubulin and its relatives. *Curr. Opin. Struct. Biol.* **16**(2), 221–229.

Oostergetel, G. T., Keegstra, W., and Brisson, A. (1998). Automation of specimen selection and data acquisition for protein crystallography. *Ultramicroscopy* **74**, 47–59.

Plitzko, J. M., Rigort, A., and Leis, A. (2009). Correlative Cryo-light microscopy and Cryo-electron tomography: From cellular territories to molecular landscapes. *Curr. Opin. Biotechnol.* **20**(1), 83–89.

Pruggnaller, S., Mayr, M., and Frangakis, A. S. (2008). A visualization and segmentation toolbox for electron microscopy. *J. Struct. Biol.* **164**(1), 161–165.

Rigort, A., Bauerlein, F.J., Leis, A., Gruska, M., Hoffmann, C., Laugks, T., Bohm, U., Eibauer, M., Gnaegi, H., Baumeister, W., and Plitzko, J.M. (2010). Micromachining tools and correlative approaches for cellular Cryo-electron tomography. *J. Struct. Biol.*. Feb21 epub ahead of print (in press).

Schwartz, C. L., Sarbash, V. I., Ataullakhanov, F. I., McIntosh, J. R., and Nicastro, D. (2007). Cryo-fluorescence microscopy facilitates correlations between light and cryo-electron microscopy and reduces the rate of photobleaching. *J. Microsc.* **227**(Pt 2), 98–109.

Shi, S., Sun, S., Andrews, S. B., and Leapman, R. D. (1996). Thickness measurement of hydrated and dehydrated cryosections by EELS. *Microsc. Res. Tech.* **33**(3), 241–250.

Sui, H., and Downing, K. H. (2006). Molecular architecture of axonemal microtubule doublets revealed by cryo-electron tomography. *Nature* **442**(7101), 475–478.

Suloway, C., Pulokas, J., Fellmann, D., Cheng, A., Guerra, F., Quispe, J., Stagg, S., Potter, C. S., and Carragher, B. (2005). Automated molecular microscopy: The new leginon system. *J. Struct. Biol.* **151**(1), 41–60.

Suloway, C., Shi, J., Cheng, A., Pulokas, J., Carragher, B., Potter, C. S., Zheng, S. Q., Agard, D. A., and Jensen, G. J. (2009). Fully automated, sequential tilt-series acquisition with leginon. *J. Struct. Biol.* **167**(1), 11–18.

van Driel, L.F., Valentijn, J. A., Valentijn, K. M., Koning, R. I., and Koster, A. J. (2009). Tools for correlative Cryo-fluorescence microscopy and Cryo-electron tomography applied to whole mitochondria in human endothelial cells. *Eur. J. Cell Biol.* **88**(11), 669–684.

Wade, R. H. (2007). Microtubules: An overview. *Methods Mol. Med.* **137**, 1–16.

Zovko, S., Abrahams, J. P., Koster, A. J., Galjart, N., and Mommaas, A. M. (2008). Microtubule plus-end conformations and dynamics in the periphery of interphase mouse fibroblasts. *Mol. Biol. Cell* **19**(7), 3138–3146.

CHAPTER 25

Automated Identification of Microtubules in Cellular Electron Tomography

Daniyar Nurgaliev[*,1], **Timur Gatanov**[*,1], *and* **Daniel J. Needleman**[†]

[*]Department of Physics, Harvard University, Cambridge, Massachusetts 01238

[†]Department of Molecular and Cellular Biology, School of Engineering and Applied Sciences, Center for Systems Biology, Harvard University, Cambridge, Massachusetts 01238

Abstract

We describe a method for automatically finding the location and conformations of microtubules in tomograms of high-pressure frozen, freeze substituted cells. Our approach uses two steps: a preprocessing step that finds locations in the tomograms that are likely to lie inside microtubules and a tracking step that uses the preprocessed data to identify the trajectories of individual microtubules. We test this method on a

[1]Daniyar Nurgaliev and Timur Gatanov have contributed equally to this work

978-0-12-381349-7
DOI: 10.1016/S0091-679X(10)97025-8

reconstruction of a *Caenorhabditis elegans* mitotic spindle and we compare our results with those obtained by a human expert who manually segmented the same data. At present, the method could be used to assist the analysis of large-scale tomography reconstructions. With further improvements, it may be possible to reliably segment cellular tomograms without human intervention.

I. Introduction

Electron microscopy and tomography are increasingly used to study larger and larger systems, including organelles (Marsh, 2005), whole cells (Hoog *et al.*, 2007), and perhaps, in the near future, complete organs (Briggman and Denk, 2006). This work has the potential to transform our understanding of cell and developmental biology by (1) bridging the gap in length scale between the structure of protein complexes—obtained from X-ray, NMR, and electron microscopy—and the structure of entire cells—obtained by light microscopy; (2) allowing massive data sets to be studied so the statistical significance of trends can be analyzed; and (3) enabling investigations of the spatial distribution of features which are too small to resolve by light microscopy.

There has been remarkable progress in sample preparation, data acquisition, and assembling reconstructions for these large-scale electron microscopy studies (Briggman and Denk, 2006; Hoenger and McIntosh, 2009), but the development of methods for analyzing the resulting data have lagged behind. In cellular tomography, most image segmentation—the identification of microtubules, membranes, and other features of interests—is still performed manually. While this approach can be highly accurate and has been very successful to date, the segmentation step is often the most time-consuming part of such work. As larger systems are investigated, purely manual segmentation will no longer be practical. For example, it currently takes a skilled human a few hours to track all of the microtubules in a tomographic reconstruction of a small spindle, such as from yeast. Spindles in human cells have a volume approximately 1000 times larger than spindle from yeast cells, so at this rate, it would take an expert a good part of a year to segment. Spindles from Xenopus egg extracts, another popular model system, have a volume about 10 times larger than spindles in human tissue culture cells; thus the resulting analysis of these structures is expected to take 10 times longer still. Fully automated or computer-assisted approaches have the potential to greatly speed segmentation. There have been previous attempts to automatically segment tomograms of plastic-embedded samples, but this remains a challenging problem (Jiang *et al.*, 2006; Sandberg, 2007).

In this chapter, we describe our efforts to develop an automated method of segmenting microtubules in cellular tomograms of plastic-embedded, freeze substituted samples. We mention a number of unsuccessful approaches we tried; we present an algorithm which gives satisfactory results; we demonstrate the method on a tomographic reconstruction of a *C. elegans* mitotic spindle and compare the automated results with manual segmentation performed by a human expert; finally, we detail ways our method can be improved.

II. Overview

Tomography data is intrinsically three dimensional (3D), consisting of a matrix of voxels with a gray value at each location. Microtubules are low-contrast objects that appear as two thin dark lines (Fig. 1). The resolution of tomograms is asymmetric, being much higher in the *XY* plane than in *Z*, and microtubules are difficult to see if they are not parallel to the viewing axis. As a practice data set, we use a reconstruction of a *C. elegans* mitotic spindle from (O'Toole *et al.*, 2003). The data set is a dual-axis tomogram of two serial sections that have been pasted together. The final reconstruction is a $2880 \times 1024 \times 286$ array of voxels which are 1.7 nm per side and have an intensity value ranging from 0 to 255.

Our method for finding microtubules consists of two steps: First, we identify points that are likely to be part of a microtubule. In this step, which we call preprocessing, we transform the original image into a new image, where the intensity of each voxel is a measure of our confidence that a microtubule exists at that location. Second, we connect these points into lines. We call the algorithm which performs this second task a tracker, because its goal is to track the trajectory of individual microtubules.

Both the preprocessing step and the tracker make use of a number of parameters whose values must be selected. Optimizing the choice of parameters requires some method of measuring the quality of the results. Validation of image recognition tasks, like finding microtubules in tomograms, is a very difficult problem because we do not know what the "ground truth" is—where the microtubules really are. After all, if we

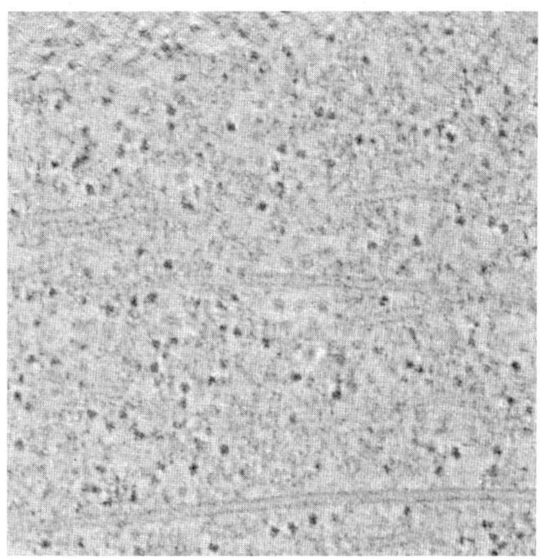

Fig. 1 A typical *XY* slice from a tomogram of a *C. elegans* spindle (O'Toole *et al.*, 2003). The dark long double lines are the walls of microtubules, the black dots are ribosomes, and the smooth region on the top left is a reconstruction artifact from the proximity of the physical edge of the sample. This region is 512×512 pixels. Each pixel is 1.7×1.7 nm.

already had an ideal way of knowing the location of all microtubules in the tomogram, we would use that as our method for finding microtubules instead of trying to develop a new algorithm. Our method of validation is to simply compare our results with those obtained manually by a human expert. We thus attempt to choose parameters that give results which are similar to those obtained by manually tracking.

We will next describe each of these steps in more detail: preprocessing, tracking, validation, and future directions.

III. Preprocessing: Finding Points in Microtubules

We have opted to use two–dimensional (2D) algorithms for our preprocessing step. The tracking algorithm then connects the preprocessed data into trajectories that follow the backbone of the microtubules in 3D. While a 3D processing step would make use of additional information, and thus would presumably be more reliable, we believe that there are a number of advantages associated with employing a simpler 2D preprocessing step. Most importantly, 2D images are much easier to work with and visualize, greatly aiding in testing and debugging the algorithm. Additionally, electron tomography has different resolution and noise structure in the *XY* and *Z* directions, so the data is already intrinsically anisotropic. Finally, 2D preprocessing is very easy to parallelize, as each *XY* slice can be analyzed by an independent computational node.

A 512×512 pixel region from an *XY* slice of a tomogram of a *C. elegans* mitotic spindle is shown in Fig. 1. The set of dark double lines are the walls of microtubules and the black circles are ribosomes. The smooth region in the top left corner is caused by errors in the reconstruction associated with joining together two tomograms from separate physical blocks. We have explored a number of different approaches for identifying which points in images like this are in microtubules. Many of our attempts failed before we developed a method we were satisfied with. We will first discuss various unsuccessful methods that we explored.

A. Unsuccessful Approaches

The walls of microtubules appear as two dark lines, so it is natural to attempt to find them by thresholding the image. Unfortunately, this simple approach produces so many false positives that it is not productive (Fig. 2). This method can be improved by using a local threshold—based on the difference between a pixel's intensity and the average intensity in its vicinity—or by first filtering out ribosomes. However, the performance is quite poor, even with these enhancements.

A common approach for finding features in an image is to use a convolution. A convolution provides a measure of how much a local region in the image matches a mask of interest. First, we need to generate a convolution mask—an average image of a microtubule. We have done this either empirically by averaging together images of microtubules found manually or by conjecturing a simple model for a microtubule, two dark lines spaced an appropriate amount. Both methods produce similar results. If we

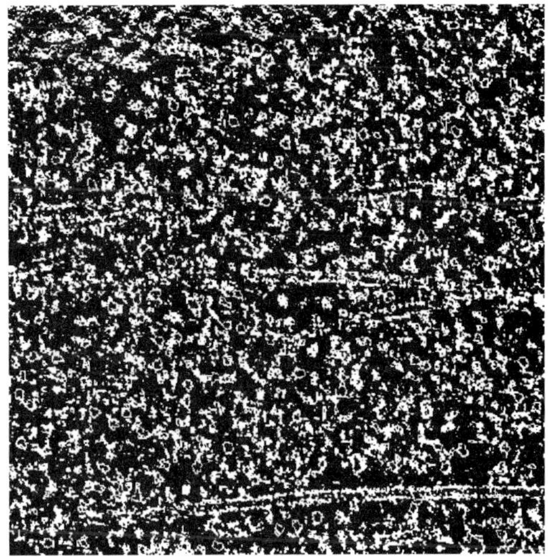

Fig. 2 Intensity thresholding. The region displayed in Fig. 1 was thresholded to highlight pixels with intensities between 100 and 150—the range present along the microtubule walls. While this procedure selects microtubule walls, there are also a very large number of false positives.

simply convolve the data with one of these masks, we are effectively looking for microtubules of a particular orientation. Microtubules in the image could be facing any direction, so we need a more flexible approach. We constructed 60 different masks by rotating one of the original masks in 3 degree increments. Each of these masks can be used to probe the image for a microtubule oriented at the corresponding angle. We created a new image by replacing the intensity of each pixel with the maximum value at the corresponding location obtained from the 60 different convolutions with the 60 different masks. By the logic outlined above, this procedure should provide an estimate of how well a particular region matches the profile of a microtubule, independent of its orientation. The resulting image is visually striking, but, unfortunately, it is not helpful for locating microtubules (Fig. 3). This simple algorithm fails because of the complexity and noise structure of the tomography slices.

An alternative method is to use more sophisticated approaches to search for lines in the image (Lindeberg, T. 1998). This will not be sufficient for finding microtubules because they are composed of pairs of lines, but it might be a reasonable starting point for a more involved procedure. A line is an object that is flat in one direction and has a local extremum in the perpendicular direction—dark lines, like those from the microtubules wall, will be a local minimum. These statements are claims about the derivatives of the image: a region in the image is a line if it has a zero first derivative and a large second derivative along one direction, and a small second derivative in the perpendicular direction.

Computing derivatives in images faces two challenges: (1) The data is not a continuous function, rather, it is composed of discrete pixels. (2) Directly taking differences

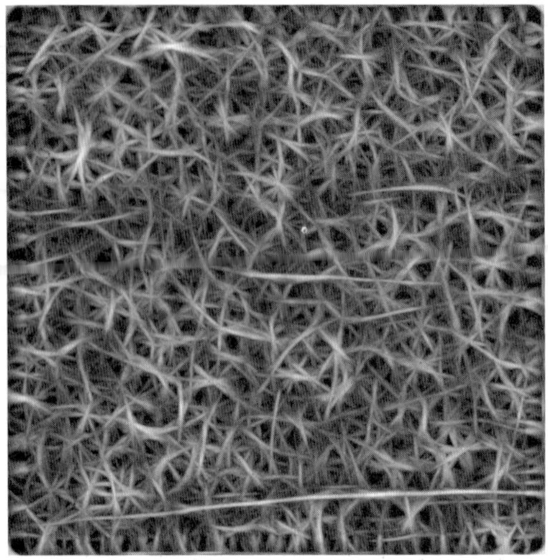

Fig. 3 Direction-independent convolution with a microtubule model. A mask to identify microtubules was empirically constructed by averaging the profile of many microtubules. The mask was oriented in 60 different directions and convolved with the region displayed in Fig. 1. Each pixel in this image corresponds to the maximum value at that location obtained from the different convolutions. Microtubules are visible as long bright lines, but the background structure is very strong and complex.

between neighboring pixels produces poor estimates of the desired derivatives because of the presence of noise. A convenient way to circumvent these difficulties is to first smooth the image by convolving it with a Gaussian kernel:

$$I_s(x,y) = \sum_{i,j} I_{ij} \frac{1}{2\pi\sigma^2} \exp\left(-\frac{(x-i)^2 + (y-j)^2}{2\sigma^2} \right) = I \times G_\sigma \tag{1}$$

where I_{ij} is the original image, $I_s(x,y)$ is the smoothed image, and σ is the width of the Gaussian kernel. An advantage of this approach is that it easily allows one to search for features at a particular length scale, set by the width of the Gaussian. Each microtubule wall in an *XY* slice of the tomogram appears as a dark line with a width of approximately 3 pixels, so it is appropriate to convolve the image with a Gaussian of width $\sigma = 3$ pixels.

The derivative of this smoothed image can then be efficiently computed by convolving the original images with the derivatives of a Gaussian. The first derivatives consist of a vector at each location whose components are the derivatives in the x and y direction:

$$I'_s = [I'_{sx}, I'_{sy}] = [I \times G_{\sigma x}, I \times G_{\sigma y}] \tag{2}$$

$$G_{\sigma x} = \frac{1}{2\pi\sigma^2}\frac{-x}{\sigma^2}\exp\left(-\frac{x^2+y^2}{2\sigma^2}\right) \quad (3)$$

$$G_{\sigma y} = \frac{1}{2\pi\sigma^2}\frac{-y}{\sigma^2}\exp\left(-\frac{x^2+y^2}{2\sigma^2}\right) \quad (4)$$

Higher order derivatives can be calculated with additional convolutions. Thus, the second derivatives form a matrix, called the hessian matrix, at each location, which is computed through further convolutions:

$$I''_s = \begin{bmatrix} I''_{sxx} & I''_{sxy} \\ I''_{syx} & I''_{syy} \end{bmatrix} = \begin{bmatrix} I * G_{\sigma xx} & I * G_{\sigma xy} \\ I * G_{\sigma yx} & I * G_{\sigma yy} \end{bmatrix} \quad (5)$$

$$G_{\sigma xx} = \frac{1}{2\pi\sigma^2}\left(-\frac{1}{\sigma^2}+\frac{x^2}{\sigma^4}\right)\exp\left(-\frac{x^2+y^2}{2\sigma^2}\right) \quad (6)$$

$$G_{\sigma xx} = \frac{1}{2\pi\sigma^2}\left(-\frac{1}{\sigma^2}+\frac{y^2}{\sigma^4}\right)\exp\left(-\frac{x^2+y^2}{2\sigma^2}\right) \quad (7)$$

$$G_{\sigma yx} = \frac{1}{2\pi\sigma^2}\frac{xy}{\sigma^4}\exp\left(-\frac{x^2+y^2}{2\sigma^2}\right) \quad (8)$$

Now points on lines can be identified as pixels where the intensity is flat in one direction—one of the eigenvalues of the hessian matrix has a small absolute magnitude—and the intensity has a local minimum in the other direction—the other eigenvalues of the hessian is large and positive, and the first derivative along that direction is zero. One complication is that, because the image is made of finite-size pixels, the first derivative will not be identically zero near the minimum. We thus need to determine if the first derivative is likely to pass through zero inside a pixel, which can be achieved by approximating the pixilated image as a continuous function. Consider the variation in intensity at a pixel along the x axis. This can be approximated as follows:

$$I_s(x,0) = I_s(0,0) + xI'_{sx}(0,0) + \frac{1}{2}x^2 I''_{sxx}(0,0) + O(x^3) \quad (9)$$

The first derivative of this function is zero at $x = -(I'_{sx}/I''_{sxx})$. If this value is less than 0.5 pixels, then the function has a local extreme inside that pixel. Furthermore, if I''_{sxx} is positive, this is a local minimum.

Running this algorithm to search for lines produces many false positives (Fig 4). The problem is that the described algorithm searches for any feature that locally looks like a line of width σ and ends out picking up linear structures present in the background.

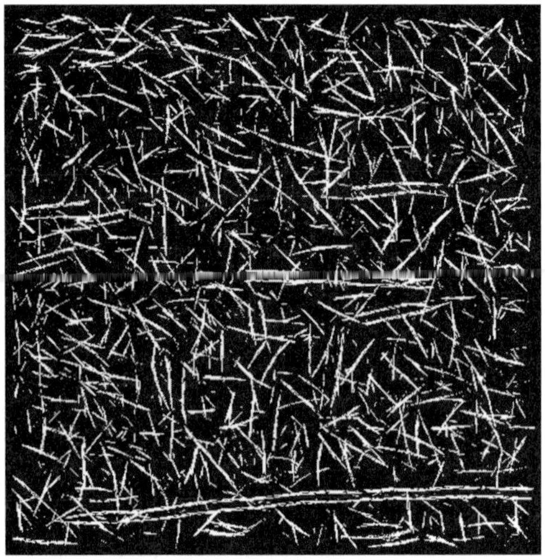

Fig. 7 The location of extended lines from the region depicted in Fig. 1, obtained from 60 different auxiliary matrices such as the one depicted in Fig. 6. See text for details.

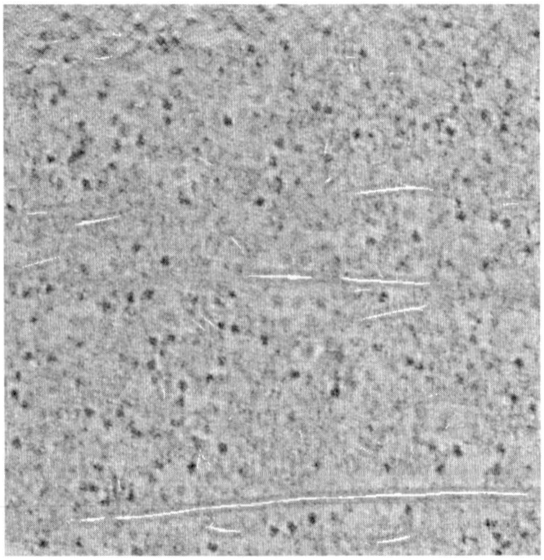

Fig. 8 The final results of the preprocessing step (white) displayed over the original image. Note that while most microtubules are well identified, false positives are also present.

The preprocessing algorithm contains several parameters that need to be set: the width of the Gaussian used for calculating the derivatives, the length L of the mask, and the two thresholds for the convolutions. Adjusting these values can make the procedure more lenient—with fewer false negatives but more false positives—or more stringent—with fewer false positives but more false negatives.

IV. Tracking: Connecting Points into Lines

The preprocessing step of our microtubule detection algorithm finds points that are likely to be inside microtubules. However, at this stage, we have not identified which points are contained in the same microtubule. This data also has errors: some points that we identify as being in microtubules might not really be in microtubules (false positives) and some points that are really in microtubules are not found by our algorithm (false negatives). Moreover, we would like to obtain information on the global properties of microtubules: How many microtubules are present in the tomogram? Where are they located? How long are they? What are their conformations? Ideally, we would like to represent each microtubule as a smooth curve that traces the microtubule's center line, so we can analyze properties of microtubules instead of properties of points that are in microtubules. Therefore, we need to connect the previously identified points into curves. We call the algorithm that performs this task a tracker, because it tracks the trajectory of individual microtubules.

Developing a tracking algorithm is challenging because the preprocessed data has a number of imperfections (Fig. 9):

- Several pixels are identified near the center line of microtubules, making the microtubule's exact position difficult to localize. These regions are only a few pixels wide in the *XY* plane, but their extent in the *Z* direction can be quite variable, and their width is not uniform along microtubules.
- There are gaps in microtubules: regions along their length where no points are found. A naive tracking algorithm might identify these gaps as true breaks, incorrectly finding multiple short microtubules where in reality only one long microtubule is present.
- Microtubules can pass close to each other and a poor tracker might falsely fuse two separate microtubules.
- Points which are clearly not associated with a microtubule are falsely identified as being in microtubules. These false positives can occur at relatively isolated locations or in clusters. Clusters of false positives are particularly difficult to distinguish from short microtubules. These clusters are often adjacent to microtubules.

Thus we need a tracking algorithm which can identify microtubules and find their trajectories despite the flaws that are present in the preprocessing step. One powerful method for tracking lines and other extended objects are "active contour models" (Kass *et al.*, 1988), in which an energy function is defined such that a local minimum corresponds to a tracked object and, ideally, the global minimum corresponds to all objects being located. When implementing active contours one has to choose an appropriate energy function. For tracking lines in high-quality images, such as the ones we obtain after a preprocessing step, active contour methods can be quite robust; thus constructing a suitable energy function is straightforward. The method is actually so powerful that it can even be used to track microtubules in the original, raw data, without the preprocessing step, but the results are less satisfactory. In the next section we will provide an introduction to active contours from a slightly more physical perspective than is standard (Kass *et al.*, 1988).

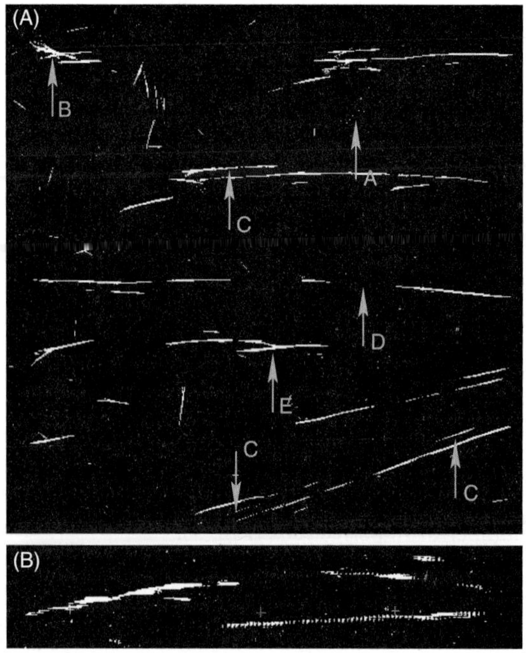

Fig. 9 (A) An *XY* slice of the preprocessing results; different types of errors and difficulties are present: A, isolated noise; B, cluster noise; C, false positives adjacent to a microtubule or between two microtubules; D, gaps (false negatives); E, close approach of two microtubules. (B) An *XZ* slice of the preprocessing results. The thickness of a microtubule in the preprocessing stage varies greatly in *Z*. Finding center the of the microtubule in *Z* is challenging.

A. Active Contour Models and Many-Body Simulation

Active contour models are a class of image analysis methods for finding extended objects—such as lines. In this approach, active contours—one-dimensional (1D) curves—are placed on the image and the energy is calculated based on the location and conformation of the contours. Then the contours are rearranged to search for the local energy minimum, which corresponds to the tracked positions of the lines. To implement this procedure, one needs to choose an appropriate energy function and an energy minimization algorithm. We shall start by considering the choice of energy function.

First, consider a single active contour designed to look for weakly bending lines in an image. An intuitive and useful choice is to select an energy functional which corresponds to an elastic string in an external potential:

$$E_{\text{contour}} = E_{\text{potential}} + E_{\text{bending}} + E_{\text{length}} = \underbrace{\int U(\vec{x}(l))dl}_{E_{\text{potential}}} + \underbrace{\int \frac{k}{R(l)^2} dl}_{E_{\text{bending}}} - \underbrace{\mu L}_{E_{\text{length}}} \quad (10)$$

The energy of the active contour consists of three components: the overlap of the contour and the image ($E_{\text{potential}} = \int U(\vec{x}(l)dl)$), how straight the contour is ($E_{\text{bending}} = \int (k/R(l)^2 dl)$), and its length ($E_{\text{length}} = -\mu L$). The first term, $E_{\text{potential}} = \int U(\vec{x}(l))dl$, represents the potential energy of the string. Here $\vec{x}(l)$ is a parametric representation of the contour, where $\vec{x}$ is the arc length along the contour and l is the position of the corresponding point. U, the potential, is a function of the image that determines which features are tracked. If $U = I$ (the intensity of the image), then the energy will be lower in the darker regions, while if $U = -I$ the energy will be lower in the brighter regions. If $U = -|\nabla I|$, the energy will be lower in the regions with high-gradients - edges. More complex functions of the intensity may also be useful. Second-order derivatives can identify local maxima or minima, and we have found that it can be helpful to include nonlocal contributions.

The second term in the energy, $E_{\text{elastic}} = \int (k/R(l)^2)dl$, represents the elastic energy of the contour. Here, $(1/R(l)) = |(d^2\vec{x}(l)/dl^2)|$ is the curvature of the contour and k is a bending modulus which weighs the relative importance of this term. If k is too small then the contour will faithfully follow the minimum of the potential, which might lead to overfitting of the data as even the local noise structure of the image will be tracked. If $k = \infty$ then only perfectly straight contours are allowed. The third term in the energy, $E_{\text{length}} = -\mu L$, favors longer contours. $L = \int dl$ is the length of the contour and μ is a stretch modulus. The role of this term is to grow the active contour until it is as long as the line to be tracked, but not any longer.

We have been discussing the energy of a single active contour, but many images of interest will have multiple lines, and the described approach must be adjusted to account for this situation. One option is to simultaneously model multiple contours and assign an energy to the entire ensemble. Part of the energy will just be the sum of the contribution of the individual contours: $\sum_i E_{\text{contour}}(i)$, where i is an index that denotes the ith contour, the sum extends over the number of contours, N, and $E_{\text{contour}}(i)$ is the energy of the ith contour of the form discussed above. As the number of lines present in the image is not known a priori, the number of contours, N, should be a variable parameter, which will contribute an additional term to the energy ($E_{\text{chempotential}}$). A natural choice is to set $E_{\text{chempotential}} = \nu N$. Here ν is a chemical potential which determines the cost of creating a new contour. This $E_{\text{chempotential}}$ term favors fewer contours and can close gaps by causing two separate contours to fuse into one. It is also undesirable if multiple contours identify the same line in the image. Therefore, the contours should not be too close to each other. This condition can be enforced by including an interaction term to the energy ($E_{\text{interaction}}$) that cause the contours to repel each other. A simple option is to define a pairwise additive potential between strings, U_{int}, that depends on the distance between points on the strings. The interaction energy, $E_{\text{interaction}}$, will be of the form $\sum_{i,j}\int_{\text{string } i}\int_{\text{string } j} U_{\text{int}}(\vec{x}(l_i)-\vec{x}(l_j))dl_i dl_j$, where the sum extends over all pairs of contours, $\vec{x}(l_i)-\vec{x}(l_j)$ is the distance between a point on contour i and a point on contour j, and the integrals cover the lengths of each contour. Different choices for U_{int} are possible, including a simple hard body interaction that prevents contours from approaching closer than a specified distance.

Taken together, these considerations lead to the multicontour energy functional:

$$\begin{aligned} E_{\text{total}} &= \sum_i E_{\text{contour}}(i) + E_{\text{interaction}} + E_{\text{chempotential}} \\ &= \sum_i E_{\text{contour}}(i) + \sum_{i,j}\int_{\text{string } i}\int_{\text{string } j} U_{\text{int}}(\vec{x}(l_i)-\vec{x}(l_j))dl_i dl_j + vN \end{aligned} \tag{11}$$

This model maps the problem of finding lines in an image onto the "physical" problem of finding the minimum potential energy of a many-body system. It would be possible to directly perform a simulation of an ensemble of strings, in which they are created, annihilated, grow and shrink, merge, and split. This very beautiful picture of a dynamic many-body system has only a few parameters: three constants k, μ, v and two energy functions U, U_{int}. Finding the lowest energy state of this system is equivalent to solving a difficult image recognition problem! This analogy allows us to apply powerful methods for finding the minimum energy of complex systems developed in physics. In addition, we can now use our physical intuition to make heuristic simplifications to improve the performance of the tracking algorithim.

B. Tracking in Practice

So far, the discussion of our tracking approach has been very abstract. A number of additional issues must be addressed when using this method to analyze actual data. A real image is not a continuous distribution of intensity values, but rather consists of discrete pixels. Similarly, we must decide how to represent the trajectory of the contours. It is natural, and computationaly efficient, to discretize them as well. A specific energy minimization algorithm must also be selected.

Contours could be represented as splines (piecewise polynomial curves) or polylines (piecewise linear curves). As long as the resulting contours overlap the image to pixel resolution (requiring an accuracy of half a pixel), these two representations will give equivalent results. While the intrinsic smoothness of splines is ascetically pleasing, polylines are more convenient to work with, so we use them instead. Moreover, we can convert polylines to splines afterward to achive subpixel resolution of the microtubule centerlines. While the length of the segments in the polylines may dynamically vary during the simulation, we still need to choose a characteristic default length that is convienient. If the segments are too short, then calulations will be inefficient, but if segments are too long the discrete nature of the polyline will cause artifacts. We would like the polyline to deviate less than half a pixel from the trajectory of the continous curve it is supposed to represent. Thus, if the smallest radius of curvature of a microtubule we expect to encounter is of order 1000 pixels, the characteristic length of the polyline segment should be less than $\sqrt{8(0.5)(1000)} \approx 60$ pixels (where the factor of 8 arises from basic geometric considerations). In practice, we have found that using a segment size of 30 pixels works well.

There are many methods for computationaly finding the minimum of a function, but our analogy with a physical system suggests that we use either Molecular Dynamics or Monte Carlo simulations. Molecular Dynamics simulations define equations of motions based on "forces" arising from the prescribed energy functional and use them to evolve the system forward from its intial configuration. For complex energy landscapes, Molecular Dynamics simulations can get stuck in local minima, making it difficult to find the true minimum of interest. Molecular Dynamics simulations have been used with active contours to track actin filaments from fluorescence microscopy (Li *et al.*, 2009a, b), but the corresponding images have a much higher signal to noise than tomograms, so the resulting energy landscape is simpler. Monte Carlo simulations, in which trial moves are made and probablistically either accepted or rejected depending on the resulting change in energy, are more efficient for finding the global minimum in high-dimensional spaces with complex energy landscapes (Metropolis *et al.*, 1953). One powerful form of Monte Carlo simulation, called Simulated Annealing (Kirkpatrick, 1983), starts by frequently accepting energetically unfavorable moves and gradually lowers the "temperature," becoming more and more stringent over time. It is difficult to know what Simulated Anealing protocol is best to use—how fast to lower the temperature and when to stop the simulation—but the technique is still very useful. We next describe an implementation that can track microtubules in tomograms, even without any preprocessing.

C. Tracking in the Original Images

In this section we will illustrate how the presented tracking methods can be used to find microtubules in tomograms, even without the benefit of a preprocessing step. Since this is only intended to demonstrate our approach, we will limit the discussion to the task of finding a single microtubule in a 2D slice. In the next section we will show how preprocessing improves the quality of the data so greatly that the tracker can be further simplified and we expand the approach to handle the full 3D, multicontour problem.

We need to specify a potential, U that will be helpful for finding microtubules. A microtubule in a 2D slice of a tomogram appears as two dark, parralel lines (Fig. 1), so it is natural to choose $U = I(\vec{x}(l) + \frac{w}{2}\vec{n}(l)) + I(\vec{x}(l) - \frac{w}{2}\vec{n}(l))$, where $\vec{n}(l)$ is a vector normal to the contour at point $\vec{x}(l)$ and w is the width of a microtubule. This results in an energy function for a contour:

$$\begin{aligned} E_{\text{contour}} &= E_{\text{potential}} + E_{\text{bending}} + E_{\text{length}} \\ &= \int \left[I(\vec{x}(l) + \frac{w}{2}\vec{x}(l)) + I(\vec{x}(l) - \frac{w}{2}\vec{n}(l)) \right] dl + \int \frac{k}{R(l)^2} dl - \mu L \end{aligned} \tag{12}$$

We next need to determine appropriate values for the three parameters w, μ, and k. By inspection, the width of a microtubule is 10 pixels, so we set $w = 10$. We want to select a value of μ such that the E_{length} term in the energy will compete with the $E_{\text{potential}}$ term, so a contour grows if it is on a microtubule and shrink if it is off one. The average image intensity is 165 and the average intensity along the walls of a microtubule is 140–150. Therefore, choosing $\mu = 155$ will have the desired effect of making $E_{\text{potential}}$ dominate if the contour is on a microtubule and E_{length} be larger if the contour is off a microtubule. As long as the bending modulus k is not too large its exact value does not seem to effect the results and we select $k = 5 \times 10^6$.

To find the global minimum we use a Monte Carlo algorithm. We start a contour in a random configuration and evolve its position through a series of steps. A trial move is made by displacing, growing, or shrinking the contour and the move is either accepted or rejected depending on how it changes the system's energy. If the new conformation lowers the energy, then the move is accepted. If the new conformation increases the energy, the move is accepted with probability $\exp(-\Delta E/T)$), where ΔE is the change in the contour's energy and T is the "temperature"—a parameter that determines how frequently energetically unfavorable moves are allowed. This process is repeated until the simulation ends. To avoid having the contour be stuck in unwanted local minimuma, we use a Simulated Anealing protocol in which we start with a high temperature, $T = 150$, and gradually decrease it, by 2% every 200 steps. When the temperature becomes very low, the contour "freezes" in its current local minimum.

The initial state of the contour is a single segment with a random position and orientation. The following types of moves were implemented in the simulation:

- Growing by one segment (at one end or another).
- Shrinking by one segment (at one end or another, if the polyline is longer than 1 segment).
- Random displacement of a single vertex of the polyline.

When the simulation starts, the segment travels in the image until it overlaps a microtubule. The contour rapidly grows along the microtubule, but will occasionally reverse direction, quickly shrinking back to the original one-segment length. Amusingly, the process is highly reminiscent of the polymerization of microtubules by dynamic instability! After annealing is complete, the contour settles in on the position of a microtubule. While the found trajectory is often very accurate, errors are present. Figure 10 shows an example where the contour suddenly shifts to the side, following one edge of the microtubule and noise instead of tracing the true microtubule backbone. This type of error occurs because the energy landscape is highly complex, and the efficiency with which local minimum are circumvented is sensitive to the annealing protocol. Another drawback of the algorithm is that it is very slow. It is therefore highly advantageous to use the preprocessed images discussed earlier—resulting in a simpler energy landscape—and to streamline the Monte Carlo simulation, greatly speeding up the algorithim.

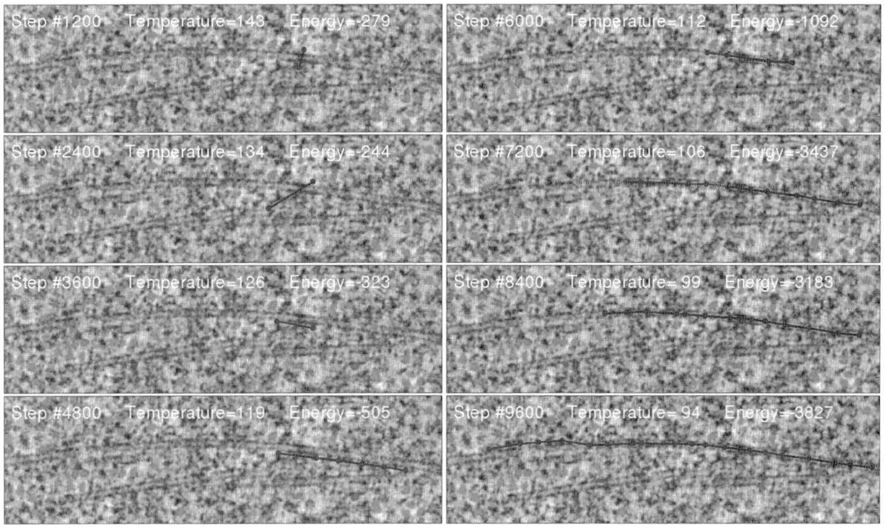

Fig. 10 Simulated annealing for tracking in the original images. Steps 1200–2400: a single segment moves randomly in the image. It does not grow when it is not on a microtubule. Step 3600: the segment overlaps with the microtubule and aligns with it. Growing from this position can lower the energy. Step 4800: the segment does grow from the position found on step 3600. However, it is not well aligned with the microtubule backbone. Step 6000: the contour undergoes a displacement causing it to more accurately follow the microtubule backbone, and the energy is substantially reduced. Steps 7200–8400: the polyline continues to grow inside the found microtubule. Step 9600: the contour has grown nearly the entire length of the microtubule. Note the region on the left: the contour jerks to the side and follows one microtubule wall and noise, instead of the true microtubule center line. This problem could be solved by increasing the bending rigidity, k, or by finding different parts of microtubules and fusing them in many-body simulation.

D. Tracking in the Preprocessed Images

In the preprocessed data, the location of microtubules are given by bright lines, so we can simply set $U(x) = -I(x)$. The resulting energy is as follows:

$$E_{\text{contour}} = E_{\text{potential}} + E_{\text{bending}} + E_{\text{length}} = -\int I(\vec{x}(l))dl + \int \frac{k}{R(l)^2}dl - \mu L \qquad (13)$$

This potential has two parameters, k and μ. As before, the end result is relatively insentive to the value of k, and we set it to $k = 5\times10^6$. μ should be intermediate between the intensity of points on microtubules and the intensity of points not on microtubules. One convenient approach to determine this level is to plot a histogram of intensity values in the preprocessed data. Then, after making an estimate of the volume fraction of microtubules, we can choose a value of μ which an appropriate fraction of intensity values are above—corresponding to points on microtubules.

The preprocessed data is of very high quality, allowing substantial simplifications of the Monte Carlo simulation. First, because most microtubules are well separated from each other, we do not explicitly perform the full many-body simulation. We simulate a single contour at a time, identifying the trajectory of each microtubule in the image independently. Second, we do not start the contours at random locations. Initial segments are added at points likely to be on microtubules—high-intensity pixels in the preprocessed images. Third, the preprocessed data is very smooth and has little noise. Therefore, the energy landscape has few unwanted local minima, and we can perform the Monte Carlo simulation at zero temperature—only accepting moves that decrease the contour's energy. With these considerations in mind, our resulting algorithm is as follows:

1. Find a high-intensity pixel in the image, this point is likely to be on a microtubule. Start a new, one-segment contour at this location at an orientation that minimizes the energy.
2. Grow the contour one segment at a time with a conformation that minimizes the energy.
3. If adding the segment results in an increase in energy, then this indicates that that portion of the contour is unlikely to be on a microtubule. This could be for two reasons: the contour might have grown past the end of the microtubule or there could be a gap in the microtubule caused by false negatives from the preprocessing step. We first check for a gap by attempting to grow the microtubule even farther to see if it can find another region that is likely to be part of the same microtubule. If a gap is not identified, then we continuosly shrink the contour until removing length no longer decreases the energy, at which point the microtubule end has been identified.
4. Refine the trajectory of the contour by repeatedly moving each point of the polyline by a random vector and keeping the move if the contour's energy is lowered.
5. Remove the found microtubule from the image by setting $I(x)$ to zero at all points within 5 pixels of the contour's final position.

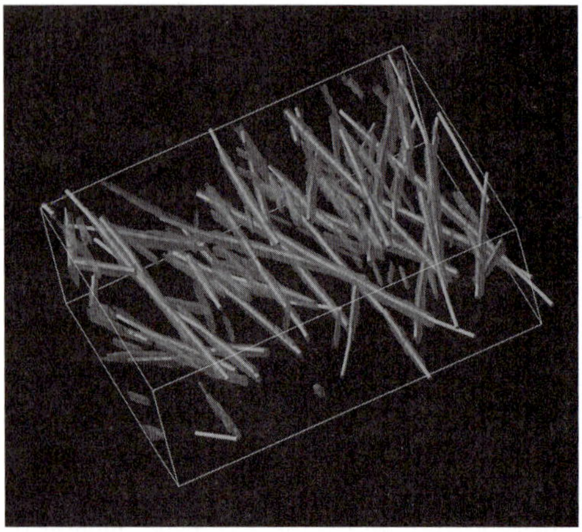

Fig. 11 Tracker's performance. The gray regions are points likely to be inside microtubules generated from the preprocessing step. The tubes mark the continuous lines found by the tracking algorithm.

6. Start again at step 1 in the new updated image and repeat until all points that are likely to be in a microtubule—those above a predetermined intensity value—have been interrogated.

Contours that are too short are most likely caused by noise and are discarded. The remaining contours give the locations and conformations of the tracked microtubules. This tracking algorithim is rapid and gives results that appear to be very good (Fig. 11).

V. Validation and Future Work

Simple visual inspection indicates that our algorithm accurately finds many microtubules in the tomogram. However, it is highly desirable to have a more systematic means of testing the success of our approach for automated microtubule identification. Having an objective validation scheme is important for optimizing the parameters used in the algorithm, discovering what type of errors are present so we can introduce new steps to minimize them, and understanding how reliable the results are. We therefore compared the output of our method with previously obtained results from a human expert who manually identified microtubules in this data set (O'Toole *et al.*, 2003) (see Fig. 12)

While many microtubules manually identified were found with our automated approach, we discovered a number of errors. The most common mistakes are as follows:

1. Some microtubules that were found manually were entirely missed by our algorithm. Many of these missed microtubules were located near the centrosome, where the noise structure is different than in other parts of the tomogram. Other missed microtubules pass through the *XY* planes at very steep angles.
2. Our automatically identified microtubules tend to be shorter than the manually identified ones: our algorithm has difficulty accurately finding the end of microtubules.
3. Long microtubules are sometimes incorrectly identified as multiple, short microtubules by our method. These errors arise because our tracking algorithm does not sufficiently correct for gaps in microtubules present in the preprocessing step.

We are in the process of developing ways to avoid these errors and we are attempting to use automated approaches to optimize the values of the parameters in our algorithm. One difficulty is that there are also errors in the manual data: some microtubules found by our algorithm were missed by the human expert and our automated method more accurately identifies the precise location of microtubule center lines.

Despite the mistakes that are present and the need for further optimization, our approach performs quite well for the tomogram under study. One concern for future applications is that the parameters in the algorithm may have to be adjusted for

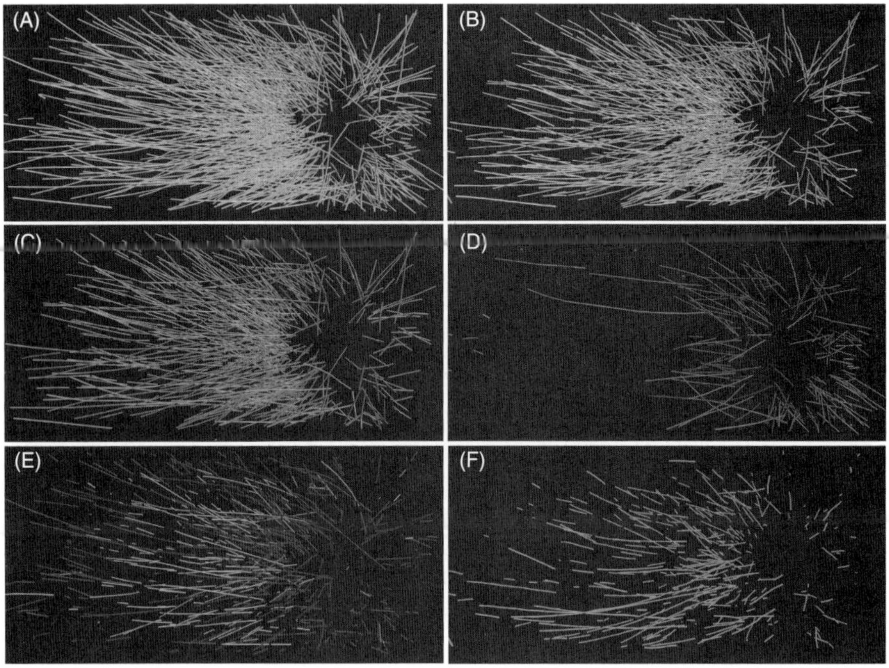

Fig. 12 Comparison of manual and automatic tracking: (A) All microtubules found by a human expert (O'Toole *et al.*, 2003). (B) All microtubules found by our algorithm. (C) Parts of microtubules found both by manual and by automatic tracking. (D) Microtubules missed completely in automatic tracking. Many are concentrated near the centrosome where the noise structure is different. Some are oriented at steep angles to *XY* plane. (E) Missing ends and gaps in the automatically identified microtubules. (F) Parts of microtubules found by our automatic algorithm and absent in the manual tracking result. Some of these are false positives, but others are true microtubules that were missed by the human expert.

different biological samples and different microscope settings. A hint that this procedure might not be straightforward is that the success of the preprocessing step can even vary within a sample: our algorithm performed worse near centrosomes, which exhibit a different background than other regions in the reconstruction. We are currently exploring the use of machine learning techniques to overcome this difficulty with preprocessing. Our preliminary work using Support Vector Machines (SVMs) is extremely promising. We create a training set by having a user click on several regions containing microtubules and several regions where microtubules are absent. With enough examples, SVMs are excellent at recognizing microtubules. Furthermore, the process of training is interactive: a user can improve the performance of the SVM by noting areas where it makes mistakes.

At the present stage of development, our automated method is not sufficiently reliable to replace manual tracking by a human expert. It could be used to provide an initial guess for the location of microtubules, which can then be corrected by a human. While this computer-assisted approach would still be cumbersome, it should be much faster than currently used, entirely manual methods. As our method is improved and new approaches are developed, it may eventually become possible to segment tomograms without human intervention. Such a development would greatly facilitate the analysis of large-scale tomographic reconstructions of cells.

Acknowledgments

We thank Eileen O'Toole for supplying the *C. elegans* tomography data and the manual microtubule identification that were used to test our automated segmentation. Figures 11 and 12 were generated with VMD (Humphrey *et al.*, 1996).

References

Briggman, K. L., and Denk, W. (2006). Towards neural circuit reconstruction with volume electron microscopy techniques. *Curr. Opin. Neurobiol.* **16**, 562–570.

Gonzalez, R. C., and Woods, R. E. (2007). "Digital Image Processing." Pearson Prentice Hall, New Jersey.

Hastings, W. K. (1970). Monte Carlo sampling methods using Markov chains and their applications. *Biometrika* **57(1)**, 97–109.

Hoenger, A., and McIntosh, J. R. (2009). Probing the macromolecular organization of cells by electron tomography. *Curr. Opin. Cell Biol.* **21**, 89–96.

Hoog, J. L., Schwartz, C., Noon, A. T., O'Toole, E. T., Mastronarde, D. N., McIntosh, J. R., and Antony, C. (2007). Organization of interphase microtubules in fission yeast analyzed by electron tomography. *Dev. Cell* **12**, 349–361.

Humphrey, W., Dalke, A., and Schulten, K. (1996). "VMD—visual molecular dynamics". *J. Mol. Graph.* **14**, 33–38.

Jiang, M., Qiang, J., and McEwen, B. F. (2006). Model-base automated extraction of microtubules from electron tomography volume. *IEEE Trans. Inf. Technol. Biomed.* **10(3)**, 608–617.

Jiang, M., Qiang, J., and McEwen, B. F. (2006). Automated extraction of fine features of kinetochore microtubules and plus-ends from electron tomography volume. *IEEE Trans. Image Process.* **15(7)**, 2035–2048.

Kass, M., Witkin, A., and Terzopoulos, D. (1988). Snakes: Active contour models. *Int. J. Comput. Vis.* **1(4)**, 321–331.

Kirkpatrick, S., Gelatt, C. D., and Vecchi, M. P. (1983). Optimization by Simulated Annealing. *Science. New Series.* **220(4598)**, 671–680.

Kremer, J. R., Mastronarde, D. N., and McIntosh, J. R. (1996). Computer visualization of three-dimensional image data using IMOD. *J. Struct. Biol.* **116(1)**, 71–76.

Li, H., Shen, T., Smith, M. B., Fujiwara, I., Vavylonis, D., and Huang, X. (2009a). Automated actin filament segmentation, tracking and tip elongation measurements based on open active contour models. *In* "ISBI'09: Proceedings of the Sixth IEEE international conference on Symposium on Biomedical Imaging," IEEE Press. (http://www.ncbi.nlm.nih.gov/pmc/articles/PMC2804988/)

Li, H., Shen, T., Vavylonis, D., and Huang, X. (2009b). Actin filaments tracking based on particle filters and stretching open active contour models. *Med. Image Comput. Comput. Assist. Interv.* **12**(Pt 2), 673–681.

Lindeberg, T. (1998). Edge detection and ridge detection with automatic scale selection. *Int. J. Comput. Vis.* **30(2)**, 117–154.

Marsh, B. J. (2005). Lessons from tomographic studies of the mammalian Golgi. *Biochim. Biophys. Acta* **1744(3)**, 273–292.

Metropolis, N., Rosenbluth, A. W., Rosenbluth, M. N., Teller, A. H., Teller, E., *et al.* (1953). Equation of state calculations by fast computing machines. *J. Chem. Phys.* **21(6)**, 1087–1092.

O'Toole, E. T., McDonald, K. L., Mantler, J., McIntosh, J. R., Hyman, A. A., and Muller-Reichert, T. (2003). Morphologically distinct microtubule ends in the mitotic centrosome of *Caenorhabditis elegans*. *J. Cell Biol.* **163(3)**, 451–456.

Sandberg, K. (2007). Methods for image segmentation in cellular tomography. (McIntosh Ed, J. R., ed.), *Methods in Cell Biology.* **79**, 769–798.

CHAPTER 26

Quality Control in Single-Molecule Studies of Kinesins and Microtubule-Associated Proteins

Gary J. Brouhard

Department of Biology, McGill University, Montréal, Québec, Canada H3A 1B1

Abstract

Commercial microscopes capable of single-molecule experiments have made it simple for researchers to adopt these powerful techniques. This chapter is meant to help newcomers assess whether their data is of sufficient quality to warrant time-intensive analysis. Two problems can hamper single-molecule experiments: (1) non-specific aggregation of the proteins of interest and (2) detection thresholds from a poor microscope setup. I outline four steps that researchers can take to overcome these problems and convince themselves that they are observing bona fide single molecules.

978-0-12-381349-7
DOI: 10.1016/S0091-679X(10)97026-X

I. Introduction

Single-molecule experiments have taken hold of the microtubule field. Ten years ago, only a few laboratories published articles purporting to observe individual kinesins or microtubule-associated proteins (MAPs) by fluorescence (Vale *et al.*, 1996). These labs used custom-built microscopes outfitted for total internal reflection fluorescence (TIRF) microscopy (Axelrod *et al.*, 1984). Now, tens of labs worldwide make use of the technique. There is every reason to expect that the technology will continue to propagate, as commercial TIRF microscopes have matured to the point where custom-built systems are no longer strictly required. The goal of this chapter is to provide a guide to newcomers interested in pursuing single-molecule studies with regard to one particular problem: how does a researcher know that she is detecting a single molecule by fluorescence? What is a single molecule?

Researchers are able to observe single molecules in cells (Cai *et al.*, 2007), but a large share of contemporary research is oriented toward *in vitro* experiments that reconstitute the interaction of kinesins or MAPs with microtubules. In these experiments, microtubules are adhered to a cover glass surface and imaged by TIRF. Fluorescent kinesins or MAPs are introduced and also visualized by TIRF (Fig. 1). Other authors have provided thoughtful and thorough descriptions of the practice of setting up a single-molecule microscope (Stuurman and Vale, 2005), immobilizing microtubules to a cover glass surface (Gell *et al.*, 2010), and performing several types of experiments (Selvin and Ha, 2008). This guide begins at the point where data are coming from the camera. Are the data of sufficient quality to warrant further (time-intensive!) analysis or should they be discarded?

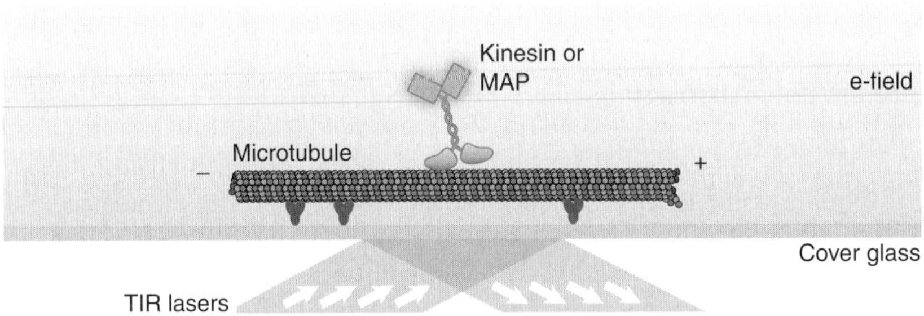

Fig. 1 Single-molecule assay for kinesins and MAPs. Stabilized, fluorescently labeled microtubules (red) are adhered to a cover glass surface by antibodies (dark blue). Excitation by TIRF allows detection of single molecules (e.g., kinesin-1-GFP, shown in green) in the evanescent field (e-field, orange). The figure is not to scale.

II. Problems in Single-Molecule Detection

What prevents a researcher from detecting a single molecule by fluorescence? Two problems may arise. First, the protein species at hand may not, in fact, be comprised of single molecules, but rather of oligomers of some sort. Second, the microscopy system in use may not be sufficiently sensitive to detect the small number of photons emitted by single fluorophores.

A. Non-specific Aggregation

In vitro single-molecule experiments make use of purified recombinant proteins. When expressed in bacteria, recombinant proteins often aggregate into aberrant oligomeric species. The high rate of production from phage-derived promoters gives rise to large numbers of unfolded polypeptide chains emerging from ribosomes. If the rate of folding is too slow, hydrophobic residues from two or more of these chains will form non-covalent bonds and the proteins will then aggregate. These aggregates often form inclusion bodies (Williams *et al.*, 1982). Large inclusion bodies are easily removed from bacterial lysates by centrifugation, but textbook protocols for this clarification step do not use *g*-forces large enough to pellet small aggregates of kinesins and MAPs. Indeed, it is nearly impossible to separate out aggregates that differ in mass from the non-aggregated species by only a few hundred kDa (e.g., false dimers and trimers). Using insect- or mammalian-cell expression systems will provide chaperone-assisted folding, but the rates of translation are still able to overwhelm the folding machinery in some cases.

The real problem, however, arises from the freezing and long-term storage of purified proteins. In 1932, F.F. Nord used measurements of surface tension and solution viscosity to show that freezing caused aggregation of egg albumin, gelatin, gum arabic, and sodium oleate (Nord and Von Ranke-Abonyi, 1932). Nord confirmed these observations with light-scattering measurements in 1949 (Bier and Nord, 1949). Today, it is the widespread experience of protein scientists that freezing leads to non-specific protein aggregation. The extent of aggregation can be alleviated by good buffers, which is why additives such as glycerol are imperative (Simpson *et al.*, 2009). Each recombinant protein will have a different propensity for aggregation during expression and/or freezing, and in some cases the problem is severe.

This persistent bogeyman of protein aggregation creates a lasting problem for single-molecule experiments (Fig. 2). If aggregates are known to form, how can the researcher be sure that she is working with a monodisperse, homogeneous species? An inhomogeneous population can significantly complicate the results of single-molecule experiments. The aggregated protein often dies and activities that might otherwise have been discovered are lost. In other cases, the aggregated molecules display aberrant or exaggerated properties, as when bundles of kinesin-1 show extended run lengths and inhomogeneous velocities. Other false-positive results would be (a) an increased lifetime of microtubule interaction, (b) stationary binding where lattice diffusion might occur, (c) lattice diffusion where very transient associations should dominate, and (d)

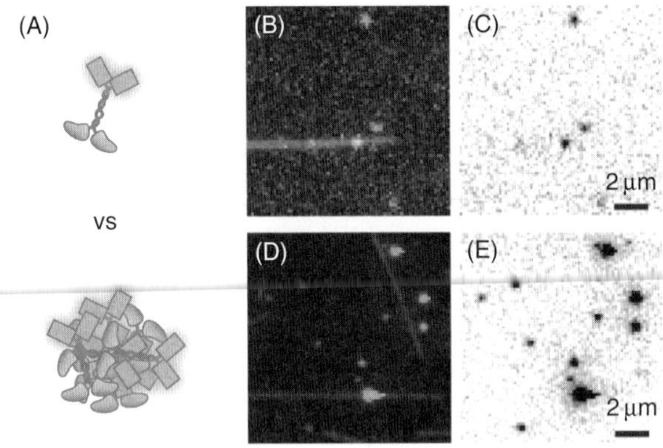

Fig. 2 Non-specific aggregation hampers single-molecule experiments. (A) Schematic representation of a single kinesin-1 dimer (top) versus a non-specific aggregate (bottom). (B) A single molecule of MCAK-His6-eGFP (green) interacting with a microtubule (red). Two other single molecules are visible on the nearby surface. (C) Inverted gray-scale image of the green channel showing single MCAK-His6-eGFP molecules. Scale bar 2 μm. (D) Under poor conditions, MCAK-His6-eGFP forms aggregates (green) and interacts with microtubules (red), as evidenced by inhomogeneous signals. (E) Inverted gray-scale image of the green channel showing aggregates of MCAK-His6-eGFP. Scale bar 2 μm.

the interpretation of aggregates as bona fide oligomers. It is thus apparent that protein aggregation must be convincingly ruled out in order for precise and accurate measurements to be made.

B. Detection Thresholds

A single fluorophore may emit only a few thousand detected photons before irreversible photobleaching occurs. This is a low signal. An original limitation to observing single molecules was a lack of camera sensitivity, as measured by the quantum efficiency of the sensor: the percentage of impinging photons that are converted into electrical signals. The lack of sensitivity was further complicated by significant noise within the electronics. In short, the signal-to-noise ratio was too low. A signal-to-noise ratio of 5, the Rose criterion, can be regarded as a generic threshold for the reliable detection of a single molecule. Low signal-to-noise ratios have largely been solved by the advent of commercially available back-illuminated electron-multiplying CCD cameras (EMCCDs). These cameras have a quantum efficiency of 90% or more, and the sensor is cooled to –80°C or lower, which substantially reduces electronics noise. The newest of these cameras are essentially perfect instruments: under ideal conditions, they are limited only by photon shot noise, the theoretical noise floor. These cameras are expensive but essential. A good camera, however, is not sufficient. Optical components out of alignment, poor filter sets, and faulty sample preparation can lead to a persistent detection threshold. The danger of a detection threshold is that only the aggregates make it over the limit, and thus single molecules are never detected.

III. Quality Control Steps

What follows are four steps that a researcher can take to convince herself that intensity signals coming from the camera are in fact single molecules.

A. Step 1. Good Initial Conditions

The starting point of every *in vitro* single-molecule project is high-quality recombinant kinesins or MAPs: monodisperse, homogenous proteins purified to 99% (Fig. 3A). The requirement is akin to that faced by structural biologists, with the exception that high concentrations of the proteins are not required. A final concentration of 1 μM is sufficient, as single-molecule experiments take place in the nanomolar regime. Higher concentrations of protein do minimize freezing-induced aggregation, so it is nevertheless worthwhile to optimize the purification protocol where possible.

The initial protein conditions can be tested by size exclusion chromatography. Not only does this facilitate purification but monodisperse proteins will also elute from the size exclusion column with a well-defined Gaussian profile in the absorbance trace (Fig. 3B). Aggregates will elute early, although care must be taken to design a protocol and select a column with adequate resolution for the separation of small aggregates

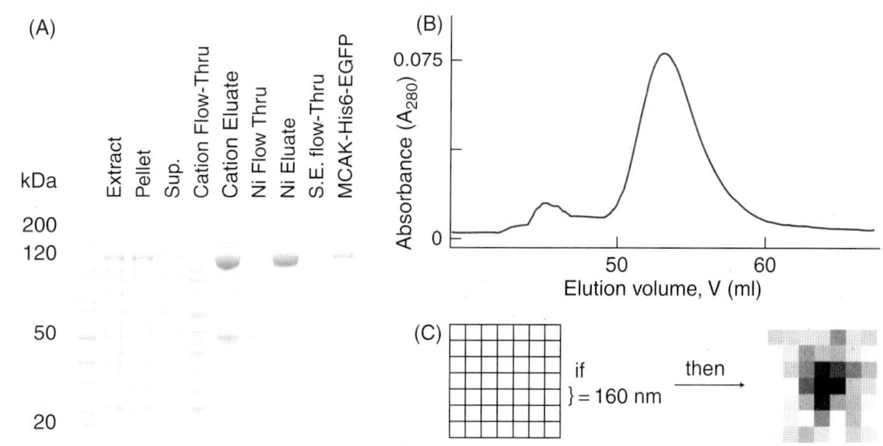

Fig. 3 Initial steps in quality control. (A) SDS-PAGE gel MCAK-His6-eGFP purification showing cation exchange, Ni-affinity, and size exclusion chromatography steps. MCAK-eGFP appears as a 115 kDa band throughout the purification. The Ni column eluate is > 90% pure, and the size exclusion column successfully removes the remaining contaminants (rightmost lane, labeled MCAK). (B) Absorbance trace for the elution of MCAK-eGFP from the size exclusion column. MCAK elutes as a single peak with a Gaussian profile. The higher weight contaminant visible in the SDS-PAGE gel elutes as an earlier, smaller peak. (C) Size-based analysis of fluorescence intensity signals. If the image pixel size is 160 nm (left), then images of single molecules will cover an approximately 3 × 3 pixel grid (right, inverted gray-scale image of a single MCAK-GFP).

(e.g., false dimers and trimers). Protein purity can be assessed by overloading of a Coomassie-stained SDS-PAGE gel (Fig. 3A) or by mass spectrometry for the hard core.

While a good three-step protein purification can take all day, it is nevertheless advisable to test a sample of the protein in the single-molecule microscope immediately after purification, without freezing. Data from freshly purified protein can then be used for comparison to samples that have been pushed through a freeze-thaw cycle to determine if freezing has had a negative impact. Freshly purified protein is often stable at 4°C for 48 h or more, which is adequate time to perform a large number of single-molecule experiments. In some cases, it is preferable to rely exclusively on freshly purified protein (Helenius *et al.*, 2006); the increase in data quality is well worth the additional labor of repeated purifications.

The purified proteins must be fluorescent, either through tagging with a genetically encoded fluorophore (e.g., eGFP) or by chemical cross-linking of a dye molecule (e.g., Cy3). For genetically encoded fluorophores, the monodispersity is assessed during the purification. Chemical cross-linking adds a post-purification step that also creates a propensity for aggregation or non-specific loss of activity. The trade-off, of course, is significantly brighter and more stable fluorescence. Monodispersity is evaluated after the labeling reaction and can coincide with the separation of unconjugated dye molecules (Gell *et al.*, 2010).

B. Step 2. Preliminary Analysis of Signals

After a good purification of the recombinant proteins (and the labeling step if necessary), the purified kinesins or MAPs are introduced into the reaction chamber with surface-immobilized microtubules. As data are coming from the cameras, non-specific aggregates are often immediately apparent (Fig. 2D and E). A monodisperse species will produce a monodisperse field of fluorescence intensity signals, although some leeway must be given for, e.g., inhomogeneities in the illumination field (Fig. 2B and C). If non-specific aggregation is present in the sample, the fluorescence intensity signals will include a range of objects of varying brightness (Fig. 2E).

A single molecule approximates a point source of light. Its image, therefore, is a diffraction-limited spot or Airy disk defined by the point spread function (PSF) of the single-molecule microscope. A typical PSF has a radius (full maximum to first full minimum) of approximately 250 nm and is approximated by a two-dimensional Gaussian distribution (Thomann *et al.,* 2002). Compare this width with the pixel size of the camera, and one can immediately get a sense for what a single molecule should look like. For example, the sensors (pixel elements) of an Andor iXon EMCCD camera have a physical size of $16 \times 16\ \mu m$. Coupled to a $100\times$ objective, one expects image pixels of 160×160 nm. If a 250 nm wide PSF was centered on one such pixel, we expect the image of the PSF to extend slightly over a 3×3 pixel grid (Fig. 3C). If the fluorescence signals appear much larger than this, they cannot be single molecules. For very bright fluorophores, the secondary and tertiary peaks of the Airy pattern may begin to contribute to and widen the image, but this is rare, especially for GFP and its derivatives. Large spatial signals are telltale of aggregation.

C. Step 3. Two-Step Photobleaching

A traditional approach to quality control in single-molecule experiments has been a demonstration of two-step photobleaching for dimeric kinesins. This technique was used by Vale *et al.* in the first observations of single kinesins (Vale *et al.*, 1996) and continues to serve as an important control. The fluorescence intensity of a molecule is plotted as a function of time, and an appropriate trace is found with two plateaus followed by a complete loss of signal, indicating that two discrete photobleaching events have occurred. The intensities of the two plateaus are ideally twofold in difference, and the molecule is thus declared a dimer.

There are three limitations to this approach. First, it does not apply to monomeric kinesins and MAPs or to proteins that were chemically labeled with a variable number of fluorophores. Of course, if one observes two-step photobleaching for a supposedly monomeric protein, this is cause for concern.

A second limitation arises when the signal-to-noise ratio of the fluorescence intensity signals is low. Figure 4A shows an idealized two-step photobleaching trace (red curve). The signal-to-noise ratio in this trace is 20, and individual plateaus and plateau intensity values are readily identifiable. Ideally, these plateau intensity values are a "calibration" of the brightness of a single fluorophore under the imaging conditions at hand. If the signal-to-noise ratio decreases to 5, however, the intensity traces become noisy (Fig. 4A, orange curve). In these cases, a step-finding algorithm would be applied to identify the plateaus and the plateau intensity values (Kerssemakers *et al.*,

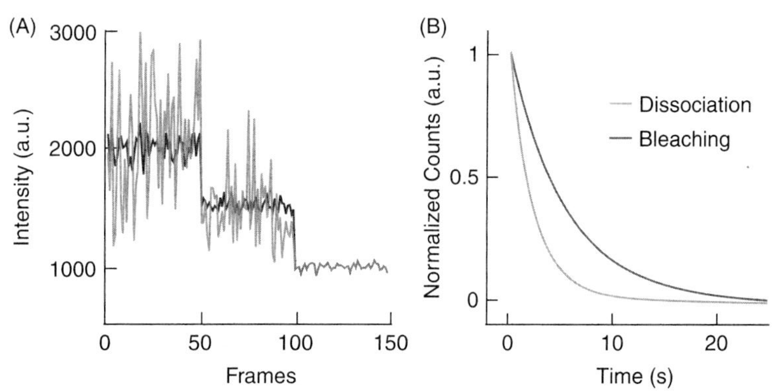

Fig. 4 Two-step photobleaching and characterization of bleaching times. (A) A simulated fluorescence intensity trace showing two-step photobleaching. At a signal-to-noise ratio of 20, two plateaus are clearly visible at 2000 and 1500 intensity units (red trace). At a signal-to-noise ratio of 5, the two plateaus are more difficult to discern (orange trace). (B) Comparison of photobleaching and dissociation from the microtubule lattice for XMAP215-GFP. XMAP215 dissociates from the microtubule lattice with a characteristic constant, $\tau_D = 2.5$ s. Single XMAP215-GFP molecules photobleach with a time constant, $\tau_B = 5.5$ s.

2006). Assuming the algorithm is unbiased and does not accept user input for the number of predicted steps, the uncertainty in the assigned plateaus and plateau intensity values will still be significant.

The third limitation is that two-step photobleaching is difficult to observe while the kinesin or MAP is interacting with the microtubule. This is because the lifetime of the interaction of the kinesin or MAPs with the microtubule may be substantially shorter than the timescale of photobleaching for its fluorophore(s) (Fig. 4B). In fact, researchers usually go to great lengths to avoid photobleaching. The greater the success in this domain, the less likely is the observation of two-step photobleaching during an interaction. For example, the average lifetime of XMAP215's interaction with the microtubule lattice is 2.5 s (Brouhard *et al.*, 2008) (Fig. 4B, green curve). The timescale for photobleaching of a single GFP in the presence of antifade reagents has been measured as 5.5 s (Fig. 4B, blue curve). One can see that it is more likely that XMAP215 dissociates than bleaches during the observation window. For this reason, two-step photobleaching experiments are carried out independently with surface-adsorbed molecules, as originally performed by Vale *et al.* In this case, the kinesin or MAP is allowed to adsorb passively onto the surface, although in principle specific antibodies could be used. Because the fluorophores are directly at the surface, the intensity of the evanescent field is at its peak. It is therefore likely that the molecules will appear brighter than if they are interacting with a microtubule held aloft by antibodies.

D. Step 4. Comparison to Known Standards

Finally, as always, a positive control is essential. In other words, a researcher can test the microscopy assay and compare the fluorescence intensity of her signals to the intensity of a known single-molecule standard. An excellent example of such a standard is kinesin 1-GFP; its velocity and processive run lengths are well established, with nearly identical results measured by numerous labs under a range of experimental conditions. The velocity and processive run length of kinesin-1 are also robust to small variations in buffer conditions. Therefore, if one performs an experiment with kinesin-1-GFP and measures velocity of 800 nm/s and an average run length of 1 μm, the population of molecules is certainly single kinesin-1's. If kinesin-1 is subject to aggregation, the run length will increase substantially (Klumpp and Lipowsky, 2005), so it is straightforward to determine if problematic aggregates are present. Therefore, by observing a population of kinesin-1-GFP, one can quickly ascertain whether the kinesin-1 itself is well behaved and whether single molecules are being observed.

If both are true, the intensity distribution of the kinesin-1-GFP molecules can be compared to the intensity distribution of the protein of interest (assuming the molecule is also a GFP-tagged dimer) (Fig. 5). Furthermore, the mean fluorescence intensity of the kinesin-1-GFP molecules serves as a "calibration" for the intensity of GFP-tagged proteins. This calibration should approximately match the fluorescence intensity values measured in the surface adsorption experiments used to test for two-step photobleaching, although again the position of the fluorophores in the evanescent field is different. Of course, other positive control single-molecule standards may be more appropriate to

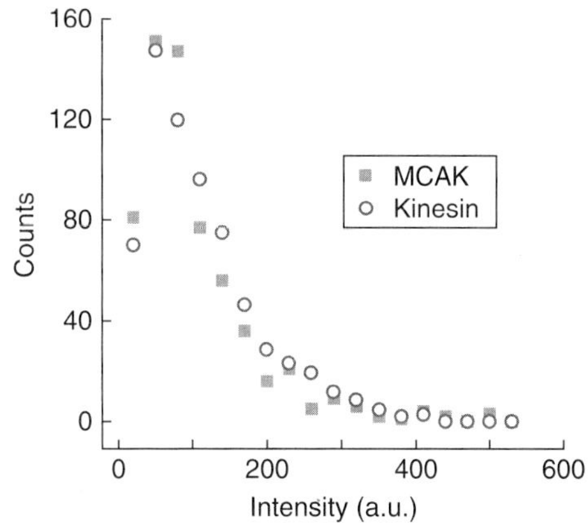

Fig. 5 Comparison to known standards. Kinesin-1-GFP was assayed and single-molecule conditions were verified by an analysis of the distribution of processive run lengths (not shown). The distribution of kinesin-1-GFP intensities (red circles) was compared to the distribution of MCAK-GFP intensities (green squares). The two distributions do not show a statistically significant deviation.

a particular project. A good standard will have been tested quantitatively in multiple labs, exhibit stereotypical properties, and be robust to experimental conditions. The observation of single molecules of a known protein standard is indeed the best confirmation that a new TIRF microscopy assay is functional.

IV. Summary

Although TIRF microscopes and EMCCD cameras are commercially available, single-molecule experiments remain a technical challenge. Non-specific aggregation of recombinant proteins and a poor microscope setup leading to detection thresholds are the two problems that require quality control. This guide has provided steps for quality control of the data from the single-molecule microscope to help separate the wheat from the chaff. Single-molecule experiments require (1) monodisperse, homogeneous, highly purified fluorescent proteins, (2) a microscope system with a sensitive, low-noise EMCCD camera, (3) uniform fluorescence intensity signals with an apparent size equivalent to a diffraction-limited spot, (4) two-step photobleaching for dimeric molecules, and (5) quantitative comparison to known fluorescent single-molecule standards. Good luck!

Acknowledgments

I thank L. Cassimeris and P. Tran for the opportunity to write this chapter; J. Helenius, an indefatigable lab mate, for his teamwork on experiments with MCAK; J. Stear, a man of distinction and charm, for his teamwork on XMAP215; C. Gell, S. Bechstedt, and M. Wieczorek for discussions and reading of the manuscript; S. Wolfson for editing; and my postdoctoral advisor, J. Howard, for his example of discerning and keen science.

References

Axelrod, D., Burghardt, T. P., and Thompson, N. L. (1984). Total internal reflection fluorescence. *Annu. Rev. Biophys. Bioeng.* **13**, 247–268.

Bier, M., and Nord, F. F. (1949). Aggregation phenomena in egg albumin solutions as determined by light scattering measurements. *Proc. Natl. Acad. Sci. U.S.A.* **35**(1), 17–23.

Brouhard, G. J., Stear, J. H., Noetzel, T. L., Al-Bassam, J., Kinoshita, K., Harrison, S. C., Howard, J., and Hyman, A. A. (2008). Xmap215 is a processive microtubule polymerase. *Cell* **132**(1), 79–88.

Cai, D., Verhey, K. J., and Meyhofer, E. (2007). Tracking single kinesin molecules in the cytoplasm of mammalian cells. *Biophys. J.* **92**(12), 4137–4144.

Gell, C., Bormuth, V., Brouhard, G., Cohen, D., Diez, S., Friel, C., Helenius, J., Nitzsche, B., Petzold, H., Ribbe, J., Schaeffer, E., Stear, J. H., et al., . (2010). Microtubule dynamics reconstituted in-vitro and imaged by single-molecule fluorescence microscopy. *Methods Cell Biol.* **95**, 221-245.

Helenius, J., Brouhard, G., Kalaidzidis, Y., Diez, S., and Howard, J. (2006). The depolymerizing kinesin mcak uses lattice diffusion to rapidly target microtubule ends. *Nature* **441**(7089), 115–119.

Kerssemakers, J. W., Munteanu, E. L., Laan, L., Noetzel, T. L., Janson, M. E., and Dogterom, M. (2006). Assembly dynamics of microtubules at molecular resolution. *Nature* **442**(7103), 709–712.

Klumpp, S., and Lipowsky, R. (2005). Cooperative cargo transport by several molecular motors. *Proc. Natl. Acad. Sci. U.S.A.* **102**(48), 17284–17289.

Nord, F. F., and Von Ranke-Abonyi, O. M. (1932). Cryolysis of lyophilic colloids, and its bearing on the mechanism of enzyme action. *Science* **75**(1932), 54–55.

Richard, J., Simpson, Peter, D., Adams, and Erica Golemis, A. Basic Methods in Protein Purification and Analysis: A Laboratory Manual." (2009). "Cold Spring Harbor Laboratory Press, Cold Spring Harbor, NY.

Selvin, P., and Ha, T. (2008) "Single-Molecule Techniques: A Laboratory Manual.". Cold Spring Harbor Laboratory Press, Cold Spring Harbor, NY.

Stuurman, N., and Vale, R. (2005). Imaging single molecules using total internal reflection fluorescence microscopy. *In* "Live Cell Imaging: A Laboratory Manual" (R. Goldman and D. Spector, eds.). Cold Spring Harbor Laboratory Press, Cold Spring Harbor, NY.

Thomann, D., Rines, D. R., Sorger, P. K., and Danuser, G. (2002). Automatic fluorescent tag detection in 3d with super-resolution: Application to the analysis of chromosome movement. *J. Microsc.* **208**(Pt 1), 49–64.

Vale, R. D., Funatsu, T., Pierce, D. W., Romberg, L., Harada, Y., and Yanagida, T. (1996). Direct observation of single kinesin molecules moving along microtubules. *Nature* **380**(6573), 451–453.

Williams, D. C., Van Frank, R. M., Muth, W. L., and Burnett, J. P. (1982). Cytoplasmic inclusion bodies in Escherichia coli producing biosynthetic human insulin proteins. *Science* **215**(4533), 687–689.

SUBJECT INDEX

D

E

F

G

H

I

K

L

N

R

S

T

U

V

W

VOLUMES IN SERIES

Founding Series Editor
DAVID M. PRESCOTT

Volume 1 (1964)
Methods in Cell Physiology
Edited by David M. Prescott

Volume 2 (1966)
Methods in Cell Physiology
Edited by David M. Prescott

Volume 3 (1968)
Methods in Cell Physiology
Edited by David M. Prescott

Volume 4 (1970)
Methods in Cell Physiology
Edited by David M. Prescott

Volume 5 (1972)
Methods in Cell Physiology
Edited by David M. Prescott

Volume 6 (1973)
Methods in Cell Physiology
Edited by David M. Prescott

Volume 7 (1973)
Methods in Cell Biology
Edited by David M. Prescott

Volume 8 (1974)
Methods in Cell Biology
Edited by David M. Prescott

Volume 9 (1975)
Methods in Cell Biology
Edited by David M. Prescott

Volume 10 (1975)
Methods in Cell Biology
Edited by David M. Prescott

Volume 11 (1975)
Yeast Cells
Edited by David M. Prescott

Volume 12 (1975)
Yeast Cells
Edited by David M. Prescott

Volume 13 (1976)
Methods in Cell Biology
Edited by David M. Prescott

Volume 14 (1976)
Methods in Cell Biology
Edited by David M. Prescott

Volume 15 (1977)
Methods in Cell Biology
Edited by David M. Prescott

Volume 16 (1977)
Chromatin and Chromosomal Protein Research I
Edited by Gary Stein, Janet Stein, and Lewis J. Kleinsmith

Volume 17 (1978)
Chromatin and Chromosomal Protein Research II
Edited by Gary Stein, Janet Stein, and Lewis J. Kleinsmith

Volume 18 (1978)
Chromatin and Chromosomal Protein Research III
Edited by Gary Stein, Janet Stein, and Lewis J. Kleinsmith

Volume 19 (1978)
Chromatin and Chromosomal Protein Research IV
Edited by Gary Stein, Janet Stein, and Lewis J. Kleinsmith

Volume 20 (1978)
Methods in Cell Biology
Edited by David M. Prescott

Advisory Board Chairman
KEITH R. PORTER

Volume 21A (1980)
Normal Human Tissue and Cell Culture, Part A: Respiratory, Cardiovascular, and Integumentary Systems
Edited by Curtis C. Harris, Benjamin F. Trump, and Gary D. Stoner

Volume 21B (1980)
Normal Human Tissue and Cell Culture, Part B: Endocrine, Urogenital, and Gastrointestinal Systems
Edited by Curtis C. Harris, Benjamin F. Trump, and Gray D. Stoner

Volume 22 (1981)
Three-Dimensional Ultrastructure in Biology
Edited by James N. Turner

Volume 23 (1981)
Basic Mechanisms of Cellular Secretion
Edited by Arthur R. Hand and Constance Oliver

Volume 24 (1982)
The Cytoskeleton, Part A: Cytoskeletal Proteins, Isolation and Characterization
Edited by Leslie Wilson

Volume 25 (1982)
The Cytoskeleton, Part B: Biological Systems and *In Vitro* Models
Edited by Leslie Wilson

Volume 26 (1982)
Prenatal Diagnosis: Cell Biological Approaches
Edited by Samuel A. Latt and Gretchen J. Darlington

Series Editor
LESLIE WILSON

Volume 27 (1986)
Echinoderm Gametes and Embryos
Edited by Thomas E. Schroeder

Volume 28 (1987)
***Dictyostelium discoideum*: Molecular Approaches to Cell Biology**
Edited by James A. Spudich

Volume 29 (1989)
Fluorescence Microscopy of Living Cells in Culture, Part A: Fluorescent Analogs, Labeling Cells, and Basic Microscopy
Edited by Yu-Li Wang and D. Lansing Taylor

Volume 30 (1989)
Fluorescence Microscopy of Living Cells in Culture, Part B: Quantitative Fluorescence Microscopy—Imaging and Spectroscopy
Edited by D. Lansing Taylor and Yu-Li Wang

Volume 31 (1989)
Vesicular Transport, Part A
Edited by Alan M. Tartakoff

Volume 32 (1989)
Vesicular Transport, Part B
Edited by Alan M. Tartakoff

Volume 33 (1990)
Flow Cytometry
Edited by Zbigniew Darzynkiewicz and Harry A. Crissman

Volume 34 (1991)
Vectorial Transport of Proteins into and across Membranes
Edited by Alan M. Tartakoff

Selected from Volumes 31, 32, and 34 (1991)
Laboratory Methods for Vesicular and Vectorial Transport
Edited by Alan M. Tartakoff

Volume 35 (1991)
Functional Organization of the Nucleus: A Laboratory Guide
Edited by Barbara A. Hamkalo and Sarah C. R. Elgin

Volume 36 (1991)
***Xenopus laevis:* Practical Uses in Cell and Molecular Biology**
Edited by Brian K. Kay and H. Benjamin Peng

Series Editors

LESLIE WILSON AND PAUL MATSUDAIRA

Volume 37 (1993)
Antibodies in Cell Biology
Edited by David J. Asai

Volume 38 (1993)
Cell Biological Applications of Confocal Microscopy
Edited by Brian Matsumoto

Volume 39 (1993)
Motility Assays for Motor Proteins
Edited by Jonathan M. Scholey

Volume 40 (1994)
A Practical Guide to the Study of Calcium in Living Cells
Edited by Richard Nuccitelli

Volume 41 (1994)
Flow Cytometry, Second Edition, Part A
Edited by Zbigniew Darzynkiewicz, J. Paul Robinson, and Harry A. Crissman

Volume 42 (1994)
Flow Cytometry, Second Edition, Part B
Edited by Zbigniew Darzynkiewicz, J. Paul Robinson, and Harry A. Crissman

Volume 43 (1994)
Protein Expression in Animal Cells
Edited by Michael G. Roth

Volume 71 (2003)
Neurons: Methods and Applications for Cell Biologist
Edited by Peter J. Hollenbeck and James R. Bamburg

Volume 72 (2003)
Digital Microscopy: A Second Edition of Video Microscopy
Edited by Greenfield Sluder and David E. Wolf

Volume 73 (2003)
Cumulative Index

Volume 74 (2004)
Development of Sea Urchins, Ascidians, and Other Invertebrate Deuterostomes: Experimental Approaches
Edited by Charles A. Ettensohn, Gary M. Wessel, and Gregory A. Wray

Volume 75 (2004)
Cytometry, 4th Edition: New Developments
Edited by Zbigniew Darzynkiewicz, Mario Roederer, and Hans Tanke

Volume 76 (2004)
The Zebrafish: Cellular and Developmental Biology
Edited by H. William Detrich, III, Monte Westerfield, and Leonard I. Zon

Volume 77 (2004)
The Zebrafish: Genetics, Genomics, and Informatics
Edited by William H. Detrich, III, Monte Westerfield, and Leonard I. Zon

Volume 78 (2004)
Intermediate Filament Cytoskeleton
Edited by M. Bishr Omary and Pierre A. Coulombe

Volume 79 (2007)
Cellular Electron Microscopy
Edited by J. Richard McIntosh

Volume 80 (2007)
Mitochondria, 2nd Edition
Edited by Liza A. Pon and Eric A. Schon

Volume 81 (2007)
Digital Microscopy, 3rd Edition
Edited by Greenfield Sluder and David E. Wolf

Volume 82 (2007)
Laser Manipulation of Cells and Tissues
Edited by Michael W. Berns and Karl Otto Greulich

Volume 83 (2007)
Cell Mechanics
Edited by Yu-Li Wang and Dennis E. Discher

Volume 84 (2007)
Biophysical Tools for Biologists, Volume One: *In Vitro* Techniques
Edited by John J. Correia and H. William Detrich, III

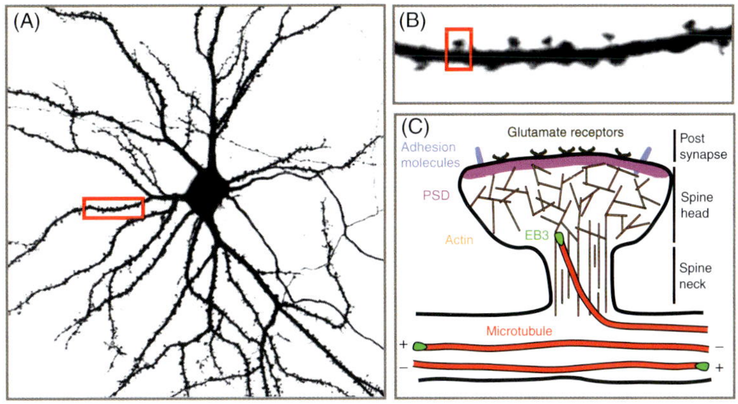

Plate 1 (Figure 1 on page 114 of this volume)

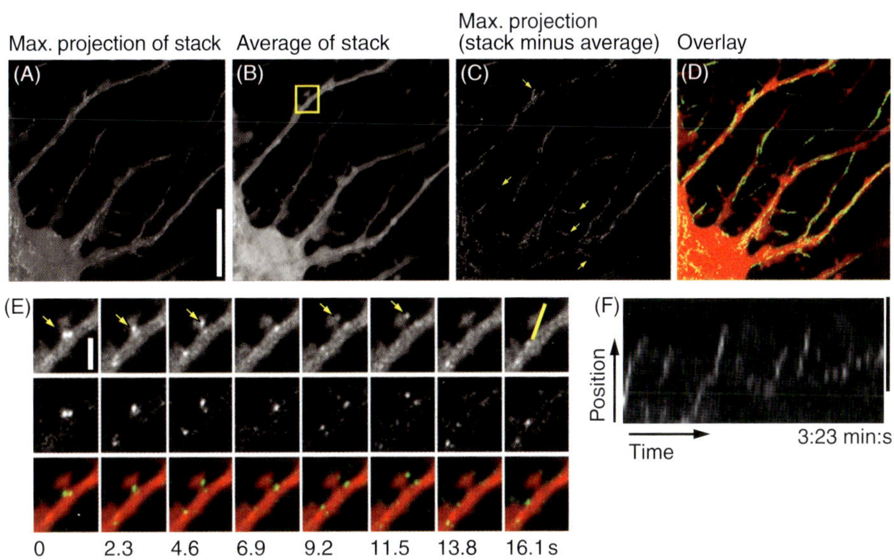

Plate 2 (Figure 3 on page 129 of this volume)

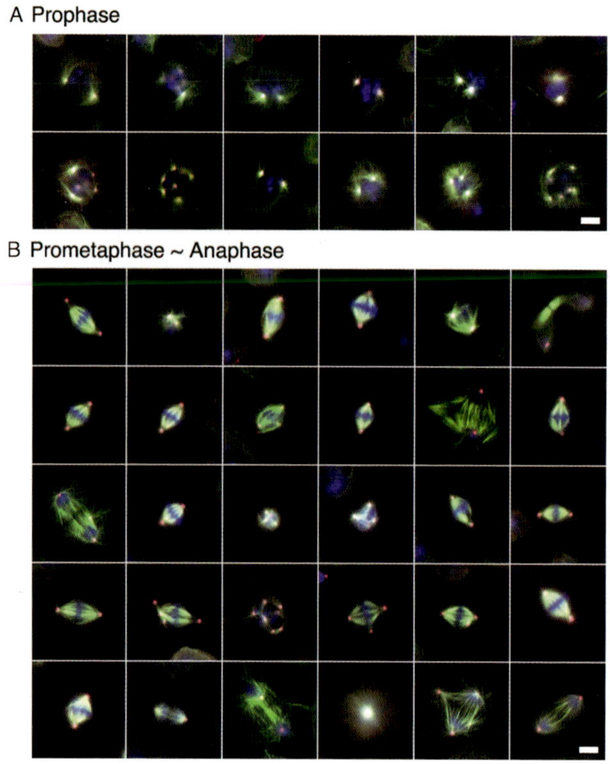

Plate 8 (Figure 3 on page 267 of this volume)

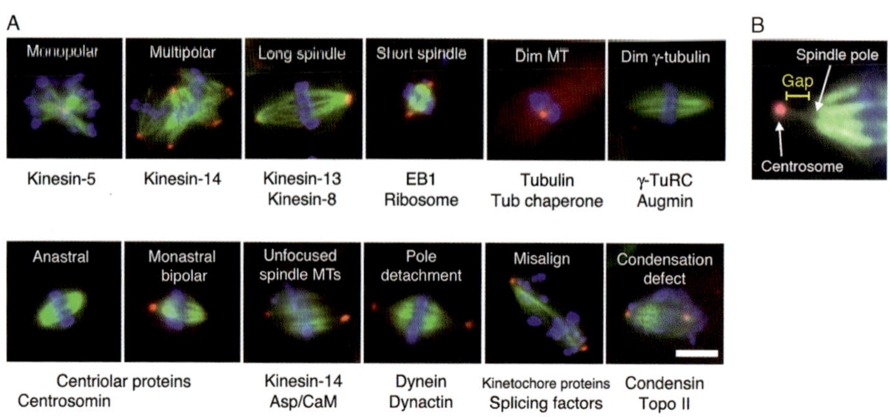

Plate 9 (Figure 4 on page 268 of this volume)

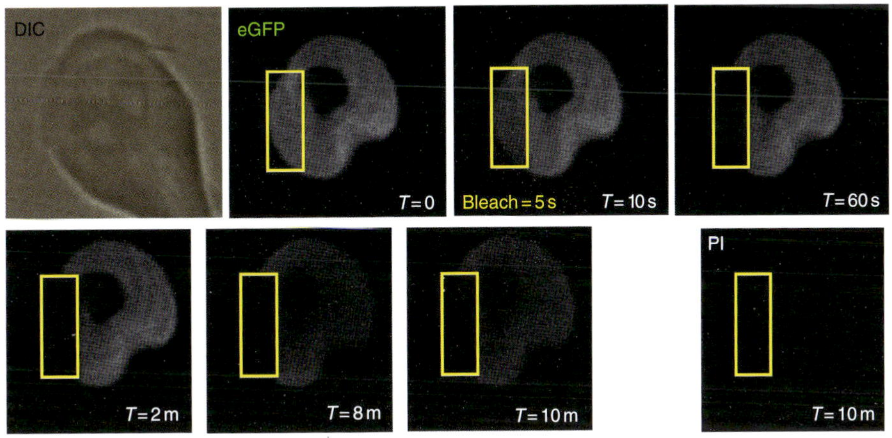

Plate 10 (Figure 4 on page 330 of this volume)

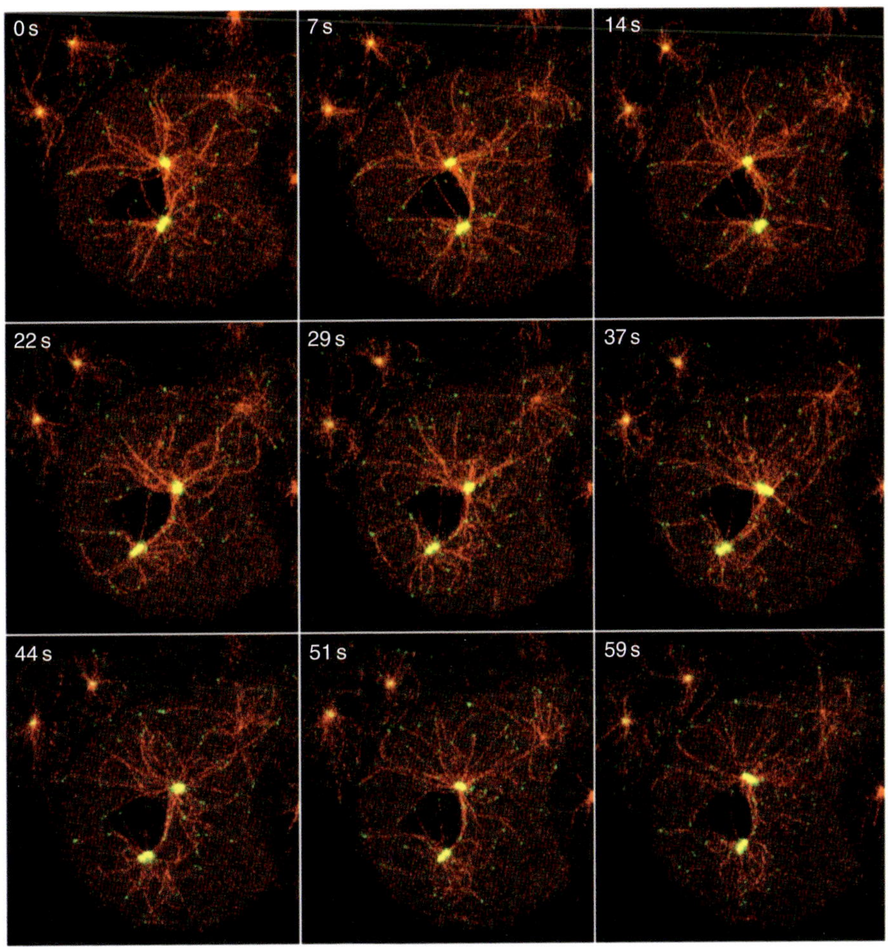

Plate 11 (Figure 3 on page 350 of this volume)

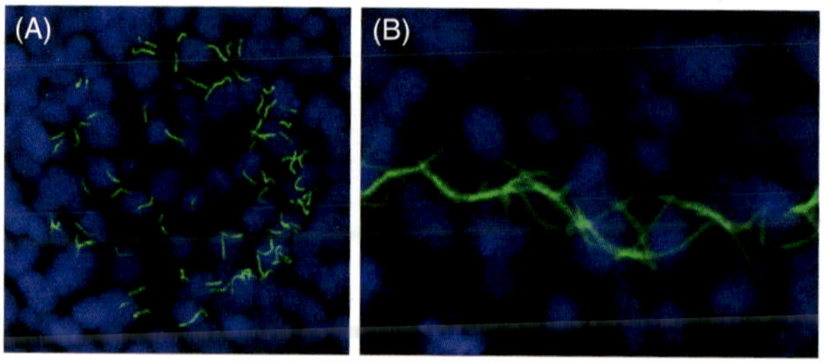

Plate 12 (Figure 2 on page 420 of this volume)

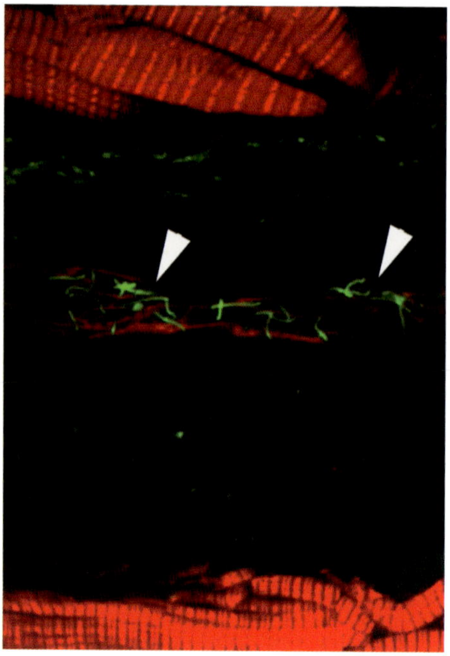

Plate 13 (Figure 3 on page 424 of this volume)

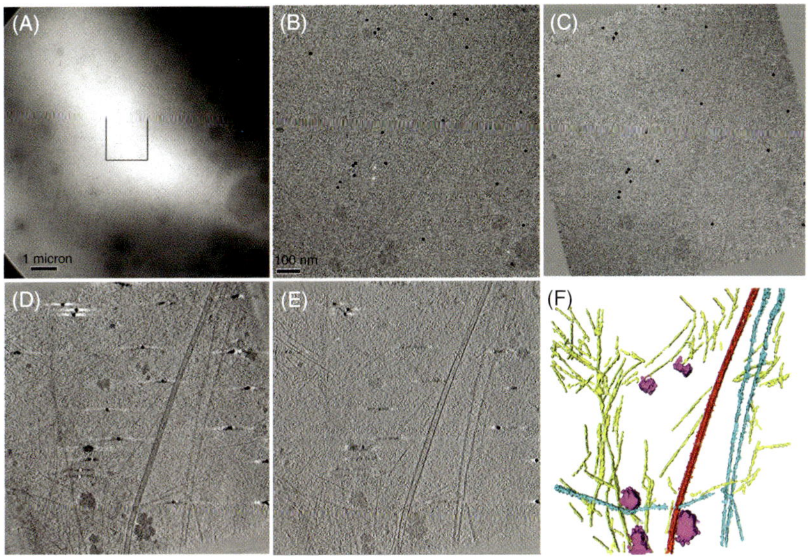

Plate 14 (Figure 4 on page 467 of this volume)

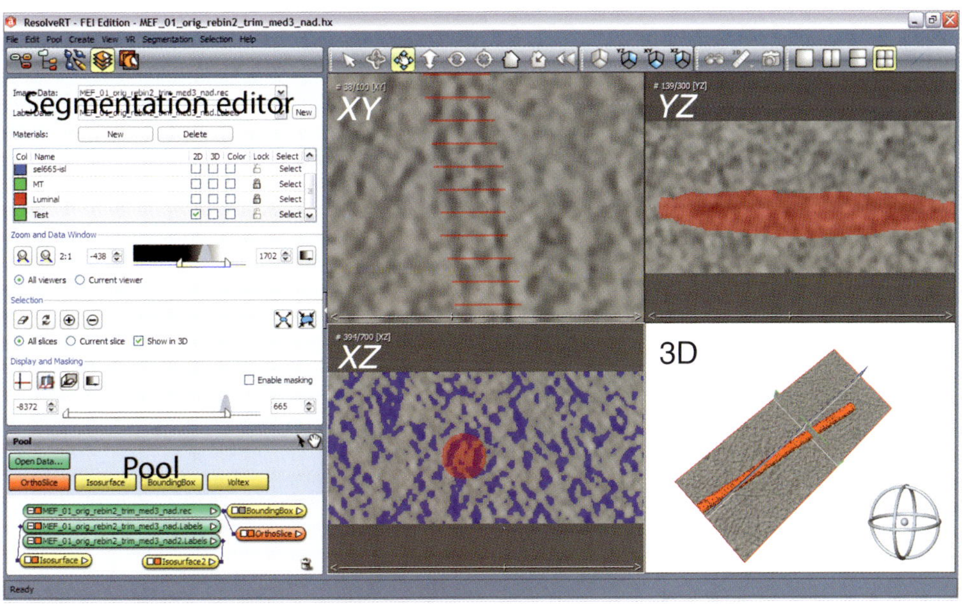

Plate 15 (Figure 6 on page 469 of this volume)

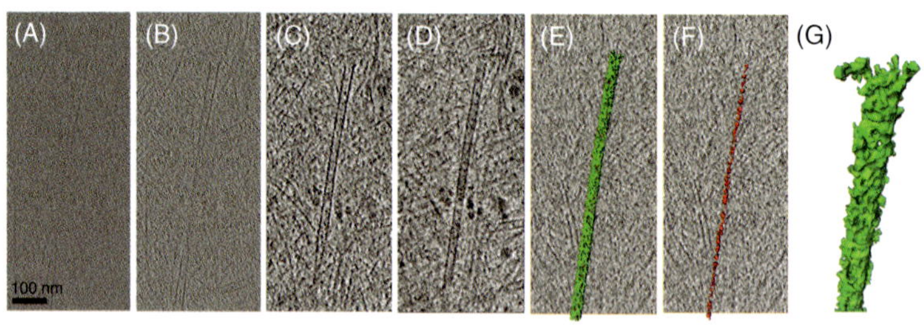

Plate 16 (Figure 7 on page 469 of this volume)